Toxicology of Amphibian Tadpoles

Editors

Eduardo Alves de Almeida
Department of Natural Sciences
Regional University of Blumenau
Blumenau, SC, Brazil

Juliane Silberschmidt Freitas
Department of Agrarian and Natural Sciences
Minas Gerais State University (UEMG)
Ituiutaba, MG, Brazil

CRC Press is an imprint of the
Taylor & Francis Group, an **informa** business
A SCIENCE PUBLISHERS BOOK

Cover credit: Photo taken by Tiago Leite Pezzuti

First edition published 2024
by CRC Press
2385 NW Executive Center Drive, Suite 320, Boca Raton FL 33431

and by CRC Press
4 Park Square, Milton Park, Abingdon, Oxon, OX14 4RN

Library of Congress Cataloging-in-Publication Data (applied for)

ISBN: 978-0-367-54967-1 (hbk)
ISBN: 978-0-367-54969-5 (pbk)
ISBN: 978-1-003-09139-4 (ebk)

DOI: 10.1201/9781003091394

Typeset in Times New Roman
by Radiant Productions

Preface

Amphibians are the group of vertebrates with the greatest number of species under population declines and extinctions in the world. Among the main causes of this threat are habitat degradation, invasive species, infectious diseases, climate changes, and environmental pollution. This group of animals is especially affected by aquatic pollutants because they have permeable skin, cutaneous breathing, and a biphasic lifecycle, i.e., with a larval phase totally aquatic. Effects of pollutants in amphibians' tadpoles or adults have been poorly studied yet, especially considering the effects of xenobiotic mixtures, the influence of abiotic factors on pollutant effects and specific habits and physiological characteristics of numerous different species. Several publications exist in the scientific literature regarding ecotoxicological effects and consequences of environmental pollutants in adult amphibians, including one specialized book. Similar studies on tadpoles have also increased in recent years, but to date, there is no published book dedicated specifically to the polluting effects on amphibians during their larval phase. Thus, the present book represents the first effort to gather information about ecotoxicological effects of aquatic pollutants in tadpoles. In order to best contextualize the toxicological aspects of environmental contaminant effects in these organisms, the book first brings recent data about the alarming decline of amphibian populations worldwide. The second chapter focuses on the need to understand the different classifications of the developmental stages of tadpoles from the egg to the climax of metamorphosis in ecotoxicological studies, since morphological and physiological changes that occur in these organisms affect the mode of action of different intoxicants. A general overview of main toxicokinetic pathways of toxic substances in tadpoles is then explored. Finally, the effects of potentially toxic substances such as pesticides, pharmaceuticals and metals on tadpoles' physiology (e.g., effects on endocrine and immune systems, generation of oxidative stress, malformations, genotoxicity, and other general biomarkers) are discussed in the remaining chapters.

Contents

1

How Pollutants are Affecting Amphibian Tadpoles

Relationship with Anthropogenic Pollutants and Perspectives for the Future

Renan Nunes Costa,[1,*] *Fernanda Franco*,[2] *Mirco Solé*,[3] *Denise de Cerqueira Rossa-Feres*[4] and *Fausto Nomura*[5]

1. Introduction

Mass extinction events in Earth's history due to natural events (e.g., massive volcanic eruptions, asteroid collision) significantly reduced biological diversity quickly. In modern times, classified by many researchers as Anthropocene, anthropic action is responsible for the sixth event of species mass extinction (Barnosky et al. 2011, Kolbert 2014, Ceballos et al. 2015, Ceballos et al. 2020). Simply put, our species (i.e., *Homo sapiens*) has directly or indirectly devastated millions of years of evolution, reducing the biological diversity similar to the impact of an asteroid.

Amphibians lead the race as the group with the most species suffering from population declines and extinctions, being the most threatened vertebrates in the world (Stuart et al. 2004, Gallant et al. 2007, Alroy 2015). About a fifth of amphibian species are extinct or threatened with extinction (Ceballos et al. 2020). Among the main causes of amphibian declines and extinctions are habitat

[1] Departamento de Ciências Biológicas, Universidade do Estado de Minas Gerais-UEMG, Praça dos Estudantes, 23, Santa Emília, 36800-000, Carangola, Minas Gerais, Brasil.

[2] Departamento de Biociências e Tecnologia, Universidade Federal de Goiás-UFG, Rua 235, Setor Leste Universitário, s/n. 74605-050, Goiânia, Goiás, Brasil. Email: fec_franco@ufg.br

[3] Departamento de Ciências Biológicas, Universidade Estadual de Santa Cruz-UESC, Rodovia Jorge Amado, km 16, 45662-900, Ilhéus, Bahia, Brasil. Email: msole@uesc.br

[4] Departamento de Zoologia e Botânica, Universidade Estadual Paulista-UNESP, Rua Cristóvão Colombo, 2265. 15054-000, São José do Rio Preto, São Paulo, Brasil. Email: deferes@gmail.com

[5] Departamento de Ecologia, Universidade Federal de Goiás-UFG, Av. Esperança, s/n. 74690-900, Goiânia, Goiás, Brasil. Email: faustonomura@ufg.br

* Corresponding author: renan.costa@uemg.br

loss, fragmentation and degradation, invasive species, infectious diseases, climate changes, and environmental pollution and contamination (Hayes et al. 2010, Blaustein et al. 2011, Alroy 2015). We can evaluate the isolated impacts of each threat factor, however, they can act simultaneously and lead to additive and/or synergistic effects on amphibian attributes, increasing their extinction potential (e.g., Blaustein et al. 2003, 2011).

Environmental pollution and contamination can be observed worldwide and is very common in freshwater environments. It is seen as one of the most significant factors causing amphibian declines (e.g., Driss et al. 2015, Lambert and Wagner 2018, Schuler and Relyea 2018, Pietrzak et al. 2019). This group is especially affected by freshwater pollution and contamination because of some specific characteristics that increase their susceptibility (Blaustein and Kiesecker 2002, Blaustein et al. 2003, Schiesari et al. 2007, Hayes et al. 2010). Among them are permeable skin, cutaneous breathing, and a biphasic lifecycle, which is a particular concern to species with indirect development. Moreover, the majority of amphibians have external fertilization and embryonic development, with aquatic eggs not protected by a shell and tadpoles confined to the interior of water bodies (Bishop et al. 1999, Gallant et al. 2007, Schiesari et al. 2007, Mann et al. 2009, Allentoft and O'Brien 2010). Most amphibian species deposit their spawn in small water bodies near roads, industries, cities, or agricultural landscapes, which are usually contaminated by chemicals (Bridges and Boone 2003, Baker et al. 2013, Lenhardt et al. 2015). For example, pesticides used in Brazilian croplands are carried to water bodies more often during the rainy seasons. It coincides with many amphibian species' reproductive period, during which they use temporary ponds in agricultural landscapes (Moreira et al. 2012, Schiesari and Corrêa 2016). Thus, tadpoles are exposed to several kinds of pollutants and contaminating substances, such as heavy metals, pesticides, fertilizers, microplastics, pharmaceutical products, domestic sewage, industrial waste, emerging organic contaminants, among others.

However, despite many ecotoxicological studies published every year, the full effects and impacts of pollutants on amphibians (adults and tadpoles) are not fully understood and it still can be underestimated. Most studies with tadpoles were conducted under laboratory conditions because it is easier to get variables under control (e.g., Relyea 2005, Costa and Nomura 2016, Freitas et al. 2016). Also, studies carried out in outdoor mesocosms have received great attention because they allow both controlling variables and simulate natural environments (e.g., Boone et al. 2007, Hua and Relyea 2014, Jones et al. 2021). As conducted in a laboratory or outdoor mesocosm, these studies have found a range of lethal and sublethal effects on tadpoles exposed to different levels of contamination.

Even the generally low concentrations tested (probably lower than concentrations reached in water bodies) are responsible for a significant reduction in tadpole survival, reflecting changes in the population parameters under natural conditions. Low tadpoles' survival reduces the number of adult individuals that will reach the reproductive phase. This effect contributes to population decline and can lead to local extinction, especially for rare or endemic species and those with few and/or small populations. Moreover, the application of ecologically relevant concentrations of contaminants, does not guarantee that these effects may be applied to natural habitats because of several other environmental factors (e.g., substrate type, water temperature, and pH) can increase or decrease the lethality and residence time of a given chemical (Giesy et al. 2000, Boone and Bridges 2003, Relyea and Hoverman 2006, Relyea 2012). Probably because of this, few studies measured and monitored contaminants' effects on tadpoles in natural water bodies (e.g., Agostini et al. 2020) and, even less, the synergistic impact from other contributing stressors. While tons of chemicals are illegally dumped into aquatic environments, their impacts on tadpoles' attributes remain only partially known. Thus, pollutants and contaminants silently threaten amphibians (Fig. 1) and their contribution to population declines remains underestimated (Brühl et al. 2013, Agostini et al. 2020).

The consequences of the decline in amphibian populations are spread across food webs, affecting the dynamics of communities and the functioning of freshwater ecosystems (Relyea and Hoverman 2006, Arribas et al. 2015). Tadpoles are an important food resource for a range of

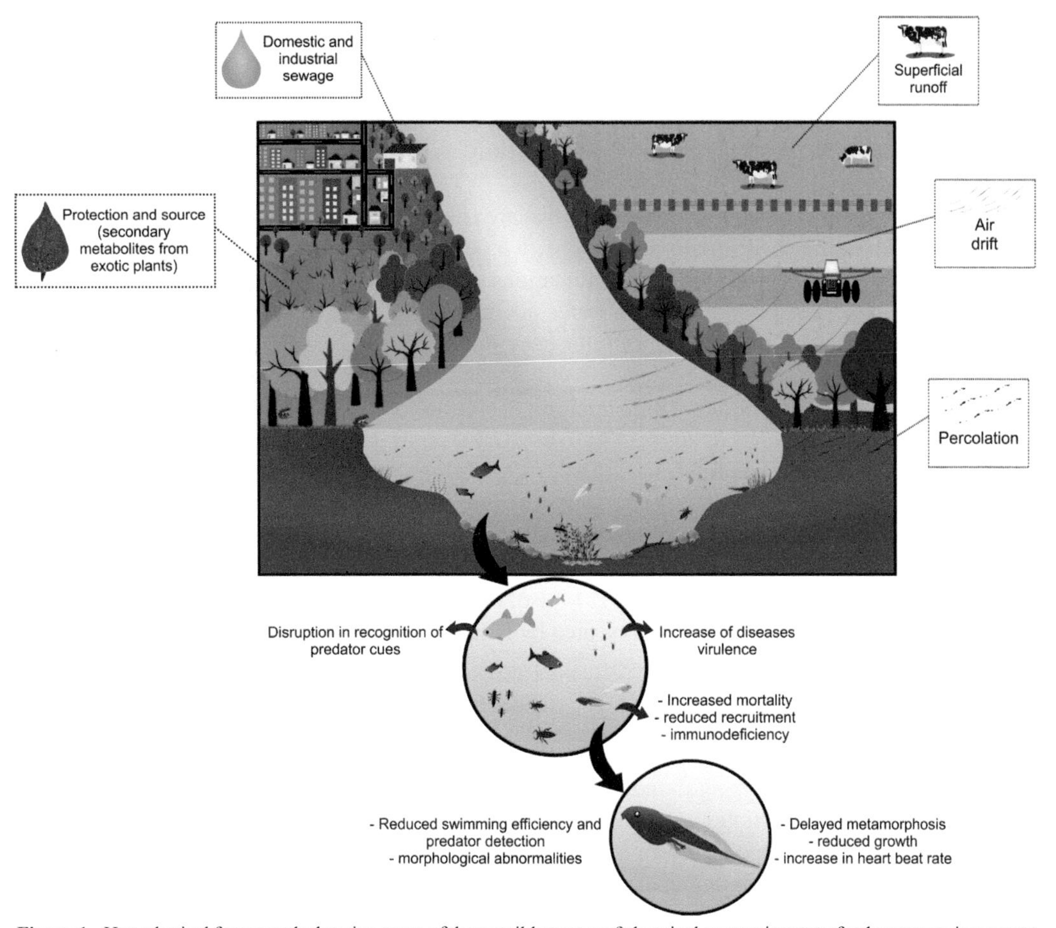

Figure 1. Hypothetical framework showing some of the possible routes of chemical contaminants to freshwater environments and its consequence for amphibian larvae. Industrial and domestic sewage are the common routes of contaminants (e.g., detergents, heavy metal, endocrine disruptors) to the freshwater environment. Even treated sewage can be a source of contamination when the sewage treatment was not designed to filter some contaminants (e.g., hormones). Superficial runoff could carry several types of contaminants used in infrastructure maintenance (such as salt used to avoid the road to freeze) and human activities (such as several types of agrochemical inputs). Chemical contaminants can also reach non-target areas like freshwater environments due to air drift during mechanized application with tractors and airplanes in large industrial farms. Percolation of chemical contaminants could eventually reach groundwater, increasing the importance of forests as a filter for contaminants and protection of aquatic habitats. Eventually, man-made interventions could change forest composition with the introduction of exotic plants for production or arborization, which could be a source of toxic secondary metabolites that native amphibians were naive to. All these contaminants can affect the amphibian larvae at the community (e.g., impairing the ability to recognize predator cues, increasing pathogen virulence, changing competitive ability), populational (e.g., increasing mortality, reducing recruitment) and individual (e.g., reducing growth, delaying metamorphosis, reducing swimming efficiency, reducing immune response to diseases, morphological abnormalities) levels.

vertebrate and invertebrate organisms, and an essential link in predator-prey interaction networks. The scarcity of tadpoles in natural environments denies a significant food resource for several predator species, generating effects on subsequent trophic levels. However, tadpoles also have an essential impact as consumers, with a diversified diet, functioning as herbivores, detritivores, and/or carnivores, consuming a range of algae, plants, fungi, pollen grains, bacteria, protists, insect larvae, frog eggs, and even other tadpoles (Altig et al. 2007). Tadpoles often occur in high densities, and their feeding activity accelerates the decomposition of organic matter and releases a significant amount of nutrients into freshwater habitats (e.g., Iwai et al. 2009, Connelly et al. 2011). Because of their role as primary and secondary consumers, tadpoles' feeding activity resonates throughout food webs, affecting zooplankton diversity, nutrient availability, and macrophyte biomass (Arribas et al.

2015). Therefore, understanding the influence of chemical contaminants through aquatic food webs is of great importance, since tadpoles are essential elements in the aquatic ecosystems' functioning because of their critical functional roles related to nutrient cycling, energy flow, and bioturbation (Hocking and Babbitt 2014, Montaña et al. 2019). The scarcity of tadpoles can result in the reduction or loss of these ecosystem functions and, consequently, in the loss of ecosystem services provided by freshwater environments.

Aiming to understand how knowledge in this area has advanced over time, and also to identify gaps and areas that need further investigations, we performed a systematic search in the Web of Science database using the following combinations of keywords: (ecotox* OR tox* OR pollut* OR "water pollut*" OR bioindicat* OR water quality OR "endocrine disrupt*" OR disrupt* OR pesticid* OR herbicid* OR agrochem* OR "heavy metal*" OR insecticid* OR nanopartic* OR contaminant* OR fungicid* OR bactericid* OR Organochlor* OR Organophosp* OR Carbamat* OR Pyrethroid* OR Terpenoid* OR Saponin* OR Phenolic* OR Tannin* OR Alkaloid*) AND (Anur* OR Amphib*) AND (Frog* OR Toad* OR Treefrog*). Moreover, we substituted the keywords Frog*, Toad*, Treefrog* to Tadpole* OR "Anura larvae*" in the systematic search to evaluate Tadpoles' application in Ecotoxicological studies. We restricted our search to the Web of Science database, considered the most comprehensive scientific journals' index, to avoid duplicate entries. We used the keywords reported by Schiesari et al. (2007) and for the chemical pollutants' "families" we used Weldeslassie et al. (2018). Also, we restricted our search from 2007 to date, once there is a previous systematic review presented by Schiesari et al. (2007).

2. The History so Far

We retrieved 1,928 papers published using keywords directed to adults, and 909 publications when we substitute keywords referring to tadpoles in our search criteria (136 entries were removed from our search because these articles were not related to tadpoles or to contaminants after inspecting the articles abstract). Thus, tadpoles ecotoxicological studies represent a little more than half the papers published since 2007 considering ecotoxicological studies with amphibians. This result illustrates the importance of tadpoles in ecotoxicological studies, unlike those observed in other areas, such as Ecology, Zoology, and Ethology. We created a cloud of words with the titles and keywords of all articles retrieved in our search, and the word "tadpole" is in the spotlight (Fig. 2). The larval stage, which in general does not receive much attention in zoological or ecological studies, shines in ecotoxicological studies.

The general interest in tadpoles as animal models appears to have increased over time, considering the number of papers published since 2007. In the last three years (2018, 2019, 2020) nearly 80 papers were published each year using tadpoles as animal models in experimental approaches (Fig. 3). This interest could be a consequence of the abundance in which tadpoles can be found in nature and the ease to collect them, together with the relatively simple facilities needed to house tadpoles in laboratories. Also, tadpoles show high phenotypic plasticity and respond fast to different ecological contexts and contaminants (e.g., Relyea 2004, Steiner and Van Buskirk 2008, Fusco and Minelli 2010, Nomura et al. 2011, Costa and Nomura 2016). The existence of systematic protocols for research based on anuran larvae, like the Frog Embryo Teratogenesis Assay Xenopus (FETAX; Dumont et al. 1983) protocol (e.g., Hu et al. 2015) and the Amphibian Metamorphosis Assay (AMA; OECD 2009) (e.g., Coady et al. 2010), is another factor that explains the great use of anuran larvae in ecotoxicological studies.

Finally, the appearance of emergent diseases, like chytridiomycosis, caused by the fungus *Batrachochytrium dendrobatidis* (Bd), and the well documented amphibian species decline could help to draw attention to amphibians and to increase the interest about the interaction of several stressors, like diseases and contaminants, and to understand causes underlying amphibian species declines (e.g., Davidson and Knapp 2007, Boone et al. 2007, Rohr et al. 2008a, Heyes et al. 2010,

Figure 2. Word cloud based on the frequency of words used in the keywords (above) and title (below) of published ecotoxicological studies focusing on amphibians in general (right side) and tadpoles in particular (left side) since 2007.

Rollins-Smith et al. 2011). Thus, although we could not differentiate between the ease of use and collection from the sense of urgency to understand the threats for this group, it is probably more accurate to assume that both aspects contributed to the increasing use of tadpoles in ecotoxicological studies.

3. Where This Study Started From: Comparison With the Previous Knowledge

Astoundingly, almost 15 years later our results related to biogeographic and taxonomic research gaps about the effects of chemical pollutants on anurans, agree with those obtained by Schiesari et al. (2007). Fortunately, our results also point out that this gap is getting filled, since some countries from tropical regions contributed to an expressive percentage with ecotoxicological studies (Fig. 4). The majority of the studies were published by authors from the USA, but countries like Brazil, Argentina, Canada, China, and Australia had also a large output of articles about Ecotoxicology using tadpoles as animal models (Fig. 4). Probably, the study by Schiesari et al. (2007) is a possible driver for the interest in studies in this area, especially in Brazil and Argentina.

It is interesting to note that India also appears in the Top10 countries with papers published since 2007' Schiesari et al. publication (Fig. 5). The increase in ecotoxicological studies with amphibian

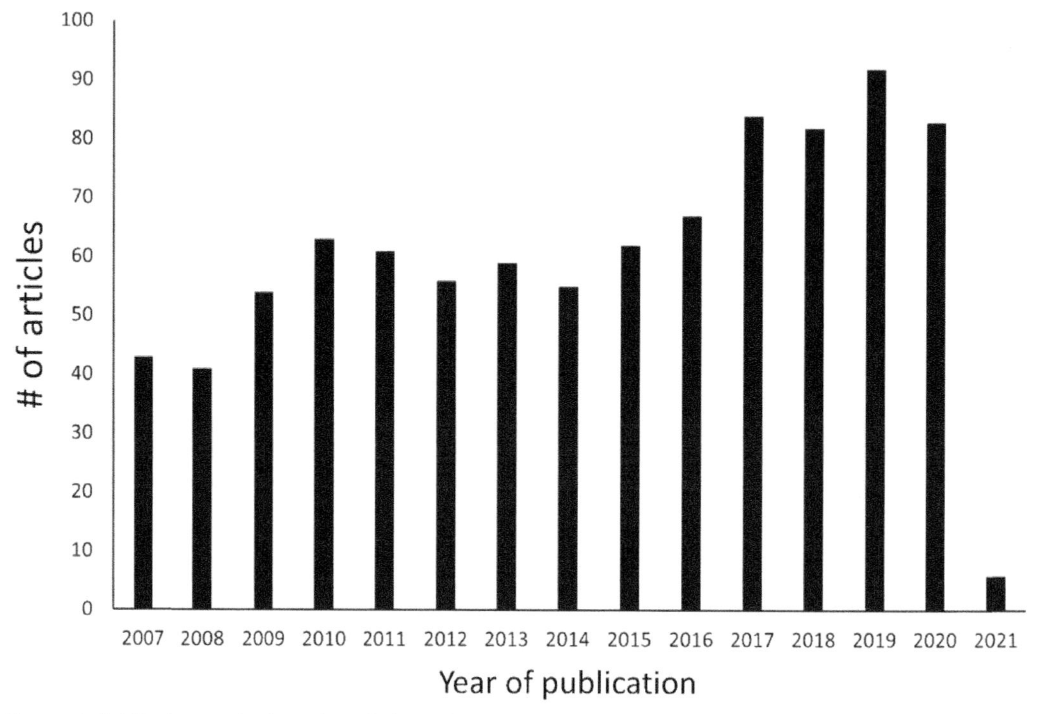

Figure 3. Publication trends along time (2007 to 2021) about ecotoxicological studies focusing on tadpoles. Journal abbreviations follow the Web of Science format.

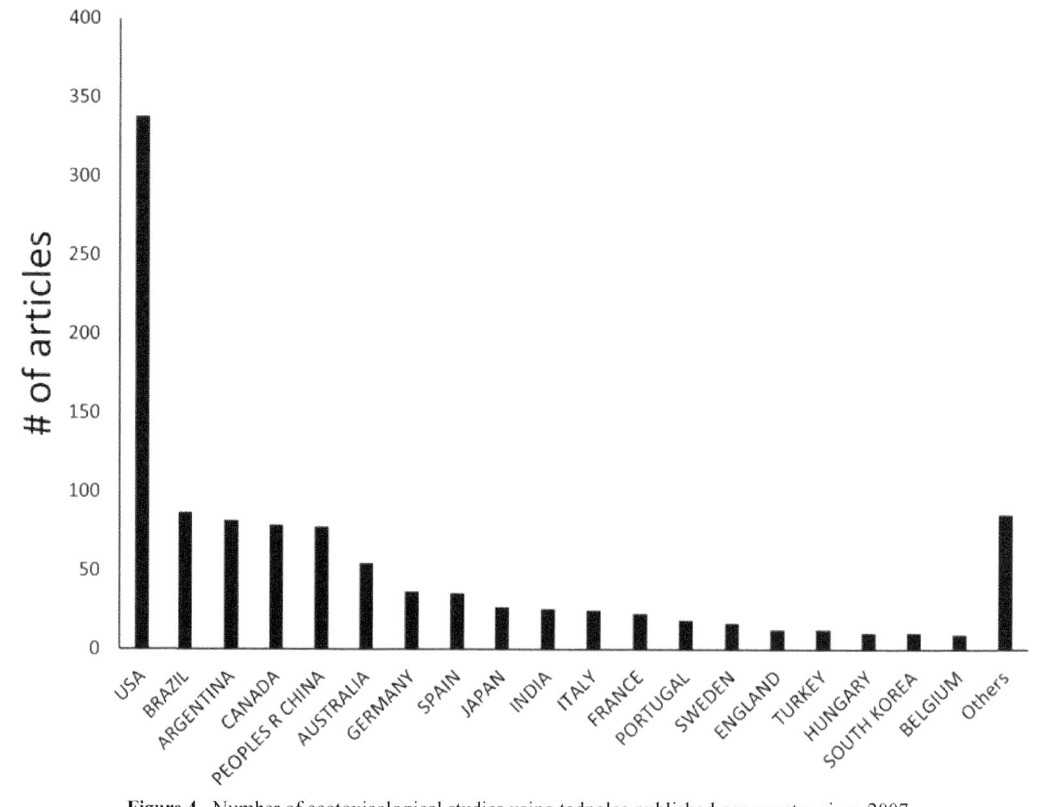

Figure 4. Number of ecotoxicological studies using tadpoles published per country since 2007.

larvae in Neotropical, Indomalayan and Australasian countries contributes to the reduction of some spatial gaps pointed out by Schiesari et al. (2007). This contributes to gather information from strategic regions, once these countries are major consumers of agrochemicals and because of the greater importance of agricultural activities on the gross domestic product (GDP) of these regions (Bain et al. 2017; Benbrook 2016). Moreover, the increase in pesticide use is much greater for each increase in GDP that this use promotes (Hedlund et al. 2020). The importance of pesticides for economic growth in agricultural countries is proportional to the environmental impacts that it causes and increases faster than the population growth of such countries (Hedlund et al. 2020). Thus, there is no expectation for a reduction in pesticide use, on the contrary, this use is estimated to increase much more.

Among the most productive authors in this field, the top four feature one researcher from the USA (R. A. Relyea) and three from Argentina (R. C. Lajmanovich, P. M. Peltzer and A. M. Attademo) (Fig. 5). Although we see women with prominent positions considering the number of articles, they lead only one-third of the thirty most cited articles, showing a possible sex bias in this field, which has to be overcome. Considering that Brazil, China, and Australia are among the countries with a large output of articles published but have no authors in the most productive list, this can indicate that these countries may have more research teams producing Ecotoxicological products using tadpoles.

In addition to the biogeographical bias, Schiesari et al. (2007) also detected a narrow universe of amphibian representatives in ecotoxicological studies. Only ten species were represented in about half of the 1,997 ecotoxicological studies retrieved by them. Besides the matter of the small number of representatives of the diversity of amphibians per se, an additional concern was that most of these few species are widely distributed and occur at higher latitudes, being more generalistic and broadly tolerant to environmental factors, and probably more robust to chemical contamination (Schiesari et al. 2007). It is important to notice that the comparison between our results with those of Schiesari et al. (2007) is not straightforward, because it was not possible to use the same keywords

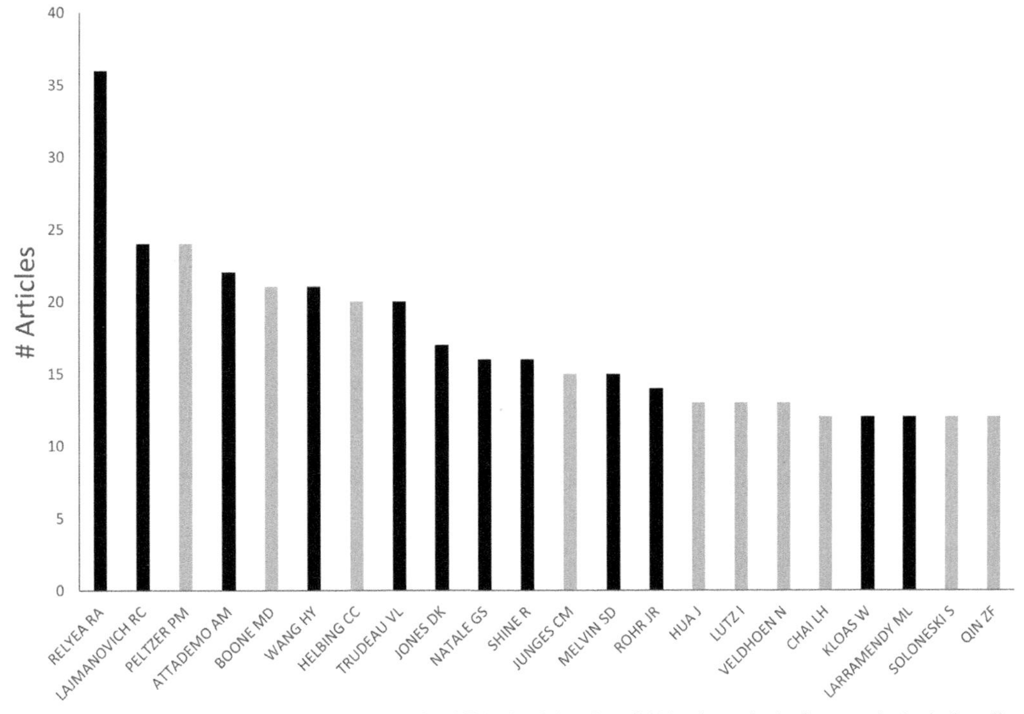

Figure 5. List of the 22 authors with more than 12 published articles since 2007 using tadpoles in ecotoxicological studies. Black = male authors. Grey = females authors.

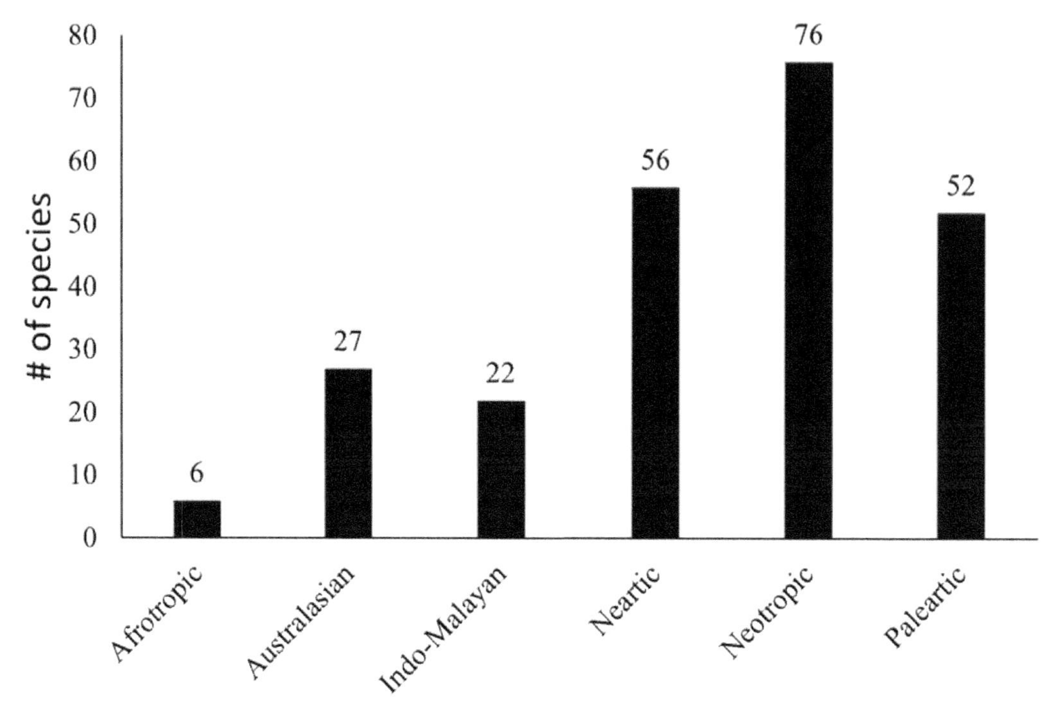

Figure 6. Number of amphibian species by ecoregions used in ecological studies. When the study focused on non-native species, we considered the ecoregion of origin of the species and not the local where the study was developed.

and we focused on larvae. Considering these differences, our results evidenced an increasing focus on amphibian species from tropical regions (Fig. 6), which harbor the greatest anuran diversity in the world (Segalla et al. 2021) and also the greatest number of species categorized as threatened (IUCN 2021). This is an important gap that deserves prioritization in ecotoxicological studies since most results, which are still based on species from the northern hemisphere, can only be applied to predictions about the exposure of amphibian species to pollutants in tropical regions with restrictions. However, we assume that coping with the higher anuran diversity in neotropics could be challenging. Many threatened anuran species were geographically rare, with low populational abundance and difficult to collect. Also, these populational traits could make difficult the expedition of license permits by the environmental regulatory agencies once most of the experiments result in the death of individuals used. Also, in general, threatened species have specific habitat requirements that prove to be challenging to replicate in laboratory conditions. The lack of standard protocols with species from high diversity areas is an important bias for ecotoxicological studies. On the other hand, focusing on amphibian species from tropical regions is of great importance and urgency. Increasing the diversity of species used in ecotoxicological studies would allow the evaluation of the evolutive effects of agrochemicals and the species' response to them. Moreover, it will contribute to understanding the consequences of the ecosystemic role of tadpoles since agrochemical can influence their functional traits. Finally, in the massive agrochemicals consuming regions, this approach is basal to drive conservation plans and actions.

4. Main Vehicles for Knowledge Dissemination

The main output for information about tadpoles as model organisms in ecotoxicological studies are articles, considering all article formats (articles published in scientific journals, as book chapters, proceeding papers, early access, and data papers), with 778 (from 909 products in total) articles published since 2007, followed by reviews, with 21 published since 2007. However, as usual,

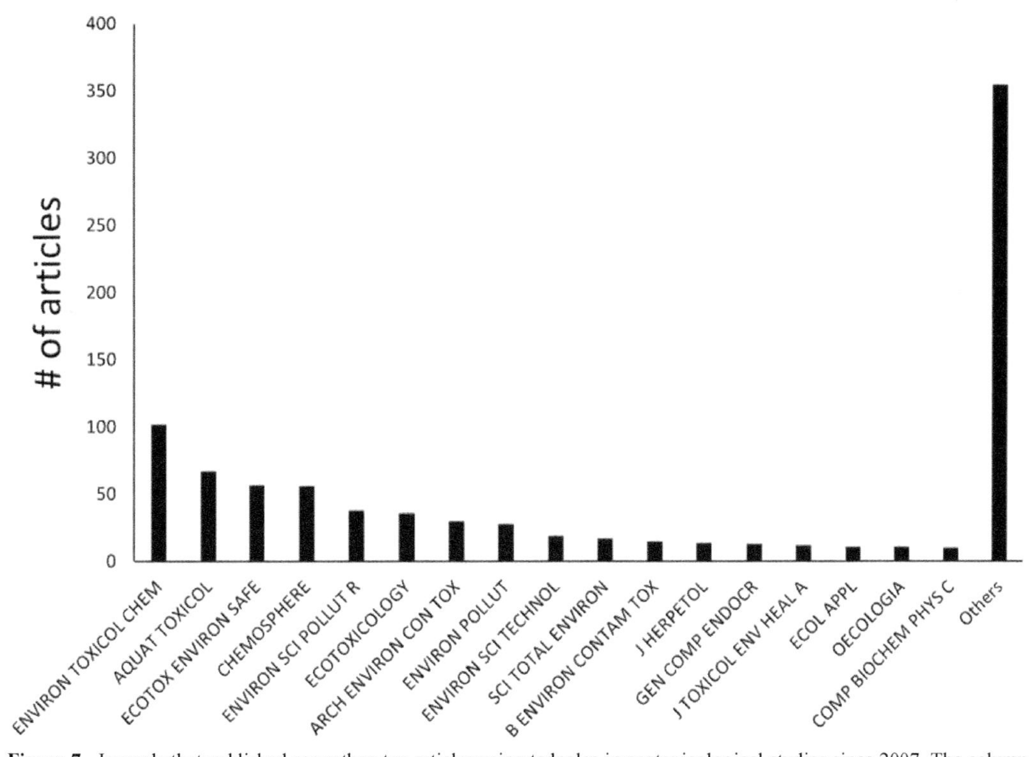

Figure 7. Journals that published more than ten articles using tadpoles in ecotoxicological studies since 2007. The column "others" represents the sum of articles published in 229 journals.

reviews receive much more attention than articles, with reviews being cited three times more often (total articles citations = 14,523; mean citation per article = 16.37; total reviews citations = 1,088; mean citation per review = 51.81). The most cited products had a more general approach and were not directly aimed at tadpoles but investigated the effects of chemical contaminants on the aquatic environment and aquatic organisms (e.g., Solomon et al. 2008, Mann et al. 2009, Hayes et al. 2010).

The journals that published the largest number of articles using tadpoles in ecotoxicological studies were Environmental Toxicology and Chemistry, Aquatic Toxicology, Ecotoxicology and Environmental Safety, Chemosphere, and Environmental Science and Pollution Research (Fig. 7). The greater frequency of publications in toxicology and environmental contamination journals may suggest that ecotoxicological experiments using tadpoles still receive little attention and are not published in multi- and interdisciplinary science journals (the so-called "mega journals", *sensu* Binfield 2014). Also, most of the publications that we retrieved are not ecotoxicological studies in the scope defined by Relyea and Hovermann (2006). These authors noted that ecologists and toxicologists have used tools and jargons forged in their own "bubbles" of knowledge, which includes their own or favorite journals to carry on the discussion. As stated by Relyea and Hovermann (2006) an ecotoxicological study "*naturally implies a hybrid of ideas and approaches from ecology and toxicology*".

5. Approaches, Trends and the Complexity of Interactions Between Contaminants and Other Stressors

Glyphosate and atrazine appear as important chemical contaminants, being the principal contaminants investigated in almost half of the 30 most cited articles (Table 1), reflecting the fact that herbicides represent about 60% of the pesticides used worldwide (Benbrook 2016, Duke and Dayan 2018).

Table 1. The thirty most cited articles published since 2007 using amphibian larvae in ecotoxicological studies.

Article	Document type	Times cited	DOI	Contaminant
Mann et al. (2009)	Review	350	10.1016/j.envpol.2009.05.015	Several
Rohr et al. (2008a)	Article	297	10.1038/nature07281	Atrazine
Hayes et al. (2010)	Article	230	10.1242/jeb.040865	Several
Hayes et al. (2011)	Review	177	10.1016/j.jsbmb.2011.03.015	Atrazine
Solomon et al. (2008)	Review	171	10.1080/10408440802116496	Atrazine
Rollins-Smith et al. (2011)	Article	138	10.1093/icb/icr095	Several
Boone et al. (2007)	Article	135	10.1890/1051-0761(2007)017 [0291:MSIACE]2.0.CO;2	Several
Rohr et al. (2008b)	Article	126	10.1890/07-1429.1	Atrazine, Carbaryl, Ghyphosate
Davey et al. (2008)	Article	122	10.1289/ehp.10131	Arsenic
Relyea and Jones (2009)	Article	119	10.1897/09-021.1	Glyphosate
Hinther et al. (2011)	Article	116	10.1021/es1041942	Thyroxine
Nations et al. (2011)	Article	112	10.1016/j.chemosphere.2011.01.061	Nanomaterials
Hayes et al. (2009)	Article	104	10.1007/s10886-009-9608-6	Bufadienolides
Davidson and Knapp (2007)	Article	89	10.1890/06-0181	Pesticides (general)
Relyea (2012)	Article	88	10.1890/11-0189.1	Glyphosate
Sparling and Fellers (2009)	Article	87	10.1897/08-336.1	Endosulfan, Chlorpyrifos
Costa et al. (2008)	Article	86	10.1007/s10646-007-0178-5	Glyphosate
Lajmanovich et al. (2011)	Article	83	10.1007/s00244-010-9578-2	Glyphosate
Heimeier et al. (2009)	Article	83	10.1210/en.2008-1503	Xenoestrogen
Brunelli et al. (2009)	Article	83	10.1016/j.aquatox.2008.09.006	Endosulfan
Carlsson et al. (2009)	Article	82	10.1897/08-524.1	Pharmaceutical industries effluents
Hogan et al. (2008)	Article	82	10.1016/j.ygcen.2008.03.011	Ethinylestradiol
Jones et al. (2009)	Article	81	10.1897/09-033.1	Endosulfan
Fini et al. (2007)	Article	77	10.1021/es0704129	Plastifiers, flame retardants (method)
Oka et al. (2008)	Article	76	10.1016/j.aquatox.2008.02.009	Atrazine
Belden et al. (2010)	Article	75	10.1002/etc.297	Strobilurin, azole, pyraclostrobin, azoxystrobilin
Moore et al. (2012)	Article	70	10.1016/j.ecoenv.2011.11.025	Glyphosate
Relyea and Hoverman (2008)	Article	70	10.1111/j.1600-0706.2008.16933.x	Malathion
Rohr et al. (2009)	Article	69	10.1007/s00442-008-1208-6	Atrazine

A common approach of the most cited articles is regarding how chemical pollutants lead to lethal and sublethal effects on tadpoles and how pollutants could affect the strength of interactions in food webs, as an explanatory mechanism for amphibian declines. Concerning lethal and sublethal effects of chemical pollutants, Moutinho et al. (2020) demonstrated that herbicides used in sugarcane crops (glyphosate, ametryn, 2,4-D, metribuzin and acetochlor), although employed in doses recommended by the manufacturers, can cause lethal and sublethal effects on tadpoles of *Boana pardalis*. The lethal effects reached 76% and 68% of mortality following prolonged exposure to ametryn and acetochlor, respectively. Ametryn significantly reduced activity rates and slowed tadpoles' developmental and growth rates, which may have important consequences by extending the larval period, increasing the risk of desiccation in the temporary and semi-permanent ponds where these tadpoles develop (Moutinho et al. 2020).

More recently, studies have focused on how chemical pollutants could affect the strength of interactions in food webs, including tadpoles. Hua and Relyea (2019) evaluated the direct and indirect consequences of the insecticide permethrin in complex aquatic mesocosms with 3 trophic levels and 13 animal species. Focusing on herbivorous tadpoles, permethrin contamination leads to extinction or population decrease of *Dryophytes versicolor*, *Lithobates pipiens* and *L. clamitans*. As an indirect effect, the reduction of herbivorous tadpoles contributed to the increase of periphyton abundance and a consequent mass increase of surviving tadpoles.

Chemical contaminants can affect amphibian populations by a multitude of pathways (Heyes et al. 2010). Davidson and Knapp (2007) exemplified the complexity of interactions between chemical contaminants and other ecological factors or stressors. They investigated the decline of anuran populations reported for the mountain yellow-legged frog (*Amerana muscosa*). This species experienced a severe population decline throughout its distribution in the Sierra Nevada range, which was associated with an invasive fish species, both by correlative (e.g., Bradford et al. 1998, Knapp 2005) and experimental (Vredenburg 2004) studies. However, frog populations continued to decline even in sites where invasive fishes were totally removed or from sites invasive fishes had never reached (Davidson and Knapp 2007). Pesticides drifting during applications and carried out by wind to the *Amerana muscosa* habitats, even in small concentrations, could explain the declining populations in sites where fishes were absent (Davidson and Knapp 2007). Moreover, the authors did not exclude further interactions with diseases and other chemicals. Indeed, they highlight that the attempt of finding causal factors to explain the amphibian declining phenomenon should take into account multiple stressors and their interactions, instead of pointing to a single cause (Davidson and Kapp 2007).

As though this wide network of interactions was not enough, Jones et al. (2009) reported that some chemical contaminants, endosulfan in particular, can have lag effects, i.e., starting to cause mortality days after exposure. This is particularly important when we consider that the main tool to evaluate the toxicity of contaminants are the LC50 curves. Once the lag effects could vary from three to ten days (Relya 2009, Jones et al. 2009), LC50 curves calculated for 48h or 96h could underestimate the toxicity and the mortality risk for non-target organisms, like tadpoles. Since chemical compounds do not need to be lethal to cause toxicity effects, this evidences that LC50 values without any other evaluation are underestimating potential toxic effects on amphibians (e.g., Freitas et al. 2019). Finally, even invasive species could be considered a source of chemical "contamination", once native species could be exposed to new or higher levels of chemical defenses inserted in their communities by invasive species (e.g., Hayes et al. 2009, Wijethunga et al. 2016). This system was investigated by Hayes et al. (2009), when they investigated the effect of cane toad toxins on native Australian frog species.

The directions about pesticide effects on freshwater systems pointed by Relyea and Hovermann (2006) summarize the approaches and trends discussed so far, especially about the importance of considering synergistic interactions between pesticides, abiotic factors and biotic stressors. Moreover, Relyea and Hovermann (2006) highlighted some important directions to ecotoxicological studies for questions as: (i) pesticide effects on oviposition and habitat use, (ii) pesticide application time and frequency, (iii) population dynamics, (iv) evolution of pesticide resistance, and (v) effects of pesticide mixtures. It is important to mention that recently, some studies (e.g., Cothran et al. 2013, Hua et al. 2015, Jones et al. 2018) started the investigation on evolution of pesticide resistance.

5.1 Interactions Among Chemical Contaminants and Abiotic Factors

Freshwater environments receive a range of contaminants that have different modes of action, and that interact among themselves and with other stressors in several different pathways. Studies assessing interactions between chemical contaminants are recent. For example, Dimitrie and Sparling (2014) evaluated the combined effects of two insecticides, endosulfan and chlorpyrifos, on survival, growth, and development of *Pseudacris regilla* tadpoles. Both insecticides, one organophosphorus and other organochlorine, have different modes of action and their interactions can vary depending on their concentrations and on the endpoints being assessed. But Dimitrie and Sparling (2014) found that interactive effects of these two insecticides facilitated the development of axial malformations. Another study assessed the toxicity of equitoxic and non-equitoxic binary mixtures of glyphosate and cypermethrin pesticides to *Rhinella arenarum* tadpoles (Brodeur et al. 2014). Equitoxic and non-equitoxic mixtures were significantly synergic, and the magnitude of the synergy varied from twofold to nine times depending on the different combinations of commercial products of these pesticides. These results evidenced the importance and urgency in understanding the mechanisms behind the synergy between/among chemical contaminants. In a wider approach, Hanlon and Parris (2014) tested the individual and combined effects of Roundup® and Sevin® and the pathogenic fungus *Batrachochytrium dendrobatidis* (Bd) on *Dryophytes versicolor* tadpoles in three phases of development (early, mid, and late). Besides having demonstrated different responses of tadpoles to the chemicals depending on the larval developmental phase, authors showed that Roundup ameliorated the effects of Bd on survival compared with tadpoles exposed to Bd alone. Sevin® reduced the mass of new metamorphs compared to Roundup® and reduced snout-vent length compared to all other treatments (Hanlon and Parris 2014). These results evidenced that chemical contaminants can attenuate the effects of Bd on tadpoles, but their effects vary according to the larval phase in which tadpoles are exposed. In another experiment, Hanlon et al. (2015) exposed *Lithobates sphenocephalus* tadpoles to fungicide thiophanate-methyl that can kill *Batrachochytrium dendrobatidis* (Bd) in the environment and clear Bd infections within anuran hosts. Tadpoles were exposed to Bd-infection and fungicide in mesocosms inoculated with aliquot of concentrated plankton suspension. Fungicide altered zooplankton diversity and caused mortality to all tadpoles in fungicide-exposed tanks. In tanks free of fungicide, Bd-exposed tadpoles metamorphosed smaller than Bd-unexposed in high-density treatments, and at the same size in low-density treatments. From these results, we can understand that using fungicide to control Bd is worse than not using it, at least for *Lithobates sphenocephalus* tadpoles. Thus, the nature of the contaminant and how it affects the environment together with the larval stage are important factors to determine the outcome of Bd infections.

Although several studies warn of the importance of evaluating the effects of chemical contaminants in different environmental conditions (e.g., Ficken and Byrne 2013, Babini et al. 2018), taking into account environmental parameters such as soil type and conductivity, temperature is, by far, the most considered environmental parameter. Aside from being easy to control in meso and microcosm experiments, variations in water temperature can have important effects on the responses of tadpoles to contaminants. Freitas et al. (2016) verified that *Lithobates catesbeianus* tadpoles treated at high temperatures (34°C) and exposed to diuron and its metabolite 3,4-dichloroaniline (3,4-DCA) presented changes in gene expression and plasma 3,3′,5-triiodothyronine (T3) concentrations, accelerating their metamorphosis. Also, temperature interacts with the herbicide Tebuthiuron, that at higher temperatures (over 32°C) lead to a decrease in the activities of ethoxyresorufin-O-deethylase (EROD) and superoxide dismutase (SOD), but increased the activities of glutathione S-transferase (GST), glutathione peroxidase (GPx), glucose 6-phosphate dehydrogenase (G6PDH), and acetylcholinesterase (AChE) (Grott et al. 2021). Likewise, temperature modulated the effects of the herbicides clomazone (Gamit®) and sulfentrazone (Boral®SC) on biochemical responses of *Physalaemus nattereri* and *Rhinella schneideri* tadpoles (Freitas et al. 2017a,b), and of glyphosate in tadpoles of *Bufo bufo* (Baier et al. 2016), as follow: (i) antioxidant enzymes, including catalase (CAT), superoxide dismutase (SOD), and glucose-6-phosphate dehydrogenase (G6PDH) had their activities increased by clomazone in *P. nattereri* tadpoles treated at high temperatures (32 and 36°C); (ii) G6PDH was increased in most *P. nattereri* groups exposed to 36°C, and also the biotransformation enzyme glutathione-S-transferase (GST) had more evident alterations in *P. nattereri* only at high temperatures (Freitas et al. 2017b); (iii) clomazone also increased the activity of SOD and GST at higher temperatures in *R. schneideri* tadpoles, but with a time lag of 3 days; (iv) lipid peroxidation was induced in both species exposed to clomazone only at higher temperatures (Freitas et al. 2017a,b). Integrated biomarker response (IBR) index indicated a synergic effect of temperature and sulfentrazone for tadpoles of *P. nattereri*, while the IBR was mainly influenced by temperature in *R. schneideri* (Freitas et al. 2017a,b).

Considering another chemical contaminant, Green and Salice (2020) demonstrated that elevated chloride levels, associated with increased temperature and variation in food quality decreased survival of *Lithobates sylvaticus* tadpoles. Baier et al. (2016) demonstrated that glyphosate effects were temperature-dependent, affecting the growth parameters, tail deformation and mortality of *Bufo bufo* tadpoles. Reinforcing the importance of considering temperature in assessment of the effects of contaminants on anuran tadpoles, Abel et al. (2020) demonstrated that *Lithobates catesbeianus* tadpoles have effective mechanisms to avoid or compensate possible inhibitions of their esterases by carbaryl. The enzyme carboxilesterase (CbE) had greater activity in tadpoles' groups exposed to the highest concentration of carbaryl (100 ug/L) at the two lowest experimental temperatures (20 and 25°C), and the acetilcolinesterase (AChE) had less activity at the highest temperature (30°C). Another important interaction is between chloride ion and temperature, since Green and Salice (2020) showed that environmentally relevant concentrations of chloride, temperatures, and the protein content of the diet negatively impacted the survival and the timing of metamorphosis of *Lithobates sylvaticus* tadpoles. On the contrary, no synergistic interactions between temperature regimes and the fungicide chlorothalonil was found in experiments with *Agalychnis callidryas* tadpoles (Alza et al. 2016). All these studies claim attention to the importance of considering the temperature effects on the action of chemicals and evaluating the effects of contaminants on tadpoles of different species, since they can present species-specific responses, as those presented above to *P. nattereri* and *R. schneideri*.

Some studies investigated the interactive effects of road salt (NaCl) and other contaminants. Santos et al. (2013) evaluated the influence of salinity on the copper (Cu) effect on *Pelophylax perezi*. Cu alone had a higher lethal toxicity to tadpoles and, although significant interactions between Cu and salinity have influenced the activities of CAT and lactate dehidrogenase enzymes, a consistent pattern of interactive effects was not observed. Stoler et al. (2017) evaluated the interaction of road salt and the commonly applied insecticide carbaryl on an aquatic food web. Although no synergistic effects were found, they evidenced a complex result of indirect community interactions and direct toxicity. Both of these studies evidence the difficulty as well as the importance of analyzing interactive effects between different stressors.

5.2 Chemical Contaminants Effects on Interspecific Interactions (Competition, Parasitism, and Predation)

The effects of chemical contaminants can take even more varied pathways. They can follow a direct pathway, due to their direct effect on the survivorship of tadpoles, or follow indirect pathways, as the immunosuppressive effects, for example, which increase the pathogen virulence and disease rates (Rohr et al. 2008a).

Rohr et al. (2008a) found that atrazine could positively influence the abundance of trematodes that parasite tadpoles, reducing their survivorship. But also, the effect of atrazine in the abundance of trematodes is larger when found together with phosphate. Rohr et al. (2008a) found that both atrazine and phosphate interact to increase the amount of periphyton in the environment. This occurs because the herbicide atrazine kills the phytoplankton, thus increasing the water clarity and luminosity penetrance, while phosphate increases the nutrient that is available for the periphyton in the environment. With an increased abundance and diversity of periphyton, the abundance and diversity of snails also increases, the latter being the intermediate host of trematodes. Moreover, sublethal effects of chemical contaminants could directly increase the susceptibility of tadpoles to parasite and disease infections, as reported for *Lithobates clamitans* exposed to atrazine, regarding trematode infections (Rohr et al. 2008b) and for *Duttaphrynus melanostictus*, exposed to four pesticides (chlorpyrifos, dimethoate, glyphosate, and propanil), regarding trematode *Acanthostomum burminis* infections (Jayawardena et al. 2017). However, Rohr et al. (2008b) drew attention to the possibility that atrazine could elicit environmental changes that increase the density of trematodes at the same time that it reduces the immune efficiency of tadpoles. Posteriorly, Rohr et al. (2013) found an increased mortality in *Osteopilus septentrionalis* tadpoles when exposed to both atrazine and the parasitic fungus *Batrachochytrium dendrobatidis*. More recently, Milotic et al. (2018) analyzed the effects of microcystins, a cyanobacterial blooms secondary compound, which can have both direct and indirect negative effects on larval amphibians by increasing their mortality and also their susceptibility to *Echinostoma* sp., a trematode parasite.

Chemical contaminants affect not only host-parasite interactions, but also change the strength and direction of competition. Sparling and Fellers (2009), for example, found that *Pseudacris regilla* tadpoles were more tolerant to endosulfan and chlorpyrifos when compared to *Rana boylii*, which helps to explain the decline observed in *R. boylii* populations while *P. regilla* populations seemed stable. *Amerana boylii* showed sublethal effects to chlorpyrifos, due to the smaller size and longer time to metamorphosis, which did not occur to *P. regilla*. Such sublethal effects could further reduce the competition efficiency of *R. boylii* when occurring together with resistant species, like *P. regilla*. Thus, Sparling and Fellers (2009) evidenced that species-specific differences in sensitivity to contaminants could be a mechanism that controls density mediated or trait mediated interactions, as interspecific competition. Shuman-Goodier et al. (2017) evaluated the effect of the exposure to a common herbicide, butachlor, on the competitive interactions between tadpoles of an invasive amphibian (*Rhinella marina*) and of a native one *Fejervarya vittigera*. Their results revealed that competition had a strong effect on the development of both species, but contrary to their predictions, the development and survival of endemic tadpoles was not affected by butachlor, whereas invasive *R. marina* tadpoles were affected across several endpoints including gene expression (Krüppel-like factor 9 - *klf9*; and thyroid hormone receptor beta - *thrb*), body size, and survival. Finally, in a more integrative approach, Kerby and Sih (2015) experimentally evaluated the toxicity of carbaryl on each one of the competitive species pair (*Pseudacris regilla* and *Amerana boylii*), on their competitive interactions, and on their interaction with a non-native crayfish predator (*Pacifastacus leniusculus*). On one hand, *Amerana boylii* was more susceptible to pesticide exposure than *P. regilla*, which reduced their ability to compete, and also increased their mortality by 50% when exposed to predators. On the other hand, tadpoles of *P. regilla* exhibited no change. These results draw our attention to the use of multiple species tests that focus on key species interactions.

Tadpoles' predation can also be affected by contaminants, since they can disrupt the ability of tadpoles to answer to predator cues. For example, Boone et al. (2007) found that treatments with fertilizer and carbaryl not only had a positive effect on body mass of anuran larvae, but also increased the mortality of tadpoles when exposed to carbaryl plus predators. Troyer and Turner (2015) evaluated the anti-predator response of anuran tadpoles kept in water collected from an impaired stream, from a treated wastewater effluent and in de-chlorinated tap water. The response to predator cues by tadpoles kept in stream water and wastewater effluent took 20% and 25% more time than those of tadpoles kept in tap water, evidencing that chemically altered environments can disturb the chemical perception of predators. Polo-Cavia et al. (2016) showed that *Pelobates cultripes* tadpoles exposed to two common contaminants, remained unresponsive to predator cues when either humic acid or ammonium nitrate was added to the water, even at low concentrations. Miko et al. (2017) exposed *Rana dalmatina* and *Bufo bufo* tadpoles to predator chemical cues under varying concentrations of glyphosate-based herbicide, finding that the herbicide was moderately toxic to tadpoles, but the presence of predator chemical cues did not affect its lethality in either species. Moreover, they draw attention that, instead of following a strict standardization that can drive to different results depending on the origin of species and other conditions, the introduction of variation into the design and replicate entire experiments may be highly beneficial to ecotoxicological experiments. Further expanding the approach, Hanlon and Relyea (2013) investigated the effects of low nominal concentrations of malathion and endosulfan on activity and survival of *Lithobates catesbeianus* and *L. clamitans* tadpoles when exposed to four predator treatments (no predators, *Belostoma flumineum, Notophthalmus viridescens* and *Anax junius*). The interactive effects of the chemicals on the survival of tadpoles of both species were perceived after 48 hrs. For *L. catesbeianus*, all predators reduced the amount of tadpole mortality when exposed to endosulfan. For *L. clamitans*, additive negative effects occurred, except that newts increased tadpole mortality when exposed to endosulfan. Contrary to these findings, some studies found no evidence that contaminants (chlorpyrifos and glyphosate, Cothran et al. 2013; and atrazine, Davis et al. 2012) lead to a performance cost of tadpoles when facing competition and fear of predation.

A more recent approach is the assessment of several environmental contaminants on species richness and food webs. For instance, Ficken and Byrne (2013) found that species richness was negatively correlated with sediment concentrations of six heavy metals (copper, nickel, lead, zinc, cadmium and mercury) and also with wetland water electrical conductivity (a proxy for salinity) and concentrations of orthophosphate. Regarding food webs, Cothran et al. (2011) and Groner and Relyea (2011) demonstrated a trophic cascade effect of agrochemicals. Cothran et al. (2011) showed that, at a high concentration, malathion was lethal to cladocerans, who are primary consumers and to dragonflies, a top predator. These lethal effects initiated a trophic cascade effect, i.e., density-mediated indirect effects. The decreasing of dragonfly foraging efficiency resulted in increased tadpole survival, which decreased periphyton, a food resource to tadpoles. Similar results were found by Groner and Relyea (2011) for malathion and carbaryl. Both insecticides caused comparable trophic cascades that affected zooplankton and phytoplankton abundances; however, their effects on *Lithobates pipiens* tadpoles differed, mostly under high concentrations of insecticides, hampering the generality of tadpoles' responses.

5.3 Chemical Contaminants Effects on Tadpole Physiology

Some of the most cited studies investigated how environmental contaminants could directly affect tadpoles by analyzing how their physiological routes could be modified or damaged by these exogenous components. Feminization induced by synthetic estrogen was investigated by Hogan et al. (2008), who found a female biased sex-ratio in *Lithobates pipiens* tadpoles exposed in the early developmental stages, due to endocrine disruption. Hayes et al. (2011) performed a literature review to link atrazine as the cause of feminization in vertebrates, and how it could affect anurans since its larval stages, besides other indirect effects. The review of Hayes et al. (2011) is diametrically opposite to the findings of Solomon et al. (2008), who did not find any evidence that atrazine affects the reproduction in fish, amphibians or reptiles. In common, both articles did not use meta-analytical tools to review the results in literature, but used guidelines to evaluate the causative agents of diseases, like Koch's postulates and the Bradford–Hill guidelines, in Solomon et al. (2008), or Hill criteria, in Hayes et al. (2011).

Davey et al. (2008) used *Xenopus laevis* tadpoles as animal models and found that even small levels of arsenic could disrupt gene regulation via thyroid hormone receptors, inducing tail malformations. Another example is the study of Carlsson et al. (2009), which submitted *Xenopus tropicalis* tadpoles to effluents of a pharmaceutical industry in India. By assessing how these pharmaceutical contaminants affect tadpole metamorphosis, Carlsson et al. (2009) investigated the effects on the thyroid hormone system, although they did not find any direct negative effect of the contaminant that could be mediated by the thyroid system. However, similar action routes were studied by Hinther et al. (2011) for bactericides (Methyl triclosan and triclocarban), and by Shuman-Goodier et al. (2017) for butachlor, an herbicide, that found evidence that the negative effects of the contaminants were mediated by the disruption of the thyroid hormone.

Studying how chemical contaminants disrupt thyroid hormone receptors in frogs is particularly interesting because the action routes of contaminants can be the same for other vertebrate groups, including mammals (Davey et al. 2008). An example of the usefulness of such similarities is the study of Heimeier et al. (2009), who used tadpoles to investigate the effects on development triggered by bisphenol A contamination, a chemical used in the plastic industry, instead of uterus-enclosed mammalian embryos. Once the physiological control of tadpole metamorphosis uses the same routes and hormones that control the postembryonic period in mammals, the results of bisphenol A in tadpoles can be used to predict the response in mammals.

Other physiological routes can be disrupted by the actions of chemical contaminants as well and tadpoles are excellent models for such studies (e.g., Lajmanovich et al. 2011). Costa et al. (2008) found that glyphosate in sub-lethal doses induces an increase in reactive oxygen species and oxidative stress in tadpoles. Further, they observed that adrenergic stimulation stress resulted in tachycardia heartbeats and hyperactive tadpoles. Costa et al. (2008) suggest that, although energetically costly, such modulation in cardiac performance is a mechanism to cope with glyphosate intoxication. More critical, they show that even small concentrations of herbicides could affect tadpole performance and, in consequence, negatively affect their survival and population dynamics. Another physiological route that can be disrupted by the actions of chemical contaminants was evidenced by Bókony et al. (2017). The authors showed that chronic exposure to a glyphosate-based herbicide affected the production of bufadienolides, making tadpoles of *Bufo bufo* more toxic.

The sensitivity to physiological effects of environmental contaminants could be different among anuran species. Sparling and Fellers (2009) reported that *Rana boylii* had a steeper rate of decline in the cholinesterase activity than *Pseudacris regilla* when tadpoles of both species were exposed to the insecticide chlorpyrifos. Leite et al. (2010) analyzed the effect of diazinon, another insecticide, in tadpoles of *Scinax fuscovarius* and found that tadpoles from earlier developmental stages seemed to be less responsive to organophosphate pesticides. However, the acetylcholinesterase (AChE) activity was sensitive to diazinon in both earlier and later tadpoles' development stages, making this enzyme a good organophosphate pesticide biomarker. The effect of the insecticide fipronil in antioxidant defense systems was also tested in tadpoles of *Scinax fuscovarius* (Margarido et al. 2013), in which it caused the inhibition of glutathione-S-transferase (GST), an enzyme involved in the detoxification process.

Nanomaterials could also affect larvae development at low concentrations. Nations et al. (2011) found that ZnO, TiO2, Fe2O3, and CuO nanomaterials (20–100 nm), despite not increasing the mortality of *Xenopus laevis* embryos, induced gastrointestinal, spinal and other developmental abnormalities. Also, the toxicity of microplastic polyethylene was investigated by Araújo et al. (2020) who found different histopathological changes on *Physalaemus cuvieri* tadpoles (e.g., blood vessel dilation, infiltration, hypertrophy and hyperplasia), as well as changes in hepatocyte nuclei size, volume and shape.

5.4 Morphological and Behavioral Changes Induced by Chemical Contaminants

Other important effects of chemical contaminants are the morphological and behavioral changes. Several amphibian malformations can be induced by chemicals. Endosulfan, for example, induced mouth and skeletal malformations in *Bufo bufo* tadpoles submitted to chronic exposure (Brunelli et al. 2009), and caused failure to develop one or both forelimbs, stunted hindlimb growth and axial malformation in tadpoles of *Fejervarya* spp. (Devi and Gupta 2013). Lajmanovich et al. (2003)

observed craniofacial and mouth deformities, eye abnormalities and bent curved tails of *Scinax nasicus* tadpoles exposed to a glyphosate formulation. Tadpoles of *Rhinella arenarum* exposed to ciprofloxacin antibiotic and a glyphosate-based herbicide (individually or combined) presented a high frequency of morphological abnormalities, such as abnormal eye shape, abnormal gill chambers, bilateral asymmetry, edema, visceral hypotrophy, gut abnormal coiling, mouth alteration and abnormal tail (Boccioni et al. 2020).

However, not all morphological changes induced by chemical contaminants are malformations. Grott et al. (2021) demonstrated that tebuthiuron (TBU) and temperature disturbed the metabolic homeostasis of *Lithobates catesbeianus* tadpoles after 16 days of exposure, causing substantial alterations in the liver morphology, determined by the severity scores from histological analyses. Through the fluctuating asymmetry (FA) analysis, Costa and Nomura (2016) observed deviations in the symmetry of bilateral morphological traits (position of the nostrils and eyes width) in *Physalaemus cuvieri* tadpoles exposed to Roundup Original®, a glyphosate-based herbicide. The FA is a sensitive biomarker of environmental stress and reflects the impacts of stressors on developmental stability of individuals (Palmer and Strobeck 1986, Beasley et al. 2013). Often, morphological changes induced by contamination affect tadpole attributes related to swimming activity, food acquisition, and predator or food detection, which can result in increased risk of predation and reduced competitive potential (e.g., Denoël et al. 2012, Hanlon and Relyea 2013, Costa and Nomura 2016).

In addition, some morphological changes could be similar to morphological responses induced by chemical cues of natural predators of tadpoles, which suggest that glyphosate is activating the same physiological routes used by tadpoles to respond to predators (Relyea 2012, Middlemis Maher et al. 2013). Relyea (2012) detected these types of morphological responses in mesocosm experimental setups, where tadpoles of *Lithobates pipiens* and *Lithobates sylvaticus* were exposed to both glyphosate-based herbicide and predators (adult newts of *Notophthalmus viridescens* or *Anax junius* dragonflies). Herbicide exposure induced the same deeper tails on tadpoles of both species than when tadpoles were exposed to dragonflies (Relyea 2012). This should explain why herbicides could increase mortality when combined with predators in laboratory experiments.

Behavioral changes induced by chemical contaminants include microhabitat use, activity degree and swim speed changes. The very early behavioral changes in the swimming activity of *Bufo bufo* tadpoles exposed to endosulfan, detected before any other morphological effect, denotes its neurotoxic effect on tadpoles behavior (Brunelli et al. 2009). Also, a significant decrease in the swimming performance was observed in tadpoles of *Rhinella humboldti*, *R. marina*, *Boana xerophylla* and *Engystomops pustulosus*, obtained from the embryos exposed to cypermethrin, evidencing a neurological action of cypermethrin that produced a high tadpole lethality and sublethal effects on developing embryos (Velasquez et al. 2017). Similar results were obtained for *Rhinella arenarum* tadpoles exposed to chlorpyrifos, whose toxicity decreased the maximum swimming distance and, additionally, this behavioral change was not compensated at high temperatures, which enhanced the swimming performance in the control treatment (Quiroga et al. 2019).

Again, the interaction between chemical cues of predator presence and chemical contaminants usually results in increased tadpole mortality (e.g., Relyea and Mills 2001, Relyea 2005). For example, Moore et al. (2015) submitted tadpoles of *Lithobates sylvaticus* to predation risk and observed that tadpoles failed to express anti-predator responses when submitted to a glyphosate formulation. The authors suggested that the absence of anti-predator responses is associated with a complete or partial deactivation of the alarm cue system of tadpoles caused by herbicide exposure. As observed by Tierney et al. (2006), an exposure to glyphosate formulation impaired the olfactory system of tadpoles, which can affect the recognition of predator cues. Opposite to this, Relyea (2012) found that behavioral changing drives to lower lethality in a mesocosm experiment. In such conditions, chemical cues from predator presence induced a microhabitat shift in tadpoles, changing their position from the surface to the bottom of the mesocosm, where the concentrations of glyphosate were lower. Glyphosate concentration is stratified in ponds, following temperature stratification of natural aquatic environments, being more concentrated near the surface than at the bottom (Jones et al. 2010, 2011). Therefore, by driving tadpoles to the bottom of the mesocosm, chemical cues of predators, as paradoxically as it is, facilitate tadpoles' survivorship to chemical contaminants (Relyea 2012).

Relyea (2012) draws attention to the degree of realism of laboratory results and on how those interactions function when transported to more complex environments. The mesocosms used in the Relyea (2012) experiments had enough depth to allow glyphosate stratification to occur, which is not observed in more usual laboratory experiments. From the tadpoles' behavior perspective this makes sense—shifting habitat is a defense mechanism that is less costly than immobility (Mamede and Nomura 2021), and thus is probably more prone to be elicited.

A range of strategies and behavioral responses of tadpoles exposed to contaminants are reported, which do not have a pattern and appear to vary according to the species and/or types of contaminants tested. For example, Moreira et al. (2019) observed that tadpoles of *Lithobates catesbeianus* exposed to environmentally relevant concentrations of the herbicide Diuron showed a reduction in swimming activity. Additionally, tadpoles showed a significant increase in pesticide avoidance behavior, displacing from their microhabitats in search of more favorable areas (i.e., uncontaminated or less contaminated. Similarly, Moore et al. (2015) observed a movement reduction of *Lithobates sylvaticus* tadpoles in response to injured conspecific cues. A set of behavioral changes was found by Denoël et al. (2013) for tadpoles of *Rana temporaria*. They showed that contaminated tadpoles traveled shorter distances, swam less often, and at a lower mean speed, and occupied a less peripheral position than control tadpoles. On the other hand, Egea-Serrano et al. (2011) observed an increase in activity and a change in the microhabitat use in tadpoles of *Pelophylax perezi* exposed to nitrogen compounds. These increases in activity can be associated with vertical movements in the water column to reach the surface and breathe air, seeking to reduce the effects of hypoxia and/or suffocation (Marco and Blaustein 1999, Hoffmann 2010).

Finally, Sievers et al. (2019) conducted a systematic review and a meta-analysis to evaluate behavioral changes of amphibians exposed to contaminants. Among several results, they highlighted that most studies focused on pesticides (mainly insecticides and herbicides), with fewer on metals, fertilizers, pharmaceuticals, salinity and others. Also, 81% of studies used tadpoles as a model. Considering the behavioral endpoints observed in tadpoles, most studies measured activity levels, followed by speed, feeding and hiding behaviors, abnormal swimming patterns, predator escape responses and surface activity. Meta-analysis results highlighted that contaminants (i) increase the rates of abnormal swimming of tadpoles, mainly when exposed to insecticides, (ii) reduce the activity levels when exposed to insecticides and metals, (iii) reduce the ability to exhibit predator escape responses when exposed to herbicides and insecticides, and (iv) reduce the feeding rates, when exposed to insecticides. These variations in activity patterns and differential use of environments (e.g., microhabitat preferences, Annibale et al. 2019, 2020) may have implications for ecological interactions with predators and competitors.

5.5 Differences in Sensitivity Between Taxonomic Groups and Species

Aside from possible multiple interactions among chemical contaminants themselves and between them and other stressors, driving to different pathways in which contaminants can affect tadpoles, Relya and Jones (2009) found that sensitivity between taxonomic groups differs. Sensitivity to glyphosate in formulation with polyethoxylated tallowamine (POEA) was higher for anurans when compared to salamanders (Relya and Jones 2009). Sensitivity to chemical contamination could vary even within anuran species, as demonstrated by Sparling and Fellers (2009) for *Pseudacris regilla* and *Amerana boylii* tadpoles, being the first more tolerant to endosulfan and chlorpyrifos when compared to *Amerana boylii*. Also, different levels of sensitivity were observed between *Physalaemus nattereri* and *Rhinella schneideri* exposed to heat stress and clomazone (Gamit®) (Freitas et al. 2017a), as well heat stress and sulfentrazone (Boral 500SC®) (Freitas et al. 2017b).

In addition to the natural levels of interspecific tolerance (i.e., species with more or less resilience to contaminant levels), exposure to contaminants could induce this tolerance as well. Hua et al. (2013) collected embryos and hatchlings of *Lithobates sylvaticus* from populations located close to and far from agricultural fields and exposed them to sublethal concentrations of carbaryl. This exposure induced higher tolerance to a subsequent lethal concentration of carbaryl later in life. Interestingly, this induced tolerance was observed only in populations far from agricultural fields. The authors explain that populations far from agricultural fields might occasionally experience insecticides and suggested an induced tolerance due to genetic assimilation (GA), a process initiated by phenotypic plasticity. Later, Hua et al. (2015) tested the GA hypothesis and found that induced tolerance of *Lithobates sylvaticus* populations was consistent with pesticides' evolutionary responses. They observed that tadpoles living closer to agricultural fields were more tolerant (constitutive tolerance) than populations living far from agriculture (induced tolerance). In accordance with this result, Cothran et al. (2013) observed constitutive tolerance of exposed *Lithobates sylvaticus* tadpoles to chlorpyrifos and demonstrated that populations collected closer to agriculture were more tolerant than populations collected far from agricultural areas. However, even contaminant-tolerant populations can experience negative effects of prolonged exposure to sublethal levels of contaminants. The prolonged exposure to chronic contamination levels can lead to various sublethal effects, such as morphological, physiological, genetic and/or behavioral changes that reduce tadpole's performance (e.g., Leite et al. 2010, Jones et al. 2010, Relyea 2012, Moore et al. 2015, Costa and Nomura 2016).

We could expect that the taxonomic distance could be a good predictor of the difference in sensitivity to chemical contaminants among species, which could help to explain part of the general decline in amphibian populations worldwide. However, despite Jones et al. (2009) having presented a hypothesis that this sensibility to chemical contaminants is phylogenetically structured, there is data still lacking to test, or even to support this claim. In one of the rare studies employing more integrated approaches, Chiari et al. (2015) used phylogenetic comparative methods to evaluate the sensitivity to copper sulfate, a commonly used pesticide, under different temperature scenarios. The authors show that this sensitivity exhibits a strong phylogenetic signal when controlled for experimental temperature. Similar findings have already been demonstrated by Hammond et al. (2012), who found significant phylogenetic signals in the sensitivity to endosulfan and in the time lag effects on tadpole mortality. These results emphasize the importance of considering the evolutionary perspective to make accurate predictions of amphibian sensitivity to chemical contaminants.

6. Conclusions and Perspectives

Most ecotoxicological assays were carried out under controlled conditions in the laboratory, but they had a low degree of realism when compared to the experiments carried out in outdoor mesocosms (e.g., Mikó et al. 2015) or in the field (e.g., Boone and James 2005, Relyea 2012). Sparling and Fellers (2009) also addressed this issue and drew attention that the results obtained in the laboratory were not always representative of real-world consequences. They call attention, for example, that environmental contaminants' behavior is not yet fully understood and tadpoles, and all other living organisms, are submitted to a plethora of different chemicals and other anthropogenic and non-anthropogenic stressors (e.g., predators, competitors, UV, diseases, temperature) that could interact in an additive and/or a synergistic way. Because of that, even non-lethal effects of chemical contaminants could trigger a cascade of physiological and ecological changes that could culminate with the species decline or extinction.

In contrast, field experiments are conducted in large-scale environments, at the cost of accuracy due to the great number of uncontrolled variables (Boone and James 2005, Mann et al. 2009). Thus, the use of standard laboratory tests may have to be reconsidered and their benefits carefully weighed against the difficulties of performing experiments under more natural conditions. In this way, the use of outdoor mesocosms seems to be the most powerful experimental technique because it allows a greater control of many variables within the experimental system and allows more realism (see review in Boone and James 2005).

Regardless of the chosen methodology, researchers must be aware of methodological-specific limitations, especially when searching for causal links or explanations for amphibian population declines. It is a common hypothesis to expect the worst scenarios when designing experiments with chemical contaminants and such expectations could lead to unrealistic experiments, as with high concentrations of chemicals in treatments because "tadpoles did not show the expected mortality in pilot trials". The trick here is that mortality is not the only possible negative outcome that a researcher should expect when investigating chemical contaminants' effects.

Finally, considering that the sensitivity to chemical contamination differs among anuran species, it is important that the tests on the effects of contaminants on tadpoles include as many species as possible (e.g., Ficken and Byrne 2013) and also species from different evolutive lineages.

Acknowledgments

The authors would like to thank Fabiane Annibale for the graphic development of Figure 1, and Daniela P. Mejía for suggestions and careful language editing. DCRF (# 304760/2021-8), FN (# 301232/2018-0) and MS (# 309365/2019-8) are CNPq fellows. RNC is "Pesquisador Produtividade da UEMG – PQ/UEMG - edital 10/2022 e edital 08/2021".

References Cited

Abel, G., S.C. Grott, D. Bitschinski, N.G. Istael, S.P. Silva, F.E. Carneiro and E.A. de Almeida. 2020. Influence of temperature on the effects of the insecticide carbaryl in bullfrog tadpoles (*Lithobates catesbeianus*). Revinter 13: 52–62.

Agostini, M.G., I. Roesler, C. Bonetto, A.E. Ronco and D. Bilenca. 2020. Pesticides in the real world: The consequences of GMO-based intensive agriculture on native amphibians. Biol. Conserv. 241: 108355.

Allentoft, M.E. and J. O'Brien. 2010. Global amphibian declines, loss of genetic diversity and fitness: a review. Diversity 2: 47–71.

Alroy, J. 2015. Current extinction rates of reptiles and amphibians. P. Natl. Acad. Sci. USA 112(42): 13003–13008.

Altig, R., M.T. Whiles and C.L. Taylor. 2007. What do tadpoles really eat? Assessing the trophic status of an understudied and imperiled group of consumers in freshwater habitats. Freshwater Biol. 52: 386–395.

Alza, C.M., M.Z. Donnelly and S.M. Whitfield. 2016. Additive effects of mean temperature, temperature variability, and chlorothalonil to red-eyed treefrog (*Agalychnis callidryas*) larvae. Environ. Toxicol. Chem. 35(12): 2998–3004.

Annibale, F.S., V.T.T. de Sousa, C.E. de Sousa, M.D. Venesky, D.C. Rossa-Feres, F. Nomura and R.J. Wassersug. 2019. Influence of substrate orientation on tadpoles' feeding efficiency. Biol. Open 8: 037598.

Annibale, F.S., V.T.T. de Sousa, C.E. de Sousa, M.D. Venesky, D.C. Rossa-Feres, R.J. Wassersug and F. Nomura. 2020. Smooth, striated, or rough: how substrate textures affect the feeding performance of tadpoles with different oral morphologies. Zoomorphology 139: 97–110.

Araújo, A.P.C., A.R. Gomes and G. Malafaia. 2020. Hepatotoxicity of pristine polyethylene microplastics in neotropical *Physalaemus cuvieri* tadpoles (Fitzinger, 1826). J. Hazard. Mater. 386: 121992.

Arribas, R., C. Díaz-Paniagua, S. Caut and I. Gomez-Mestre. 2015. Stable isotopes reveal trophic partitioning and trophic plasticity of larval amphibian guild. PLoS ONE 10: 1–19.

Babini, M.S., C.D. Bionda, Z.A. Salinas, N.E. Salas and A.L. Martino. 2018. Reproductive endpoints of *Rhinella arenarum* (Anura, Bufonidae): populations that persist in agroecosystems and their use for the environmental health assessment. Ecotox. Environ. Safe. 154: 294–301.

Baier, F., E. Gruber, T. Hein, E. Bondar-Kunze, M. Ivankovic, A. Mentler, C.A. Bruhl, B. Spangl and J.G. Zaller. 2016. Non-target effects of a glyphosate-based herbicide on Common toad larvae (*Bufo bufo*, Amphibia) and associated algae are altered by temperature. PeerJ. 4: e2641.

Bain, C., T. Selfa, T. Dandachi and S. Velardi. 2017. 'Superweeds' or 'Survivors'? Framing the problem of glyphosate resistant weeds and genetically engineered crops. J. Rural Stud. 51: 211–21.

Baker, N.J., B.A. Bancroft and T.S. Garcia. 2013. A meta-analysis of the effects of pesticides and fertilizers on survival and growth of amphibians. Sci. Total Environ. 449: 150–156.

Barnosky, A.D., N. Matzke, S. Tomiya, G.O.U. Wogan, B. Swartz, T.B. Quental et al. 2011. Has the Earth's sixth mass extinction already arrived? Nature 471: 51–57.

Beasley, D.A.E., A. Bonisoli-Alquati and T.A. Mousseau. 2013. The use of fluctuating asymmetry as a measure of environmentally induced developmental instability: a meta-analysis. Ecol. Indic. 30: 218–226.

Belden, J., S. McMurry, L. Smith and P. Reilley. 2010. Acute toxicity of fungicide formulations to amphibians at environmentally relevant concentrations. Environ. Toxicol. Che. 29: 2477–2480.

Benbrook, C.M. 2016. Trends in glyphosate herbicide use in the United States and globally. Environ. Sci. Eur. 28(3): 1–15.

Binfield, P. 2014. Novel scholarly journal concepts. pp. 155–163. *In*: Bartling, S. and S. Friesike [eds.]. Opening Science: the Evolving Guide on How the Internet is Changing Research, Collaboration and Scholarly Publishing. SpringerOpen, London, United Kingdom.

Bishop, C.A., N.A. Mahony, J. Struger, P. Ng and K.E. Pettit. 1999. Anuran development, density and diversity in relation to agricultural activity in the Holland River Watershed, Ontario, Canada (1990 – 1992). Environ. Monit. Assess. 57: 21–43.

Blaustein, A.R. and J.M. Kiesecker. 2002. Complexity in conservation: lessons from the global decline of amphibian populations. Ecol. Lett. 5(4): 597–608.

Blaustein, A.R., J.M. Romansic, J.M. Kiesecker and A.C. Hatch. 2003. Ultraviolet radiation, toxic chemicals and amphibian population declines. Divers. Distrib. 9: 123–140.

Blaustein, A.R., B.Z. Han, R.A. Relyea, P.T.J. Johnson, J.C. Buck, S.S. Gervasi and L.B. Kats. 2011. The complexity of amphibian population declines: understanding the role of cofactors in driving amphibian losses. Ann. N. Y. Acad. Sci. 1223: 108–119.

Boccioni, A.P.C., R.C. Lajmanovich, P.M. Peltzer, A.M. Attademo and C.S. Martinuzzi. 2020. Toxicity assessment at different experimental scenarios with glyphosate, chlorpyrifos and antibiotics in *Rhinella arenarum* (Anura: Bufonidae) tadpoles. Chemosphere. 273: 128475.

Bókony, V., Z. Mikó, A.M. Móricz, D. Krüzselyi and A. Hettyey. 2017. Chronic exposure to a glyphosate-based herbicide makes toad larvae more toxic. Proc. R. Soc. B 284(1858): 1–7.

Boone, M.D. and C.M. Bridges. 2003. Effects of carbaryl on green frog (*Rana clamitans*) tadpoles: timing of exposure versus multiple exposures. Environ. Toxicol. Chem. 22: 2695–2702.

Boone, M.D. and S.M. James. 2005. Use of aquatic and terrestrial mesocosms in ecotoxicology. Appl. Herpetol. 2: 231–257.

Boone, M.D., R.D. Semlitsch, E.E. Little and M.C. Doyle. 2007. Multiple stressors in amphibian communities: effects of chemical contamination, bullfrogs, and fish. Ecol. Appl. 17: 291–301.

Bradford, D.F., S.D. Cooper, T.M. Jenkins Jr., K. Kratz, O. Sarnelle and A.D. Brown. 1998. Influences of natural acidity and introduced fish on faunal assemblages in California alpine lakes. Can. J. Fish. Aquat. Sci. 55: 2478–2491.

Bridges, C.M. and M.D. Boone. 2003. The interactive effects of UV-B and insecticide exposure on tadpole survival, growth and development. Biol. Conserv. 113: 49–54.

Brodeur, J.C., M.B. Poliserpi, M.F. D'Andrea and M. Sanchez. 2014. Synergy between glyphosate- and cypermethrin-based pesticides during acute exposures in tadpoles of the common South American toad *Rhinella arenarum*. Chemosphere 112: 70–76.

Brühl, C.A., T. Schmidt, S. Pieper and A. Alscher. 2013. Terrestrial pesticide exposure of amphibians: an underestimated cause of global decline? Sci. Rep. 3: 1–4.

Brunelli, E., I. Bernabò, C. Berg, K. Lundstedt-Enkel, A. Bonacci and S. Tripepi. 2009. Environmentally relevant concentrations of endosulfan impair development, metamorphosis and behaviour in *Bufo bufo* tadpoles. Aquat. Toxicol. 91: 135–142.

Carlsson, G., S. Örn and D.G.J. Larsson. 2009. Effluent from bulk drug production is toxic to aquatic vertebrates. Environ. Toxic. Chem. 28(12): 2656–2662.

Ceballos, G., P.R. Ehrlich, A.D. Barnosky, A. García, R.M. Pringle and T.M. Palmer. 2015. Accelerated modern human–induced species losses: entering the sixth mass extinction. Sci. Adv. 1(5): 1–5.

Ceballos, G., P.R. Ehrlich and P.H. Raven. 2020. Vertebrates on the brink as indicators of biological annihilation and the sixth mass extinction. P. Nalt. Acad. Sci. USA 117 (24): 13596–13602.

Chiari, Y., S. Glaberman, N. Seren, M.A. Carretero and I. Capellini. 2015. Phylogenetic signal in amphibian sensitivity to copper sulfate relative to experimental temperature. Ecol. Appl. 25(3): 596–602.

Coady, K., T. Marino, J. Thomas, R. Currie, G. Hancock, J. Crofoot, L. McNalley, L. McFadden, D. Geter and G. Klecha. 2010. Evaluation of the amphibian metamorphosis assay: exposure to the goitrogen methimazole and the endogenous thyroid hormone l-thyroxine. Environ. Toxicol. Chem. 29(4): 869–880.

Connelly, S., C.M. Pringle, M.R. Whiles, K.R. Lips, S. Kilham and R. Brenes. 2011. Do tadpoles affect leaf litter decomposition in neotropical streams? Freshwater Biol. 56: 1863–1875.

Costa, M.J., D.A. Monteiro, A.L. Oliveira-Neto, F.T. Rantin and A.L. Kalinim. 2008. Oxidative stress biomarkers and heart function in bullfrog tadpoles exposed to Roundup Original. Ecotoxicology 17: 153–163.
Costa, R.N. and F. Nomura. 2016. Measuring the impacts of Roundup Original® on fluctuating asymmetry and mortality in a Neotropical tadpole. Hydrobiologia 765: 85–96.
Cothran, R.D., F. Radarian and R.A. Relyea. 2011. Altering aquatic food webs with a global insecticide: arthropod-amphibian links in mesocosms that simulate pond communities. J. N. Am. Benthol. Soc. 30(4): 893–912.
Cothran, R.D., J.M. Brown and R.A. Relyea. 2013. Proximity to agriculture is correlated with pesticide tolerance: evidence for the evolution of amphibian resistance to modern pesticides. Evol. Appl. 6: 832–841.
Davey, J.C., A.P. Nomikos, M. Wungjiranirun, J.R. Sherman, L. Ingram, C. Batki et al. 2008. Arsenic as an endocrine disruptor: arsenic disrupts retinoic acid receptor–and thyroid hormone receptor–mediated gene regulation and thyroid hormone–mediated amphibian tail metamorphosis. Environ. Health Persp. 116(2): 165–172.
Davidson, C. and R.A. Knapp. 2007. Multiple stressors and amphibian declines: dual impacts of pesticides and fish on Yellow-Legged frogs. Ecol. Appl. 17(2): 587–597.
Davis, M.J., J.L. Purrenhage and M.D. Boone. 2012. Elucidating predator-prey interactions using aquatic microcosms: complex effects of a crayfish predator, vegetation, and atrazine on tadpole survival and behavior. J. Herpetol. 46(4): 527–534.
Denoël, M., B. D'Hooghe, G.F. Ficetola, C. Brasseur, E. De Pauw, J.P. Thomé and P. Kestemont. 2012. Using sets of behavioral biomarkers to assess short-term effects of pesticide: a study case with endosulfan on frog tadpoles. Ecotoxicology 21: 1240–1250.
Denoël, M., S. Libon, P. Kestemont, C. Brasseur, J.F. Focant and E. De Pauw. 2013. Effects of a sublethal pesticide exposure on locomotor behavior: a video-tracking analysis in larval amphibians. Chemosphere 90(3): 945–951.
Devi, N.N. and A. Gupta. 2013. Toxicity of endosulfan to tadpoles of *Fejervarya* spp. (Anura: Dicroglossidae): mortality and morphological deformities. Ecotoxicology 22: 1395–1402.
Dimitrie, D.A. and D.W. Sparling. 2014. Joint toxicity of chlorpyrifos and endosulfan to pacific treefrog (*Pseudacris regilla*) tadpoles. Arch. Environ. Contam. Toxicol. 67(3): 444–452.
Driss, R., H. Imhof, W. Sanchez, J. Gasperi, F. Galgani, B. Tassin and C. Laforsch. 2015. Beyond the ocean: contamination of freshwater ecosystems with (micro-) plastic particles. Environ. Chem. 12(5): 539–550.
Duke, S.O. and F.E. Dayan. 2018. Herbicides. *In*: eLS. John Wiley & Sons, Ltd: Chichester. DOI: 10.1002/9780470015902.a0025264.
Dumont, J., T.W. Schultz, M. Buchanan and G. Kao. 1983. Frog embryo teratogenesis assay-*Xenopus* (FETAX) - a short-term assay application to complex environmental mixtures. pp. 393–405. *In*: Waters, M.D., S.S. Sandhu, J. Lewtas, L. Claxton, N. Chernoff and S. Nesnow [eds.]. Symposium on the Application of Short-term Bioassays in the Analysis of Complex Environmental Mixtures III. Plenum Press, New York, USA.
Egea-Serrano, A., M. Tejedo and M. Torralva. 2011. Behavioral responses of the Iberian waterfrog, *Pelophylax perezi* (Seoane 1885), to three nitrogenous compounds in laboratory conditions. Ecotoxicology 20: 1246–1257.
Ficken, K.L.G. and P.G. Byrne. 2013. Heavy metal pollution negatively correlates with anuran species richness and distribution in south-eastern Australia. Austral Ecol. 38(5): 523–533.
Fini, J.B., S. Le Mevel, N. Turque, K. Palmier, D. Zalko, J.P. Cravedi and B.A. Demeneix. 2007. An *in vivo* multiwell-based fluorescent screen for monitoring vertebrate thyroid hormone disruption. Environ. Sci. Technol. 41: 5908–5914.
Freitas, J.S., A.J. Kupsco, G. Diamante, A.A. Felício, E.A. de Almeida and D. Schlenk. 2016. Influence of temperature on the thyroidogenic effects of Diuron and its metabolite 3,4-DCA in tadpoles of the American bullfrog (*Lithobates catesbeianus*). Environ. Sci. Technol. 50: 13095–13104.
Freitas, J.S., A.A. Felício, F.B. Teresa and E.A. de Almeida. 2017a. Combined effects of temperature and clomazone (Gamit®) on oxidative stress responses and B-esterase activity of *Physalaemus nattereri* (Leiuperidae) and *Rhinella schneideri* (Bufonidae) tadpoles. Chemosphere 185: 548–562.
Freitas, J.S., F.B. Teresa and E.A. de Almeida. 2017b. Influence of temperature on the antioxidant responses and lipid peroxidation of two species of tadpoles (*Rhinella schneideri* and *Physalaemus nattereri*) exposed to the herbicide sulfentrazone (Boral 500SC®). Comp. Biochem. Physiol. C Toxicol. Pharmacol. 197: 32–44.
Freitas, J.S., L. Girotto, B.V. Goulart, L.O.G. Alho, R.C. Gebara, C.C. Montagner, L. Schiesari and E.L.G. Espíndola. 2019. Effects of 2,4-D-based herbicide (DMA® 806) on sensitivity, respiration rates, energy reserves and behavior of tadpoles. Ecotox Environ Safe 82: 109446.
Fusco, G. and A. Minelli. 2010. Phenotypic plasticity in development and evolution: facts and concepts. Phil. Trans. R. Soc. B 365: 547–556.
Gallant A.L., R.W. Klaver, G.S. Casper and M.J. Lannoo. 2007. Global rates of habitat loss and implications for amphibian conservation. Copeia 4: 967–979.
Giesy J.P., S. Dobson and K.R. Solomon. 2000. Ecotoxicological risk assessment for Roundup® herbicide. Rev. Environ. Contam. T. 167: 35–120.

Grenn, F.B. and C.J. Salice. 2020. Increased temperature and lower resource quality exacerbate chloride toxicity to larval *Lithobates sylvaticus* (wood frog). Environ. Pollut. 266: 115188.

Groner, M.L. and R.A. Relyea. 2011. A tale of two pesticides: how common insecticides affect aquatic communities. Freshwater Biol. 56(11): 2391–2404.

Grott, S.C., D. Bitschinski, N.G. Israel, G. Abel, S.P. da Silva, T.C. Alves, D. Lima, A.C.D. Bainy, J.J. Mattos, E.B. da Silva, C.A.C. de Albuquerque and E.A. de Almeida. 2021. Influence of temperature on biomarker responses and histology of the liver of American bullfrog tadpoles (*Lithobates catesbeianus*, Shaw, 1802) exposed to the herbicide Tebuthiuron. Sci. Total Environ. 771: 144971.

Hammond, J.I., D.K. Jones, P.R. Stephens and R.A. Relyea. 2012. Phylogeny meets ecotoxicology: evolutionary patterns of sensitivity to a common insecticide. Evol. Appl. 5(6): 593–606.

Hanlon, S.M. and R.A. Relyea. 2013. Sublethal effects of pesticides on predator–prey interactions in amphibians. Copeia 4: 691–698.

Hanlon, S.M. and M.J. Parris. 2014. The interactive effects of chytrid fungus, pesticides, and the exposure timing on gray treefrog (*Hyla versicolor*) larvae. Environ Toxicol Chem. 33(1): 216–222.

Hanlon, S.M., K.J. Lynch, J.L. Kerby and M.J. Parris. 2015. The effects of a fungicide and chytrid fungus on anuran larvae in aquatic mesocosms. Environ. Sci. Pollut. Res. 22(17): 12929–12940.

Hayes, R.A., M.R. Crossland, M. Hagman, R.J. Capon and R. Shine. 2009. Ontogenetic variation in the chemical defenses of cane toads (*Bufo marinus*): toxin profiles and effects on predators. J. Chem. Ecol. 35: 391–399.

Hayes, T.B., P. Falso, S. Gallipeau and M. Stice. 2010. The cause of global amphibian declines: a developmental endocrinologist's perspective. J. Exp. Biol. 213: 921–933.

Hayes, T.B., L.L. Anderson, V.R. Beasley, S.R. Solla, T. Iguchi, H. Ingraham et al. 2011. Demasculinization and feminization of male gonads by atrazine: consistent effects across vertebrate classes. J. Steroid Biochem. Mol. Biol. 127: 64–73.

Hedlund, J., S.B. Longo and R. York. 2020. Agriculture, pesticide use, and economic development: a global examination (1990–2014). Rural Sociol. 85(2): 519–544.

Heimeier, R.A., B. Das, D.R. Buchholz and Y. Shi. 2009. The xenoestrogen bisphenol A inhibits postembryonic vertebrate development by antagonizing gene regulation by thyroid hormone. Endocrinology 150(6): 2964–2973.

Hinther, A., C.M. Bromba, J.E. Wulff and C.C. Helbing. 2011. Effects of triclocarban, triclosan, and methyl triclosan on thyroid hormone action and stress in frog and mammalian culture systems. Environ. Sci. Technol. 45: 5395–5402.

Hocking, D.J. and K.J. Babbitt. 2014. Amphibian contribution to ecosystem services. Herpetol. Conserv. Bio. 9(1): 1–17.

Hoffmann, H. 2010. Cyanosis by methemoglobinemia in tadpoles of *Cochranella granulosa* (Anura: Centrolenidae). Rev. Biol. Trop. 58(4): 1467–1478.

Hogan, N.S., P. Duarte, M.G. Wade, D.R.S. Lean and V.L. Trudeau. 2008. Estrogenic exposure affects metamorphosis and alters sex ratios in the northern leopard frog (*Rana pipiens*): identifying critically vulnerable periods of development. Gen. Comp. Endocr. 156: 515–523.

Hu, L., J. Zhu, J.M. Rotchell, L. Wu, J. Gao and H. Shi. 2015. Use of the enhanced frog embryo teratogenesis assay-*Xenopus* (FETAX) to determine chemically-induced phenotypic effects. Sci. Total Environ. 508: 258–265.

Hua, J., N.I. Morehouse and R.A. Relyea. 2013. Pesticide tolerance in amphibians: induced tolerance in susceptible populations, constitutive tolerance in tolerant populations. Evol. Appl. 6: 1028–1040.

Hua, J., D.K. Jones, B.M. Mattes, R.D. Cothran, R.A. Relyea and J.T. Hoverman. 2015. The contribution of phenotypic plasticity to the evolution of insecticide tolerance in amphibian populations. Evol. Appl. 8(6): 586–596.

Hua, J. and R.A. Relyea. 2019. The effect of a common pyrethroid insecticide on wetland communities. Environ. Res. Commun. 1(1): 015003.

IUCN 2021. The IUCN red list of threatened species. Version 2021-1. https://www.iucnredlist.org ISSN 2307-8235.

Iwai, N., R.G. Pearson and R.A. Alford. 2009. Shredder–tadpole facilitation of leaf litter decomposition in a tropical stream. Freshwater Biol. 54: 2573–2580.

Jayawardena, U.A., J.R. Rohr, P.H. Amerasinghe, A.N. Navaratne and R.S. Rajakaruna. 2017. Effects of agrochemicals on disease severity of *Acanthostomum burminis* infections (Digenea: Trematoda) in the Asian common toad, *Duttaphrynus melanostictus*. BMC Zool. 2(13): 1–10.

Jones, D.K., J.I. Hammond and R.A. Relyea. 2009. Very highly toxic effects of endosulfan across nine species of tadpoles: lag effects and family-level sensitivity. Environ. Toxicol. Chem. 28(9): 1939–1945.

Jones, D.K., J.I. Hammond and R.A. Relyea. 2010. Roundup® and amphibians: the importance of concentration, application time, and stratification. Environ. Toxicol. Chem. 29(9): 2016–2025.

Jones, D.K., J.I. Hammond and R.A. Relyea. 2011. Competitive stress can make the herbicide Roundup® more deadly to larval amphibians. Environ. Toxicol. Chem. 30(2): 446–454.

Jones, D.K., E.K. Yates, B.M. Mattes, W.D. Hintz, M.S. Schuler and R.A. Relyea. 2018. Timing and frequency of sublethal exposure modifies the induction and retention of increased insecticide tolerance in wood frogs (*Lithobates sylvaticus*). Environ. Chem. Toxicol. 37: 2188–2197.

Jones, D.K., J. Hua, B.M. Mattes, R.D. Cothran, R.A. Relyea and J.T. Hoverman. 2021. Predator-and competitor-induced responses in amphibian populations that evolved different levels of pesticide tolerance. Ecol. Appl. 31: e02305.

Kerby, J.L. and A. Sih. 2015. Effects of carbaryl on species interactions of the foothill yellow legged frog (*Rana boylii*) and the Pacific treefrog (*Pseudacris regilla*). Hydrobiologia 746(1): 255–269.

Knapp, R.A. 2005. Effects of nonnative fish and habitat characteristics on lentic herpetofauna in Yosemite National Park, USA. Biol. Conserv. 121: 265–279.

Kolbert, E. 2014. The Sixth Extinction: An Unnatural History. Henry Holt and Company, 336p.

Lajmanovich, R.C., M.T. Sandoval and P.M. Peltzer. 2003. Induction of mortality and malformation in *Scinax nasicus* tadpoles exposed by glyphosate formulations. B. Environ. Contam. Tox. 70: 612–618.

Lajmanovich, R.C., A.M. Attademo, P.M. Peltzer, C.M. Junges and M.C. Cabagna. 2011. Toxicity of four herbicide formulations with glyphosate on *Rhinella arenarum* (Anura: Bufonidae) tadpoles: B-esterases and glutathione S-transferase Inhibitors. Arch. Environ. Contam. Toxicol. 60: 681–689.

Lambert, S. and M. Wagner. 2018. Microplastics are contaminants of emerging concern in freshwater environments: an overview. pp. 1–23. *In*: Wagner, M. and S. Lambert. [orgs]. Freshwater Microplastic. Emerging Environmental Contaminants? The Handbook of Environmental Chemistry. SpringerOpen, Berlin/Heidelberg, Germany.

Leite, P.Z., T.C.S. Margarido, D. Lima, D.C. Rossa-Feres and E.A. Almeida. 2010. Esterase inhibition in tadpoles of *Scinax fuscovarius* (Anura, Hylidae) as a biomarker for exposure to organophosphate pesticides. Environ. Sci. Pollut. Res. 17: 1411–1421.

Lenhardt, P.P., C.A. Brühl and G. Berger. 2015. Temporal coincidence of amphibian migration and pesticide applications on arable fields in spring. Basic. Appl. Ecol. 16(1): 54–63.

Mamede, J.L. and F. Nomura. 2021. *Dendropsophus minutus* (Hylidae) tadpole evaluation of predation risk by fishing spiders (*Thaumasia* sp.: Pisauridae) is modulated by size and social environment. J. Ethol. 39: 217–223.

Mann M.R., R.V. Hyne, C.B. Choung and S.P. Wilson. 2009. Amphibians and agricultural chemicals: review of the risks in a complex environment. Environ. Pollut. 157: 2903–2927.

Marco, A. and A.R. Blaustein. 1999. The effects of nitrite on behavior and metamorphosis in cascades frogs (*Rana cascadae*). Environmental Toxicology and Chemistry 18(5): 946–949.

Margarido, T.C.S., A.A. Felício, D.C. Rossa-Feres and E.A. Almeida. 2013. Biochemical biomarkers in *Scinax fuscovarius* tadpoles exposed to a commercial formulation of the pesticide fipronil. Mar. Environ. Res. 91: 61–67.

Middlemis Maher, J., E.E. Werner and R.J. Denver. 2013. Stress hormones mediate predator-induced phenotypic plasticity in amphibian tadpoles. Proc. R. Soc. B 280: 20123075.

Mikó Z., J. Ujszegi, Z. Gál, Z. Imrei and A. Hettyey. 2015. Choice of experimental venue matters in ecotoxicology studies: comparison of a laboratory-based and an outdoor mesocosm experiment. Aquat. Toxicol. 167: 20–30.

Miko, Z., J. Ujszegi, Z. Gal and A. Hettyey. 2017. Standardize or diversify experimental conditions in ecotoxicology? A case study on herbicide toxicity to larvae of two anuran Amphibians. Arch. Environ. Con. Tox. 73(4): 562–569.

Milotic, M., D. Milotic and J. Koprivnikar. 2018. Exposure to a cyanobacterial toxin increases larval amphibian susceptibility to parasitism. Parasitol. Res. 117(2): 513–520.

Montaña, C.G., S.D.G.T.M. Silva, D. Hagyari, J. Wager, L. Tiegs, C. Sadeghian et al. 2019. Revisiting "what do tadpoles really eat?" A 10-year perspective. Freshwater Biol. 64: 2269–2282.

Moore, H., D.P. Chivers and M.C.O. Ferrari. 2015. Sub-lethal effects of Roundup™ on tadpole anti-predator responses. Ecotox. Environ. Safe. 111: 281–285.

Moore, L.J., L. Fuentes, J.H. Rodgers, W.W. Bowerman, G.K. Yarrow, W.Y. Chao and W.C. Bridges. 2012. Relative toxicity of the components of the original formulation of Roundup (R) to five North American anurans. Ecotox. Environ. Safe. 78: 128–133.

Moreira, J.C., F. Peres, A.C. Simões, W.A. Pignati, E.C. Dores, S.N. Vieira, C. Strüssmann and T. Mott. 2012. Groundwater and rainwater contamination by pesticides in an agricultural Region of Mato Grosso State in Central Brazil. Ciência e Saúde Coletiva 17(6): 1557–1568.

Moreira, R.A., J.S. Freitas, T.J.S. Pinto, L. Schiesari, M.A. Daam, C.C. Montagner et al. 2019. Mortality, spatial avoidance and swimming behavior of bullfrog tadpoles (*Lithobates catesbeianus*) exposed to the herbicide diuron. Water Air Soil Pollut. 230: 125.

Moutinho, M.F., E.A. Almeida, E.L.G. Espíndola, M.A. Daam and L. Schiesari. 2020. Herbicides employed in sugarcane plantations have lethal and sublethal effects to larval Boana pardalis (Amphibia, Hylidae). Ecotoxicology 29: 1043–1051.

Nations, S., M. Wages, J.E. Cañas, J. Maul, C. Theodorakis and G.P. Cobb. 2011. Acute effects of Fe2O3, TiO2, ZnO and CuO nanomaterials on *Xenopus laevis*. Chemosphere 83(8): 1053–1061.

Nomura, F., V.H.M. do Prado, F.R. da Silva, R.E. Borges, N.Y.M. Dias and D.C. Rossa-Feres. 2011. Are you experienced? Predator type and predator experience trade-offs in relation to tadpole mortality rates. J. Zool. 284: 144–150.

OECD (Organization for Economic Co-operation and Development). 2009. Test No. 231: Amphibian Metamorphosis Assay, OECD Guidelines for the Testing of Chemicals, Section 2: Effects on Biotic Systems, OECD Publishing.

Oka, T, O. Tooi, N. Mitsui, M. Miyahara, Y. Ohnishi, M. Takase, A. Kashiwagi, T. Shinkai, N. Santo and T. Iguchi. 2008. Effect of atrazine on metamorphosis and sexual differentiation in *Xenopus laevis*. Aquat. Toxicol. 87: 215–226.

Palmer, A.R. and C. Strobeck. 1986. Fluctuating asymmetry: measurement, analysis, patterns. Annu. Rev. Ecol. Syst. 17: 391–421.

Pietrzak, D., J. Kania, G. Malina, E. Kmiecik and K. Wątor. 2019. Pesticides from the EU First and Second watch lists in the water environment. Clean-Soil Air Water 47(7): 1–13.

Polo-Cavia, N., P. Burraco and I. Gomez-Mestre. 2016. Low levels of chemical anthropogenic pollution may threaten amphibians by impairing predator recognition. Aquat. Toxicol. 172: 30–35.

Quiroga, L.B., E.A. Sanabria, M.V. Fornes, D.A. Bustos and M. Tejedo. 2019. Sublethal concentrations of chlorpyrifos induce changes in the thermal sensitivity and tolerance of anuran tadpoles in the toad *Rhinella arenarum*? Chemosphere 219: 671–677.

Relya, R.A. 2009. A cocktail of contaminants: how mixtures of pesticides at low concentrations affect aquatic communities. Oecologia 159: 363–376.

Relyea, R.A. and N.E. Mills. 2001. Predator-induced stress makes the pesticide carbaryl more deadly to gray treefrog tadpoles (*Hyla versicolor*). P. Nalt. Acad. Sci. USA 98: 2491–2496.

Relyea, R.A. 2004. Fine-tuned phenotypes: tadpole plasticity under 16 combinations of predators and competitors. Ecology 85(1): 172–179.

Relyea, R.A. 2005. The lethal impacts of roundup and predatory stress on six species of North American tadpoles. Arch. Environ. Con. Tox. 48: 351–357.

Relyea, R.A. and J. Hoverman. 2006. Assessing the ecology in ecotoxicology: a review and synthesis in freshwater systems. Ecol. Lett. 9: 1157–1171.

Relyea, R.A. and J. Hoverman. 2008. Interactive effects of predators and a pesticide on aquatic communities. Oikos 117: 1647–1658.

Relyea, R.A. and D.K. Jones. 2009. The toxicity of Roundup® Max to 13 species of larval amphibians. Environ. Toxicol. Chem. 28(9): 2004–2008.

Relyea, R.A. 2012. New effects of Roundup® on amphibians: Predators reduce herbicide mortality; herbicides induce antipredators morphology. Ecol. Appl. 22(2): 634–647.

Rohr, J.R., A.M. Schotthoefer, T.R. Raffel, H.J. Carrick, N. Halstead, J.T. Hoverman, C.M. Johnson, L.B. Johnson, C. Lieske, M.D. Piwoni, P.K. Schoff and V.R. Beasley. 2008a. Agrochemicals increase trematode infections in a declining amphibian species. Nature 455: 1235–1240.

Rohr, J.R., T.R. Raffel, S.K. Sessions and P.J. Hudson. 2008b. Understanding the net effects of pesticides on amphibians trematode infection. Ecol. Appl. 18(7): 1743–1753.

Rohr, J.R., A. Swan, T.R. Raffel and P.J. Hudson. 2009. Parasites, info-disruption, and the ecology of fear. Oecologia 159: 447–454.

Rohr, J.R., T.R. Raffel, N.T. Halstead, T.A. McMahon, S.A. Johnson, R.K. Boughton and L.B. Martin. 2013. Early-life exposure to a herbicide has enduring effects on pathogen-induced mortality. Proc. R. Soc. B 280: 1–7.

Rollins-Smith, L.A., J.P. Ramsey, J.D. Pask, L.K. Reinert and D.C. Woodhams. 2011. Amphibian immune defenses against chytridiomycosis: impacts of changing environments. Integr. Comp. Biol. 51(4): 552–562.

Santos, B., R. Ribeiro, I. Domingues, R. Pereira, A.M.V.M. Soares and I. Lopes. 2013. Salinity and copper interactive effects on perez's frog *Pelophylax perezi*. Environ. Toxicol. Chem. 32(8): 1864–1872.

Schiesari, L. and D.T. Corrêa. 2016. Consequences of agroindustrial sugarcane production to freshwater biodiversity. GCB Bioenergy 8: 644–657.

Schiesari, L., B. Grillitsch and H. Grillitsch. 2007. Biogeographic biases in research and their consequences for linking amphibians declines to pollution. Conserv. Biol. 21(2): 465–471.

Schuler, M.S. and R.A. Relyea. 2018. A review of the combined threats of road salts and heavy metals to freshwater systems. BioScience 68(5): 327–335.

Segalla, M.V., B. Berneck, C. Canedo, U. Caramaschi, C.A.G. Cruz, P.C.A. Garcia, T. Grant, C.F.B. Haddad, A.C.C. Lourenço, S. Mângia, T. Mott, L.B. Nascimento, L.F. Toledo, F.P. Werneck and J.A. Langone. 2021. List of Brazilian Amphibians. Herpetol. Brasil. 10(1): 121–216.

Shuman-Goodier, M.E., G.R. Singleton and C.R. Propper. 2017 Competition and pesticide exposure affect development of invasive (*Rhinella marina*) and native (*Fejervarya vittigera*) rice paddy amphibian larvae. Ecotoxicology 26(10): 1293–1204.

Sievers, M., R. Hale, K.M. Parris and S.D. Melvin. 2019. Contaminant-induced behavioural changes in amphibians: A meta-analysis. Science of The Total Environment 693: 133570.

Solomon, K.R., J.A. Carr, L.H.D. Preez, J.P. Giesy, R.J. Kendall, E.E. Smith and G.J.V.D. Kraak. 2008. Effects of atrazine on fish, amphibians, and aquatic reptiles: a critical review. Crit. Rev. Toxicol. 38: 721–772.

Sparling, D.W. and G.M. Fellers. 2009. Toxicity of two insecticides to California, USA, anurans and its relevance to declining amphibian populations. Environ. Toxicol. Chem. 28(8): 1696–1703.

Steiner, U.K. and J. Van Buskirk. 2008. Environmental stress and the costs of whole-organism phenotypic plasticity in tadpoles. J. Evolution. Biol. 21: 97–103.

Stoler, A.B., B.M. Walker, W.D. Hintz, D.K. Jones, L. Lind, B.M. Mattes, M.S. Schuler and R.A. Relyea. 2017. Combined effects of road salt and an insecticide on wetland communities. Environ. Toxicol. Chem. 36(3): 771–779.

Stuart S.N., J.S. Chanson, N.A. Cox, B.E. Young, A.S. Rodrigues, D.L. Fischman and R.W. Waller. 2004. Status and trends of amphibian declines and extinctions worldwide. Science 306: 1783–1786.

Tierney, K.B., P.S. Ross, H.E. Jarrard, K. Delaney and C.J. Kennedy. 2006. Changes in juvenile coho salmon electro-olfactogram during and after short-term exposure to current-use pesticides. Environ. Toxicol. Chem. 25: 2809–2817.

Troyer, R.R. and A.M. Turner. 2015. Chemosensory perception of predators by larval amphibians depends on water quality. PLoS One 10(6): e0131516.

Velasquez, T.M.T., L.M.H. Munoz and M.H.B. Bautista. 2017. Acute toxicity of the insecticide cypermethrin (Cypermon (R) 20 EC) on four species of Colombian anurans. Acta Biol. Colomb. 22(3): 340–347.

Vredenburg, V.T. 2004. Reversing introduced species effects: experimental removal of introduced fish leads to rapid recovery of a declining frog. P. Natl. Acad. Sci. USA 101: 7646–7650.

Weldeslassie, T., Naz, H., Singh, B. and M. Oves. 2018. Chemical contaminants for soil, air and aquatic ecosystem. pp. 1–22. *In*: Oves, M., M. Zain Khan and I.M.I. Ismail [eds.]. Modern Age Environmental Problems and their Remediation. Springer, Cham. Berlin/Heidelberg, Germany.

Wijethunga, U., M. Greenlees and R. Shine. 2016. Far from home: responses of an American predator species to an American prey species in a jointly invaded area of Australia. Biol. Invasions 18(6): 1645–1652.

2

The Importance of Tadpole Staging in Ecotoxicological Studies

Suelen Cristina Grott, Camila Fatima Rutkoski and *Eduardo Alves de Almeida**

1. Introduction

Amphibians are organisms with permeable skin, eggs without shell, and have a life cycle usually divided into aquatic and terrestrial phases. Most species undergo a process of metamorphosis, in which there is complete morphological remodeling of the aquatic larvae into an adult terrestrial tetrapod (Duellman and Trueb 1986). Metamorphosis process is usually divided into three periods: pre-metamorphosis, pro-metamorphosis and climax (Misra and Dash 1984). Pre-metamorphosis refers to a period when the genesis and early growth and development of tadpoles occur in the absence of thyroid hormones (TH). During pro-metamorphosis, the hindlimbs undergo morphogenesis, as evidenced by toe differentiation and by the rapid and extensive hindlimb growth (McDiarmid and Altig 1999). The metamorphic climax is the period of major changes, which culminate in the loss of most larval characters with rapid differentiation from tadpole to adult. In this stage, there is complete resorption of the tail and development of physiological structures and functions that are essential for adult terrestrial life (Hall and Larson 1998, Mcdiarmid and Altig 2000).

In order to differentiate the developmental stages of amphibians, staging allows us to recognize certain morphological landmarks useful for comparing the sequence of developmental events from fertilization to adulthood (Mcdiarmid and Altig 1999). In ecotoxicological studies, the staging of animals is essential for a correct interpretation of the effects of environmental pollutants, since at each stage of development animals present different physiological characteristics. For example, hormonal levels such as cortisol, THs and growth hormone vary substantially from one developmental stage to another, generating profound changes in the metabolism of the animals, which are also associated with cell proliferation and apoptosis, resulting in extensive organ remodeling. As a consequence, the effects of pollutants can be quite variable throughout the development of tadpoles.

Centro de Estudos em Toxicologia Aquática, Universidade Regional de Blumenau, Rua São Paulo, 3366, Bloco Q, setor Q101, Blumenau, SC, 89030-000, Brazil. Emails: suelencgtt@gmail.com; cfrutkoski@furb.br

* Corresponding author: eduardoalves@furb.br

2. Main Proposals for Classification of Developmental Stages of Tadpoles

In order to better describe the development of amphibians, several classification methods were created to standardize each stage of development in different species. One of the first studies to present a division of amphibian development into stages was published by Pollister and Moore (1937) for the species *Rana sylvatica*, where 23 stages were described from fertilization to the larval stage. Later, the study by Weisz (1945) divided *Xenopus laevis* development into 23 different periods. Subsequently, several other authors sought to develop staging tables, including the classification proposed by Taylor and Kollros (1946), Conte et al. (1952), Nieuwkoop and Faber (NF classification) (1956), Gosner (1960), Etkin (1968), Shumway (1940), Limbaugh and Volpe (1957), Daudin (1802), Iwasawa and Futagami (1992), Shimizu and Ota (2003) and Segura–Solís and Bolaños (2009).

Among the works mentioned above, the study by Gosner 1960 is one of the most used in scientific literature, as it presents a more general table, which can be applied to most anuran families. The system includes 46 stages, from the fertilized embryo (stage 1) to complete metamorphosis (stage 46). In the Gosner classification, only external anatomy characters are evaluated for staging, such as the size of the hind limbs and the position of the eyes in relation to the nostril. The color, size and number of eggs vary widely between species and are not addressed in the Gosner classification. The sequential changes, from fertilization to cleavage, blastula and gastrula, are similar between species and are present in the staging table. Stages 1 to 20 correspond to the development of the egg after fertilization. After stage 17 the embryo is already practically formed and in stage 20 hatching normally occurs. This first stage is called embryonic phase and it corresponds to the period in which individuals are protected by a thin gelatinous layer (Mcdiarmmid and Altig 1999). From stage 21 to 25, there is the "hatchling" phase, which includes young organisms that have hatched from the egg, with well-developed mouth and gills (Mcdiarmmid and Altig 1999). The larval stage is represented by stages from 26 to 41, which is the longest period of development due to the growth of hind and fore limbs, necessary for the transition from the aquatic to the terrestrial phase (Mcdiarmmid and Altig 1999). Stages between 26 to 40 are mostly examined by the morphology and length of the hind limbs. From stages 26 to 30, the specific identification is given by the length/diameter ratio of the limb bud that is growing. In the stage 31, the limb is paddle-shaped and the later stages up to 37 are characterized by the appearance of individual fingers. Proportional differences in finger length are seen from stage 38 to 40, accompanied by changes in the appearance of the "metatarsal and subarticular tubercles". Very noticeable changes in metamorphosis occur after stage 40, as the tadpole begins to decrease its total length through tail resorption and the larval mouthparts start disappearing. The skin of the forelimbs becomes transparent at stage 41 and the cloacal tail usually disappears at or shortly after this stage (Mcdiarmmid and Altig 1999). The last stage is the metamorphosis phase, which begins at stage 42 and ends at stage 46, where individuals reach adulthood (Mcdiarmmid and Altig 1999). At stage 42, lower limbs appear and until stage 46, morphological changes in the head are observed, identified mainly by changes in the mouth. At stage 46, a frog is considered a juvenile and may differ from its fully adult form.

Different classifications of amphibian development have also been proposed by other authors. Etkin (1968), for example, divided the stages of development into three phases: pre-metamorphosis, pro-metamorphosis, and climax. Pre-metamorphosis includes larval stages that tadpoles can reach even if they lack the thyroid gland or thyroid hormone (TK stage I-X). Pro-metamorphosis includes the stages in which the growth of the tail and hind legs is hormonally stimulated (TK stage XI–XIX). Finally, the climax includes all the stages with tail resorption (from beginning to end) (stage TK XX–XXV). Nieuwkoop and Faber (1956) divided the developmental stages of *Xenopus laevis* into 66 stages, which corresponds to NF 1 to NF 66. *X. laevis* is a species native to sub-Saharan Africa (Tinsley et al. 1996) and strictly aquatic in all their stages of development (Carreño and Nishikawa 2010).

The classification used by Nieuwkoop and Faber (1956) is widely disseminated in studies with tadpoles.

Pollister and Moore (1937), as well as Shumway (1940), in turn, proposed a staging classification based on *Rana sylvatica* development. In this study, authors divided the initial development of the species (fertilization to the larval stage) into 23 different stages, using as parameters the external morphology of the species and the age (hours), under a constant temperature of 18°C. The larval stage to metamorphosis was divided by Taylor and Kollros (1946) into 25 stages distinguished by Roman numerals from the embryonic stages. These larval stages were based, as far as possible, on the emergence of new external morphological structures, or on readily detectable changes in structures already present. Taylor and Kollros (1946) stages (TK stages) begin with the emergence of the hind limbs (I to V), which are rudimentary structures and essentially no more than a simple bud. Stages VI and X comprise the tadpoles in which the bud is first transformed into a fin, and then the fingers are separated from each other. It is worth mentioning here that tadpole fin is a very simple structure, formed by overlapping layers of skin (different from those observed in fish, which present bone or cartilaginous tissue). Next, the authors determined the paw stages, or pre-metamorphic stages, which correspond to stages between XI to XVII. Then, "metamorphic" stages advance from XVIII to XXV, which begin with regression from the tail to the cloaca, and end when tail resorption is completed. Finally, stage XXV represents the fully developed juvenile frog. Although this developmental classification was initially described to differentiate the species *Rana pipiens*, it has also been applied to several other species, such as *Lithobates catesbeiana* (Heerema et al. 2018), *Rana (R.) rugosa* (Oike et al. 2017), *Pelobates syriacus* (Katz et al. 2003), and *Rana nigromaculata* (Kashiwagi et al. 2005).

According to what has already been described, the process of metamorphosis is generally divided into three important periods in the development of tadpoles: pre-metamorphosis, pro-metamorphosis and climax. The main proposals used in this classification are summarized in Fig. 1.

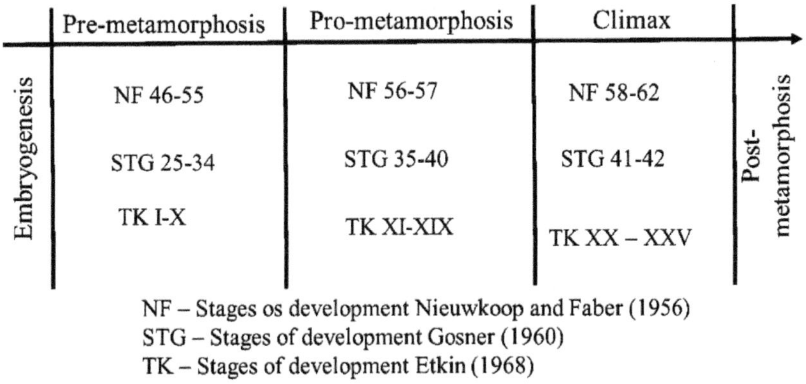

Embryogenesis	Pre-metamorphosis	Pro-metamorphosis	Climax	Post-metamorphosis
	NF 46-55	NF 56-57	NF 58-62	
	STG 25-34	STG 35-40	STG 41-42	
	TK I-X	TK XI-XIX	TK XX – XXV	

NF – Stages os development Nieuwkoop and Faber (1956)
STG – Stages of development Gosner (1960)
TK – Stages of development Etkin (1968)

Figure 1. Principal proposals for classifying tadpole development stages.

3. Species with their Own Staging Table

Species differ from each other in terms of their developmental patterns. However, stages of development were not well characterized (or not yet) for many of them. The use of a staging system allows comparisons between tadpoles of different size and developmental period that belong to different species with similar developmental stages. However, these developmental descriptions are not yet available for several groups of anurans, requiring further studies and improvement of classification systems. This is essential to better adapt (or standardize) the staging of the species (Shimizu and Ota 2003). In this sense, some authors have suggested species-specific tables for staging development in tadpoles.

The species *Bryobatrachus nimbus*, also known as the Australian moss frog, for example, is the only Tasmanian frog whose life cycle is entirely terrestrial (Mitchell and Swain 1996). In this species, females lay small clutches of eggs in nests built in mosses, lichens or peat during the breeding season, which corresponds to spring and/or summer (Rounsevell et al. 1994). Since *B. nimbus* life cycle is terrestrial, the authors Mitchell and Swain (1996) divided the development of the species into 17 stages, from oviposition (stage 1) to tail emergence (stage 7), capsule hatching (stage 10), emergence of hind and fore limbs (stages 11 to 15) and by complete resorption of the tail and formation of the adult individual (stages 16 and 17). The description of this staging evidenced that development in this tadpole species was not predominantly intracapsular, as the researchers had previously believed. Instead, the larvae hatch from the capsule two to three months after oviposition and continue to develop in the nest within a homogeneous fluid derived from the capsule. The capsule size in *B. nimbus* is nearly double that of any other Australian frog, while the larval lifespan is the longest known for any terrestrial breeding Australian species. Metamorphosis occurs approximately 12 months after oviposition (Mitchell and Swain 1996).

On the other hand, the species *Gastrotheca riobambae*, which also has a terrestrial cycle, has a peculiar developmental system, in which adult females have a permanent dorsal pouch where eggs are kept during incubation until the embryo reaches the larval period (tadpole) (del Pino 1989). On hatching, tadpoles are released from the pouch into standing water, where they complete metamorphosis. For this reason, authors staged the species into 25 developmental stages, starting with fertilization (stage 1) and close in stage 25, when the tadpole is born with the emergence of its forelimbs (similar to Gosner stage 33).

Anuran species that do not have tadpole's stages are also common in some genera, and therefore do not fit into traditional staging tables. Examples are the species *Oreobates barituensis* (Goldberg et al. 2012), *Eleutherodactylus coqui* (Townsend and Stewart 1985), and four species within the genus *Pipa* (*P. aspera, P. arrabali, P. pipa* and *P. snethlageae*) (Araújo et al. 2016). These species present a direct terrestrial development, in which adults lay eggs on land that hatch directly into sub-adult frogs. *E. coqui* embryos, for example, do not emerge as tadpoles, but develop directly into frogs between 17 and 26 days after laying eggs. The absence of a tadpole stage allows nesting to take place in dry or waterless places, such as tree holes and curled palm petioles. The staging of the *E. coqui* species is divided into 15 stages, based on changes in the external morphology of the embryos, formation of external gills, progressive pigmentation of the eyes, growth and differentiation of limbs, development of endolymphatic calcium deposits, expansion of a pigmented wall of the body, growth of a large, vascularized tail, and changes in embryonic behavior.

In addition to the tables cited above, other authors have divided amphibian development into different tables, from fertilization to adult formation. Some examples are the studies carried out by Ba-Omar et al. (2014) who described the development of the species *Bufo arabicus*, which was based on the Gosner table (1960). Limbaugh and Volpe's (1957) staging table was developed to assess the development of the species *Bufo valliceps*, which divides the species' development into 46 stages and was based on those described by Shumway (1940) and Taylor and Kollros (1946), described for *Rana pipiens*. Xiong et al. (2010) performed the staging of the species *Odorrana tormota*, popularly known as concave-eared torrent frog. The species is found in isolated places in China and presents peculiar characteristics, among them, the presence of tympanic membranes highly unusual in adult males, which are recessed from the surface of the body and allow the production and perception of ultrasonic frequencies. The staging proposed by the authors is divided into 32 stages, including the early embryonic period (1–25) and larval period (26–32), from the tadpole stage to the tail absorption stage. Fabrezi et al. (2009) analyzed the development of the species *Pseudis platensis* (Hylidae) and found that it did not fit into the existing tables, mainly because the species has a larval stage with tadpoles larger than adults, making necessary the establishment of a specific staging method. For this, the total development of the species was divided into 13 stages, following the description by Gosner, with some changes: stages I–IV corresponds to the larval stage, stages V–XII includes the metamorphosis period, and stage XIII is the adult stage.

Iwasawa and Futagami (1992) described the staging of the frog *Hyla japonica* based on the stages of Gosner (1960). This species occurs in Japan, Korea, China, Northern Mongolia, the Russian Far East and as far south as Lake Baikal (Amphibia Web 2017). The animals usually occupy mixed broadleaf forests, savannas, forest steppes, meadows and swamps, but they can also inhabit unforested areas such as bushy river valleys and large city settlements (Amphibia Web 2017). Development comprises 46 stages, very similar to that proposed by Gosner (1960), except by the more detailed embryonic stage (stages 8–11) and the development of gills during the larval stages 21 and 22.

Shimizu and Ota (2003) studied *Microhyla ornata* staging based on the classification by Gosner (1960) and by Iwasawa and Futagami (1992). The species occurs in Taiwan, India, Japan, as well as Southeast China and Asia, inhabiting ground amid leaves and grasses (Amphibia Web 2002). The development was divided into 45 stages, as follows: blastula cleavage, stages 1–10; gastrula, stages 11–14; neurula, stages 15–18; tail bud, stages 19–21; external gill, stages 22–28; hindlimb bud, stages 29–33; hindlimb formation, stages 34–41; and metamorphosis, stages 42–45 (Shimizu and Ota 2003).

In addition to the studies already mentioned, several other species have their development studied and documented, such as the species *Polypedates teraiensis* (Chakravarty et al. 2011), *Hoplobatrachus tigerinus* (Khan 1969, Agarwal and Niazi 1977), *Fejervarya limnocharis* (Roy and Khare 1978) *Euphlyctis cyanophlyctis* (Kumar 1982), *Hyla annectans* (Ao and Bordoloi 2001), *Rotunda arenophryne* (Anstis et al. 2007), *Myobatrachus gouldii* (Anstis et al. 2007), *Philautus silus* (Kerney et al. 2007), *Euphlyctis cyanophlyctis* (Lalremsanga and Hooroo 2013).

4. Effects of Contamination in Development of Amphibians

Tadpoles are experiencing a population decline worldwide due to adverse environmental pressures, such as the loss of habitats, contamination of aquatic environments, infectious diseases, climate changes and increased incidence of ultraviolet radiation, due to decreased ozone layer (Lips et al. 2005, Collins 2010, Alton and Franklin 2017). These factors can act synergistically with abiotic factors, such as changes in temperature, pH, food availability and oxygen, potentializing the negative effects on amphibians (Hooper et al. 2013).

The final destination of most anthropogenic pollutants is the aquatic environment (Santos et al. 2010), which may cause physiological disturbances in non-target organisms, such as tadpoles, eventually causing disturbances (delay or acceleration) in their development (Gomez-Mestre et al. 2013, Peltzer et al. 2013, Rissoli et al. 2016, Carvalho et al. 2017).

The ability to accelerate development in response to negative environmental conditions is a consensus in the scientific literature (Morand et al. 1997, Murata and Yamauchi 2005, Brande-Lavridsen et al. 2010). Although this may bring some benefits to the animal (e.g., faster formation of adult individuals that are capable to leave more rapidly disturbed aquatic environments), it may also compromise the completion of metamorphosis, eventually generating adults with altered body size and/or incomplete or immature reproductive organs (Semlitsch 1988). These premature adults may also present altered physiological performance being not competitive enough to survive in nature (Hayes 1995, Loman 1999, Burraco et al. 2017, Márquez-García et al. 2009, Goater et al. 1993). The acceleration of tadpoles' development in response to pollutants has been reported in different species of anurans (Newman 1989, Freitas et al. 2017, Grott et al. 2022). Exposure to different pollutants can also lead to developmental delay (Foster et al. 2010, Li et al. 2016, Navarro-Martín et al. 2014). Developmental delay can affect the survival of tadpoles in natural environments, since this process often occurs in ephemeral ponds that normally dry during the summer, increasing vulnerability to predation and desiccation (Mochizuki et al. 2012). Much more details and examples of pollutant effects on growth and development of tadpoles are discussed in Chapter 12.

5. Conclusions and Perspectives

One important step when planning ecotoxicological experiments using tadpoles is the identification and selection of adequate developmental stages. The toxicity of different substances to tadpoles can vary according to their developmental stages, as important morphological and physiological alterations occur during the course of metamorphosis. Experimental groups (on ecotoxicological studies) should be composed by animals at same or closely similar stages of development, thus allowing better interpretation of results and less variation on evaluated endpoints due to dissimilar responses of the organisms with large discrepancies on their developmental stage. It also permits to evaluate if contaminants alter the time required to reach metamorphosis, since numerous studies have reported both acceleration or delayed effects of toxicants on tadpole`s development. In this context, following one of the classifications already proposed and established in the scientific literature allows better comparisons between different studies on tadpoles, in addition to allowing the identification of the most or the least susceptible stages of tadpoles to environmental contaminants.

Acknowledgements

We thank "Conselho Nacional de Desenvolvimento Científico e Tecnológico, CNPq", "Fundação de Amparo à Pesquisa e Inovação do Estado de Santa Catarina, FAPESC" and "Coordenação de Aperfeiçoamento de Pessoal de Nível Superior, CAPES" (001) for the financial support.

References Cited

Agarwal, S.K. and I.A. Niazi. 1977. Normal table of developmental stages of the Indian bullfrog, *Rana tigrina* Daud. (Ranidae, Anura, Amphibia). Proc. nat. Acad. Sci., India B, 47: 7992.

Alton, L.A. and C.E. Franklin. 2017. Drivers of amphibian declines: effects of ultraviolet radiation and interactions with other environmental factors. Clim. Chang. Responses 4: 1–26.

AmphibiaWeb. 2002. *Microhyla ornata*: Ornate Narrow-Mouthed Toad < https://amphibiaweb.org/species/2180 > Universidade da Califórnia, Berkeley, CA, EUA. Accessed Oct 8, 2021.

AmphibiaWeb. 2009. Incilius aucoinae < https://amphibiaweb.org/species/6409 > Universidade da Califórnia, Berkeley, CA, EUA. Accessed Oct 8, 2021.

AmphibiaWeb. 2017. *Hyla japonica*: Japanese Tree Frog <https://amphibiaweb.org/species/832> University of California, Berkeley, CA, USA. Accessed Oct 8, 2021.

Anstis, M., J.D. Roberts and R. Altig. 2007. Direct development in two myobatrachid frogs, *Arenophryne rotunda* Tyler and *Myobatrachus gouldii* Gray, from western Australia. Rec. West Aust. Mus. 23: 259–271.

Ao, J.M. and S. Bordoloi. 2001. Development of *Hyla annectans* Jerdon, 1870 from Nagaland, India. J. Bombay nat. Hist. Soc. 98: 169178.

Araújo, O.G., C.F. Haddad, H.R. Silva and L.A. Pugener. 2016. A simplified table for staging embryos of the pipid frog *Pipa arrabali*. An. Acad. Bras. Cienc. 88: 1875–1887.

Ba-Omar, T., I. Ambu-Saidi, S. Al-Bahry and A. Al-Khayat. 2004. Embryonic and larval staging of the Arabian Toad, *Bufo arabicus* (Amphibia: Bufonidae). Zool. Middle East 32: 47-56.

Brande-Lavridsen, N., J. Christensen-Dalsgaard, and B. Korsgaard. 2010. Effects of ethinylestradiol and the fungicide prochloraz on metamorphosis and thyroid gland morphology in *Rana temporaria*. Open Zool. J. 3: 1.

Burraco, P., C. Díaz-Paniagua and I. Gomez-Mestre. 2017. Different effects of accelerated development and enhanced growth on oxidative stress and telomere shortening in amphibian larvae. Sci. Rep. 7: 1–11.

Carreño, C.A. and K.C. Nishikawa. 2010. Aquatic feeding in pipid frogs: the use of suction for prey capture. J. Exp. Biol. 213: 2001–2008.

Carvalho, C.S., H.S.M. Utsunomiya, T. Pasquoto, R. Lima, M.J. Costa and M.N. Fernandes. 2017. Blood cell responses and metallothionein in the liver, kidney and muscles of bullfrog tadpoles, *Lithobates catesbeianus*, following exposure to different metals Environ. Pollut. 221: 445–452.

Chakravarty, P., S. Bordoloi, S. Grosjean, A. Ohler and A. Borkotoki. 2011. Tadpole morphology and table of developmental stages of *Polypedates teraiensis* (Dubois, 1987). Alytes 27: 85–115.

Collins, J.P. 2010. Amphibian decline and extinction: what we know and what we need to learn. Dis. Aquat. 92: 93–99.

Conte, E., D. Sirlin and J. Leo. 1952. Pattern series of the first embryonary stages in *Bufo arenarum*. Anat. Rec. 112: 125–135.

Daudin, F.M. 1802. Histoire naturelle des rainettes, des grenouilles et des crapauds. Paris: Bertrandet, Libraire Levrault.

del Pino, E.M. and B. Escobar. 1981. Embryonic stages of *Gastrotheca riobambe* (Fowler) during maternal incubation and comparison of development with other marsupial frogs. J. Morphol. 167: 277–295.

del Pino, E.M. 1989. Modifications of oogenesis and development in marsupial frogs. Development 107: 169–187.

Duellman, W.E. and L. Trueb. 1986. Biology of amphibians. McGraw-Hill, New York.

Etkin, W. 1968. Hormonal control of amphibian metamorphosis. pp. 313–34.8 *In*: Etkin, W. and L.J. Gilbert [eds.]. Metamorphosis: A Problem in Developmental Biology. New York: Appleton.

Fabrezi, M., S.I. Quinzio and J. Goldberg. 2009. Giant tadpole and delayed metamorphosis of *Pseudis platensis* Gallardo, 1961 (Anura, Hylidae). J. Herpetol. 43: 228–243.

Freitas, J.S., A.A. Felício, F.B. Teresa and E.A. de Almeida. 2017. Combined effects of temperature and clomazone (Gamit®) on oxidative stress responses and B-esterase activity of *Physalaemus nattereri* (Leiuperidae) and *Rhinella schneideri* (Bufonidae) tadpoles. Chemosphere 185: 548–562.

Foster, H.R., G.A. Burton, N. Basu and E.E. Werner. 2010. Chronic exposure to fluoxetine (Prozac) causes developmental delays in *Rana pipiens* larvae. Environ. Toxicol. Chem. 29: 2845–2850.

Goater, C.P., R.D. Semlitsch and M.V. Bernasconi. 1993. Effects of body size and parasite infection on the locomotory performance of juvenile toads, *Bufo bufo*. Oikos 66: 129–136.

Goldberg, J., F.V. Candioti and M.S. Akmentins. 2012. Direct-developing frogs: ontogeny of *Oreobates barituensis* (Anura: Terrarana) and the development of a novel trait. Amphib-reptil 33: 239–250.

Gomez-Mestre, I., S. Kulkarni and D.R. Buchholz. 2013. Mechanisms and consequences of developmental acceleration in tadpoles responding to pond drying. PloS one 8: e84266.

Gosner, K.L. 1960. A simplified table for staging anuran embryos and larvae with notes on identification. Herpetologica 16: 183–190.

Grott, S.C., N. Israel, D. Lima, D. Bitschinski, G. Abel, T.C. Alves, E.B. Silva, C.A.C. Albuquerque, J.J. Mattos, A.C.D. Bainy and E.A. de Almeida. 2022. Influence of temperature on growth, development and thyroid metabolism of American bullfrog tadpoles (*Lithobates catesbeianus*) exposed to the herbicide tebuthiuron. Environ. Toxicol. Pharmacol. 94: 103910.

Hall, J.A. and J.H. Larsen Jr. 1998. Postembryonic ontogeny of the spadefoot toad, *Scaphiopus intermontanus* (Anura: Pelobatidae): skeletal morphology. J. Morphol. 238: 179–244.

Hayes, T.B. 1995. Interdependence of corticosterone and thyroid hormones in larval toads (*Bufo boreas*). I. Thyroid hormone-dependent and independent effects of corticosterone on growth and development. J. Exp. Zool. 271: 95–102.

Heerema, J.L., K.W. Jackman, R.C. Miliano, L. Li, T.S. Zaborniak, N.C. Veldhoen and C. Helbing. 2018. Behavioral and molecular analyses of olfaction-mediated avoidance responses of *Rana (Lithobates) catesbeiana* tadpoles: Sensitivity to thyroid hormones, estrogen, and treated municipal wastewater effluent. Horm. Behav. 101: 85–93.

Hooper, M.J., G.T. Ankley, D.A. Cristol, L.A. Maryoung, P.D. Noyes and K.E. Pinkerton. 2013. Interactions between chemical and climate stressors: A role for mechanistic toxicology in assessing climate change risks. Environ. Toxicol. Chem. 32: 32–48.

Iwasawa, H. and N. Kawasaki. 1979. Normal stages of development of the Japanese green frog *Rhacophorus arboreus* (Okada et Kawano). Jap. J. Herpetol. 8: 22–35.

Iwasawa, H. and J. Futagami. 1992. Normal stages of development of a tree frog, Hyla japonica Gunther. Jap. J. Herpetol. 14: 129–142.

Kashiwagi, K., T. Shinkai, E. Kajii and A. Kashiwagi. 2005. The effects of reactive oxygen species on amphibian aging. Comp. Biochem. Physiol. C Toxicol. Pharmacol. 140: 197–205.

Katz, U., A. Rozman and S. Gabbay. 2003. Skin epithelial transport and structural relationships in naturally metamorphosing *Pelobates syriacus*. J. Exp. Zool. A Comp. Exp. Biol. 298: 1–9.

Kerney, R., M. Meegaskumbura, K. Manamendra-Arachchi and J. Hanken. 2007. Cranial ontogeny in *Philautus silus* (Anura: Ranidae: Rhacophorinae) reveals few similarities with other direct-developing anurans. J. Morphol. 268: 715–725.

Khan, M.S. 1965. A normal table of *Bufo melanostictus* Schneider. Biologia 11: 1–39, Lahore.

Khan, M.S. 1969. A normal table of *Rana tigrina* Daudin 1. Early development (stages 127). Pakistan J. Sci. 21: 3650.

Kumar, A. 1982. Studies on certain aspects of ecology, development and experimental breeding of *Rana cyanophlyctis*, Schneider. PhD Thesis, NEHU.

Lalremsanga, H.T. and R.N.K. Hooroo. 2013. Observation on the normal development of Indian skipping frog, *Euphlyctis cyanophlyctis* (Anura: Dicroglossidae). Traditional Knowledge, 200.
Limbaugh, B.A. and P. Volpe. 1957. Early development of the gulf coast toad, *Bufo valliceps* Wiegmann. Am. Mus. Novit. 1842: 1–32.
Lips, K.R., P.A. Burrowes, J.R. Mendelson and G. Parra-Olea. 2005. Amphibian declines in Latin America: Widespread population declines, extinctions, and impacts 1. Biotropica: Conserv. Biol. 37: 163–165.
Li, M., S. Li, T. Yao, R. Zhao, Q. Wang and G. Zhu. 2016. Waterborne exposure to triadimefon causes thyroid endocrine disruption and developmental delay in *Xenopus laevis* tadpoles. Aquat. Toxicol. 177: 190–197.
Loman, J. 1999. Early metamorphosis in common frog *Rana temporaria* tadpoles at risk of drying: an experimental demonstration. Amphib-reptil 20: 421–430.
Mcdiarmid, R.W. and R. Altig. 1999. Tadpoles: The Biology of Anuran Larvae. The University of Chicago Press, Chicago and London, 444 pp.
Mishra, P.K. and M.C. Dash. 1984. Larval growth and development of a tree frog, *Rhacophorous maculatus* (Gray). Trop. Ecol. 25: 203–207.
Mitchell, N. and R. Swain. 1996. Terrestrial development in the Tasmanian frog, *Bryobatrachus nimbus* (Anura: Myobatrachinae): larval development and a field staging table. Pap. Proc. 130: 75–80.
Mochizuki, K., T. Goda and K. Yamauchi. 2012. Gene expression profile in the liver of *Rana catesbeiana* tadpoles exposed to low temperature in the presence of thyroid hormone. Biochem. Biophys. Res. Commun. 420: 845–850.
Morand, A., P. Joly and O. Grolet. 1997. Phenotypic variation in metamorphosis in five anuran species along a gradient of stream influence. C. R. Acad. Sci. 320: 645–652.
Murata, T. and K. Yamauchi. 2005. Low-temperature arrest of the triiodothyronine-dependent transcription in *Rana catesbeiana* red blood cells. Endocrinology 146: 256–264.
Nieuwkoop, P.D. and J. Faber. 1967. Normal Table of *Xenopus laevis* (Daudin). North Holland Publishing Co, Amsterdam, 252 pp.
Navarro-Martín, L., C. Lanctôt, P. Jackman, B.J. Park, K. Doe, B.D. Pauli and V.L. Trudeau. 2014. Effects of glyphosate-based herbicides on survival, development, growth and sex ratios of wood frogs (*Lithobates sylvaticus*) tadpoles. I: Chronic laboratory exposures to VisionMax®. Aquat. Toxicol. 154: 278–290.
Newman, R.A. 1989. Developmental plasticity of *Scaphiopus couchii* tadpoles in an unpredictable environment. Ecology 70: 1775–1787.
Nieuwkoop, P.O. and J. Faber. 1956. Normal table of *Xenopus laevis* (Daudin). Amsterdam: North Holland Publ.
Oike, A., M. Kodama, S. Yasumasu, T. Yamamoto, Y. Nakamura, E. Ito and M. Nakamura. 2017. Participation of androgen and its receptor in sex determination of an amphibian species. PLoS One 12: e0178067.
Peltzer, P.M., R.C. Lajmanovich, A.M. Attademo, C.M. Junges, M.C. Cabagna-Zenklusen, M.R. Repetti, M.E. Sigrist and H. Beldoménico. 2013. Effect of exposure to contaminated pond sediments on survival, development, and enzyme and blood biomarkers in veined treefrog (*Trachycephalus typhonius*) tadpoles. Ecotoxicol. Environ. Saf. 98: 142–51.
Poilister, A.W. and J.A. Moore. 1937. Moore tables for the normal development of *Rana sylvatica*. Anat. Rec. 68: 489–496.
Rissoli, R.Z., F.C. Abdalla, M.J. Costa, F.T. Rantin, D.J. McKenzie and A.L. Kalinin. 2016. Effects of glyphosate and the glyphosate based herbicides Roundup Original® and Roundup Transorb® on respiratory morphophysiology of bullfrog tadpoles. Chemosphere 156: 37–44.
Rong-Chuan, X., J. Jian-Ping, F. Liang, W. Bin and Y. Chang-Yuan. 2010. Embryonic development of the concave-eared torrent frog with its significance on taxonomy. Zool. Res. 31: 490–498.
Rossa-Feres, D.D.C., R.J. Sawaya, J. Faivovich, J.G.R. Giovanelli, C.A. Brasileiro, L. Schiesari, J. Alexandrino and C.F.B. Haddad. 2011. Anfíbios do Estado de São Paulo, Brasil: conhecimento atual e perspectivas. Biota Neotrop. 11: 47–66.
Rounsevell, D.E., D. Ziegeler, P.B. Brown, M. Davies and M.J. Littlejohn. 1994. A new genus and species of frog (Anura: Myobatrachinae) from soutlern Tasmania. Trans. R. Soc. S. Aust. 118: 171–185.
Roy, D. and M.K. Khare. 1978. Normal table of development of *Rana limnocharis* Weigmann. Proc. nat. Acad. Sci. India 48: 5–16.
Santos, L.H., A.N. Araújo, A. Fachini, A. Pena, C. Delerue-Matos and M.C.B.S.M. Montenegro. 2010. Ecotoxicological aspects related to the presence of pharmaceuticals in the aquatic environment. J. Hazard. 175: 45–95.
Segura-Solís, S. and F. Bolaños. 2009. Desarrollo embrionario y larva del sapo *Incilius aucoinae* (Bufonidae) en Golfito, Costa Rica. Rev. Biol. Trop. 57: 291–299.
Semlitsch, R.D. 1988. Time and size are metamorphosis related to adult fitness. pdf. Ecology 69: 184–192.

Shimizu, S. and H. Ota. 2003. Normal development of *Microhyla ornata* the first description of the complete embryonic and larval stages for the Microhylid Frogs (Amphibia: Anura). Curr. Herpetol. 22: 73–90.

Shumway, W. 1940. Stages in the normal development of *Rana pipiens* I. External form. Anat. Rec. 78: 139–147.

Taylor, C.A. and J.J. Kollros. 1946. Stages in the normal development of *Rana pipiens* larvae. Anat. Rec. 94: 7–23.

Tinsley, R.C., C. Loumont and H.R. Kobel. 1996. Geographical distribution and ecology. The Biology of *Xenopus* (ed. by Tinsley, R.C. and H.R. Kobel), pp. 33–39. Oxford University Press, Oxford.

Townsend, D.S. and M.M. Stewart. 1985. Direct development in *Eleutherodactylus coqui*: a staging table. Copeia 423–436.

Weisz, P. B. 1945. The normal stages in tile development of the South African clawed toad, *Xenopus laevis*. Anat. Rec. 93: 161–169.

Xiong, R.C., J.P. Jiang, L. Fei, B. Wang and C.Y. Ye. 2010. Embryonic development of the concave-eared torrent frog with its significance on taxonomy. Int. J. Zoo. Res. 31: 490–498.

3

Toxicokinetic Pathways of Environmental Contaminants in Amphibian Tadpoles

Daniel Schlenk[1] and *Eduardo Alves de Almeida*[2,*]

1. Introduction: General Principles of Toxicokinetics

Toxic effects of environmental pollutants on tadpoles are dependent on the route and duration of exposure, as well as toxicokinetics. To cause any toxic effect, xenobiotics should be in contact with a molecular target within the organism for a determined period of time. The chemical characteristics of the xenobiotic and its environmental availability influences toxicokinetics and the magnitude of toxic effect produced by the toxicant. To be absorbed by the organism, the toxicant should pass through different biological barriers, such as the skin, or the epithelium of the respiratory or digestive surfaces. Whatever the route of exposure, the compound should be able to passively or actively pass through cellular membranes or between cells that are part of these biological barriers to be absorbed. If the absorbed compound can be transported to the circulatory system, blood circulation can distribute the toxicant throughout the animal's body, taking it to target organs where the compound can either accumulate or be metabolized by specialized enzymes, and potentially excreted from the cell. Unaltered molecules of the toxicant or their metabolic by-products can return to the blood circulation and potentially get eliminated by the kidney into the urine during the renal filtration process. Alternatively, the compound may be eliminated by diffusion or specialized efflux mechanisms through the skin or the gill surface, or directed to the intestine via bile from the liver through the gall bladder. Biliary elimination occurs generally within the feces, although in some cases metabolites may be altered by gut microflora and reabsorbed through enterohepatic circulation. A general overview of the toxicokinetic process of toxic substances in tadpoles is depicted in Fig. 1.

[1] Department of Environmental Sciences, University of California, Riverside 900 University Ave, 92521 Riverside, California, United States. Email: daniel.schlenck@ucr.edu

[2] Department of Natural Sciences, Regional University of Blumenau, Rua São Paulo 3366, Bloco Q, Blumenau, SC, Brazil, CEP 89030-000.

* Corresponding author: eduardoalves@furb.br

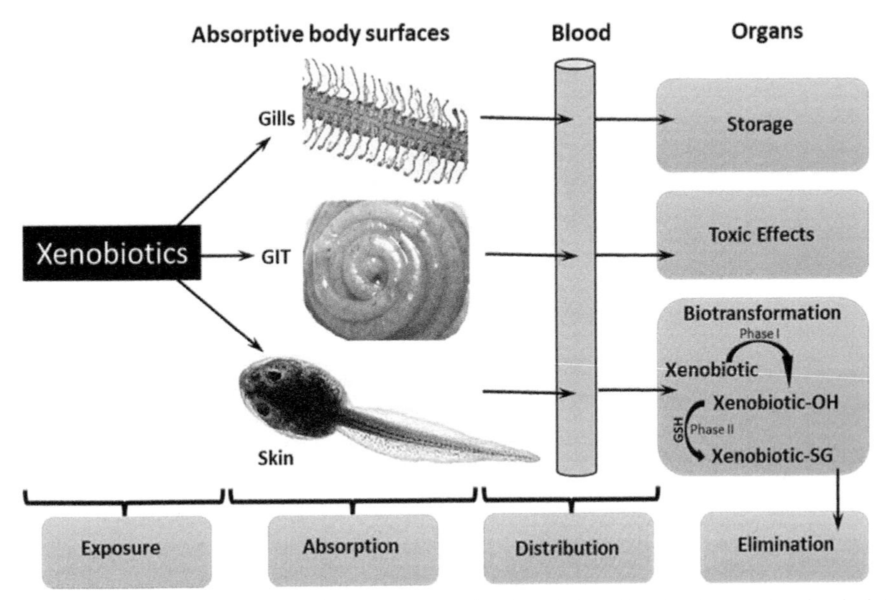

Figure 1. Schematic representation of xenobiotic toxicokinetics in tadpoles. Xenobiotics can be absorbed from the environment through the gills, skin or the gastrointestinal tract and then be distributed to different organs by the blood stream. Once reaching different organs, the xenobiotic can be retained (stored), biotransformed into more or less reactive metabolites, interact with cellular components causing toxic effects, and/or get eliminated from cells and from the organism.

2. Absorption of Toxicants in Tadpoles

The first step of toxicokinetics involves the absorption of the toxic compound from the environment. To be absorbed, the organism should be in contact with the toxic agent, and the toxicant should be available in an absorbable form. In larval tadpoles, the main organs that are typically responsible for toxicant uptake from the environment are the skin, the gills and/or the gastrointestinal tract (GIT).

Generally, the longer the animal is exposed to the toxicant in an absorbable form, greater the potential intoxication. Consequently, environmentally persistent compounds in general have a greater likelihood of exposure and potential absorption. Toxicants can be absorbed by four main mechanisms: passive diffusion, facilitated diffusion, active transport or vesicle-mediated transport (endocytosis) (Fig. 2). Protein-mediated transport (facilitated diffusion and active transport) of xenobiotics is most frequent when the absorption process occurs through the GIT and the gills. However, the skin of tadpoles does not possess a high number of transporters for toxicant uptake. Numerous membrane transporters for different nutrients are present in the intestine and gill epithelium, and can often mediate the transport of some toxicants, depending on the similarity of their structures with organic compounds that have affinity to the transporter. Both tissues also have very specific transporters for metals which are often hijacked by non-essential metals. However, with regard to organic toxicants, due to the high specificity of most membrane transporters to chemical structures of nutrients, the large majority of absorption process of xenobiotics through the GIT and gill primarily occurs by passive diffusion. Transport of toxicants by protein transporters does not depend on only having an affinity for the transporter but often energy and or conformational change of the transporter is required so that the toxicant can be transported from the outside to inside of the cell. Moreover, after being transported, the toxicant should be released by the transporter into the cytosol. For this reason, the vast majority of toxic substances are often relatively lipid soluble, a characteristic that favors passive diffusion through cellular membranes.

Surfaces of the skin tend to have limited concentrations of protein transporters, and any absorption processes that occurs in these organs are often dependent solely on passive diffusion. As passive diffusion is the predominant transport process for the absorption of majority of organic

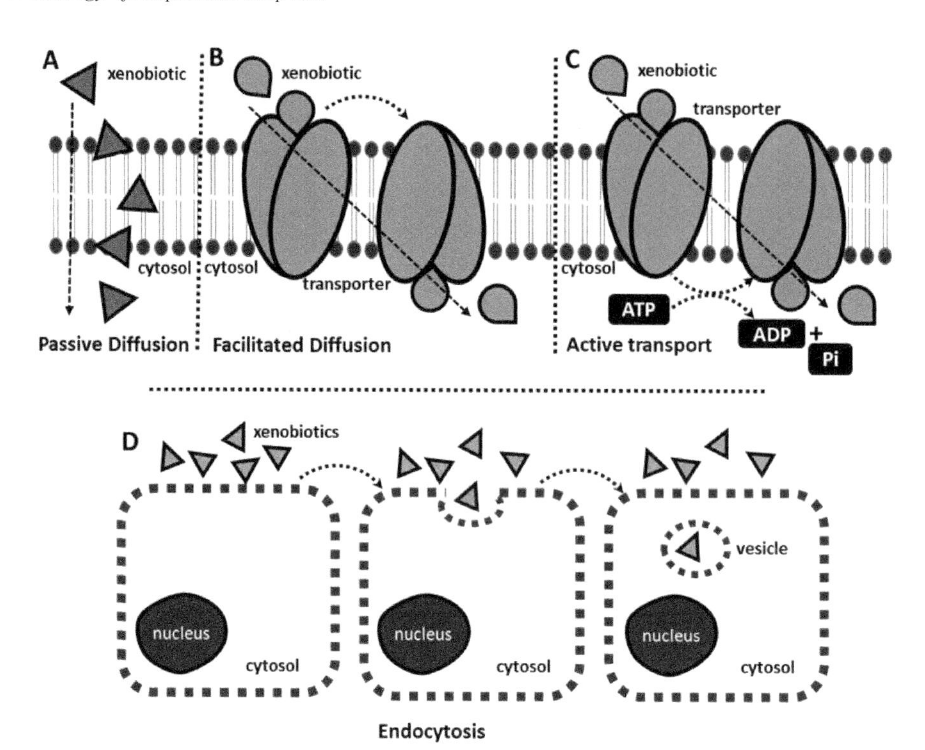

Figure 2. Main mechanisms of transference of xenobiotics across cellular membranes: passive diffusion (A), facilitated diffusion (B), active transport (C) and endocytosis (D). Vertical and diagonal dotted arrows indicate the direction of xenobiotic transport. *Simple diffusion can also occur through membrane pores or ion channels, a process named "filtration diffusion". Filtration diffusion depends on concentration gradient, the size of the compound and its capacity to interact with the interior of the pore/channel. This process is particularly important for the absorption of metals; i.e., Pb^{2+}, which is able to pass through some Ca^{2+} channels.

xenobiotics, only molecules with considerable lipophilicity will meet the requirements to cross the cellular membrane. In general, lipophilic molecules are also metabolized in the membranous smooth endoplasmic reticulum (SER) where biotransformation often occurs.

The route of exposure is also important for absorption. For example, toxicants that are absorbed through the GIT in vertebrates are often directed to the liver by the hepatic-portal circulation. The passage of toxicants through the liver can increase or decrease the toxicity of the teoxicant due to its transformation into more or less toxic metabolites by hepatic enzymes. When more reactive metabolites are generated, they can directly react with cellular components of the hepatocytes causing hepatotoxicity. Alternatively, these metabolites can be excreted to the systemic circulation and reach other organs in a more toxic form, causing systemic effects. Excretion can occur through the bile canaliculus into the gallbladder, where bile is stored and eventually discharged into the GIT and incorporated into the excreta or feces. Smaller molecules may also be excreted back into the circulatory system and transported to the kidney where elimination can occur.

On the other hand, if absorption occurs at the respiratory epithelium or the skin, the toxicant is more likely to be distributed to different target organs in its unmodified form unless transportation to the liver occurs. If the toxicity depends on the formation of reactive metabolites through biotransformation processes, absorption at these sites may lead to more limited toxicity than if the compound is absorbed in the GIT.

Most studies regarding toxicant absorption in amphibians have been focused on adults. However, the absorptive surfaces of tadpoles can be much more susceptible to absorption of toxicants than those from adults due to their structural simplicity (Uchiyama and Konno 2006). The jelly capsule that protects the amphibian embryos in aquatic environments before hatching can be very permeable to toxic compounds potentially causing embryotoxicity. Furthermore, the absorption through maternal transfer to embryos can also account for intoxication of tadpoles in the environment.

2.1 Absorption Through the Skin

The skin possesses a specialized epithelium that isolates the organism from the external environment and prevents the loss of water and other endogenous materials. It usually consists of four layers: the hypodermis, dermis, epidermis and stratum corneum. The latter covers the body surface and represents the main barrier to percutaneous absorption (Lillywhite 2006, Hadley 1991). The skin in amphibians tends to have higher permeability than other vertebrates because it is physiologically involved in respiration and in osmoregulation (Uchiyama and Konno 2006). In general, the permeability of the epithelium is proportional to its thickness, and the stratum corneum is roughly 10–20 times thicker in mammals (~ 20 μm) than in amphibians (~ 2 μm in *Rana pipiens*) (Carrer et al. 2008, Farquar and Palade 1965). This difference implies that organic xenobiotics can diffuse into amphibians more effectively than in mammals, depending on the chemical's hydrophobicity. For example, Quaranta et al. (2009) estimated that the percutaneous passage of xenobiotics such as atrazine, paraquat and glyphosate in the skin of adult *R. esculenta* (adults) was one or two orders of magnitude faster than in pigs. Willens et al. (2006a,b) also examined the percutaneous absorption of the insecticide malathion by adult anuran skin *in vitro*. Using a diffusion cell model, the total absorption of ventral and dorsal skin of the American bullfrog (*Lithobates catesbeianus*) was 81 and 69%, and for the cane toad (*Bufo marinus*) 83 and 77%. In contrast, with the harvested perfused anuran pelvic limb (HPAPL) model, 46% absorption of the total malathion dose administered was measured in frog skin (pelvic limb). Moreover, Mendez et al. (2009) assessed the uptake of atrazine by adult American toads (*Bufo americanus*) *in vivo*, exposing them to atrazine-treated soil. Dehydrated toads that were exposed to soil spiked with C^{14}-atrazine-treated water rapidly absorbed atrazine as the animal rehydrated through the pelvic patch.

Similar studies on the rates of toxicant absorption through the skin in tadpoles are not available in scientific literature, but considering the fact that the skin of tadpoles is less developed compared to that of adults, it can be expected that the process of percutaneous passage of pollutants in tadpoles may be more rapid than in adults. The skin of tadpoles is subjected to a profound remodeling process during metamorphosis. When anuran tadpole metamorphoses from an aquatic larva to a terrestrial frog, the conversion of the larval skin into the adult stratified skin is one of the most important changes to adapt to the dry environment (Nakajima et al. 2005). Therefore, it can be also expected that the capacity of the skin of tadpoles to absorb pollutants from water varies as changes in the type and number of skin cells and structures (i.e., the appearance of keratin) is altered during development. In contrast to the adult, the tadpole skin is very similar to mammalian fetal skin, and consists of two to three cell layers in contact with an acellular collagen lamella, with no dermis (Brown and Cai 2007). When metamorphosis starts, the increase in circulating thyroid hormones induces the apoptosis of this larval skin cells (Schreiber and Brown 2003) and the basal cells to form a germinative epithelium that is characteristic of adult vertebrate skin. The more complex skin of adult amphibians has an underlying keratinized dermis and skin glands that appear for the first time at metamorphosis climax, possibly decreasing its permeability. Willens et al. (2006a) noted that a 2 to 3 cell layer thickness in the dorsal skin of *L. catesbeianus* with more glandular tissues resulted in a 25 to 30% reduction in permeability to the insecticide malathion, when compared to the dorsal skin of *Bufo marinus* with 1 to 2 cell layers and lower density of glandular tissues. Therefore, alterations in skin during metamorphosis possibly results in changes in the capacity of the skin to function as a barrier for toxicant absorption, thus resulting in differential susceptibility to intoxication according to the developmental stage of the amphibian.

2.2 Absorption Through the Gastrointestinal Tract (GIT)

The GIT of vertebrates is the location where ingested food is digested and the resulting nutrients get absorbed by the organism. Therefore, the cells tend to be very specialized in digestive secretion and absorption of nutrients. Food taken in through the mouth is digested to extract nutrients and absorb energy, and the non-absorbed waste is further expelled at the anus as feces. Numerous toxic chemicals that eventually are dissolved in water or adhered to food particles and are ingested by tadpoles can, therefore, be effectively absorbed by the GIT. Lanctôt et al. (2017b) showed that tadpoles exposed to metal-contaminated water absorbed and bioaccumulated 2–3 times more Cd and Zn when food particles were present in water compared to tadpoles exposed in clean water.

Similar to the skin, the GIT also undergoes a profound remodeling during metamorphosis in tadpoles. Like typical herbivores, tadpoles have a long, highly coiled, narrow-bore, thin-walled intestine consisting of a single layer of epithelial cells with a long colon, whereas adult animals have a short, wide-bore, thick-walled intestine typical of carnivores. In fact, during the final stages of development in pro-metamorphic tadpoles, primary (larval) epithelial cells are completely removed through apoptosis and replaced by secondary (adult) epithelial cells that form a more complex multi-folded epithelium (Shi and Ishizuya-Oka 1996, Ishizuya-Oka 2011). At the climax of metamorphosis, the intestines of tadpoles also shorten by 60 to 90% depending on the species (Pretty et al. 1995). The mesenchyme and muscle layers become thicker, and the epithelium folds into the typical crypts and villi that characterize all adult vertebrate intestines. Indeed, by the time the gut has completed shortening, the cross-sectional diameter has also decreased by about one-third (Schreiber et al. 2005).

Besides the changes in the length and morphology, the process of intestine remodeling in tadpoles also results in alterations in nutrient uptake due to changes in the amount, variety and affinities of membrane transporters. For example, the glucose transporter has a higher capacity in tadpoles than in adults, corresponding to the higher carbohydrate content of the tadpole diet. Conversely, the proline transporter has a higher capacity in adults than in tadpoles, corresponding to the higher protein content of the adult diet (Toloza and Diamond 1990a,b). Alterations in number, variety and specificity of membrane transporters along the GIT can, thus, change the type and range of substances that could eventually be absorbed through protein-mediated transport. In addition, although the intestine decreases in length during metamorphosis, it folds and develops villi and microvilli structures, which in addition to the brush border of the epithelial cells, increases the surface area for absorption. These alterations are consistent with the increasing absorptive capacity of the GIT as tadpoles reach later stages of development and complete the metamorphosis process (Hourdry et al. 1996).

Thus, toxicant uptake by tadpoles through the GIT can vary substantially as the morphology and function of the intestine changes during development, and possibly alter susceptibility to intoxication. As a rule, toxicant uptake by passive diffusion in the GIT is dependent on the membrane composition, but also on the number and thickness of epithelial cells. Indeed, for weak acids and bases, whose absorption will depend on the partitioning of ionized and unionized forms, absorption by passive diffusion can also vary as the pH of the GIT changes along development. Pre-metamorphic larvae do not have a complex gastric pouch (or stomach) as seen in adults, but the wall in the gastric region of most tadpoles forms a thickened sheath, which produces a secretion with a pH close to 3.5, with no substantial production of digestive enzymes. During metamorphosis climax, the glandular sheath regresses and is replaced by a large permanent stomach associated with functional changes, such as the abundant production of pepsin (pH optimum 2.3) and hydrochloric acid by the gastric serous cells (Hourdry et al. 1996).

Furthermore, bile salts may influence the absorption process of contaminants in the intestine. Bile salts can enhance xenobiotic penetration through various biological membranes by interacting with phospholipids in cell membranes and increasing its permeability (Moghimipour et al. 2015). Anderson et al. (1979) observed that a higher number of bile salt molecules appear as *L. catesbeianus* tadpoles pass from pro-metamofic to post-metamorfic stages, due to the activity of new enzymes involved in biliary salt synthesis. Therefore, absorption rates through the intestine may increase as the production of salt bile increases in tadpoles.

Feeding behavior can also affect the rate of toxicant uptake by the GIT in different tadpoles. For example, embryos of some anurans have a sufficient yolk reserve and reach a late developmental stage without feeding (Altig and Johnston 1989, Burggren and Just 1992). In these species, hatching results in free-living nonfeeding tadpoles that stay near the nest until metamorphosis (Altig and Johnston 1989). Most tadpole species cease eating during the climax of metamorphosis and start to make use of the organic reserves from the fat tissue and from the tail cells that are reabsorbed. Thus, during these periods of limited feeding, absorption of toxicants may be impaired through the GIT.

Feeding behavior can also influence the absorption of toxicants through the gills. Numerous tadpoles' species are nonselective filter feeders extracting suspended particles in open water through gill filters (Seale 1987). These tadpoles are characterized by large gill chambers and well-developed filter systems that retain and adhere food particles to mucous traps located on the underside of the ventral velum (Hourdry et al. 1996). Although particle retention on the gill filters can be advantageous for feeding, it can also allow toxicants adhered to the food to have contact with gill filaments, allowing toxicants to be absorbed by the gill surface.

Another feeding behavior that may have some importance for chemical absorption through the GIT is coprophagy. For some tadpole species most parts of ingested material that passes through the gut appears untouched on elimination (Savage 1962) with only ~ 10% of the organic matter extracted from the food on the passage through the GIT (Altig and McDearman 1975). Tadpoles permitted to practice coprophagy can continue to extract calories still present in the reingested feces (Gromko et al. 1973) and consequently grow larger than tadpoles prevented from reingesting their feces (Steinwascher 1978). Coprophagy for some tadpole species can represent a crucial strategy for obtaining nutrients during development, especially in environments with scarce food availability. However, if toxic chemicals are present in the feces due to excretion after detoxification or because they were not absorbed after ingestion, the coprophagic habit can possibly enhance the degree of absorption.

2.3 Absorption Through the Gills

In most tadpole species, branchial ventilation is comparable to that of teleost fishes, and is responsible for about 40% of O_2 uptake from water. Due to a large skin area to body mass ratio, the presence of the tail, and thin highly vascularized skin, 60% of O_2 uptake in these animals is cutaneous. However, in some species such as *Rana temporaria, Bufo bufo* and *Litoria ewingii* cutaneous gas exchange seems to be not so important in the tadpole phase, because of limited skin vascularization (McIndoe and Smith 1984). As development proceeds, gills degenerate and the appearance of lungs assume importance in O_2 uptake in air-breathing adults, but the skin often remains the major organ of O_2 uptake until metamorphosis is nearly complete. Immediately after metamorphosis, O_2 uptake by the lungs is elevated to 80% of total consumed O_2. Similarly, nearly 40% of CO_2 excretion in aquatic tadpoles occurs through the gills, with this process replaced ~ 90% by the lungs and 10% in the skin of adults (Burggren and West 1982).

In tadpoles, the rate and efficacy of absorption of toxic compounds through the gills is undistinguishable from the skin, therefore, the contribution of gill absorption is often estimated in total with the uptake through skin. However, the differences in both cutaneous and branchial vascularization among different species, as previously addressed, may lead to different uptake rates of organic xenobiotics. The more vascularized the absorptive surface, the more rapid is the internal dispersion of the toxicant molecules through the blood circulation, maintaining the gradient of concentration of the xenobiotic between the external and the internal environment. In turn, the regression of the gills, being replaced by the adult lungs also reduces the rate of toxicant uptake by the gills during the metamorphosis, which may also reflect on distinct capacities of the gills to absorb contaminants depending on the stage of development.

Although the morphological and functional characteristics of the gill and of the skin epithelium can enhance permeability to lipophilic contaminants, ventilation for branchial respiration may be of even greater importance for enhanced uptake of contaminants through the gills. The gills of tadpoles are actively ventilated by the alternating action of buccal and pharyngeal pumps (Gradwell 1973, 1973) at a considerable energy cost. An eventual ventilation of the skin due to swimming movements in tadpoles does not seems to be as constant or efficient. Ventilation of the skin capillaries due to exposure to water flow was also not proven to be as extensive as the exposure of gill capillaries. Gill ventilation promotes higher water flux through the gill filaments, which may increase the rate of contaminants in contact with the respiratory surface, possibly increasing its absorption in comparison to the skin. This could be particularly important in situations in which tadpoles need to increase ventilation, such as in response to hypoxic or anoxic environments, or events in which tadpoles have high energetic and oxygen demands, such as escapes from predator attempts.

The gills are also an important route for metal uptake. Mechanisms by which metals are absorbed by tadpoles through the gills are not well characterized, but given the structural and physiological similarities between the gills of tadpoles and freshwater fishes, it is likely that the processes of metal uptake is similar. The gills are essential for osmoregulation, being involved in ammonia elimination and in the balance of Na^+, Cl^- and H^+ ions through specialized active and passive transporters and channels. In general, freshwater vertebrates with gills extract needed NaCl from the medium via parallel, antiport systems involving exchange of intracellular H^+ or NH_4^+ for external Na^+ and internal HCO_3^- (or possibly $0H^-$) in exchange for external Cl^-. Channels and transporters for the uptake of essential metals (e.g., Cu and Zn) are also present in the epithelium of the gill, which in some cases are also capable to transport other metallic ions.

Hogstrand et al. (1996) showed that Zn^{2+} and possibly Cd pass across the apical membrane of the chloride cells of the rainbow trout gills via Ca^{2+} voltage-insensitive channels. Moreover, Glynn et al. (1994) showed that the calcium-channel blocker verapamil caused a concentration-dependent decrease in gill uptake of Cd and Hg in *Brachydanio rerio*, and that exposure to Cd or Hg in the presence of 0.2 mM Ca^{2+} showed a lower branchial uptake of the metals than the controls, which suggest an additional passive influx of Cd and Hg through calcium channels. Comhaire et al. (1994) also suggested that Co ions are absorbed in the gills of *Cyprinus carpio* through Ca^{2+}-channels. Zn, Cu, but especially Cd, can displace Ca^{2+} from Ca^{2+}-ATPase on the gill surface, enhancing the influx of these toxic metals in the organism (Zia and McDonald 1994). Moreover, loss of calcium from the membrane and tight junctions interconnecting the apical areas of the upper cell layers of the gill epithelium due to displacement from other metals often increase the permeability of the gills, which can result in increased uptake of xenobiotics (McDonald et al. 1991, McWilliams 1982).

Copper is an essential metal for living organism, and for this reason the apical membrane of the gill epithelia of fishes contains Cu-specific channels and Cu leak through epithelial Na^+ channels when external Na^+ concentration is low (Handy et al. 2002). Thus, it can be expected that increased concentration of Cu in water would increase its influx through these channels. However, Grossel and Wood (2002) showed that gill Cu concentration stabilizes over time in rainbow trout exposed to elevated external concentrations of Cu, suggesting the establishment of an equilibrium between uptake and elimination from the gill tissue. At the highest copper concentration, a biphasic accumulation pattern was observed, with an initial increase followed by a marked drop in branchial Cu levels. The authors suggested that elevated accumulation of Cu in gill cells may initiate a stimulation of extrusion mechanisms such as the copper-specific ATPases and/or cause a reduction in the apical entry of copper, resulting in a protective response against intoxication when elevated copper concentrations are present. Despite the lack of additional information in the scientific literature, possible similar protective mechanisms against intoxication by other metals should not be disregarded.

Considering the importance of protein-mediated transport for metal uptake through the gills, their affinity to different metals, transport efficacy, variety and number of copies in cell membranes are limiting factors that will drive the rate by which different metals are absorbed by the gills. Despite a total lack on information about these transporters in the gill of tadpoles, it is likely that changes occur during development until the animal completes metamorphosis. However, Dietz and Alvarado (1974) studied the Na^+ and Cl^- fluxes across gill chamber epithelium of *L. catesbeianus* tadpoles at various stages (TK VI to XVII), and reported no significant differences in transfer rates between the different stages. This suggests that ion transporters in general are similarly active during all stages of development of tadpoles. However, further studies are required for confirmation.

2.4 Absorption Through Maternal Transfer

Although the gills, skin and GIT are the most common sites for exposure in tadpoles to toxicants in the environment, amphibian larvae can also absorb toxicants from maternal transfer during embryogenesis (Kadokami et al. 2004, Bergeron et al. 2010), ultimately leading to reduced hatching success and offspring viability (Kotyzova and Sundeman 1998, Hopkins et al. 2006, Bergeron et al. 2011). Hopkins et al. (2006), for example, reported that narrow-mouth toads (*Gastrophryne carolinensis*) inhabiting an industrial area transferred significant amounts of selenium and strontium to their eggs, causing a 11% reduction in the hatching success. In surviving larvae from the contaminated site, the frequency of developmental abnormalities increased, and abnormal swimming was also highly reduced, compromising the survival of the tadpoles. Bergeron et al. (2011) showed that larvae of female American toads exposed to Hg presented delayed growth and produced smaller adults at the climax of metamorphosis than larvae from unexposed females. Metts et al. (2012) also demonstrated that previous maternal exposure of toads (*Bufo terrestris*) to coal combustion wastes reduced larval survival to metamorphosis up to 57% compared to larvae of unexposed females, and was related to the maternal transfer of contaminants to the embryos. These authors also showed that females collected from the coal combustion waste-contaminated area had 27% reduced offspring viability and were able to transfer elevated levels of Cu, Pb, Se, and Sr to their eggs compared to females from the reference site (Metts et al. 2013).

2.5 Additional Absorption Barriers in Tadpoles

As previously mentioned, the jelly coat that cover the embryos during the early development of amphibians can be permeable to toxic compounds present in the environment. In natural environments, this jelly structure helps the eggs to adhere to the vegetation, protects the embryos against mechanical disturbances, UV-B radiation and act as a possible protective agent against some chemicals (Grant and Licht 1995, Berrill et al. 1997, Jung and Walker 1997, Marquis 2006). For example, Licht (1985) showed that embryos exposed to DDT without protective jelly capsules accumulated 10 times more DDT than eggs with jelly capsules. Marquis et al. (2006a) also demonstrated that *R. temporaria* embryos with jelly coat tend to suffer a lower mortality rate than embryos without jelly after exposure to different PAHs (naphthalene, phenanthrene and pyrene).

The analysis of the molecular components of the jelly coats of different amphibian species reveals that it is composed mainly of carbohydrates and proteins that are species specific, with a remarkable heterogeneity in their structures (Salthe 1963). This heterogeneity implies differential capacities of the jelly coat to protect the embryos among species. Greulich et al. (2002) exposed *Bombina bombina* eggs to ^{14}C-isoproturon, a phenylurea herbicide, and found that 2% of the herbicide was absorbed by the larvae in 30 min, and 4.8% in 48 h, suggesting low protective action of the jelly coat. Absorption of TCDD into both eggs and tadpoles of American toads, leopard frogs and green frogs also occurred rapidly (within approx. 2 h), with TCDD being found primarily in the embryo proper (the jelly capsule contained 4% TCDD relative to the embryo) (Jung and Walker 1997). These results imply that the protection provided by jelly coat against the intoxication can vary from one species to another according to their composition, but also that this protection can be dependent on the chemical properties of the toxicant.

Another physiological response to environmental toxicants in tadpoles is the secretion of mucus through the skin and gills. Mucogenic activity during larval development has been shown to protect the epidermis against exogenous stressors (Fox 1992). Excess levels of mucus in skin has been reported in the skin of both fish and amphibians exposed to organochlorine insecticides (Fenoglio et al. 2009, Zaccone et al. 1989). Similarly, elevated mucous secretion has been observed in the gill epithelium as a result of various perturbations in aquatic environments (Bernabò et al. 2008, 2013a, Brunelli et al. 2010, Brunelli and Tripepi 2005). In tadpoles, Bernabò et al. (2013b) noted increased mucus secretion in the skin of different species exposed to the insecticide endosulfan. It is thus conceivable that high mucus-secreting activity could be an immediate and nonspecific response to environmental toxicants, eventually providing an extra barrier against absorption.

3. Distribution of Toxicants in Tadpoles

The presence of internal barriers is also important for toxicant absorption and its distribution to different organs of the body. Once internalized, the toxic agent can reach different organs through the blood circulation. But depending on the cellular and acellular components of the different tissues that compose the organ, its absorption capacity can be substantially different. The composition of the cell membranes in terms of lipid and protein, the presence, complexity and thickness of extracellular matrix components and glycocalyx, number and types of influx and efflux proteins in cellular membrane, and the characteristics of the vascular structure of the tissues are examples of factors that may alter the feasibility of a toxicant to be more or less distributed to different organs.

A typical internal barrier in vertebrates that prevents most toxicants from reaching the nervous system is the blood-brain barrier (BBB). The BBB is a component of the vascular system of the brain and acts as a selective interface between the blood and the brain. The BBB separates the circulation from the brain, allowing for the protection from xenobiotic uptake and the regulation of serum factors and nutrient uptake. The BBB is a combination of a physical barrier (i.e., due to the presence of specialized tight junctions between endothelial cells) in addition to the action of selective transporters that mediate the transit of substances between the blood and neurons. It is also a metabolic barrier, containing and releasing enzymes locally, including some isoforms of cytochrome P450 (Abbott et al. 2006, Ghosh et al. 2011). Although little is known about the BBB function in anuran amphibians, some studies suggest that it may be under-developed in tadpoles compared to adults. However, Andino et al. (2016) showed that the BBB in *X. laevis* pre-metamorphic tadpoles (3 weeks-old) is fully functional, therefore providing some degree of protection against intoxication of the nervous system.

After being absorbed, toxic substances that reach the blood stream can be distributed to different organs of the tadpole. Accordingly, xenobiotic molecules will be delivered at higher amounts to those organs that are more vascularized, such as the brain, liver and kidneys. As most organic xenobiotics must have a relatively high lipophilicity to permit it to cross the cellular membranes of the absorptive body surfaces, much of the absorbed substances will not easily dissolve in aqueous phase of the blood, needing the presence of circulating lipophilic substances to serve as vehicles, such as lipoproteins or even the cellular membranes of blood cells. The free (unbound) fraction of a toxicant in plasma is usually the portion that exerts a toxic effect. Binding of toxicants in plasma proteins and distribution though the blood stream is particularly important for those tadpoles with higher body length and mass, in which the internal distribution of the toxicant cannot be only dependent on simple diffusion process among blood fluids. For very small tadpoles, passive diffusion may contribute significantly for toxicant distribution. For example, *Scinax luizotavioi* tadpoles at Gosner stage 37 are ~ 5 mm in length and ~ 2 mm in width (Bertoluci et al. 2007), which may favor a higher rate of absorption and distribution of lipophilic toxicants through the body fluids by simple diffusion.

Since distribution may be dependent on the presence of lipoproteins, the distribution and disposal of lipophilic substances through the circulation in tadpoles varies according to the capacity of the liver in producing and releasing these proteins into the circulation.

Similar to other organs in tadpoles, the liver is also remodeled at metamorphosis. When tadpoles metamorphose to terrestrial frogs they change from ammonia to urea excretion, and the genes of hepatic enzymes of the urea cycle are only upregulated at metamorphosis (Paik and Cohen 1960, Atkinson et al. 1996). Albumin synthesis by the liver is also up-regulated in the liver late in tadpole life (Moskaitis et al. 1989), a process that is mediated by thyroid hormones. Thus, distribution and disposal of contaminants that are dependent on binding to lipoproteins such as albumin in tadpoles vary according to the capacity of the liver to produce such proteins along development. If protein binding is reduced due to lower available lipoproteins, a greater free fraction of the xenobiotic may be available, which may increase toxicity.

4. Biotransformation in Tadpoles

The physicochemical properties of organic xenobiotics play an important role in the absorption of the chemical with lipophilic materials being more readily taken up by cells. The ability of organisms to metabolically transform the chemical structures of xenobiotics into more soluble and excretable derivatives is critical to prevent xenobiotics from causing toxicity (Ertl and Winston 1998). When an organism absorbs a toxicant from the environment, the efficiency of metabolic enzymes may influence whether the toxicant bioaccumulates in the body, how it is distributed in the organism, how it becomes more or less toxic to target organs, and the potential routes for its the elimination from the body. The metabolic capacity of organisms varies substantially among different species and even among different organs of the same individual; thus, different susceptibilities can be observed depending on the studied species and/or on the stage of development of the tadpole.

Numerous studies have characterized the presence of biotransformation enzymes in amphibians. The newt *Pleurodeles waltl*, for example, was shown to be able to convert the polyaromatic hydrocarbon (PAH) benzo[a]pyrene (BaP) into water-soluble products (Marty et al. 1989). Another study showed that leopard frogs (*Rana pipiens*) are capable to metabolize and eliminate different polychlorinated biphenyls (PCBs) during all life stages, with metamorphs presenting the higher metabolic activity (Leney 2005).

Biotransformation is an enzymatic cellular process, often divided in two phases (Fig. 3). During phase I, the structure of the toxicant can be modified by both cytosolic SER enzymes. SER enzymes have been primarily characterized in the microsomal fraction of cellular homogenates. They are prepared in the laboratory by differential centrifugation at 100,000 g, and are generated as the pellet. The resulting supernatant possesses the cytosolic enzymes. Cytosolic phase I enzymes include several reductases, hydrolases, oxidases and esterases, while examples of microsomal Phase I enzymes include the cytochrome P450 monooxygenase family (CYPs) and flavin monooxygenases (FMOs). Phase II biotransformation pathways involve conjugation reactions of modified or unmodified toxicant molecules with endogenous compounds, such as glucuronic acid, sulphate and the tripeptide glutathione (GSH), by the actions of different transferases found both in the microsomes and cytosol (e.g., glucuronosyl transferases, sulphate transferases and glutathione *S*-transferases). Both phase I and phase II metabolism often result in more water-soluble metabolites, and the conjugation reactions of phase II target the toxicant molecule for specialized membrane efflux proteins, facilitating their elimination from cells.

Although phase I and II enzymes are present in cells at constitutive baseline levels, the expression of genes of these enzymes as well as their catalytic activities can be substantially induced in response to a great variety of contaminants, through different cellular signaling pathways. One of the most studied signaling mechanism includes the Aryl Hydrocarbon Receptor (AhR), a transcript factor that binds to lipophilic xenobiotics, such as PAHs and PCBs. After binding, the complex is translocated to the nucleus as a transcription factor and promotes the transcription of numerous

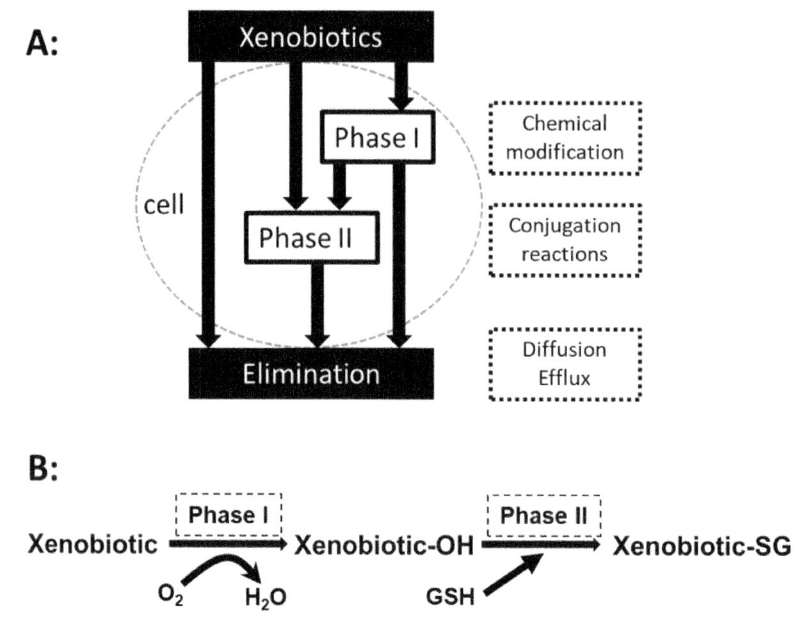

Figure 3. General scheme of cellular biotransformation processes. **A:** Xenobiotics that enter the cell can be directly eliminated without metabolization, or passes through phase I and/or phase II metabolic reactions before being eliminated. **B:** Phase I causes a modification of the chemical structure of the xenobiotics, while phase II is related to conjugation reactions with endogenous molecules such as glutathione (GSH).

genes, among them, several phase I and phase II enzymes and also aryl hydrocarbon receptor repressors (AhRR), which can modulate negatively the responsiveness of AhR to typical inducers in negative feedback mechanism (Fig. 4).

Leney et al. (2006a) revealed that green frogs and leopard frogs possess metabolic enzymes for biotransformation of PCBs during each of three distinct life stages: tadpoles, metamorphs and adults. The highest metabolism of different PCBs was found in metamorphs, followed by an intermediate activity in tadpoles, and lowest activity in adults. One possible explanation for higher activity of the CYP system in metamorphs was related to a maintenance of homeostasis of endogenous substances. As discussed by the authors, metamorphic transformation in amphibians is controlled by different hormones; therefore, high levels of CYP enzymes could be present during metamorphosis simply because of an increased need for hormone metabolism. Thus, these enzymes would represent a convenient advantage against PCB intoxication, being able to metabolize it. A second potential explanation for higher CYP systems in metamorphosis is that these enzymes would be produced naturally during metamorphosis to metabolize accumulated plant toxins from the diet. Tadpoles are primarily herbivorous; thus, they are exposed to phytotoxins from their diet and require CYP enzymes to metabolize and excrete these toxins. Therefore, it is plausible that CYP activity naturally increases during metamorphosis to break down these compounds. Thus, the metabolism of PCBs may simply be a side effect of this natural enzyme activity. The reduction of CYP systems after metamorphosis would then be a reflection of the transformation of the herbivorous juveniles into the carnivorous adult which require less CYP.

Another possibility is that the PCBs themselves are upregulating biotransformation enzymes, although with less efficacy in adults and pre-metamorphic organisms. Biotransformation enzymes are present in amphibians even at early life stages, and in some cases, actually are higher in larvae than in adults. For example, the constitutive activity of BaP hydroxylase in adult newts was only approximately 60% that of the larvae (Marty et al. 1989). This contradiction could be related to the fact that biotransformation systems can be more or less effective depending on the toxic molecule considered, or that the AhR system in tadpoles would be less responsible.

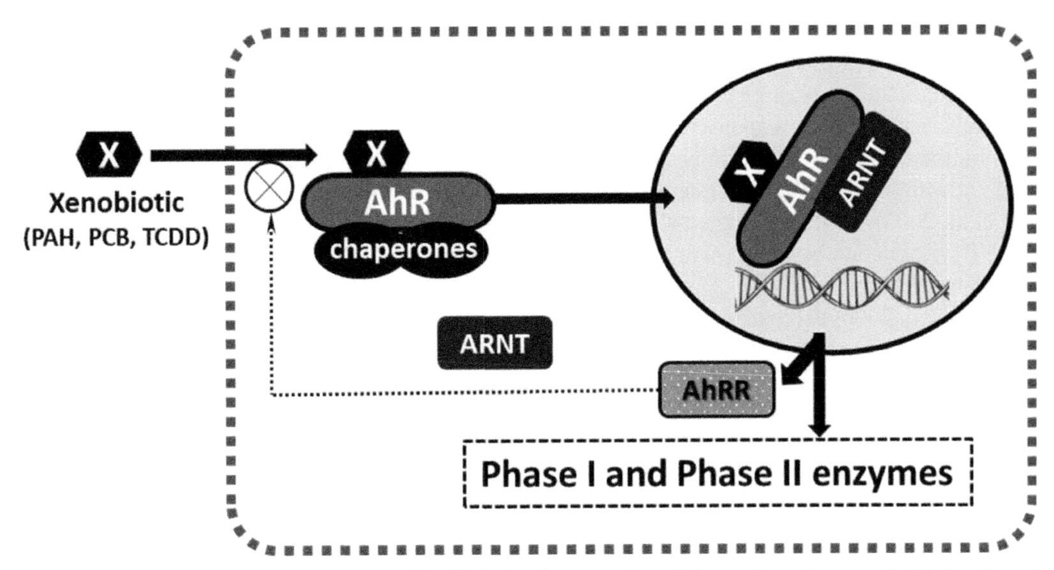

Figure 4. Most lipophilic xenobiotics are capable of inducing the expression of biotransformation genes by binding the aryl hydrocarbon receptor (AhR). The interaction of the xenobiotic with AhR causes the release of chaperones complexed with the receptor, allowing AhR to dimerize with the aryl hydrocarbon receptor nuclear translocator (ARNT). The complex formed by the AhR bound to the xenobiotic molecule and ARNT then migrates to the nucleus and binds to the promotor region of a gene cluster of biotransformation enzymes, promoting its mRNA transcription. The expression of an aryl hydrocarbon receptor repressor (AhRR) can be also induced, which reduce the responsiveness of AhR to typical inducers.

CYP enzymes present in larval stages are inducible, meaning that their abundance and/or activity can increase in response to some xenobiotics (Jönsson et al. 2011). However, it has been shown that the AhR receptors in tadpoles are substantially less responsible to some classic agonists, such as dioxins (Lavine et al. 2005, Elskus 2005, Kazzaz et al. 2020). The earliest developmental stages of tadpoles often exhibit very low responsiveness to AhR agonists such as benzo[a]pyrene (BaP), 2,3,7,8-tetrachlorodibenzo-*p*-dioxin (TCDD) and Aroclor 1254. For example, Sestak et al. (2018) showed that *Xenopus laevis* tadpoles exposed to BaP did not have the same sensitivity to cardiac impairment in comparison to fish. Compared with other vertebrates, frogs are up to 1000-fold less sensitive to TCDD-induced mortality (Beatty et al. 1976, Jung and Walker 1997). On the other hand, responsiveness to AhR agonists and consequent CYP induction seems to increase in later life stages, as previously mentioned (Fujita et al. 1999, Leney et al. 2006c). Metamorphic tadpoles of the green frog (*Rana clamitans*) and northern leopard frog (*R. pipiens*) display a greater capacity for PCB metabolism and elimination than tadpoles that have not commenced metamorphosis (Leney et al. 2006b), and CYP1A genes are clearly induced in *X. laevis* adult liver following 3-methylcholanthrene exposure (Fujita et al. 1999)

The low responsiveness of early frog life stages stands in contrast to mouse embryos, in which TCDD-inducible CYP1A1 expression can be measured in blastocysts (Wu et al. 2002), and to zebrafish embryos, in which CYP1A induction by TCDD can be observed within twenty-four hours post fertilization (Tanguay et al. 1999). AhR1α and AhR1β, the two AhR paralogs in *Xenopus laevis*, bind TCDD with more than 25-fold lower affinity than mouse AhR (Lavine et al. 2005), which seems to be related to the structure of the AhR in frogs. Odio et al. (2013) identified three residues with side chains protruding into the putative ligand-binding cavity (N325, A354, and A370) and demonstrated that changing them to their mouse-like counterparts substantially enhanced TCDD binding by frog AHR1β. They also demonstrated that a single mutation in a residue that controls the binding cavity characteristics of the high affinity chicken AhR (Karchner et al. 2006) increased TCDD binding by the frog receptor. Interestingly, all the three Amphibian Orders (Anura, Caudata and Gymnophiona) presents these same three modified amino acid residues in the ligand-binding domain, suggesting that AhRs with low TCDD affinity represent a characteristic that evolved in a common ancestor of all three extant amphibian groups (Shoots et al. 2015, Kazzaz et al. 2020).

However, the weak response to AhR agonists in early life stages of frogs cannot be explained only by the relatively low affinity AhRs, since the same AhRs are expressed at reasonably high levels at both earlier CYP1A-refractory stages and later, more responsive stages (Lavine et al. 2005, Ohi et al. 2003). Also, it has been shown that fish from populations that have acquired resistance to CYP1A induction by dioxin-like PCBs remain responsive to PAHs (Courtenay et al. 1999), even though both compound classes exert their effects through the AhR (Fernandez-Salguero et al. 1996, Shimizu et al. 2000, Elskus 2005).

Zimmermann et al. (2008) tested the hypothesis that an AhRR would be contributing to the lower responsivity of AhR receptors. The AhR repressor is a member of the AhR protein family that is involved in the modulation (attenuation) of AhR-driven transcriptional activation (Mimura et al. 1999). AhR inducers upregulate AhRR mRNA, suggesting that this repressor protein acts in a negative feedback loop to down-regulate AhR activity (Fig. 4) (Karchner et al. 2002, Mimura et al. 1999). However, AhR1α was shown to be a higher AhRR inducer when compared to AhR1β in *X. laevis* cell lines treated with 6-formylindolo[3,2-b]carbazole (Freeburg et al. 2017). AhRR mRNA was expressed at low levels in embryos and was induced approximately 2-fold following TCDD exposure (42 ng/g wet weight). However, mRNA constitutive expression of AhRR was gradually increased during the course of tadpole development, and was more than 3-fold higher in prometamorphic and metamorphic climax tadpoles. Thus, attenuated transcriptional activation of AhR target genes and low TCDD toxicity in *X. laevis* embryos was also not explained by possible constitutive, high-level expression of AhRR.

Rowatt et al. (2003) hypothesized that a lack of coordinate expression or functional interaction between AhR and ARNT proteins during early life stages of tadpoles may contribute to the xenobiotic insensitivity of AhR, but their results indicated that both *X. laevis* ARNT1 and ARNT2 mRNAs are present well before the onset of CYP1A inducibility. Nevertheless, these authors also mention that only ARNT1 is constitutively expressed in early tadpoles, with increasing concomitant mRNA expression of ARNT2 at later stages. Also, ARNT1 was predominately cytosolic in tadpoles, needing to dimerize with SIM1 proteins to migrate to the nucleus, which differs from other vertebrates, where ARNT1 is located in the nucleus. ARNT2, in contrast, seems to be located in the nucleus of *X. laevis*, as in other vertebrates. Consequently, the lower responsivity of AhR to classic agonists may be a result of its lower affinity to the chemical structures of the contaminants, interference of AhRRs, and differential mechanisms of interaction with the translocator protein. Nevertheless, even though CYP1 induction is lower compared to other vertebrates, the induction of biotransformation enzymes in tadpoles occurs, and does provide some degree of protection against intoxication. However, CYP induction in most studies in tadpoles apparently only occurs at relatively high contaminant concentration (Gillardin et al. 2008).

4.1 Phase I and Phase II Enzymes in Tadpoles

There are numerous studies in the scientific literature reporting alterations in the enzymatic activity and or mRNA expression of phase I and phase II enzymes in tadpoles, specially CYPs and GSTs. Examples of some relevant studies showing alterations and responses of biotransformation enzymes to distinct xenobiotics in different tadpole species are listed in Table 1. All those results are consistent with an active and responsive biotransformation system in tadpoles when facing xenobiotic exposures.

The *Xenopus tropicalis* genome shows at least a single gene in each of the four cytochrome P450 1 (CYP1) subfamilies that occur in most vertebrates (CYP1A, CYP1B1, CYP1C1, and CYP1D1) (Jönsson et al. 2011). This and other anuran species also express two CYP1A mRNAs (CYP1A6

Table 1. Relevant publications on pollutant effects on biotransformation enzymes of different species of tadpoles.

Tadpole species	Stage of development	Toxic agents	Effects on biotransformation enzymes	Reference
Rana temporaria	Gs 35 – 36	b-naphthoflavone EE2 $CdCl_2$	Increased CYP1A mRNA levels Unaltered CYP1A mRNA levels Unaltered CYP1A mRNA levels	Carlsson and Tydén 2018
Xenopus laevis	NF 51 – 54 NF 36 – 64 Gs 30 – 34 NF 64/65	TCDD Glyphosate Methidathion Aroclor 1254	Increased CYP1A6 mRNA levels Increased CYP1A mRNA levels Decreased GST activity Increased GST activity Increased CYP1A mRNA	Taft et al. 2018 Zimmermann et al. 2008 Güngördü et al. 2016 Jelaso et al. 2005
Xenopus tropicalis	4 or 15 dpf	PCB126	Increased CYP1A, CYP1B1 and CYP1C1 mRNA levels Increased EROD activity Unaltered CYP1D1 and AHR mRNA levels	Jönsson et al. 2011
Physalaemus gracilis	Gs 25	Chlorpyrifos, Fipronil or Cypermethrin	Increased GST activity	Rutkoski et al. 2020 Rutkoski et al. 2021
Boana pulchella	Gs 29	Chlorpyrifos	Increased GST activity	Barreto et al. 2020
Rhinella schneideri	Gs 27 – 30 Gs 27 – 30	Sulfentrazone Clomazone	Increased GST activity Increased GST activity	Freitas et al. 2017a Freitas et al. 2017b
Physalaemus nattereri	Gs 27 – 30 Gs 27 – 30 Gs 30 – 32	Sulfentrazone Clomazone Fipronil	Increased GST activity Increased GST activity Decreased GST activity	Freitas et al. 2017a Freitas et al. 2017b Boscolo et al. 2017
Scinax fuscovarius	Gs 29 – 32	Fipronil	Decreased GST activity	Margarido et al. 2013
Pelophylax ridibundus	Gs 30 – 34	Glyphosate and methidathion	Increased GST activity	Güngördü et al. 2016
Bufotes viridis	Gs 30 – 34	Glyphosate Methidathion	Decreased GST activity Increased GST activity	Güngördü et al. 2016
Aquarana catesbeiana	Gs 26 – 30 Kollros VI–VIII Gs 26 – 27	Sulfentrazone or glyphosate Atrazine or EE2 Tebuthiuron	Increased GST in liver; decreased GST in tail. Increased brain CYP19 mRNA Unaltered GST and SULT mRNA levels Unaltered GST and EROD activities	Wilkens et al. 2019 Gunderson et al. 2011 Grott et al. 2021
Melanophryniscus admirabilis	Gs 25 – 30	Sulfentrazone and glyphosate	Increased GST	Silva et al. 2021

Table 1 contd. ...

...Table 1 contd.

Tadpole species	Stage of development	Toxic agents	Effects on biotransformation enzymes	Reference
Rana nigromaculata	Gs 25	Cypermethrin	Increased GST	Xu and Huang 2017
Bombina variegata	Gs 25	Cypermethrin	Increased GST	Greulich and Pflugmacher 2004
Eupemphix nattereri	Gs 29 – 33	Fipronil, fipronil sulfone or fipronil sulfide	Unaltered GST activity	Gripp et al. 2017
Rana arvalis	Gs 25 Gs 20 – 21	Cypermethrin TCDD	Increased GST Increased EROD activity	Greulich and Pflugmacher 2004 Van der Brink et al. 2003
Silurana tropicalis	NF 12 – 46	Bitumen	Increased CYP1A mRNA	Lara-Jacobo et al. 2019
Boana pardalis	Gs 26	Glyphosate, 2,4D or Metribuzin Ametryn or Acetochlor	Unaltered GST activity Increased GST activity	Moutinho et al. 2020
Pleurodeles walt	GD 53	Benzo(a)Pyrene	Increased EROD activity	Marty et al. 1998
Rana pipiens	n.s.	Phenobarbital	Increased EROD, PROD and aldrin epoxidase activities	Khan et al. 1998
Rhinella arenarum	Gs 23 – 25 Gs 36 – 38 Gs 26 – 30 Gs 26 – 28	Ciprofloxacin Enrofloxacin Glyphosate Dimethoate Antiretrovirals (lamivudine, stavudine, zidovudine and nevirapine)	Increased GST activity Decreased GST activity Decreased GST activity Increased GST activity Increased GST activity	Peltzer et al. 2017 Lajmanovich et al. 2011 Martinuzzi et al. 2020 Fernández et al. 2020
Odontophrynus americanus	Gs 26 – 30	Pyriproxyfen	Increased GST activity	Lajmanovich et al. 2019
Elachistocleis bicolor	Gs 30 – 34	Dicamba	Decreased GST activity	Attademo et al. 2021
Scinax nasicus	Gs 30 – 34	Dicamba	Increased GST activity	Attademo et al. 2021
Scinax squalirostris	Gs 26 – 30	Glyphosate or glufosinate ammonium	Increased GST activity	Lajmanovich et al. 2022

n.e.: not specified

and CYP1A7) with relatively low conservation (55 to 63%) in amino acid sequence compared to mammalian forms. CYP1A6 and CYP1A7 are inducible in tadpoles by organic contaminants, such as TCDD (Zimmermann et al. 2008, Taft et al. 2018, Iwamoto et al. 2012).

Temporal trends for basal expression of CYP1A, CYP1B1, CYP1C1, and CYP1D1, as well as AhR and ARNT2, were examined by Jönsson et al. (2011) over the first 16 days of development in tadpoles. The six genes displayed various trends for increasing levels of expression with development. Expression of CYP1C1 peaked at 4 days post fertilization (dpf) whereas CYP1A, CYP1B1, and CYP1D1 expression peaked at 8 dpf. Constitutive expression of the AhR and ARNT2 genes showed increasing trends up to day 4, at which time the levels started to decrease. For AhR the level of expression stabilized between day 8 and day 16 post-fertilization whereas the CYP1s and ARNT2 decreased after peaking. These same authors also studied the responsiveness of CYP1 genes to 3,3',4,4',5-pentachlorobiphenyl (PCB126) and β-naphthoflavone (βNF) in two age groups of tadpoles, 4 and 14 dpf. Both age groups responded to PCB126 and βNF with CYP1 induction, but the strongest response was observed in the older tadpoles. Curiously, none of the compounds significantly affected AhR expression, as previously discussed. Moreover, Colombo et al. (1996) reported no apparent constitutive expression of CYP1A in 8 days post fertilization *X. laevis* tadpoles, but the enzyme was induced by 1-day exposure to BaP solution. Zimmermann et al. (2008) also observed a substantial increase in mRNA transcripts of *X. leavis* after exposure to TCDD, with a higher induction in NF 62-64 than in NF 36-37 and NF 52–55 tadpoles.

A NADPH-P450 reductase (NPR) was also purified from hepatic microsomes of adult *Xenopus laevis*, which was most abundantly expressed in the kidney, followed by the liver, lung, and heart, and had limited expression in the brain (Mori et al. 2006). Interestingly, the level of NPR protein was almost the same from embryos to larvae (GK stages 2 to 35).

Overall, amphibian tadpoles are equipped with important phase I enzymes that can be induced under exposure to xenobiotics, especially in older tadpoles. Since biotransformation enzymes are higher in older than younger tadpoles, protection against toxicity in older tadpoles may be present, but only when the xenobiotic is biotransformed into a less reactive metabolite. In contrast, for those chemicals that exert their toxicity due to metabolic activation, an increase in toxicity may be observed as the tadpole develops. For example, Jelaso et al. (2003) showed that the toxicity of Aroclor 1254 to *Xenopus laevis* increases substantially as the tadpoles advance through their developmental stages. These authors found that younger tadpoles produce very low levels of CYP1A1, and that the expression of this enzyme gradually increases during the development of the animal, thus being capable to transform PCBs into more reactive metabolites, which have more negative effects to tadpoles. A similar pattern was observed in fish by Guiney et al. (1997), which showed that young fish do not express CYP1A1 (p4501A1) protein and were more resistant to 2,3,7,8-tetrachlorodibenzo-p-dioxin (TCDD) exposure than older fish.

Regarding Phase II enzymes, Di Ilio et al. (1992) identified six different cytosolic GST isoforms in *Bufo bufo* embryos (GST I-VI), which were attributed to the pi family of GSTs. They also purified and characterized a GST isoenzyme (BbGST P1-1) which is continuously expressed at high levels up to the end of development of *B. bufo* tadpoles, declines to very low levels in adult toad liver, and then replaced by a novel isoform, BbGST P2-2 (Bucciarelli et al. 1999). It has been found that BbGSTP2-2 is more efficient in scavenging organic hydroperoxides, and the appearance of BbGSTP2-2 in adults would provide protection against reactive oxygen species that occur following transition to adult life. During this transition, *B. bufo* leaves the aquatic environment to live predominantly in the terrestrial environment, thus being exposed to higher oxygen concentration causing increased oxidative stress. In fact, exposure to H_2O_2 induces expression of BbGSTP2 at both transcriptional and translational levels, confirming its protective action against oxidative stress (Amicarelli et al. 2004).

De Luca et al. (2003) also identified a novel Pi-class GST in *X. laevis*, namely XlGSTP1-1, which has shown one of the highest specific activities towards 1-chloro-2,4-dinitrobenzene obtained with any known GST. In another study, these same authors also identified a Mu-class GST in the same species, XlGSTM1-1, having a narrow spectrum of substrate specificity, and was less effective in conjugating 1-chloro-2,4-dinitrobenzene (De Luca et al. 2002).

Despite having a great variety of described GST classes, the responses of these specific isoforms to different contaminants in tadpoles is still unclear. Most studies evaluating pollutant effects on GSTs have been focused primarily on the activity of the enzyme toward the CDNB substrate. Studies on other phase II enzymes in tadpoles, to our knowledge are also absent, although Grott et al. (2021) recently reported unaltered mRNA levels for a sulfotransferase in *Aquarana catesbeianus* after exposure to the herbicide tebuthiuron.

GST activity also varies according to the species and the stage of development in tadpoles. In a study exposing *Bombina variegata* tadpoles to the herbicide isoproturon Greulish et al. (2002) noted that soluble GST activity was stage dependent with tadpoles at Gosner stages 24 and 25 having higher GST activity. GST activity in tadpoles at stages 20, 21, 22, 23, 26 and 27 were unchanged by treatment. Sacchetta et al. (1997), also identified a Pi-class GSTIII subunit in embryos of *Bufo bufo* that was continuously expressed from embryos to the end of tadpole development, but then declined to very low levels in adult toad liver. Similar to phase I enzymes, variations in phase II enzymes from species to species and along development can also change the magnitude of the toxic effects of xenobiotics to tadpoles, as these enzymes are capable to generate more or less reactive conjugated-derivatives.

5. Elimination of Toxicants in Tadpoles

Elimination of xenobiotics usually occurs through the feces, urine or by the respiratory surface of the gills. Similar to absorption processes, the preferable route of elimination will depend on several physiological conditions and on the chemical properties of the toxicant molecule. For example, depending on its lipophilicity, xenobiotics present in blood circulation can diffuse through the gill epithelium or even the skin to the external environment, since both the gills and the skin are highly vascularized.

Toxicants in blood can also be eliminated through renal filtration. Similar to mammals, amphibian embryos have a pronephros that develops to a mesonephros during metamorphosis, culminating in a metanephros in adults (Vize et al. 1995). The tadpole mesonephros primarily serves to eliminate waste and regulate water, ions and acid-base balance. The mesonephros of tadpoles undergoes significant change and shifts from ammonia to urea production during metamorphosis. In general, like other vertebrates, both the pro- and the meso-nephros of tadpoles are capable to filter the blood and eliminate wastes into the urine. They can also reabsorb nutrients, and carry out active tubular secretion (Vize et al. 1997). Thus, it is reasonable to conclude that toxic molecules freely available in the blood stream can be eliminated through renal filtration in tadpoles. In contrast, toxic molecules associated to plasma proteins will be retained, since plasma proteins, due to size, are not able to cross the glomeruli pores. This implies the increased capacity of the liver in producing such proteins during development may affect the capacity of the kidneys to eliminate lipophilic contaminants. Similarly, the binding of the toxicants to plasma proteins may also affect elimination by diffusion through the gills.

An additional pathway for toxicant elimination is through the GIT. As discussed previously, the liver of tadpoles is capable of xenobiotic biotransformation, generating both conjugated and non-conjugated metabolites, which could be directed to the gallbladder through liver canaliculi and biliary ducts. When the gallbladder is emptied, its contents together with these metabolites are released into the intestine, being incorporated to the forming stool. However, it is possible that conjugated metabolites can be hydrolyzed by gut microflora and the parent compounds can be reabsorbed in the gut. However, in most cases biliary metabolites are eliminated through the feces.

Cary and Karasov (2013) studied the elimination rates of different polybrominated diphenyl ethers (PBDE) congeners in leopard frog tadpoles (*Lithobates pipiens*) at different stages of development. Tadpoles had faster elimination compared to juvenile frogs. Metamorphosing frogs, however, did not eliminate PBDEs during metamorphosis. These discrepancies were attributed to metabolic changes and organ remodeling that occurs during the course of metamorphosis. The fact that most tadpole species do not feed during the metamorphosis climax may also explain why xenobiotic elimination decreases during this period; the lower food intake during metamorphosis (PBDE was given to tadpoles through the food) leads to lower toxicant uptake and, consequently, its elimination. Curiously, some PBDE congeners were eliminated much faster than other ones, indicating that the capacity of the animals to eliminate these compounds are dependent on the structure of the PBDE.

In contrast to these observations, Leney et al. (2006c) showed that metamorphosing green frogs exposed to PCBs had the highest rates of elimination, followed by tadpoles and then adult frogs. In this case, the faster PCB elimination was attributed to a gradual decrease in lipid content that occurs during tadpole development and metamorphosis (greater consumption of fat stores), thus decreasing the capacity of the organism to retain the different PCB congeners (Leney et al. 2006c). Adequate nutrient storage is essential for non-feeding metamorphic tadpoles to fuel their morphological remodeling and basic metabolism, and fat is the major fuel used during starvation in pro-metamorphic tadpoles (Wright et al. 2011, Zhu et al. 2019). As lipid is increasingly used as fuel during metamorphosis, lipophilic xenobiotics that were eventually stored in fat-storage organs are also gradually released, increasing its elimination rate from the body. Thus, discrepancies between the results of the studies in leopard frogs and green frog tadpoles could be related to the class of xenobiotics tested (PBDEs x PCBs), but also related to species specific differences of developmental physiology of the studied organism. Nevertheless, it is evident that, significant differences occur in the elimination processes of toxic substances during tadpole development, which may reflect on differential susceptibility of the organisms for bioaccumulation and toxic effects.

Jung and Walker (1997) also found that eggs and tadpoles of American toads, leopard frogs and green frogs eliminated ^{3}H-TCDD much faster than other vertebrates, with $t_{1/2}$ of 1 to 5 d (American toad), 2 to 7 d (leopard frog), and 4 to 6 d (green frog). In this study, elimination rates were also dependent on food availability, being lowest for tadpoles fed nothing, fastest for those fed a low-fat diet, and intermediate for those fed a high-fat diet. Environmental temperature seems to also influence toxicant elimination by tadpoles. Licht (1976) showed that *R. sylvatica* premetamorphic tadpoles presented faster elimination of ^{14}C-DDT at 21 °C, when compared to animals kept at 15°C.

6. Storage of Toxicants in Tadpoles

As the toxicant molecules reach different organs in the organism, it can be retained by the organ for different periods of time depending on the chemical composition and characteristics of the cells within the tissue. Chemicals may accumulate in body compartments due to protein binding, active transport processes, or due to a high affinity for a particular tissue (for example, accumulation of lipophilic xenobiotics in fat rich organs, such as the adipose tissue). Accumulation of chemicals in any tissue compartment are often in equilibrium with its free concentration in blood, making storage a dynamic process. Any removal of free chemical from the body by metabolism or excretion shifts the equilibrium such that stored chemical is released from the storage organ (Dix 2010).

Hall and Kolbe (1980) evaluated bioaccumulation of xenobiotics in tadpoles, by evaluating the capacity of *Lithobates catesbeianus* to bioaccumulate organophosphate pesticides (OP) from the aquatic environment. They based their experiments on previous information that adult frogs were capable of accumulating paraoxon *in vitro* (Potter and O'Brien 1964) and on the fact that cholinesterases (ChEs) from most amphibian species are substantially more resistant to inhibition by OP than any other terrestrial vertebrates (Hawkins and Mendel 1946, Edery and Schatzberg-Porath 1960, Andersen et al. 1977). Hall and Kolbe hypothesized that the resistance to OP inhibition may be a reflection of the accumulation of the pesticides in the body, thus decreasing its internal bioavailability to interact with ChEs. In fact, their study demonstrated that *L. catesbeianus* tadpoles were capable to magnify an average of 64 and 62 times the environmental concentrations of parathion and fenthion, respectively. However, when these authors fed ducks with OP-contaminated tadpoles, toxicity and lethality was observed in the birds, indicating the trophic transfer of the pesticides.

Lipid content in adults and tadpole is considered to be the most important determinant for bioconcentration of organic contaminants and generally ranges from 1 to 5% on a wet-weight whole-body basis (Leney et al. 2006a, Edginton and Rouleau 2005, Schuytema et al. 1991). In accordance to this, chemicals with high logKow values, such as DDT, dieldrin, fluoranthene, cypermethrin and deltamethrin showed high BCF values in tadpoles (Katagi and Ose 2014). In adult *R. temporaria* which were force-fed with ^{14}C-DDT, the amount of radioactive residues was proportional to the lipid content of each organ with the higher residues in fat body, gall bladder, liver and kidney, and for females in the ovary (Harri et al. 1970). In another study, the bioconcentration of the antibacterial triclosan was examined in tadpoles of *X. laevis*, *B. woodhousii* and *Rana sphenocephala*. A species-dependent difference in BCF (variating from 2 to 540) was observed, together with a dependence on the larval stage of tadpoles, which was associated with the animals' mass and lipid content (Palenske et al. 2010). A similar species dependence was reported for the bioconcentration of dieldrin in *X. laevis*, *L. catesbeianus* and *R. pipiens*, but to a lesser extent (Schuytema et al. 1991). Moreover, *Xenopus laevis* tadpoles accumulated the PCB Aroclor 1254 in a dose-response manner (Jelaso et al. 2003), reaching a BAF of approximately 240.

Dependent on lipid content, the bioaccumulation of many compounds varies in tadpoles as changes in lipid reserves occur during development. In *X. laevis*, thyroid hormones activate lipogenesis, increasing lipid storage during the pro-metamorphic period. This increased fat biosynthesis is important to generate energy reserves to be used during the period in which tadpoles cease eating and remodeling occurs (Sheridan 1998). After completing metamorphosis, lipid contents increase again in juveniles and adults as the animal returns to feed. Less variation in fat amount and composition occurs in adults compared to tadpoles, which supposedly increases the capacity of adults to store lipophilic xenobiotics. Consistent with this observation, *X. laevis* tadpoles exposed to radiolabeled BDE-99 showed that lower amounts of adipose tissue in tadpoles may play a major role in the lower uptake/retention of BDE-99 in comparison to adult frogs with higher fat stores (Carlsson et al. 2007).

Inorganic compounds such as metals can also accumulate in tadpoles. Snodgrass et al. (2004), for example, showed that *R. clamitans* and *R. sylvatica* tadpoles exposed to coal combustion waste-contaminated sediments accumulated significant amounts of Cd, Cu, Fe, Ni, Sr, V and Zn when compared to tadpoles kept in water with clean sand. Lanctot et al. (2017a) showed that *Limnodynastes peronii* tadpoles are capable to retain and accumulate Se after 7 days of exposure. Even after a depuration period of 27 days, Se concentrations were still at 10 to 14% of that found at the end of the exposure period. Se accumulated in the tail, muscle, connective tissue, spinal cord, notochord, gills, and even in the eyes, but Se was predominantly found in digestive and excretory tissues (liver > mesonephros > gut > gallbladder). Curiously, these authors also reported that Se accumulation in some organs varied significantly throughout development, which would be related to the transference of the contaminant from one tissue to other, as a result of transformative processes that occur during metamorphosis, such as tissue remodeling, resorption and *de novo* tissue development. As previously discussed, during the course of metamorphosis, the liver increases in size while the length of the gut decreases by 75% (Brown and Cai 2007), which explains the gradual increase of accumulative capacity of the liver and a reduced ability of the gut to accumulate toxicants. In another study, Lanctôt et al. (2016) also showed that tadpoles exposed to coal mine wastewater had elevated levels of Se, Co, Mn and As in tail and liver tissue compared to unexposed controls. Moreover, Lanctôt et al. (2017b) revealed that metal accumulation can differ markedly when tadpoles are exposed to individual elements compared to a mixture. Specifically, tadpoles exposed to a mixture of Cd, Se and Zn accumulated and retained approximately half the amount of Cd compared to tadpoles exposed to Cd alone, suggesting that Se and/or Zn may have antagonistic effects against Cd uptake.

Metal accumulation rates can also vary during tadpole development. Roe et al. (2005) showed that the concentrations of As, Cd, Cu, Ni, Pb, and Zn decreased from the larval to metamorphic stage in *Bufo terrestris* and *Rana sphenocephala* tapdoles, after being exposed to coal combustion waste. These elements were concentrated mostly in the digestive tract, and are thus likely to be eliminated or redistributed in the body during remodeling of the gut and other organ systems. Additionally, because anurans cease feeding during metamorphosis, elimination may exceed uptake rates. The combination of these factors could contribute to lower concentrations of these elements in metamorphs.

The jelly coat that covers amphibian eggs can also account as a storage site for some xenobiotics. As previously commented, Jung and Walker (1997) found 4% TCDD relative to the embryo content in the jelly coat. Similarly, wood frog (*R. sylvatica*) embryos exposed to waterborne DDT showed little DDT in the jelly capsule (8.8%) relative to the embryo (Licht 1985). Being composed of proteins, mucoproteins, and mucopolysaccharides, the jelly coat might not readily store lipophilic compounds and thus would not be expected to retain significant amounts of such compounds (Jung and Walker 1997).

7. Conclusions and Perspectives

Toxicokinetic processes of environmental contaminants have been poorly studied in tadpoles, and many gaps exist documenting the internal fate of toxicants in these organisms. In this chapter we tried to gather general information available about general physiological processes in amphibian tadpoles that can impact toxicokinetic processes of xenobiotics. All the processes involved in toxicokinetics of xenobiotic substances in most vertebrate species are similar as those in tadpoles. But since they are aquatic organisms that undergo profound remodeling processes during their development and metamorphosis to form a terrestrial adult, significant differences are expected to occur with respect to the magnitude of the toxicokinetic event considered during the stage of development of the organism. This possibly results in distinct susceptibilities of the organism to the toxic effects of the contaminant during development. Moreover, when considering the vast behavioral and development differences between amphibian species, toxicity of contaminants can also vary between species due to particular physiological processes that can eventually modify the toxicokinetics of toxic substances among these different species. This chapter also points out the necessity of more studies on this topic focused on developing tadpoles, providing a better understanding on how pollutants impact tadpoles, and possibly permitting the development of better strategies for their conservation.

Acknowledgements

We thank "Conselho Nacional de Desenvolvimento Científico e Tecnológico, CNPq", "Fundação de Amparo à Pesquisa e Inovação do Estado de Santa Catarina, FAPESC" and "Coordenação de Aperfeiçoamento de Pessoal de Nível Superior, CAPES" (001) for the financial support.

References Cited

Abbott, N.J., L. Ronnback and E. Hansson. 2006. Astrocyte-endothelial interactions at the blood-brain barrier. Nat. Rev. Neurosci. 7: 41–53.

Altig, A. and W. McDearman. 1975. Percent assimilation and clearance times of five anuran tadpoles. Herpetologica 31: 67–69.

Altig, R. and G.F. Johnston. 1989. Guilds of anuran larvae: relationship among developmental modes, morphologies and habitats. Herpetol. Monogr. 3: 81–109.

Amicarelli, F., S. Falone, F. Cattani, M.T. Alamanou, A. Bonfigli, O. Zarivi et al. 2004. Amphibian transition to the oxidant terrestrial environment affects the expression of glutathione S-transferases isoenzymatic pattern. Biochim. Biophys. Acta 1691: 181–192.

Andersen, R.A., I. Aaraas, G. Gaare and F. Fonnum. 1977. Inhibition of acetylcholinesterase from different species by organophosphorus compounds, carbamates and methylsulphonylfluoride. Gen.l Pharmacol. 8: 331–334.
Anderson, I.A., T. Briggs, G.A.D. Haslewood, S.H. Oldham and L. Tokes. 1979. Comparison of the bile salts of frogs with those of their tadpoles: bile salt changes during the metamorphosis of *Rana catesbeiana* Shaw. Biochem. J. 183: 507–511.
Andino, F.J., L. Jones, S.B. Maggirwar and J. Robert. 2016. Frog Virus 3 dissemination in the brain of tadpoles, but not in adult Xenopus, involves blood brain barrier dysfunction. Sci. Reports 6: 22508.
Atkinson, B.G., C. Helbing and Y. Chen. 1996. Reprogramming of genes expressed in amphibian liver during metamorphosis. pp. 539–566. *In*: Gilbert, L.I., J.R. Tata and B.G. Atkinson [eds.]. Metamorphosis. Academic Press, New York.
Attademo, A.M., R.C. Lajmanovich, P.M. Peltzer, A.P.C. Boccioni, C. Martinuzzi, F. Simonielo et al. 2021. Effects of the emulsifiable herbicide Dicamba on amphibian tadpoles: an underestimated toxicity risk? Environ. Sci. Pollut. Res. Int. 28: 31962–31974.
Beatty, P.W., M.A. Holscher and R.A. Neal. 1976. Toxicity of 2,3,7,8-tetrachloridibenzo-p-dioxin in larval and adult forms of *Rana catesbeiana*. Bull. Environ. Contam. Toxicol. 16: 578–581.
Bergeron, C.M., C.M. Bodinof, J.M. Unrine and W.A. Hopkins. 2010. Bioaccumulation and maternal transfer of mercury and selenium in amphibians. Environ. Toxicol. Chem. 29: 989–997.
Bergeron, C.M., W.A. Hopkins, B.D. Todd, M.J. Hepner and J.M. Unrine. 2011. Interactive effects of maternal and dietary mercury exposure have latent and lethal consequences for amphibian larvae. Environ. Sci. Technol. 15: 3781–3787.
Bernabò, I., E. Brunelli, C. Berg, A. Bonacci and S. Tripepi. 2008. Endosulfan acute toxicity in *Bufo bufo* gills: Ultrastructural changes and nitric oxide synthase localization. Aquat. Toxicol. 86: 447–456.
Bernabò, I., A. Bonacci, F. Coscarelli, M. Tripepi and E. Brunelli. 2013a. Effects of salinity stress on *Bufo balearicus* and *Bufo bufo* tadpoles: Tolerance, morphological gill alterations and Na+/K+ - ATPase localization. Aquat. Toxicol. 132-133: 119–33.
Bernabò, I., A. Guardia, D. La Russa, G. Madeo, S. Tripepi and E. Brunelli. 2013b. Exposure and post-exposure effects of endosulfan on *Bufo bufo* tadpoles: morpho-histological and ultrastructural study on epidermis and iNOS localization. Aquat Toxicol. 142-143: 164–75.
Bertoluci, J., F.S. Leite, C.C. Eisemberg and M.A.S. Canelas. 2007. Description of the tadpole of *Scinax luizotavioi* from the Atlantic rainforest of southeastern Brazil. Herpet. J. 17: 14–18.
Bucciarelli, T., P. Sacchetta, A. Pennelli, L. Cornelio, R. Romagnoli, S. Melino et al. 1999. Characterization of toad liver glutathione transferase. Biochim. Biophys. Acta 1431: 189–198.
Berrill, M., S. Bertram and B. Pauli. 1997. Effects of pesticides on amphibian embryos and larvae. Herpetol. Conserv. 1: 233–245.
Boscolo, C.N.P., A.A. Felício, T.S.B. Pereira, T.C.S. Margarido, D.C. Rossa-Feres, E.A. Almeida and J.S. Freitas. 2017. Comercial insecticide fipronil alters antioxidant enzymes response and accelerates the metamorphosis in *Physalaemus nattereri* (Anura: Leiuperidae) tadpoles. Eur. J. Zool. Res. 5: 1–7.
Brown, D.D. and L. Cai. 2007. Amphibian metamorphosis. Develop. Biol. 306: 20–33.
Brunelli, E. and S. Tripepi. 2005. Effects of low pH acute exposure on survival and gill morphology in *Triturus italicus* larvae. J. Exp. Zool. Part A 303: 946–957.
Brunelli, E., I. Bernabò, E. Sperone and S. Tripepi. 2010. Gill alterations as biomarkers of chronic exposure to endosulfan in *Bufo bufo* tadpoles. Histol. Histopathol. 25: 1519–1529.
Burggren, W.W. and N.H. West. 1982. Changing Respiratory importance of gills, lungs and skin during metamorphosis in the bullfrog *Rana catesbeiana*. Resp. Physiol. 47: 151–164.
Burggren, W.W. and J.J. Just. 1992. Developmental changes in physiological systems. pp. 467–530. *In*: Feder, M.E. and W.W. Burggren [eds.]. Environmental physiology of the Amphibians. University of Chicago Press, Chicago.
Carlsson, G., P. Kulkarni, P. Larsson and L. Norrgren. 2007. Distribution of BDE-99 and effects on metamorphosis of BDE-99 and-47 after oral exposure in *Xenopus tropicalis*. Aquat. Toxicol. 84: 71–79.
Carlsson, G. and E. Tydén. 2018. Development and evaluation of gene expression biomarkers for chemical pollution in common frog (*Rana temporaria*) tadpoles. Environ. Sci. Pollut. Res. 25: 33131–33139.
Carrer, D.C., C. Vermehren and L.A. Bagatolli. 2008. Pig skin structure and transdermal delivery of liposomes: a two photon microscopy study. J. Control. Rel. 132: 12–20.
Cary, T.L. and W.H. Karasov. 2013. Toxicokinetics of polybrominated diphenyl ethers across life stages in the northern leopard frog (*Lithobates pipiens*). Environ. Toxicol. Chem. 32: 1631–1640.
Colombo, A., P. Bonfanti, M. Ciccotelli, M. Doldi, N. Dell'Orto and M. Camatini. 1996. Induction of cytochrome P4501A isoform in *Xenopus laevis* is a valid tool for monitoring the exposure to benzo[a]pyrene. J. Aquatic Ecosys. Health. 5: 207–211.

Comhaire, S., R. Blust, L. Van Ginneken and O.L.J. Vanderborght. 1994. Cobalt uptake across the gills of the common carp, *Cyprinus carpio*, as a function of calcium concentration in the water of acclimation and exposure. Comparat. Biochem. Physiol. C 109: 63–76.

Courtenay, S.C., C.M. Grunwald, G.L. Kreamer, W.L. Fairchild, J.T. Arsenault, M. Ikonomou et al. 1999. A comparison of the dose and time response of CYP1A1 mRNA induction in chemically treated Atlantic tomcod from two populations. Aquat. Toxicol. 47: 43–69.

De Luca, A., B. Favaloro, S. Angelucci, P. Sacchetta and C. Di Ilio. 2002. Mu-class glutathione transferase from *Xenopus laevis*: molecular cloning, expression and site-directed mutagenesis. Biochem. J. 365: 685–691.

De Luca, A., B. Favaloro, A. Carletti, P. Sacchetta and C. Di Ilio. 2003. A novel amphibian Pi-class glutathione transferase isoenzyme from *Xenopus laevis*: importance of phenylalanine 111 in the H-site Biochem. J. 373: 539–545.

Dietz, T.H. and R.H. Alvarado. 1974. Na and Cl transport across gill chamber epithelium of *Rana catesbeiana* tadpoles. Am. J. Physiol. 226: 764–770.

Di Ilio, C., A. Aceto, T. Bucciarelli, B. Dragani, S. Angelucci, M. Miranda et al. 1992. Glutathione transferase isoenzymes from *Bufo bufo* embryos at an early developmental stage. Biochem. J. 283: 217–222.

Dix, K.J. 2010. Distribution and Pharmacokinetics. pp. 923–939. *In*: Krieger, R. [ed.]. Hayes' Handbook of Pesticide Toxicology, 3rd edition, Academic Press, 2010, 2000 p.

Edery, H. and G. Schatzberg-Porath. 1960. Studies on the effects of organophosphorus insecticides on amphibians. Arch. Int. Pharm. 124: 212–224.

Edinton, A.N. and C. Rouleau. 2005. Toxicokinetics of 14C-atrazine and its metabolites in stage-66 *Xenopus laevis*. Environ. Sci. Technol. 39: 8083–8089.

Elskus, A.A. 2005. The implications of low-affinity AhR for TCDD insensitivity in frogs. Toxicol. Sci. 88: 1–3.

Ertl, R.P. and G.W. Winston. 1998. The microsomal mixed-function oxidase system of amphibians and reptiles: Components, activities, and induction. Comparat. Biochem. Physiol. C 121: 85–105.

Farquhar, M.G. and G.E. Palade. 1965. Cell junctions in amphibian skin. J. Cell Biol. 26: 263–291.

Fenoglio, C., A. Grosso, E. Boncompagni, C. Gandini, G. Milanesi and S. Barni. 2009. Exposure to heptachlor: Evaluation of the effects on the larval and adult epidermis of *Rana kl. esculenta*. Aquat. Toxicol. 91: 151–160.

Fernández, L.P., R. Brasca, A.M. Attademo, P.M. Peltzer, R.C. Lajmanovich and M.J. Culzoni. 2020. Bioaccumulation and glutathione S-transferase activity on *Rhinella arenarum* tadpoles after short-term exposure to antiretrovirals. Chemosphere. 246: 125830.

Fernandez-Salguero, P., D. Hilbert, S. Rudikoff, J. Ward and F. Gonzalez. 1996. Aryl-hydrocarbon receptor–deficient mice are resistant to 2,3,7,8-tetrachlorodibenzo-p-dioxin-induced toxicity. Toxicol. Appl. Pharmacol. 140: 173–179.

Freeburg, S.H., E. Engelbrecht and W.H. Powell. 2017. Subfunctionalization of paralogous aryl hydrocarbon receptors from the frog *Xenopus Laevis*: Distinct target genes and differential responses to specific agonists in a single cell type. Toxicol. Sci. 155: 337–347.

Freitas, J.S., F.B. Teresa and E.A. Almeida. 2017a. Influence of temperature on the antioxidant responses and lipid peroxidation of two species of tadpoles (*Rhinella schneideri* and *Physalaemus nattereri*) exposed to the herbicide sulfentrazone (Boral 500SC®). Comp. Biochem. Physiol. C 197: 32–44.

Freitas, J.S., A.A. Felício, F.B. Teresa and E.A. Almeida. 2017b. Combined effects of temperature and clomazone (Gamit®) on oxidative stress responses and B-esterase activity of *Physalaemus nattereri* (Leiuperidae) and *Rhinella schneideri* (Bufonidae) tadpoles. Chemosphere 185: 548–562.

Fox, H. 1992. Figures of Eberth in the amphibian larval epidermis. J. Morphol. 212: 87–97.

Fujita, Y., H. Ohi, N. Murayama, K. Saguchi and S. Higuchi. 1999. Molecular cloning and sequence analysis of cDNAs coding for 3-methylcholanthrene-inducible cytochromes P450 in *Xenopus laevis* liver. Arch. Biochem. Biophys. 371: 24–28.

Ghosh, C., V. Puvenna, J. Gonzalez-Martinez, D. Janigro and N. Marchi. 2011. Blood-brain barrier P450 enzymes and multidrug transporters in drug resistance: A synergistic role in neurological diseases. Curr. Drug Metab. 12: 742–749.

Glynn, A.W., L. Norrgren and A. Müssener. 1994. Differences in uptake of inorganic mercury and cadmium in the gills of the zebrafish, *Brachydanio rerio*. Aquat. Toxicol. 30: 13–26.

Gillardin V., F. Silvestre, M. Dieu, E. Delaive, M. Raes, J.P. Thomé and P. Kestemont. 2009. Protein expression profiling in the African clawed frog *Xenopus laevis* tadpoles exposed to the polychlorinated biphenyl mixture aroclor 1254. Mol. Cell Proteomics. 8(4): 596–611.

Gradwell, N. 1973. On the functional morphology of suction and gill irrigation in the tadpole of Ascaphus, and notes on hibernation. Herpetologica 29: 84–93.

Grant, K.P. and L.E. Licht. 1995. Effects of ultravioletradiation on life-history stages of anurans from Ontario, Canada. Can. J. Zool. 73: 2292-2301.

Greulich, K., E. Hoque and E. Pflugmacher. 2002. Uptake, metabolism, and effects on detoxication enzymes of isoproturon in spawn and tadpoles of amphibians. Ecotoxicol. Environ. Saf. 52: 256-266.

Greulich, K. and S. Pflugmacher. 2004. Uptake and effects on detoxication enzymes of cypermethrin in embryos and tadpoles of amphibians. Arch. Environ. Contam. Toxicol. 47: 489–495.

Gripp, H.S., J.S. Freitas, E.A. Almeida, M.C. Bisinoti and A.B. Moreira. 2017. Biochemical effects of fipronil and its metabolites on lipid peroxidation and enzymatic antioxidant defense in tadpoles (*Eupemphix nattereri*: Leiuperidae). Ecotoxicol. Environ. Saf. 136: 173–179.

Gromko, M.H., F.S. Mason and S.J. Smith-Gill. 1973. Analysis of the crowding effect in *Rana pipiens* tadpoles. J. Exp. Zool. 186: 63–72.

Grosell, M. and C.M. Wood. 2002. Copper uptake across rainbow trout gills: mechanisms of apical entry. J. Exp. Biol. 205: 1179–1188.

Grott, S.C., D. Bitschinski, N.G. Israel, G. Abel, S.P. Silva, T.C. Alves et al. 2021. Influence of temperature on biomarker responses and histology of the liver of American bullfrog tadpoles (*Lithobates catesbeianus*, Shaw, 1802) exposed to the herbicide Tebuthiuron. Sci. Total Environ. 771: 144971.

Guiney, P.D., R.M. Smolowitz, R.E. Peterson and J.J. Stegeman. 1997. Correlation of 2,3,7,8-tetrachlorodibenzo-p-dioxin induction of cytochrome P4501A in vascular endothelium with toxicity in early life stages of lake trout. Toxicol. Appl. Pharmacol. 143: 256–273.

Gunderson, M.P., N. Veldhoen, R.C. Skirrow, M.K. Macnab, W. Ding, G. van Aggelen et al. 2011. Effect of low dose exposure to the herbicide atrazine and its metabolite on cytochrome P450 aromatase and steroidogenic factor-1 mRNA levels in the brain of premetamorphic bullfrog tadpoles (*Rana catesbeiana*). Aquat. Toxicol. 102: 31–38.

Güngördü, A., M. Uçkun and E. Yologlu. 2016. Integrated assessment of biochemical markers in premetamorphic tadpoles of three amphibian species exposed to glyphosate- and methidathion-based pesticides in single and combination forms. Chemosphere 144: 2024–2035.

Hadley, N.F. 1991. Integumental lipids of plants and animals: comparative function and biochemistry. Adv. Lipid Res. 24: 303–320.

Hall, R.J. and E. Kolbe. 1980. Bioconcentration of organophosphorus pesticides to hazardous levels by amphibians. J. Toxicol. Environ. Health 6: 853–60.

Handy, R.D., F.B. Eddy and H. Baines. 2002. Sodium-dependent copper uptake across epithelia: a review of rationale with experimental evidence from gill and intestine. Biochim. Biophys. Acta 1566: 104–115.

Harri, M.N.E., J. Laitinen and E.L. Valkama. 1970. Toxicity and retention of DDT in adult frogs, *Rana temporaria* L. Environ. Pollut. 20: 45–55.

Hawkins, R.D. and B. Mendel. 1946. True cholinesterases with pronounced resistance to eserine. J. Cell. Comparat. Physiol. 27: 69–85.

Hopkins, W.A., S.E. DuRant, B.P. Staub, C.L. Rowe and B.P. Jackson. 2006. Reproduction, embryonic development, and maternal transfer of contaminants in the amphibian *Gastrophryne carolinensis*. Environ. Health Perspect. 114: 661–666.

Hogstrand, C., P.M. Verbost, S.E. Bonga and C.M. Wood. 1996. Mechanisms of zinc uptake in gills of freshwater rainbow trout: Interplay with calcium transport. Reg. Int. Comparat. Physiol. 270: R1141–R1147.

Hourdry, J., A. L'Hermite and R. Ferra. 1996. Changes in the digestive tract and feeding behavior of anuran amphibians during metamorph. Physiol. Zool. 69: 219–222.

Ishizuya-Oka, A. 2011. Amphibian organ remodeling during metamorphosis: insight into thyroid hormone-induced apoptosis. Dev. Growth Differ. 53: 202–212.

Iwamoto, D.V., C.M. Kurylo, K.M. Schorling and W.H. Powell. 2012. Induction of cytochrome P450 family 1 mRNAs and activities in a cell line from the frog *Xenopus laevis*. Aquat. Toxicol. 114-115: 165–72.

Jelaso, A.M., C. DeLong, J. Means and C.F. Ide. 2005. Dietary exposure to Aroclor 1254 alters gene expression in *Xenopus laevis* frogs. Environ. Res. 98: 64–72.

Jelaso, A.M., E. Lehigh-Shirey, J. Means and C.F. Ide. 2003. Gene expression patterns predict exposure to PCBs in developing *Xenopus laevis* Tadpoles. Environ. Mol. Mutagen. 42: 1–10.

Jönsson, M.E., C. Berg, J.V. Goldstone and J.J. Stegeman. 2011. New CYP1 genes in the frog *Xenopus* (Silurana) tropicalis: induction patterns and effects of AHR agonists during development. Toxicol. Appl. Pharmacol. 250(2): 170–183.

Jung, R.E. and M.K. Walker. 1997. Effects of 2,3,7,8-Tetrachlorodibenzo-p-dioxin (TCDD) on development of anuran amphibians. Environ. Toxicol. Chem. 16: 230–240.

Kadokami, K., M. Takeishi, M. Kuramoto and Y. Ono. 2004. Maternal transfer of organochlorine pesticides, polychlorinated dibenzo-p-dioxins, dibenzofurans, and coplanar polychlorinated biphenyls in frogs to their eggs. Chemosphere 57: 383–389.

Karchner, S.I., D.G. Franks, W.H. Powell and M.E. Hahn. 2002. Regulatory interactions among three members of the vertebrate aryl hydrocarbon receptor family: AHR repressor, AHR1, and AHR2. J. Biol. Chem. 277: 6949–6959.

Karchner, S.I., D.G. Franks, S.W. Kennedy and M.E. Hahn. 2006. The molecular basis for differential dioxin sensitivity in birds: Role of the aryl hydrocarbon receptor. Proc. Natl. Acad. Sci. USA 103: 6252–6257.

Khan, M.A., S.Y. Qadri, S. Tomar, D. Fish, L. Gururajan and M.S. Poria. 1998. Induction of hepatic cytochrome P-450 by phenobarbital in semi-aquatic frog (*Rana pipiens*). Biochem. Biophys. Res. Commun. 244: 737–44.

Katagi, T. and K. Ose. 2014. Bioconcentration and metabolism of pesticides and industrial chemicals in the frog. J. Pestic. Sci. 39: 55–68.

Kazzaz, S.A., S. Giani Tagliabue, D.G. Franks, M.S. Denison, M.E. Hahn et al. 2020. An aryl hydrocarbon receptor from the caecilian *Gymnopis multiplicata* suggests low dioxin affinity in the ancestor of all three amphibian orders. Gen. Comp. Endocrinol. 299: 113592.

Kotyzova, D. and F.W. Sundeman. 1998. Maternal exposure to Cd(II) causes malformations of *Xenopus laevis* embryos. Ann. Clin. Lab. Sci. 28: 224–235.

Lajmanovich, R.C., A.M. Attademo, P.M. Peltzer, C.M. Junges and M.C. Cabagna. 2011. Toxicity of four herbicide formulations with glyphosate on *Rhinella arenarum* (anura: bufonidae) tadpoles: B-esterases and glutathione S-transferase inhibitors. Arch. Environ. Contam. Toxicol. 60: 681–689.

Lajmanovich, R.C., P.M. Peltzer, C.S. Martinuzzi, A.M. Attademo, A. Bassó and C.L. Colussi. 2019. Insecticide pyriproxyfen (Dragón ®) damage biotransformation, thyroid hormones, heart rate, and swimming performance of *Odontophrynus americanus* tadpoles. Chemosphere 220: 714–722.

Lajmanovich, R.C., A.M. Attademo, G. Lener, A.P.C. Boccioni, P.M. Peltzer, C.S. Martinuzzi et al. 2022. Glyphosate and glufosinate ammonium, herbicides commonly used on genetically modified crops, and their interaction with microplastics: Ecotoxicity in anuran tadpoles. Sci. Total Environ. 804: 150177.

Lanctôt, C., W. Bennett, S. Wilson, L. Fabbro, F.D.L. Leusch and S.D. Melvin. 2016. Behaviour, development and metal accumulation in striped marsh frog tadpoles (*Limnodynastes peronii*) exposed to coal mine wastewater. Aquat. Toxicol. 173: 218–227.

Lanctot, C.M., T. Cresswell, P.D. Callaghan and S.D. Melvin. 2017a. Bioaccumulation and biodistribution of selenium in metamorphosing tadpoles. Environ. Sci. Technol. 51: 5764–5773.

Lanctôt, C.M., T. Cresswellb and S.D. Melvina. 2017b. Uptake and tissue distributions of cadmium, selenium and zinc in striped marsh frog tadpoles exposed during early post-embryonic development. Ecotoxicol. Environ. Saf. 144: 291–299.

Lara-Jacobo, L.R., B. Willard, S.J. Wallace and V.S. Langlois. 2019. Cytochrome P450 1A transcript is a suitable biomarker of both exposure and response to diluted bitumen in developing frog embryos. Environ. Pollut. 246: 501–508.

Lavine, J.A., A.J. Rowatt, T. Klimova, A.J. Whitington, E. Dengler, C. Beck and W.H. Powell. 2005. Aryl hydrocarbon receptors in the frog *Xenopus laevis*: Two AhR1 paralogs exhibit low affinity for 2,3,7,8-tetrachlorodibenzo-p-dioxin (TCDD). Toxicol. Sci. 88: 60–72.

Leney, J.L. 2005. Toxicokinetics of PCBs and PAHs in green frogs (*Rana clamitans*) and leopard frogs (*Rana pipiens*) at various life stages. Electronic Theses and Dissertations. 4409. https://scholar.uwindsor.ca/etd/4409.

Leney, J.L., K.C. Balkwill, K.G. Drouillard and G.D. Haffner. 2006a. Determination of polychlorinated biphenyl and polycyclic aromatic hydrocarbon elimination rates in adult green and leopard frogs. Environ. Toxicol. Chem. 25: 1627–1634.

Leney, J.L., K.G. Drouillard and G.D. Haffner. 2006b. Metamorphosis increases biotransformation of polychlorinated biphenyls: a comparative study of polychlorinated biphenyl metabolism in green frogs (*Rana clamitans*) and leopard frogs (*Rana pipiens*) at various life stages. Environ. Toxicol. Chem. 25: 2971–2980.

Leney, J.L., K.G. Drouillard and G.D. Haffner. 2006c. Does metamorphosis increase the susceptibility of frogs to highly hydrophobic contaminants? Environ. Sci. Technol. 40: 1491–1496.

Licht, L.E. 1976. Time course of uptake, elimination, and tissue levels of [^{14}C]DDT in wood-frog tadpoles. Can. J. Zool. 54: 355v360.

Licht, L.E. 1985. Uptake of ^{14}C-DDT by wood frog embryos after short term exposure. Comp. Biochem. Physiol. C. 81: 117–119.

Lillywhite, H.B. 2006. Water relations of tetrapod integument. J. Exp. Biol. 209: 202–226.

Margarido, T.C.S., A.A. Felício, D. de Cerqueira Rossa-Feres and E.A. de Almeida EA. 2013. Biochemical biomarkers in *Scinax fuscovarius* tadpoles exposed to a commercial formulation of the pesticide fipronil. Mar. Environ. Res. 91: 61–67.

Marquis, O., A. Millery, S. Guittonneau and C. Miaud. 2006a. Toxicity of PAHs and jelly protection of eggs in the Common frog *Rana temporaria*. Amphibia-Reptilia 27: 472–475.

Marquis, O., A. Millery, S. Guittonneau and C. Miaud. 2006b. Solvent toxicity to amphibian embryos and larvae. Chemosphere 63: 889–892.

Martinuzzi, C.S., A.M. Attademo, P.M. Peltzer, T.M. Mac Loughlin, D.J.G. Marino and R.C. Lajmanovich. 2020. Comparative toxicity of two different Dimethoate formulations in the common toad (*Rhinella arenarum*) tadpoles. Bull. Environ. Contam. Toxicol. 104: 35–40.

Marty, J., P. Lesca, A. Jaylet, C. Ardourel and J.L. Rivière. 1989. *In vivo* and *in vitro* metabolism of benzo(a)pyrene by the larva of the newt, *Pleurodeles waltl.* Comp. Biochem. Physiol. C. 93: 213–219.

Marty, J., J.E. Djomo, C. Bekaert and A. Pfohl-Leszkowicz. 1998. Relationships between formation of micronuclei and DNA adducts and EROD activity in newts following exposure to benzo(a)pyrene. Environ. Mol. Mutagen. 32: 397–405.

McDonald, D.G., V. Cavdek and R. Ellis. 1991. Gill design in freshwater fishes: Interrelationships among gas exchange, ion regulation, and acid-base regulation. Physiol. Zool. 64: 103–123.

McIndoe, R. and D.G. Smith. 1984. Functional anatomy of the internal gills of the tadpole of *Litoria ewingii* (Anura, Hylidae). Zoomorphol. 104: 280–291.

McWilliams, P.G. 1982. The effects of calcium on sodium fluxes in the brown trout, *Salmo trutta*, in neutral and acid water. J. Exp. Biol. 96: 439–442.

Mendez, S.I.S., D.E. Tillitt, T.A.G. Rittenhouse and R.D. Semlitsch. 2009. Behavioral response and kinetics of terrestrial atrazine exposure in American toads (*Bufo americanus*). Arch. Environ. Contam. Toxicol. 57: 590–597.

Metts, B.S., K.A. Buhlmann, D.E. Scott, T.D. Tuberville and W.A. Hopkins. 2012. Interactive effects of maternal and environmental exposure to coal combustion wastes decrease survival of larval southern toads (*Bufo terrestris*). Environ. Pollut. 164: 211–218.

Metts, B.S., K.A. Buhlmann, T.D. Tuberville, D.E. Scott and W.A. Hopkins. 2013. Maternal transfer of contaminants and reduced reproductive success of southern toads (*Bufo [Anaxyrus] terrestris*) exposed to coal combustion waste. Environ. Sci. Technol. 47: 2846–2853.

Mimura, J., M. Ema, K. Sogawa and Y. Fujii-Kuriyama. 1999. Identification of a novel mechanism of regulation of Ah (dioxin) receptor function. Genes Dev. 13: 20–25.

Moghimipour, E., A. Ameri and S. Handali. 2015. Absorption-enhancing effects of bile salts. Molecules 20: 14451–14473.

Mori, T., A. Yamazaki, T. Kinoshita and S. Imaoka. 2006. Purification of NADPH-P450 reductase (NPR) from *Xenopus laevis* and the developmental change in NPR expression. Life Sci. 79: 247–251.

Moskaitis, J.E., T.D. Sargent, L.H. Smith Jr., R.L. Pastori and D.R. Schoenberg. 1989. *Xenopus laevis* serum albumin: sequence of the complementary deoxyribonucleic acids encoding the 68- and 74-kilodalton peptides and the regulation of albumin gene expression by thyroid hormone during development. Mol. Endocrinol. 3: 464–473.

Moutinho, M.F., E.A. Almeida, E.L.G. Espíndola, M.A. Daam and L. Schiesari. 2020. Herbicides employed in sugarcane plantations have lethal and sublethal effects to larval *Boana pardalis* (Amphibia, Hylidae). Ecotoxicology 29: 1043–1051.

Nakajima, K., K. Fujimoto and Y. Yaoita. 2005. Programmed cell death during amphibian metamorphosis. Sem. Cell Develop. Biol. 16: 271–280.

Odio, C., S.A. Holzman, M.S. Denison, D. Fraccalvieri, L. Bonati, D.G. Franks et al. 2013. Specific ligand binding domain residues confer low dioxin responsiveness to AHR1β of *Xenopus laevis*. Biochemistry 52: 1746–1754.

Ohi, H., Y. Fujita, M. Miyao, K. Saguchi, N. Murayama and S. Higuchi. 2003. Molecular cloning and expression analysis of the aryl hydrocarbon receptor of *Xenopus laevis*. Biochem. Biophys. Res. Commun. 307: 595–599.

Paik, W.K. and P.P. Cohen. 1960. Biochemical studies on amphibian metamorphosis: I. The effect of thyroxine on protein synthesis in the tadpole. J. Gen. Physiol. 43: 683–696.

Palenske, N.M., G.C. Nallani and E.M. Dzialowski. 2010. Physiological effects and bioconcentration of triclosan on amphibian larvae. Comparat. Biochem. Physiol. C. 152: 232–240.

Peltzer, P.M., R.C. Lajmanovich, A.M. Attademo, C.M. Junges, C.M. Teglia, C. Martinuzzi et al. 2017. Ecotoxicity of veterinary enrofloxacin and ciprofloxacin antibiotics on anuran amphibian larvae. Environ. Toxicol. Pharmacol. 51: 114–123.

Potter, J.L. and R.D. O'Brien. 1964. Parathion activation by livers of aquatic and terrestrial vertebrates. Science 144: 55–57.

Pretty, R., T. Naitoh and R.J. Wassersug. 1995. Metamorphic shortening of the alimentary tract in anuran larvae (*Rana catesbeiana*). Anat. Rec. 242: 417–423.

Quaranta, A., V. Bellantuono, G. Cassano and C. Lippe. 2009. Why amphibians are more sensitive than mammals to xenobiotics. PLoS ONE 4: e7699.

Roe, J.H., W.A. Hopkins and B.P. Jackson. 2005. Species- and stage-specific differences in trace element tissue concentrations in amphibians: Implications for the disposal of coal-combustion wastes. Environ. Pollut. 136: 353–363.

Rowatt, A.J., J.J. Depowell and W.H. Powell. 2003. ARNT gene multiplicity in amphibians: Characterization of ARNT2 from the frog *Xenopus laevis*. J. Exp. Zool. 300B: 48–57.
Rutkoski, C.F., N. Macagnan, A. Folador, V.J. Skovronski, A.M.B. do Amaral, J. Leitemperger, M.D. Costa, P.A. Hartmann, C. Müller, V.L. Loro and M.T. 2020. Hartmann. Morphological and biochemical traits and mortality in *Physalaemus gracilis* (Anura: Leptodactylidae) tadpoles exposed to the insecticide chlorpyrifos. Chemosphere 250: 126162.
Rutkoski, C.F., N. Macagnan, A. Folador, V.J. Skovronski, A.M.B. do Amaral, J.W. Leitemperger, M.D. Costa, P.A. Hartmann, C. Müller, V.L. Loro and M.T. Hartmann. 2021 Cypermethrin- and fipronil-based insecticides cause biochemical changes in *Physalaemus gracilis* tadpoles. Environ. Sci. Pollut. Res. Int. 28(4): 4377–4387.
Sacchetta, P., R. Petruzzelli, S. Melino, T. Bucciarelli, A. Pennelli, F. Amicarelli et al. 1997. Amphibian embryo glutathione transferase: Amino acid sequence and structural properties. Biochem. J. 322: 679–680.
Salthe, S.N. 1963. The egg capsules in the amphibian. J. Morphol. 113: 161–171.
Savage, R.M. 1962. The ecology and life history of the common frog (*Rana temporaria temporaria*). New York: Hafner, 1962.
Schreiber, A.M. and D.D. Brown. 2003. Tadpole skin dies autonomously in response to thyroid hormone at metamorphosis. Proc. Natl Acad. Sci. USA 100: 1769–1774.
Schreiber, A.M., L. Cai and D.D. Brown. 2005. Remodeling of the intestine during metamorphosis of *Xenopus laevis*. Proc. Natl. Acad. Sci. USA 102: 3720–3725.
Seale, D.B. 1987. Amphibia. pp. 467–552. *In*: Vernberg, P.J. and T.J. Pandian [eds.]. Vol. 2. Animal Energetics. Academic Press, New York
Sestak, M.C., J.A. Pinette, C.M. Lamoureux and S.L. Whittemore. 2018. Early exposure to polycyclic aromatic hydrocarbons (PAHs) and cardiac toxicity in a species (*Xenopus laevis*) with low aryl hydrocarbon receptor (AHR) responsiveness. bioRxiv 301846.
Sheridan, M.A. and Y.H. Kao. 1998. Regulation of metamorphosis-associated changes in the lipid metabolism of selected vertebrates. Am Zool. 38: 350–368.
Shi, Y.B. and A. Ishizuya-Oka. 1996. Biphasic intestinal development in amphibians: Embryogenesis and remodeling during metamorphosis. Curr. Top. Dev. Biol. 32: 205–235.
Shimizu, Y., Y. Nakatsuru, M. Ichinose, Y. Takahashi, H. Kume, J. Mimura et al. 2000. Benzo[a]pyrene carcinogencity is lost in mice lacking the aryl hydrocarbon receptor. Proc. Natl. Acad. Sci. USA 97: 779–782.
Shoots, J., D. Fraccalvieri, D.G. Franks, M.S. Denison, M.E. Hahn, L. Bonati et al. 2015. An aryl hydrocarbon receptor from the salamander *Ambystoma mexicanum* exhibits low sensitivity to 2,3,7,8-Tetrachlorodibenzo-p-dioxin. Environ. Sci. Technol. 49: 6993–7001.
Silva, P.R., M. Borges-Martins and G.T. Oliveira. 2021. *Melanophryniscus admirabilis* tadpoles' responses to sulfentrazone and glyphosate-based herbicides: an approach on metabolism and antioxidant defenses. Environ. Sci. Pollut. Res. 28: 4156–4172.
Schuytema, G.S., A.V. Nebeker, W.L. Griffis and K.N. Wilson. 1991. Teratogenesis, toxicity, and bioconcentration in frogs exposed to dieldrin. Arch. Environ. Contamin. Toxicol. 21: 332–350.
Snodgrass, J.W., W.A. Hopkins, J. Broughton, D. Gwinna, J.A. Baionno and B. Burger. 2004. Species-specific responses of developing anurans to coal combustion wastes. Aquat. Toxicol. 66: 171–182.
Steinwascher, K. 1978. The effect of coprophagy on the growth of *Rana catesbeiana* tadpoles. Copeia 1: 130–134.
Taft, J.D., M.M. Colonnetta, R.E. Schafer, N. Plick and W.H. Powell. 2018. Dioxin exposure alters molecular and morphological responses to thyroid hormone in *Xenopus laevis* cultured cells and prometamorphic tadpoles. Toxicol. Sci. 161: 196–206.
Tanguay, R.L., C.C. Abnet, W. Heideman and R.E. Peterson. 1999. Cloning and characterization of the zebrafish (*Danio rerio*) aryl hydrocarbon receptor. Biochim. Biophys. Acta. 1444: 35–48.
Toloza, E.M. and J.M Diamond. 1990a. Ontogenetic development of nutrient transporters in bullfrog intestine. Am. J. Physiol. 258(5 Pt 1): G760–769.
Toloza, E.M. and J.M. Diamond. 1990b. Ontogenetic development of transporter regulation in bullfrog intestine. Am. J. Physiol. 258: G770–3.
Uchiyama, M. and N. Konno. 2006. Hormonal regulation of ion and water transport in anuran amphibians. Gen. Comparat. Endocrinol. 147: 54–61.
Van den Brink, N.W., M.B.E. Lee-de Groot, P.A.F. de Bie and A.T.C. Bosveld. 2003. Enzyme markers in frogs (*Rana* spp.) for monitoring risk of aquatic pollution. Aquat. Ecosys. Health Managem. 6: 441–448.
Vize, P.D., E.A. Jones and R. Pfister. 1995. Development of the *Xenopus pronephric* system. Dev. Biol. 171: 531–540.
Vize, P.D., D.W. Seufert, T.J. Carroll and J.B. Wallingford. 1997. Model systems for the study of kidney development: Use of the pronephros in the analysis of organ induction and patterning. Dev. Biol. 188: 189–204.

Xu, P. and L. Huang. 2017. Effects of α-cypermethrin enantiomers on the growth, biochemical parameters and bioaccumulation in *Rana nigromaculata* tadpoles of the anuran amphibians. Ecotoxicol. Environ. Saf. 139: 431–438.

Willens, S., M.K. Stoskopf, R.E. Baynes, G.A. Lewbart, S.K. Taylor and S. Kennedy-Stoskopf. 2006a. Percutaneous malathion absorption by anuran skin in flow-through diffusion cells. Environ. Toxicol. Pharmacol. 22: 255–262.

Willens, S., M.K. Stoskopf, R.E. Baynes, G.A. Lewbart, S.K. Taylor and S. Kennedy-Stoskopf. 2006b. Percutaneous malathion absorption in the harvested perfused anuran pelvic limb. Environ. Toxicol. Pharmacol. 22: 263–267.

Wilkens, A.L.L., A.A.N. Valgas and G.T. Oliveira. 2019. Effects of ecologically relevant concentrations of Boral® 500 SC, Glifosato® Biocarb, and a blend of both herbicides on markers of metabolism, stress, and nutritional condition factors in bullfrog tadpoles. Environ. Sci. Pollut. Res. 26: 23242–23256.

Wright, M.L., S.E. Richardson and J.M. Bigos. 2011. The fat body of bullfrog (*Lithobates catesbeianus*) tadpoles during metamorphosis: Changes in mass, histology, and melatonin content and effect of food deprivation. Comp. Biochem. Physiol. A 160: 498–503.

Wu Q., S. Ohsako, T. Baba, K. Miyamoto and C. Tohyama. 2002. Effects of 2,3,7,8-tetrachlorodibenzo-p-dioxin (TCDD) on preimplantation mouse embryos. Toxicology 174(2): 119–129.

Zaccone, G., S. Fasulo, P. Lo Cascio, L. Ainis, M.B. Ricca and A. Licata. 1989. Effects of chronic exposure to endosulfan on complex carbohydrates and enzyme activities in gill and epidermal tissues of the freshwater catfish *Heteropneustes fossilis* (Bloch). Arch. Biol. 100: 171–185.

Zimmermann, A.L., E.A. King, E. Dengler, S.R. Scogin and W.H. Powell. 2008. An aryl hydrocarbon receptor repressor from *Xenopus laevis*: Function, expression, and role in dioxin responsiveness during frog development. Toxicol. Sci. 104: 124–134.

Zhu, W., M. Zhang, L. Chang, W. Zhu, C. Li, F. Xie et al. 2019. Characterizing the composition, metabolism and physiological functions of the fatty liver in *Rana omeimontis* tadpoles. Front. Zool. 16: 42.

Zia, S. and D.G. McDonald. 1994. Role of the gills and gill chloride cells in metal uptake in the freshwater-adapted rainbow trout, *Oncorhynchus mykiss*. Can. J. Fish. Aquat. Sci. 51: 2482–2492.

4

Hierarchical Levels of Biomarkers in Amphibian Tadpoles Exposed to Contaminants

From Enzyme Disruptions to Etho-Toxicology Studies in Argentina

Paola Mariela Peltzer,[1,2] *Ana Paula Cuzziol Boccioni,*[1,2] *Attademo Andres Maximiliano,*[1,2] *Lucila Marilén Curi,*[1,3] *María Teresa Sandoval,*[4] *Agustín Bassó,*[2] *Candela Soledad Martinuzzi,*[1,2] *Evelina Jésica León,*[1,2,5] *Rafael Fernando Lajmanovich*[6] and *Rafael Carlos Lajmanovich*[1,2,*]

1. Brief Introduction of Amphibian Monitoring in Contaminated Environments

Environmental stress source not only degrades the habitats and deteriorates morphological and physiological aspects of amphibians, but also is an important stress factor that may affect

[1] Consejo Nacional de Investigaciones Científicas Técnicas (CONICET). Buenos Aires, Argentina.

[2] Laboratorio de Ecotoxicología. Facultad de Bioquímica y Ciencias Biológicas. Universidad Nacional del Litoral. Ciudad Universitaria, Pje. El Pozo. Santa Fe (3000) Argentina. Emails: paolapeltzer@hotmail.com; anapaulacuzziolboccioni@gmail.com; mattademo@hotmail.com; bassoagustin@gmail.com

[3] Instituto de Ictiología del Nordeste (INICNE). Facultad de Ciencias Veterinarias, Universidad Nacional del Nordeste (FCV, UNNE), (3400) Corrientes Argentina. Email: lucilacuri@gmail.com

[4] Universidad Nacional del Nordeste. Facultad de Ciencias Exactas y Naturales y Agrimensura. Embriología Animal. Av. Libertad 5470 (3400). Corrientes, Argentina. Email: mtsandoval@exa.unne.edu.ar

[5] Instituto Nacional de Limnología. Ciudad Universitaria, Pje. El Pozo. Santa Fe (3000) Argentina. Email: evelinaleon93@hotmail.com

[6] Facultad de Ciencias Médicas. Ciudad Universitaria, Pje. El Pozo. Santa Fe (3000) Argentina. Email: rafaellajmanovich@gmail.com

* Corresponding author: lajmanovich@hotmail.com

communication and behavior. Amphibians "indicate" the health status of the ecosystems where they occur (Saber et al. 2017). Some authors, like Scott and Sloman (2004), coined the term "ecological death" to refer to those aquatic organisms that no longer play an environmental role or provide an environmental service, or to those that "sacrifice" a biological function to offset the damage or environmental stress. In this sense, through field records, León et al. (2019) demonstrated that males of *S. nasicus* shift their vocal structure in acoustically disturbed sites, mainly by making vocal "adjustments" in their frequencies and amplitudes to counteract the effect of noise masking. This has sublethal effects on the reproductive success, clutch size, survival of individuals and populations viability highlighting of how is important to stop the collapse of aquatic ecosystems.

In-situ methods for the recording and sampling of anurans are recommended in contaminated environments (Fig. 1). These methods gain environmental relevance because they interpret further biological data. However, laboratory-scale assays have other advantages, such as the control of certain variables.

A summary of a number of considerations to be taken into account in the diverse methods for anuran sampling in contaminated environments is presented below (Box 1).

Box 1. Points to consider before and during field sampling in contaminated environments.

Before the field sampling	During the field sampling
I. Elaborating the experimental design should answer the questions: Where and when should samples be taken? How may samples should be taken? How should they be collected? II. (a) Critically selecting sampling sites (including the control site) based on research hypotheses and objectives). (b) acquiring updated satellite images of the sampling sites. III. Determining the timing of sampling according to the objectives (reproductive period and phenology of amphibian species, season of pesticide application, crop phenology, discharge of effluents, etc.). IV. Elaborating a list of material and tools necessary for field studies (forms, maps, GPS, compasses, flashlights, cameras, recorders, containers for environmental sample collection, bags, equipment for animal marking, balance). V. Requesting the corresponding collection and bioethics permissions.	I. Recording the highest number of environmental variables: ambient and water temperature, humidity, wind direction and speed, depth of water bodies, water physical-chemical parameters (turbidity, conductivity, pH, ammonium, nitrites, nitrates, dissolved oxygen, etc.). Georeference (*GPS*). II. Using reliable tools that can be adapted to field work, e.g., commercial kits for measuring parameters. III. Efficiently storing the environmental samples, e.g., freezer or portable refrigerators. IV. Taking photographic and audiovisual records of the sites, indicating details considered relevant. V. Transporting eggs, larvae and/or adults within the 3 h after collection, in dark flask containing water of the environment where they had been collected. At the laboratory, allowing for acclimation of individuals (minimum 12 h).

1.1 Biomarkers in Ecotoxicological Studies Involving Amphibians

The responses of organisms obtained through the analysis of biomarkers can be interpreted as early warning signals of the degree of degradation or ecological health of a given environment. In general, to be used as a biomarker, a parameter should be sufficiently sensitive, specific and capable of providing an early response. However, since they are not always specific, using different types of markers allows a more accurate exploration of the effects of contaminants (Chapman 2002, Spurgeon et al. 2005).

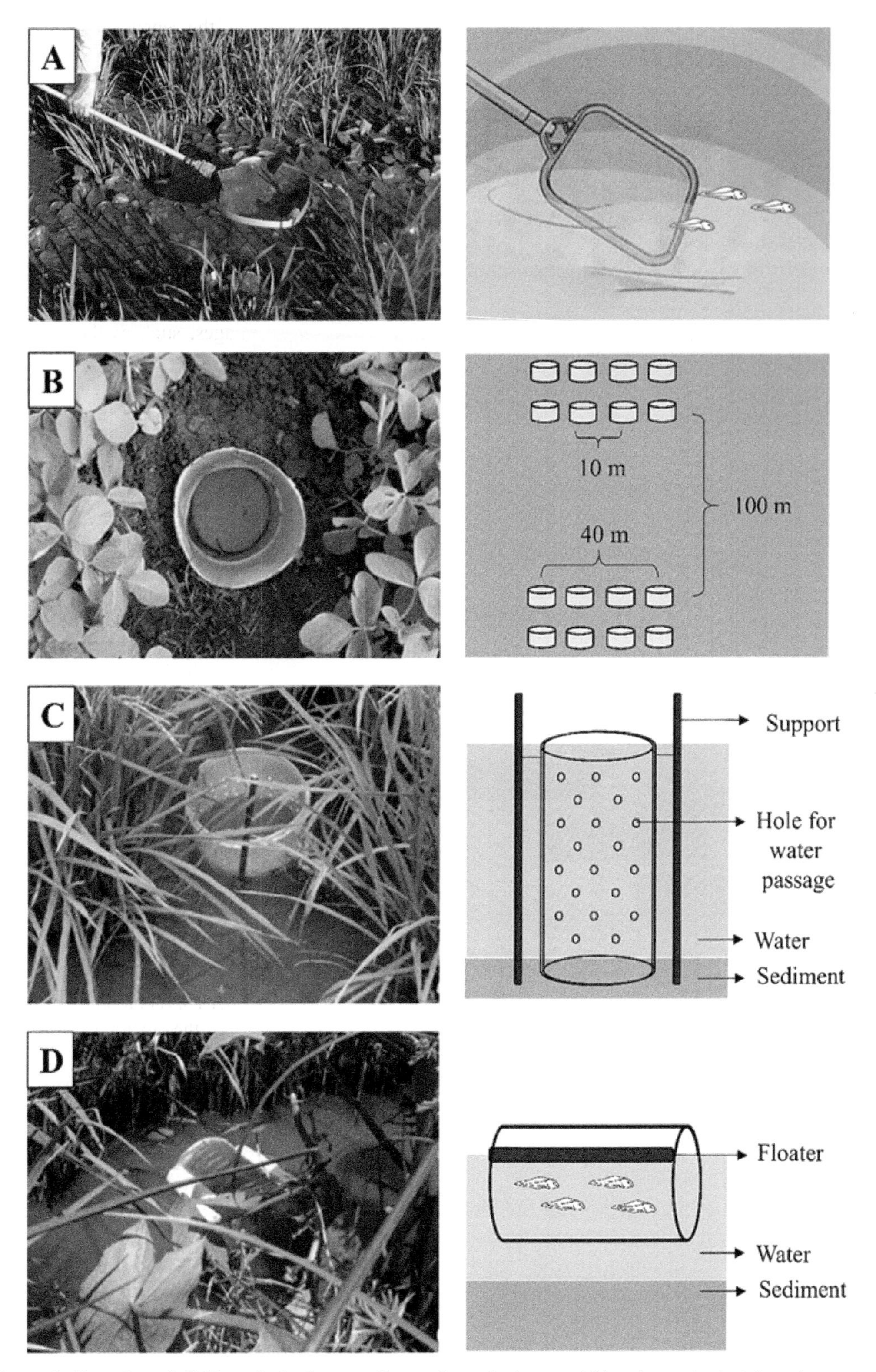

Figure 1. Examples of field methods for sampling and monitoring amphibian in contaminated environments: Scooping with net (A), live-pitfall with wet sponge at the bottom and scheme of the arrangement of the traps on the ground at the right (B), Limnocorral buried in the substrate (C), and floating limnocorral (D).

There is a great diversity of biomarker types, according to:

- TYPE OF INFORMATION. *Exposure* biomarkers (those indicating that the contaminant is present in the organism); *effect* biomarker (presence of an alteration), and *susceptibility* biomarker (organism-specific sensitivity).
- SAMPLE COLLECTION. Destructive biomarkers (those implying the animal death) and non-destructive ones
- HIERARCHICAL LEVEL. Evaluating toxic effects from the molecular scale to higher organization levels representing the entire organism allows a better understanding of the toxicity of an agent and its mechanism of action (Fig. 2).

Literature on reference values of most of the biomarkers for the wild fauna is scarce. In addition, evaluation of control individuals (individuals of the same species that are not exposed to the compounds whose effects are being studied) is recommended.

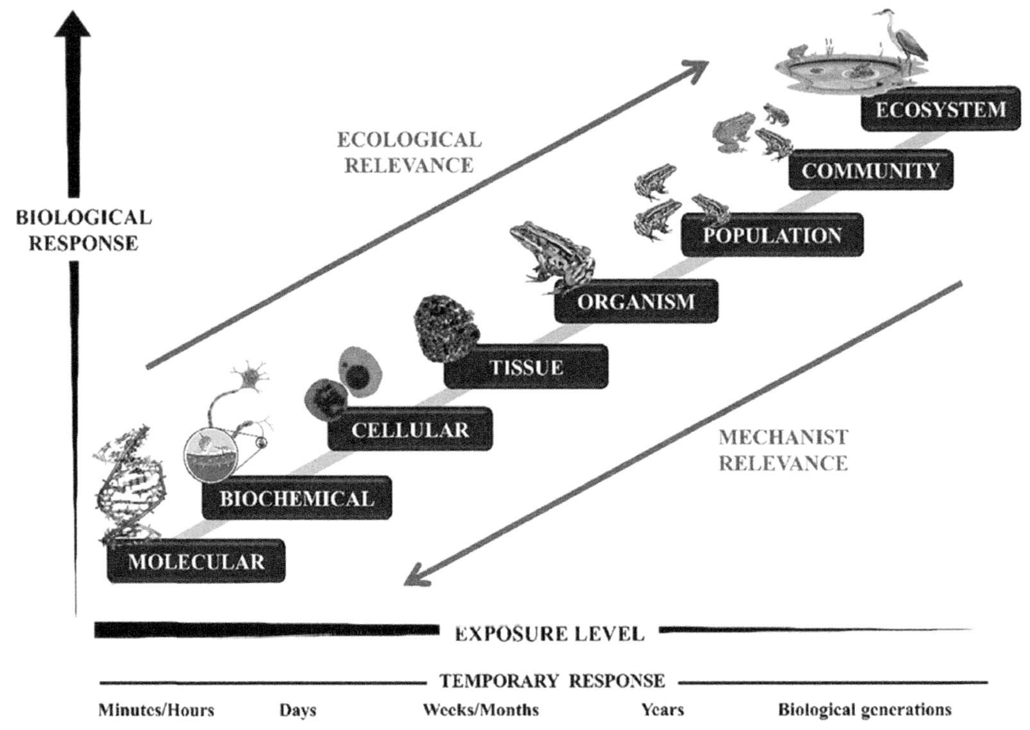

Figure 2. Hierarchical levels of ecotoxicological studies regarding ecological and mechanist relevance (modified from Spurgeon et al. 2005).

1.2 Biochemical Biomarkers

Changes at the biochemical level are usually detected before the adverse effects of contaminants can be observed at higher levels of the biological organization. For this reason, they are often applied as an useful tool for early warning and diagnosis of contaminated aquatic environments. For example, the results obtained from the analyses of different enzymatic biomarkers may evidence that amphibian populations present in agricultural environments are being exposed to environmental stress conditions caused by the intense use of agrochemicals and the presence of other toxic substances such as contaminant of emerging concerns.

Activities of acetylcholinesterase (AChE), butyrylcholinesterase (BChE), carboxylesterase (CbE) and glutathione S-transferase (GST) or glutathione reductase (GR) were characterized to understand the metabolic impact of agrochemicals in adults and tadpoles of amphibians (Attademo et al. 2007, 2011, Lajmanovich et al. 2009, 2011). Later, several studies analyzing enzyme biomarkers in tadpoles exposed to different contaminants were performed. For example, using limnocorrals and tadpoles located on rice crops. The exposure resulted in significant differences in AChE, BChE and GST enzymes of tadpoles than that of those of reference sites-natural sites (Attademo et al. 2011, 2014, 2015).

In accordance to these results, laboratory studies also revealed inhibition of AChE and BChE and induction of GST due to tadpole exposure to different pesticides (i.e., glyphosate, picloram, dicamba, pyriproxyfen, glufosinate-ammonium) and other emerging contaminants (anti-retroviral drugs and diclofenac), corroborating the pollutant effects of field studies. Likewise, the alarming neurotoxicity of contaminants such as glyphosate, glufosinate-ammonium, diclofenac, chlorpyrifos has been determined based on the measurement of cholinesterases (Peltzer et al. 2013a, 2019, Lajmanovich et al. 2019a, Paradina Fernandez et al. 2020, Attademo et al. 2021).

1.3 Hormonal Biomarkers

The thyroid hormones (TH) are essential for amphibian metamorphosis. These hormones provide information *in vivo* about the mechanisms occurring during the developmental stages (Carlsson 2019). In neotropical amphibians, data on disruption of thyroid signaling is scarce. Any alteration in the TH balance in tadpoles causes drastic structural and functional changes in larval morphology (e.g., malformations in the notochord, and defects in the heart and other organs) and on the duration of metamorphosis (Carlsson 2019).

The first data on TH levels in tadpoles of neotropical amphibians exposed to two contaminants, glyphosate and/or arsenic, and basal levels were provided by Lajmanovich et al. (2019b) for *Odontophrynus americanus* (Gosner stages 28–32, Gosner 1960). Levels of triiodothyronine (T3) and thyroxine (T4) showed a notable increase in response to contaminants. These results are crucial, since both pollutants are thyroid hormone disruptors. Other hormonal herbicides and insecticides like dicamba (Attademo et al. 2021) and pyriproxyfen (Lajmanovich et al. 2019b), respectively, also act as thyroid disruptors in amphibians (Freitas et al. 2016). More data about pollutant effects in thyroidogenesis of tadpoles can be found in chapter 5.

1.4 Genotoxicity

Typically, a micronucleus (MN) is defined as a small extranuclear chromatinic body originating from an acentric fragment or whole chromosome lost from the metaphase phase (Fig. 3). The MN test has been widely used in amphibian erythrocytes, since this cell type is easily handled and cellular dissociation is not required. When compared with other DNA damage detection techniques (e.g., Comet Assay), the MN test has some advantages: it can be performed rapidly, it is not complex or expensive, and its preparation and analysis are simpler and faster than other tests for chromosomal aberrations. This test has been used for the first time to evaluate the effect of pesticides on amphibian larvae in Argentina (Lajmanovich et al. 2005, Cabagna et al. 2006). Also, other erythrocyte nuclear abnormalities (ENA), such as lobed nuclei, binucleated cells, kidney-shaped nuclei and notched nuclei have been observed in erythrocytes of amphibian tadpoles as a consequence of exposure to contaminants in agricultural ponds (Attademo et al. 2011, Babini et al. 2016). Although the mechanism responsible for the formation of all ENA types has not been totally explained, these abnormalities are considered to be indicators of genotoxic damage and therefore may complement the scoring of MN in routine assays for genotoxicity screening in amphibian tadpoles (e.g., Lajmanovich et al. 2014). More details about genotoxicity of toxicants to tadpoles will be discussed in chapter 7.

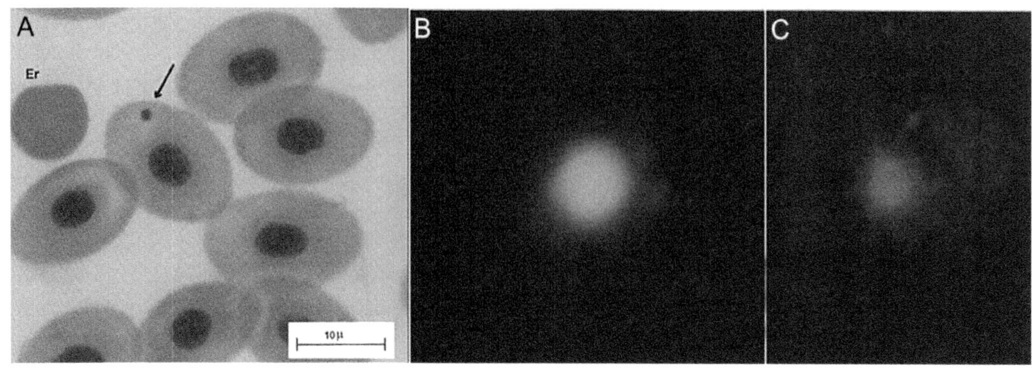

Figure 3. Detail of red blood cells (RBCs) with micronuclei (MN, arrow) and (Er, anucleate erythrocyte) observed in *Pseudis minuta* tadpoles (A). May Grunwald Giemsa, 100x. DNA damage index (DI) (based on the Comet assay) at 22 days of exposure to the NOEC-48 h/8 of glyphosate-based herbicide (GBH) and arsenite (As(III)), and the mixture of 50:50 v/v GBH-As(III); digitized images showing examples of undamaged (Type I comet) (B) and medium damage (Type III comet) RBCs (C) (400 x).

1.5 Histological Biomarkers

Histological analysis of selected organs of tadpoles may reveal sublethal effects at individual and even population levels, which can be evidenced through significant changes in the configuration, composition and appearance of some cells and tissues. Whole body sections are also useful to analyze not only the organs of interest but also their general and spatial configuration; being helpful also to confirm the presence of malformations that are observed at the macroscopic scale (Cuzziol Boccioni et al. 2020; Box 2).

Among the most common histological alteration due to pollutant exposure in tadpoles is the presence of pigmented phagocytizing cells or melanomacrophages (MMs). MMs are located in liver, spleen and kidney, and are widely studied in anurans because their changes both in number and size can indicate physiological and immunological responses to contaminants (Huespe et al. 2017, Curi et al. 2019, 2021, Cuzziol Boccioni et al. 2020; Box 2). Histopathology of the gonads are also commonly evaluated during tadpoles' development. Besides abnormalities in the general structure of gonads, their histological configuration is usually taken into account, such as the proportion of germinates cell types (Sánchez et al. 2013, 2014, Curi et al., 2019, 2021; Box 2). The analysis of reproductive organs is also useful for studying the effect of a toxic agent on the endocrine system, which is involved in the reproduction, and, ultimately, for estimating the implications of the contaminant in the reproductive success of species exposed to pollution.

The appearance and configuration of the intestinal epithelium is also another frequently evaluated histological biomarker. This tissue is sensitive to external agents and is directly associated with physiological and metabolic processes involved in the absorption of nutrients and, therefore, changes in this tissue might affect those processes (Lajmanovich et al. 2015, 2017, Cuzziol Boccioni et al. 2020; Box 2).

1.6 Cardiac Biomarkers

The heart is one of the first organs to form and the first to function during embryogenesis. Knowledge of the normal development of this organ is essential to understand the etiology of cardiac malformations and dysfunctions. In anurans, two critical periods of cardiac development were described, which would be the most susceptible to the effects of teratogens (Gaona et al. 2019): the embryonic period and the metamorphic larval period. During the embryonic development period (Gosner stage, GS 1 to 20) early cardiogenesis events occur (formation of the endocardial tube,

Box 2. Examples of histological biomarkers of amphibian used in ecotoxicological studies.

Tissue	Description	References
Whole body (tadpoles) 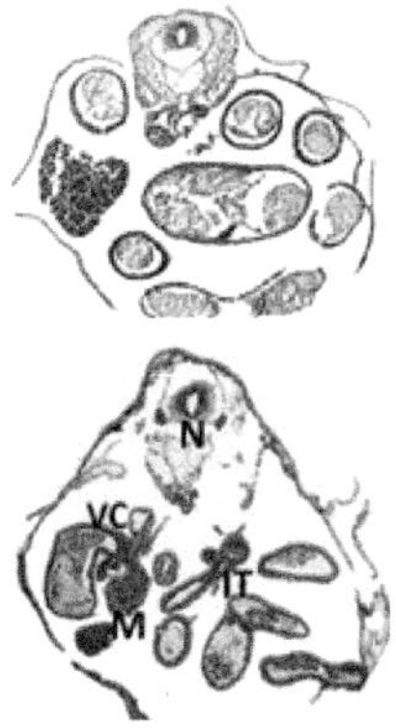	Cross section of the whole body (at abdominal level) of *Rhinella arenarum* tadpoles from control (up) and dexamethasone treatment (down). The arrows show: abnormal intestine coiling and alteration of intestinal tract (IT); visceral congestion (VC); alteration of nothocord and neural tube (N); melanomacrophages in pancreatic and hepatic tissues (M).	Cuzziol Boccioni et al. (2020)
Liver 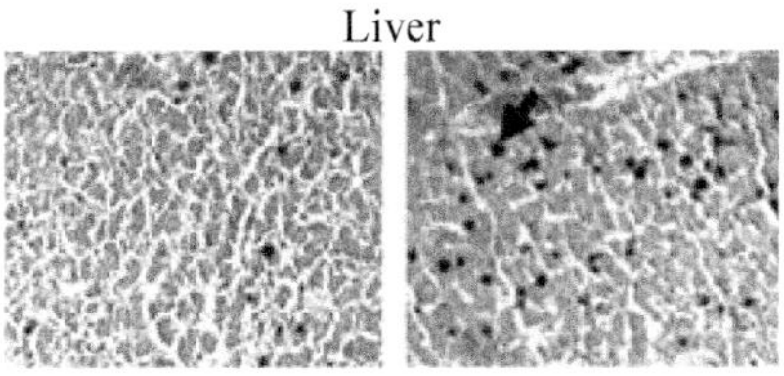	Liver sections of *Leptodactylus macrosternun* adults from control site (left) and rice field after pesticides application (right). The arrow shows the higher number of melanomacrophages respect to control.	Huespe et al. (2017)
Gonads 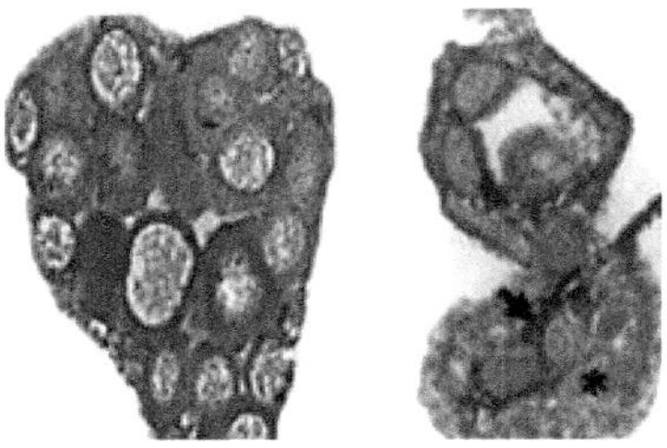	Ovary of metamorphic individuals of *Physalaemus albonotatus* from control (left) and herbicide formulation Amina Zamba ® treatment (right). The arrows show melanophores (arrow) and ovocites in different stages of maduration (*).	Curi et al. (2019)
Intestine 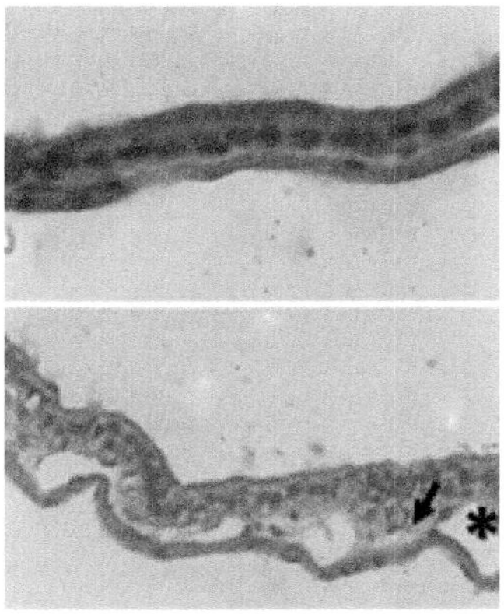	Intestinal walls *of Leptodactylus macrosternum* tadpoles from control (up) and treatment of aqueous suspension formulation of *Bacillus thuringiensis var. israelensis (down)*, with infiltration in the connective tissue underlying the epithelium (arrow) and dilation of blood vessels (*) in the treated sample.	Lajmanovich et al. (2015)

start of the looping into an S-like shape and chamber formation septation). During the metamorphic larval period (GS 42 to 46), the changes that lead to the acquisition of the final configuration in adult become evident (final displacement of the atrium to a dorsal and anterior position with respect to the ventricle and atrial septation) (Fig. 4) (Sandoval et al. 2022). This is useful baseline information for the recognition and interpretation of cardiac morphological anomalies caused by environmental xenobiotics, and could point critical stages in cardiogenesis (Fig. 5).

Several studies of anuran larvae have shown the cardiotoxic effect of some insecticides, such as chlorpyrifos, dichlorvos and diazinon (Watson et al. 2014), and pyriproxyfen (Lajmanovich et al. 2019b) and of the herbicide atrazine (Asouzu Johnson et al. 2019). Likewise, recent articles have

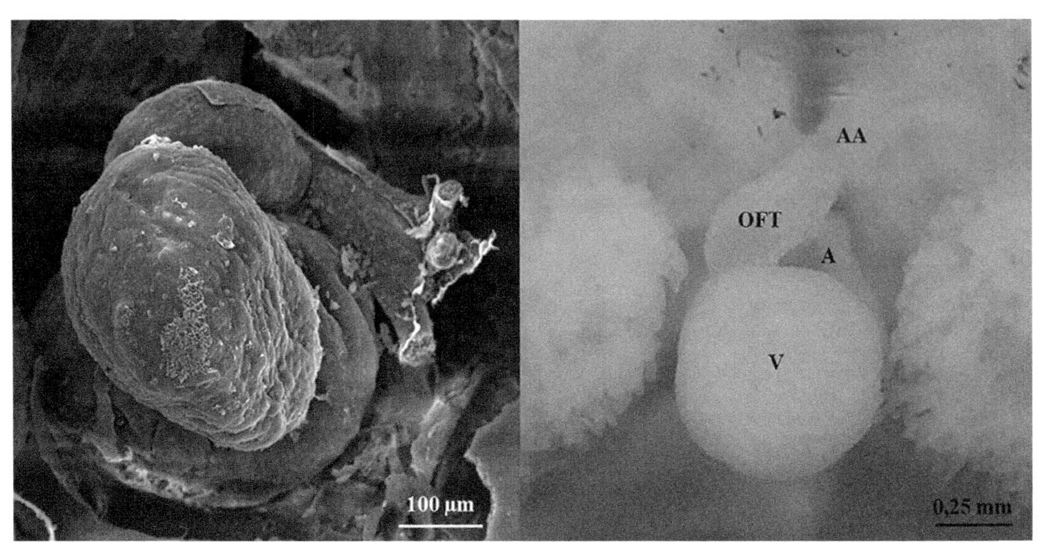

Figure 4. Heart of tadpoles of *Elachistocleis bicolor* at GS 30-31, photographed with scanning electronic and optical microscope, respectively. References: A: atrium, AA: aortic arches, OFT: outflow tract, V: ventricle.

CRITICAL PERIODS

Specification of cardiogenic mesoderm.

Formation of heart tube.

Delimitation of sinus venosus, atrium, ventricle, and outflow tract.

Onset of looping.

Onset of the heartbeat.

Development of cardiac trabeculae.

Growth of the atrium.

Final displacement of the atrium to a dorso-anterior position respect to the ventricle.

Septation of the atrium.

Anatomic and topographic changes of aortic arches.

Embryonic period **Premetamorphic** **Prometamorphic** **Metamorphic**

Larval periods

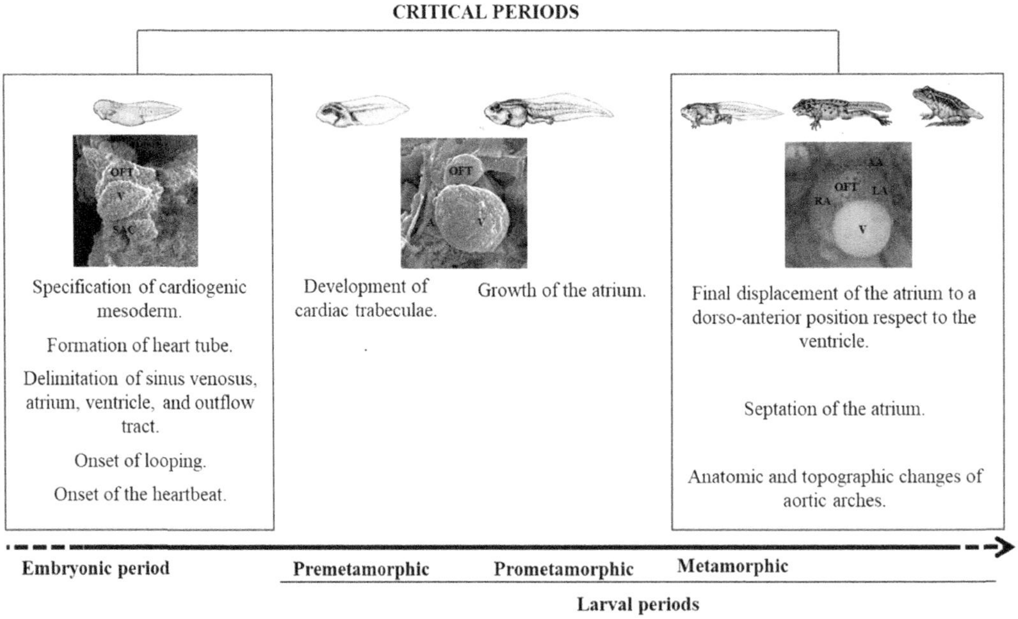

Figure 5. Critical period during cardiogenesis of anuran tadpoles. References: A: atrium, AA: aortic arches, OFT: outflow tract, V: ventricle.

described the capacity of some emerging contaminants, such as the analgesic agent Diclofenac (Peltzer et al. 2019) and the anti-inflammatory drug dexamethasone (Cuzziol Boccioni et al. 2020), to alter the cardiac frequency in larvae of *T. typhonius*, *P. albonotatus* and *R. arenarum*, native species of Argentina littoral.

1.7 Morphological Biomarkers

Among the morphological biomarkers, standard parameters like body length (termed snout-vent length in amphibians) and body weight are particularly important for the study of the organisms' response in terms of growth. In the case of bioassays involving larvae, length can be measured in small-sized individuals using digital images obtained with a stereoscopic microscope equipped with a camera, through an image analysis software or manually using specific tools (digital caliper) in larger tadpoles and in adults. To record weight (body mass) with digital scales, the excess water should be removed from the animal before measure. Growth and development rates (development inhibition), teratogenesis, and time to reach metamorphosis are also usually estimated. Morphometric indices that relate several external parameters are also frequently used in ecotoxicological evaluations for a better estimation of individual fitness (e.g., animal condition factor, hepatosomatic or gonadosomatic index, see Table 1). Several abnormalities during development could be recorded. Some examples are shown in Fig. 6. The hepatosomatic index reflects differences in energy storage and reproductive activities between males and females that can vary due to endogenous or exogenous factors (e.g., temperature, fat storage, xenobiotic, parasite infections) (Bruslè and Anadon 1996). In addition, the gonadosomatic reflects the degree of gonadal development with respect to body weight (Schmitt and Dethloff 2000) that can also vary due to exogenous stressors (see de Oliveira et al. 2016, Gondim et al. 2020, Leão et al. 2021). More examples and details on malformations in tadpoles due to pollutant will be discussed in Chapter 14.

Table 1. Morphological parameters pointed in different studies with exposed tadpoles *in situ* or lab bioassays.

Morphological parameters	Toxic compounds	References
Size	Pyretroids, 2,4D, AMPA Mixed contaminants (metals and pesticides) Ciprofloxacin, enrofloxacin	Peltzer et al. (2008) Peltzer et al. (2013a) Peltzer et al. (2017)
Animal condition factor	Clorpirifos Mixed contaminants (Rice fields)	Huespe et al. (2017) Curi et al. (2021)
Development and growth rates	Poultry litter Diclofenac Mixed contaminants (metals and pesticides)	Curi et al. (2017) Peltzer et al. (2019) Peltzer et al. (2013a)
Teratogenesis	Glyphosate Dexamethasone Agricultural occurrence	Lajmanovich et al. (2003) Cuzziol Boccioni et al. (2021) Peltzer et al. (2011)
Time to reach metamorphosis	2,4-D Atrazine	Brodeur et al. (2013) Curi et al. (2019)

1.8 Ethological Biomarkers

Amphibian behavior is the result of the integration of physiological processes, basically those determined by the information obtained from sensory organs, the nervous system and the endocrine system, mainly the pituitary gland as regulator of the core endocrine organs (thyroids, adrenal, testicles, ovaries and epiphysis). Therefore, changes in physiology, such as those produced by pesticides or pharmaceuticals (Table 2), must affect the homeostasis of important functions of the organism, and could be related to ethological alterations. These changes can be recorded through direct observation or video-recorded and quantified through specific software and subjected to

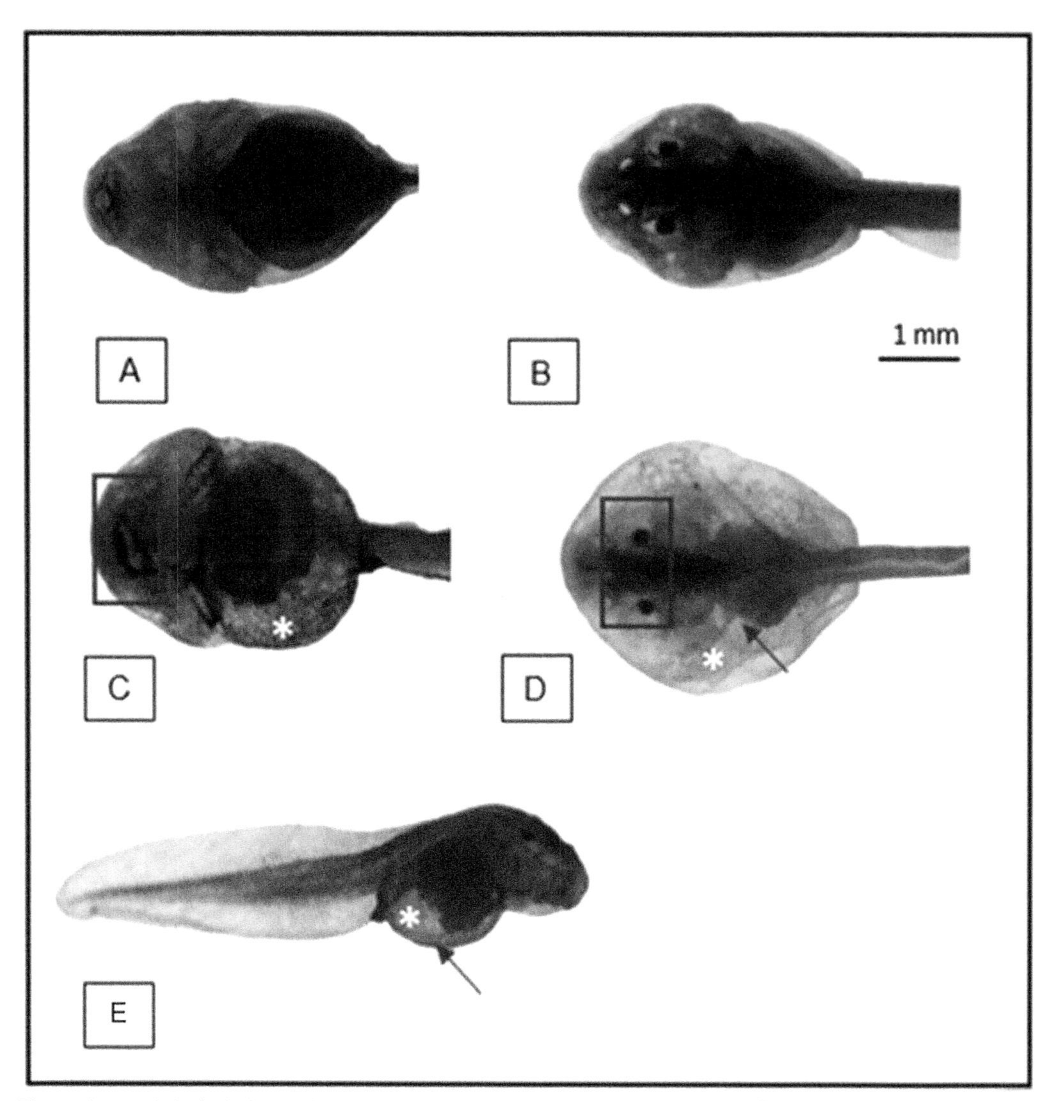

Figure 6. Morphological abnormalities in tadpoles after exposition of a contaminant. References: Normal morphology in ventral (A) and dorsal (B) view; altered oral disc (C), distance between eye, swollen body and absence well-gut development (D), ventral edema and protruded head (E).

statistical analysis. Some of the parameters studied in amphibian larvae include the total distance moved, changes in speed or general activity, and the interactions among individuals or with predators. All these changes are directly related to survival or fitness. Moreover, the lack of specificity of these biomarkers becomes an advantage when the harmful substances or the action mechanisms of a contaminant are unknown.

2. The Impact of GMO-Based Intensive Agriculture on Native Amphibians of Latin America

Habitat loss and fragmentation pressure make amphibian populations use other aquatic ecosystems that are generally immersed or near crop fields (Bishop et al. 1999, Peltzer 2006). In the last two decades, the arrival of biotechnological techniques in agriculture promoted the agricultural expansion with genetically modified organisms (GMO), mainly soybean and maize, and the recent controversial HB4 wheat. Thus, vast areas of rainforests, riparian forests and wetlands have been

Table 2. Overview of experimental design of ecotoxicological studies with ethological biomarkers on tadpoles.

Species	Stimuli	Endpoint	Results	References
Rhinella dorbignyi	Cues of fish predator, an injured conspecific (10 μg/L cypermethrin, 10 min)	Total distance moved, mean speed, global activity, number of contacts between tadpoles) in alone or in groups	Dependent of group size	Curi et al. (2022)
Boana pulchella	Glufosinate ammonium (3.55–35.5 mg/L, 48 h)	Total distance moved, mean speed	Increase of the swimming speed and mean distance at 15 mg/L; negative correlation between swimming speed and BChE activity	Peltzer et al. (2013b)
Rhinella arenarum, Physalaemus santafecinus, Leptodactylus latrans, Elachistocleis bicolor	LOEC Trifloxystrobin, Fish predator (24 h)	Predation rate	Predation rate differed among prey species; not due to funguicide exposure, but to interspecific differences in behavior	Junges et al. (2012)
Rhinella arenarum	Deltamethrin 1%+tetramethrin 0.33%, and piperonyl butoxide (48 h)	Total distance moved, mean speed, maximun speed, global activity, resting time	Uncoordinated swimming, spasmodic and erratic movement, high period of immobility	Lajmanovich et al. (2018)
Odontoprynus americanus	Pyriproxyfen (0.01–0.1 mg/L, 22 d)	Mean speed, total distance, global activity	Reduced mean speed, and global activity, negative correlation with AChE activity and T4	Lajmanovich et al. (2019b)
Trachycephalus typhonius, Physalaemus albonotatus	Diclofenac (125 and 4000 μg/L , 2–20 d)	Total distance moved, mean speed, global activity	Neurotoxic effect of DCF correlation between AChE activity and swimming parameters	Peltzer et al. (2019)

deforested, drained and burned for agricultural use in South America (especially in Brazil and Argentina) (Altieri and Nicholls 2017). In addition, an ongoing expansion of the meat agroindustry requires a massive use of pharmaceuticals and veterinary products that are discharged to rivers, streams and lakes without any treatment (Teglia et al. 2017, 2019, Rojo et al. 2019).

Our research team has done pioneer studies in these issues by showing how some pharmaceuticals (anti-inflammatory and antibiotic drugs) alter the normal development, growth and survival of tadpoles present in water bodies containing emerging contaminants, and how the mixture with microplastics and/or pesticides produce negative synergies on enzyme activities (Peltzer et al. 2019, Cuzziol Boccioni et al. 2020, 2021, Lajmanovich et al. 2022). These results are similar to those obtained by Pelusso et al. (2022) in *Rhinella arenarum* tadpoles.

For instance, the toxicity of the biological insecticide that uses a Gram + bacterium *Bacillus thuringiensis* (or *Bt*), which produces endotoxins, also known as "*Bt* toxins" or "*Cry proteins*" was also evaluated in exposed tadpoles (Lajmanovich et al. 2015). These data allowed us to assess the ecotoxicological risk to which non-target organisms are exposed after the application of those insecticides to control vectors of known diseases, such as dengue, or after consuming GMO-derived food products, such as rice, maize or *Bt*-soybean (e.g., Séralini et al. 2014, Holderbaum et al. 2015).

This biotechnological revolution of transgenic crops is undoubtedly devastating the ecosystems in Argentina and Brazil, inducing declines and/or local extinction of anuran amphibians due to contamination and alteration of their natural habitats. Most of the agrochemicals used to produce GMO are found in fat and muscle of amphibians, reptilians, birds and mammals, being the first report of pesticide bioaccumulation for Argentina (Lajmanovich et al. 2005). In addition, there are the consequences of the climatic crisis, such as the current drought affecting the Paraná-Plata Basin, the continuous fires and the deforestation, which has decimated millions of hectares of wetlands and riparian forests. The interruption of the feedback of the basins such as the Parana River through the "flying rivers" of the Amazonia seems to alert to a large-scale environmental collapse (Zemp et al. 2017).

Taking into account the numerous national and international studies, mainly in South America, that have used aquatic organisms such as amphibians as model organisms and early warning indicators, we ask the following questions: how much more evidence is needed to demonstrate that the extractive system of food production and resource exploitation drive amphibian decline and extinction as part of a global ecocide? Do we still have time to stop these ecocides, or is it possible that we have come to a point of no return?

Acknowledgements

We would like to acknowledge the editors for inviting us to participate in this work. We thank PICT 2017 N° 1069; PICT 2018 N° 3293; CAID+D 2020 N° 50620190100036LI for financing our studies. We thank to J. Brasca for the language correction. We dedicate our contribution in this chapter to the worthy science "Ciencia Digna" at the service of nature and society that goes beyond the barriers of state institutions (politic, academic and scientific) that in complicity with multinational companies in pursuit of economic interests are destroying the nature of Latin American.

References Cited

Altieri, M.A. and C.I. Nicholls. 2017. Agroecology: A brief account of its origins and currents of thought in Latin America. Agroecol. Sustain. Food Sys. 41(3-4): 231–237.

Asouzu Johnson, J., A. Ihunwo, L. Chimuka and E.F. Mbajiorgu. 2019. Cardiotoxicity in African clawed frog (*Xenopus laevis*) sub-chronically exposed to environmentally relevant atrazine concentrations: Implications for species survival. Aquat. Toxicol. 213: 105218.

Attademo, A.M., P.M. Peltzer, R.C. Lajmanovich, M. Cabagna and G. Fiorenza. 2007. Plasma B-esterases and glutathione S-transferase activities in the toad *Chaunus schneideri* (Amphibia, Anura) inhabiting rice agroecosystems of Argentina. Ecotoxicology 16: 533–539.

Attademo, A.M., M. Cabagna Zenklusen, R.C. Lajmanovich, P.M. Peltzer, C.M. Junges and A. Bassó. 2011. B-esterase activities and blood cell morphology in the Frog *Leptodactylus chaquensis* (Amphibia: Leptodactylidae) on rice agroecosystems from Santa Fe Province (Argentina). Ecotoxicology 20: 274–282.

Attademo, A.M., P.M. Peltzer, R.C. Lajmanovich, M. Cabagna Zenklusen, C.M. Junges and A. Bassó. 2014. Biological endpoints, enzyme activities, and blood cell parameters in two anuran tadpole species in rice agroecosystems of mid-eastern Argentina. Environ. Monit. Assess. 186: 635–649.

Attademo, A.M., P.M. Peltzer, R.C. Lajmanovich, M. Cabagna Zenklusen, C.M. Junges, E. Lorenzatti, C. Aró and P. Grenón. 2015. Biochemical changes in certain enzymes of *Lysapsus limellium* (Anura: Hylidae) exposed to chlorpyrifos. Ecotoxicol. Environ. Saf. 113: 287–294.

Attademo, A.M., J.C. Sanchez-Hernandez, R.C. Lajmanovich, M.R. Repetti and P.M. Peltzer. 2021. Enzyme activities as indicators of soil quality: Response to intensive soybean and rice crops. Wat. Air Soil Pollut. 232: 1–12.

Babini, M., C. Bionda, N. Salas and A. Martino, Adolfo. 2016. Adverse effect of agroecosystem pond water on biological endpoints of common toad (*Rhinella arenarum*) tadpoles. Environ. Monit. Assess. 188: 459.

Bishop, C.A., N. Mahony, J. Struger, P. Ng and K.E. Pettit. 1999. Anuran development, density and diversity in relation to agricultural activity in the Holland River watershed, Ontario, Canada (1990–1992). Environ Monit Assess 57(1): 21–43. https://doi.org/10.1023/A:1005988611661.

Brodeur, J.C., A. Sassone, G.N. Hermida and N. Codugnello. 2013. Environmentally-relevant concentrations of atrazine induce non-monotonic acceleration of developmental rate and increased size at metamorphosis in *Rhinella arenarum* tadpoles. Ecotoxicol. Environ. Saf. 92: 10–17.

Bruslé, J. and G.G. Anadon. 1996. The structure and function of fishliver. pp. 77–93. *In*: JSD, M. and H.M. Dutta [eds.]. Fish morphology. Lebanon: Horizon of New Research, Science Publishers Inc.

Cabagna, M., R.C. Lajmanovich, P.M. Peltzer, A.M. Attademo and E. Ale. 2006. Induction of micronucleus in tadpoles of *Odontophrynus americanus* (Amphibia: Leptodactylidae) by the pyrethroid insecticide cypermethrin. Environ. Toxicol. Chem. 88: 729–737.

Carlsson, G. 2019. Effect-based environmental monitoring for thyroid disruption in Swedish amphibian tadpoles. Environ. Monit. Assess. 191: 454.

Chapman, P.M. 2002. Integrating toxicology and ecology: putting the "eco" into ecotoxicology. Mar. Pollut. Bull. 44: 7–15.

Chaves, M.F., F.C. Tenório, I.L. Santos, J.C. Neto, V.W. Texeira, G.J. Moura and A.A. Texeira. 2017. Correla-tions of condition factor and gonadosomatic, hepatosomatic and lipo-somatic relations of *Leptodactylus macrosternum* (Anura: Leptodactylidae) in the Brazilian semi-arid. An. Acad. Brasil. Ci. 89: 1591–1599.

Curi, L.M., P.M. Peltzer, C.S. Martinuzzi, A.M. Attademo, S. Seib, M.F Simoniello and R.C. Lajmanovich. 2017. Altered development, oxidative stress and DNA damage in *Leptodactylus chaquensis* (Anura: Leptodactylidae) larvae exposed to poultry litter. Ecotoxicol. Environ. Saf. 143: 62–71.

Curi, L.M., P.M. Peltzer, M.T. Sandoval and R.C. Lajmanovich. 2019. Acute toxicity and sublethal effects caused by a commercial herbicide formulated with 2, 4-D on *Physalaemus albonotatus* tadpoles. Wat. Air and Soil Poll. 230(1): 1–15.

Curi. L.M., P.M. Peltzer, A.M. Attademo and R.C. Lajmanovich. 2021. Alterations in gonads and liver tissue in two neotropical anuran species commonly occurring in rice fields crops. Wat. Air Soil Pollut. 232(5): 1–18.

Curi, L.M., A.P.C. Boccioni, P.M. Peltzer, A.M. Attademo, A. Bassó, E.J. León and R.C. Lajmanovich. 2022. Signals from predators, injured conspecifics, and pesticide modify the swimming behavior of the gregarious tadpole of the dorbigny's toad, *Rhinella dorbignyi* (Anura: Bufonidae). Can. J. Zool. 100: 19–27.

Cuzziol Boccioni, A.P., P.M. Peltzer, C.S. Martinuzzi, A.M. Attademo, E.J. León and R.C. Lajmanovich. 2020. Morphological and histological abnormalities of the neotropical toad, *Rhinella arenarum* (Anura: Bufonidae) larvae exposed to dexamethasone. J. Environ. Sci. Health B 56(1): 41–53.

Cuzziol Boccioni, A.P., R.C. Lajmanovich, P.M. Peltzer, A.M. Attademo and C.S. Martinuzzi. 2021. Toxicity assessment at different experimental scenarios with glyphosaye, chrorpyrifos and antibiotics *in Rhinella arenarum* (Anura: Bufonidae) tadpoles. Chemosphere 273: 128475.

Freitas, J., A. Kupsco, G. Diamante, A. Felicio, E. Almeida and D. Schlenk. 2016. Influence of Temperature on the Thyroidogenic Effects of Diuron and Its Metabolite 3,4-DCA in Tadpoles of the American Bullfrog (*Lithobates catesbeianus*). Environ. Sci. Tech. 50(23): 13095–13104.

Gaona, L., F. Bedmar, V. Gianelli, A.J. Faberi and H. Angelini. 2019. Estimating the risk of groundwater contamination and environmental impact of pesticides in an agricultural basin in Argentina. Int. J. Sci. Environ. 16(11): 6657–6670.

Gondim, P.M., J.F.M. Rodrigues and P. Cascon. 2020. Fluctualing asymmetry and organosomatic indices in anuran population in agricultural environments in semi-arid Brazil. Herpetol. Cons. Biol. 15(2): 354–366.

Gosner, K.L. 1960. A simplified table for staging anuran embryos larvae. Herpetodologists' League 16(3): 183–190.

Holderbaum, D.F., M. Cuhra, F. Wickson, A.I. Orth, R.O. Nodari and T. Bøhn. 2015. Chronic responses of *Daphnia magna* under dietary exposure to leaves of a transgenic (Event MON810) Bt-Maize hybrid and its conventional Near-Isoline. J. Toxicol. Environ. Health A 78(15): 993–1007.

Huespe, I., M. Cabagna-Zenklusen, L.M. Curi, P.M. Peltzer, A.M. Attademo, N. Villafane and R.C. Lajmanovich. 2017. Liver melanomacrophages and gluthation s-transferase activity in *Leptodactylus chaquensis* (Anura, Leptodactylidae) as biomarkers of oxidative stress due to chlorpyrifos exposition. Acta Biol. Colomb. 22(2): 234–237.

Junges, C.M., P.M. Peltzer, R. Lajmanovich, A.M. Attademo, M. Cabagna Zenklusen and A. Bassó. 2012. Toxicity of the fungicide trifloxystrobin on tadpoles and its effect on fish-tadpole interaction. Chemosphere 87(11): 1348–1354.

Lajmanovich, R.C., M.T. Sandoval and P.M. Peltzer. 2003. Induction of mortality and malformation in *Scinax nasicus* tadpoles exposed by glyphosate formulations. Bull. Environ. Contam. Toxicol. 70: 612–618.

Lajmanovich, R.C., M. Cabagna, P.M. Peltzer, G.A. Stringhini and A.M. Attademo. 2005a. Micronucleus induction in erythrocytes of the *Hyla pulchella* tadpoles (Amphibia: Hylidae) exposed to insecticide endosulfan. Mutat. Res. 587(1-2): 67–72.

Lajmanovich, R., P. De La Sierra, F. Marino, P.M. Peltzer, A. Lenardón and E. Lorenzatti. 2005b. Determinación de residuos de organoclorados en vertebrados silvestres del litoral fluvial de Argentina. pp. 255–262. en: Aceñolaza, F.G. [ed.]. Temas de la Biodiversidad del Litoral Fluvial Argentino II, 14, INSUGEO, Miscelánea.

Lajmanovich, R.C., A.M. Attademo, P.M. Peltzer and C.M. Junges. 2009. Inhibition and recovery of brain and tail cholinesterases of *Odontophrynus americanus* tadpoles (Amphibia: Cycloramphidae) exposed to fenitrothion. J. Environ. Biol. 30: 923–992.

Lajmanovich, R.C., A.M. Attademo, P.M. Peltzer, C.M. Junges and M.C. Cabagna. 2011. Toxicity of four glyphosate formulations on *Rhinella arenarum* (anura: bufonidae) tadpoles: B-esterases and glutathione S-transferase inhibitions. Arch. Environ. Contam. Toxicol. 60: 681–689.

Lajmanovich, R.C., M. Cabagna Zenklusen, A.M. Attademo, C.M. Junges, P.M. Peltzer and A. Bassó. 2014. Induction of micronuclei and nuclear abnormalities in common toad tadpoles (*Rhinella arenarum*) treated with Liberty® and glufosinate-ammonium. Mutat. Res. 769: 7–12.

Lajmanovich, R.C., C.M. Junges, M.C. Cabagna-Zenklusen, A.M. Attademo, P.M. Peltzer, M. Maglianese, V.E. Márquez and A.J. Beccaria. 2015. Toxicity of *Bacillus thuringiensis* var. *israelensis* in aqueous suspension on the South American common frog *Leptodactylus latrans* (Anura: Leptodactylidae) tadpoles. Environ. Res. 136: 205–212.

Lajmanovich, R.C., A.M. Attademo, P.M. Peltzer, C.M. Junges and C.S. Martinuzzi. 2017. Acute toxicity of apple snail *Pomacea canaliculata*'s eggs on *Rhinella arenarum* tadpoles. Toxin Rev. 36: 45–51.

Lajmanovich, R.C., P.M. Peltzer, C.S. Martinuzzi, A.M. Attademo, A. Basso, M.I. Maglianese and C. Colussi. 2018. B-esterases and behavioral biomarkers in tadpoles exposed to pesticide Pyrethroid-TRISADA®. Toxicol. Environ. Health Sci. 10: 237–244.

Lajmanovich, R.C., P.M. Peltzer, A.M. Attademo, C.S. Martinuzzi, M.F. Simonillo, C.L. Colussi and M. Sigrist. 2019a. First evaluation of novel potential synergistic effects of glyphosate and arsenic mixture on *Rhinella arenarum* (Anura: Bufonidae) tadpoles. Heliyon 5: e02601.

Lajmanovich, R.C., P.M. Peltzer, C.S. Martinuzzi, A.M. Attademo, A. Basso and C. Colussi. 2019b. Insecticide pyriproxyfen (Dragón®) damage biotransformation, thyroid hormones, heart rate, and swimming performance of *Odontophrynus americanus* tadpoles. Chemosphere 220: 714–722.

Lajmanovich, R.C., A.M. Attademo, G. Lener, A.P. Cuzziol Boccioni, P.M. Peltzer, C.S. Martinuzzi, L. Demonte and M.R. Reppeti. 2022. Glyphosate and glufosinate ammonium, herbicides commonly used on genetically modified crops, and their interaction with microplastics: Ecotoxicity in anuran tadpoles. Sci. Total Environ. 804: 150177.

Leâo, T., M. Siqueira, S. Marcondes, L. Franco-Belussi, C. De Oliveira and C.E. Fernandes. 2021. Comparative liver morpghology associated with the hepatosomatic index in five neotropical anuran species. Anat. Rec. 304: 860–871.

León, E., P.M. Peltzer, R. Lorenzón, R.C. Lajmanovich and A.H. Beltzer. 2019. Effect of traffic noise on *Scinax nasicus* advertisement call (Amphibia, Anura). Iheringia Ser. Zool. 109: e2019007.

Oliveira, J.S., A.A. Silva and V. Silva. 2016. Phytotherapy in reducing glycemic index and testicular oxidative stress resulting from induced diabetes: A review. Braz. J. Biol. 77(1): 68–78.

Paradina Fernández, L., R. Brasca, A.M. Attademo, P.M. Peltzer, R.C. Lajmanovich and M.J. Culzoni. 2020. Bioaccumulation and glutathione S-transferase activity on *Rhinella arenarum* tadpoles after short-term exposure to antiretrovirals. Chemosphere 246: 125830.

Peltzer, P.M. 2006. La fragmentación de hábitat y su influencia en la diversidad y distribución de anfibios anuros de áreas ecotonales de los dominios fitogeográficos amazónico y chaqueño. Tesis doctoral, Universidad Nacional de La Plata.

Peltzer, P.M., R.C. Lajmanovich, J.C. Sanchez-Hernandez, M. Cabagna, A.M. Attademo and A. Bassó. 2008. Effects of agricultural pond eutrophication on survival and health status of *Scinax nasicus* tadpoles. Ecotoxicol. Environ. Saf. 70(1): 185–197.

Peltzer, P.M., R.C. Lajmanovich, L.C. Sanchez, A.M. Attademo, C.M. Junges, C.B. Bionda and A. Martino. 2011. Morphological abnormalities in amphibian populations from the mid-eastern region of Argentina. Herpetological Conserv. Biol. 6(3): 432–442.

Peltzer, P.M., R.C. Lajmanovich, A.M. Attademo, C.M. Junges, M.C. Cabagna-Zenklusen, M.C. Repetti and H. Beldoménico. 2013a. Effect of exposure to contaminated pond sediments on survival, development, and enzyme and blood biomarkers in veined tree frog (*Trachycephalus typhonius*) tadpoles. Ecotoxicol. Environ. Saf. 98: 142–151.

Peltzer, P.M., C.M. Junges, A.M. Attademo, A. Bassó, P. Grenón and R. Lajmanovich. 2013b. Cholinesterase activities and behavioral changes in *Hypsiboas pulchellus* (Anura: Hylidae) tadpoles exposed to glufosinate ammonium herbicide. Ecotoxicology 22: 1165–1173.

Peltzer, P.M., R.C. Lajmanovich, A.M. Attademo, C.M. Junges, C.M. Teglia, C.M. Martinuzzi, L. Curi, M.J. Culzoni and H.C. Goicoechea. 2017. Ecotoxicity of veterinary enrofloxacin and ciprofloxacin antibiotics on anuran amphibian larvae. Environ. Toxicol. Pharmacol. 51: 114–123.

Peltzer, P.M., R.C. Lajmanovich, C.S. Martinuzzi, A.M. Attademo, L.M. Curi and M.T. Sandoval. 2019. Biotoxicity of diclofenac on two larval amphibians: Assessment of development, growth, cardiac function and rhythm, behavior and antioxidant system. Sci. Total Environ. 683: 624–637.

Peluso, J., A. Lanuza, C. Pérez Coll and C. Aronzon, Carolina. 2022. Synergistic effects of glyphosate- and 2,4-D-based pesticides mixtures on *Rhinella arenarum* larvae. Environ. Sci. Poll. Res. 29: 1–10.

Rojo, M., D. Álvarez-Muñoz, A. Dománico, R. Foti, S. Rodriguez-Mozaz, D. Barceló and P. Carriquiriborde. 2019. Human pharmaceuticals in three major fish species from the Uruguay River (South America) with different feeding habits. Environ. Poll. 252 (Part A): 146–154.

Saber, S., W. Tito, R. Said, S. Mengistou and A. Alqahtani. 2017. Amphibians as bioindicators of the health of some wetlands in ethiopia. Egypt. J. Hosp. Med. 66: 66–73.

Sánchez, L.C., P.M. Peltzer, R.C. Lajmanovich, A.S. Manzano, C.M. Junges and A.M. Attademo. 2013. Reproductive activity of anurans in a dominant agricultural landscape from central-eastern Argentina. Rev. Mex. Biod. 84: 912–926.

Sánchez, L.C., R.C. Lajmanovich, P.M. Peltzer, A.S. Manzano, C.M. Junges and A.M. Attademo. 2014. First evidence of the effects of agricultural activities on gonadal form and function in *Rhinella fernandezae* and *Dendropsophus sanborni* (Amphibia: Anura) from Entre Ríos Province, Argentina. Acta Herpetol. 9(1): 75–88.

Sandoval, M.T., R. Gaona, L.M. Curi, F.N. Abreliano, R.C. Lajmanovich and P.M. Peltzer. 2022. Anuran heart development and critical developmental periods: A comparative analysis of three neotropical anuran species. Anat. Rec. 2222: 1–15.

Schmitt, C.J. and G.M. Dethloff. 2000. Biomonitoring of Environmental Status and Trends (BEST) Program: selected methods for monitoring chemical contaminants and their effects in aquatic ecosystems. Information and Technology Report USGS/BRD2000–0005, U.S. Geological Survey, Biological Resources Division, Columbia, Missouri, USA. 81 p

Scott, G.R. and K.A. Sloman. 2004. The effects of environmental pollutants on complex fish behaviour: Integrating behavioural and physiological indicators of toxicity. Aquat. Toxicol. 68(4): 369–392.

Séralini, G.E., E. Clair and R. Mesnage. 2014. Republished study: Long-term toxicity of a Roundup herbicide and a Roundup-tolerant genetically modified maize. Environ. Sci. Eur. 26: 14.

Spurgeon, D.J., H. Ricketts, C. Svendsen, A.J. Morgan and P. Kille. 2005. Hierarchical responses of soil invertebrates (earthworms) to toxic metal stress. Environ. Sci. Tech. 39(14): 5327–5334.

Teglia, C., P.M. Peltzer, S.N. Seib, R.C. Lajmanovich, M.J. Culzonia and H.C. Goicoechea. 2017. Simultaneous multi-residue determination of twenty-one veterinary drugs in poultry litter by modelling three-way liquid chromatography with fluorescence and absorption detection data. Talanta 167: 442–452.

Teglia, C.M., F.A. Perez, N. Michlig, M.R. Repetti, H.C Goicoechea and M.J. Culzoni. 2019. Occurrence, distribution, and ecological risk of fluoroquinolones in rivers and wastewaters. Environ. Toxicol. Chem. 38: 2305–2315.

Watson, F.L., H. Schmidt, Z.K. Turman, N. Hole, H. Garcia, J. Gregg, J. Tilghman and E.A. Fradinger. 2014. Organophosphate pesticides induce morphological abnormalities and decrease locomotor activity and heart rate in *Danio rerio* and *Xenopus laevis*. Environ. Toxicol. Chem. 33(6): 1337–45.

Zemp, D.C., C.F. Schleussner, H.M. Barbosa, M. Hirota, V. Montade, G. Sampaio, A. Staal, L. Wang-Erlandsson and A. Rammig. 2017. Self-amplified Amazon forest loss due to vegetation-atmosphere feedbacks. Nat. Commun. 13(8): 14681.

5

Effects of Pollutants on the Endocrine System of Tadpoles

Katharina Ruthsatz[1,*] and *Julian Glos*[2]

1. Introduction

Environmental contamination of aquatic ecosystems by pollutants or xenobiotics is among the major causes of the decline of amphibian populations (Hopkins 2007, Wake and Vredenburg 2008, Hayes et al. 2006, 2010). At all life stages, amphibians are highly sensitive to contaminants due to their permeable skin (Yu et al. 2013, Strong et al. 2017). Although the skin enables gas, water, and electrolyte exchange with the environment (rev. in Quaranta et al. 2009), it also facilitates the passive absorption of pollutants (Willens et al. 2006, Van Meter et al. 2014, 2015).

Amphibian larvae are particularly susceptible to aquatic contamination as they are limited in their capacity for habitat selection (Sanzo and Hecnar 2006). Furthermore, as most tadpoles are omnivorous, feeding items such as biofilm, algae, dead animals, and plants (Dodd 2010), could also allow pollutants to enter the tadpole's body by ingestion in addition to absorption through the skin (Hu et al. 2016).

Many pollutants are known to affect the endocrine system through inhibitory or stimulatory action (Mann et al. 2009, Carr and Patiño 2011), which might lead to profound effects on tadpoles, since growth, development, and metabolism are controlled by hormones (Shi 2000, Tata 2006, Brown and Cai 2007). Any imbalance of these physiological processes during the events leading to metamorphosis might impair amphibian health, performance, and survival in later life stages through carry-over effects (Ruthsatz et al. 2019, 2020a) and contribute to significant population declines (Trudeau et al. 2020).

Beside common aquatic contaminants such as pharmaceuticals, agrochemicals, industrial chemicals, and heavy metals (rev. in Mann et al. 2009, Carr and Patiño 2011, Thambirajah et al. 2019), habitat acidification, road salt exposure, and microplastics pollution have also been identified as potential disruptors of endocrine processes in amphibian larvae. This chapter reviews the

[1] Zoological Institute, Technische Universität Braunschweig, Mendelssohnstraße 4, 38106 Braunschweig, Germany.
[2] Institute of Cell and System Biology, Universität Hamburg, Martin-Luther-King-Platz 3, 20146 Hamburg, Germany.
Email: julian.glos@uni-hamburg.de
* Corresponding author: katharinaruthsatz@gmail.com

endocrine disruptive effects of wide-spread pollutants on amphibian hormonal pathways associated with metamorphosis and discusses interactive effects of multiple pollutants and stressors in the face of global change.

2. Endocrine Control of Amphibian Metamorphosis

Amphibian metamorphosis is a unique example of endocrine regulation of development (Kikuyama et al. 1993, Tata 2006) resulting in profound anatomical modifications (Paris and Laudet 2008, Tribondeau et al. 2021). There are three major changes that take place, the complete resorption of tadpole-specific organs (e.g., tail and gills), the *de novo* development of frog-specific organs (e.g., limbs and lungs), and the remodeling of existing organs into their adult forms (e.g., liver, nervous system, intestine, skin and skeleton reshaping, and ossification, reviewed in Shi 2000, Tata 2006, Brown and Cai 2007). The entire suite of molecular, biochemical, and morphological changes is orchestrated by thyroid hormones (TH, reviewed in Denver 2013). THs initiate gene expression cascades in diverse tissues that lead to cell proliferation, death, differentiation, or migration (Brown and Cai 2007).

2.1 The Drivers of Amphibian Metamorphosis: Thyroid Hormones

Two naturally occurring THs regulate amphibian metamorphosis: 3,5,3′,5′-tetraiodothyronine (T4), commonly known as thyroxine, and 3,5,3′-triiodothyronine (T3) (Shi 2000, Tata 2006). T4 is the precursor for T3 and can be converted to T3, which is the more biologically active form (Frieden 1981, Shi 2000). Thyroid activity increases significantly during larval period, peaks at metamorphic climax, and declines thereafter until an adult level of activity is reached (Dodd and Dodd 1976, Kikuyama et al. 1993, rev. in Denver 2013). Plasma concentration and whole-body content of T3 and T4 increase coincidentally with other measures of thyroid activity such as expression of thyroid hormone receptors (Yaoita and Brown 1990, Tata 2006, Denver 2009).

The endocrine system of amphibians is generally organized like most vertebrates, with hierarchic structures for several endocrine feedback mechanisms (Kloas et al. 2009a, Fig. 1). During metamorphosis, thyroid activity and thus TH levels are regulated via the hypothalamic-pituitary-thyroid (HPT) axis (Shi 2000): the corticotropin-releasing factor (CRF) from the hypothalamus regulates the release of thyroid-stimulating hormone (TSH) from the pituitary (Denver 1997a,b, Carr and Patiño 2011). The release of THs from thyroid follicles is regulated by pituitary TSH (Denver 1997a,b). Circulating THs negatively influence the activity of the hypothalamus and the pituitary (Shi 2000). Thus, the activity of the hypothalamic-pituitary-thyroid axis is regulated by negative feedback.

2.2 TH Action can be Modulated by other Hormones

TH action can be modulated by other hormones. Important modulators include prolactin, glucocorticoid hormones (CRF, adrenocorticotropin, aldosterone, and corticosterone (CORT)), and gonadal steroids (testosterone and estradiol) (Gray and Janssens 1990). In the growing larva, the pituitary hormone prolactin as well as gonadal steroids are suggested to enhance larval growth while blocking the actions of THs on metamorphosis (Etkin 1968, White and Nicoll 1981, Denver 1997, Beachy et al. 1999, Huang and Brown 2000). In contrast, glucocorticoids are known to potentiate TH-induced metamorphosis (Kikuyama et al. 1993, Denver 1997b, 1998, Glennemeier and Denver 2002).

2.3 Cross Talk between Endocrine Axes

The corticosteroids CORT and aldosterone are the primary stress hormones in amphibians and also play important roles in metamorphosis. Perception of environmental stressors by the central nervous

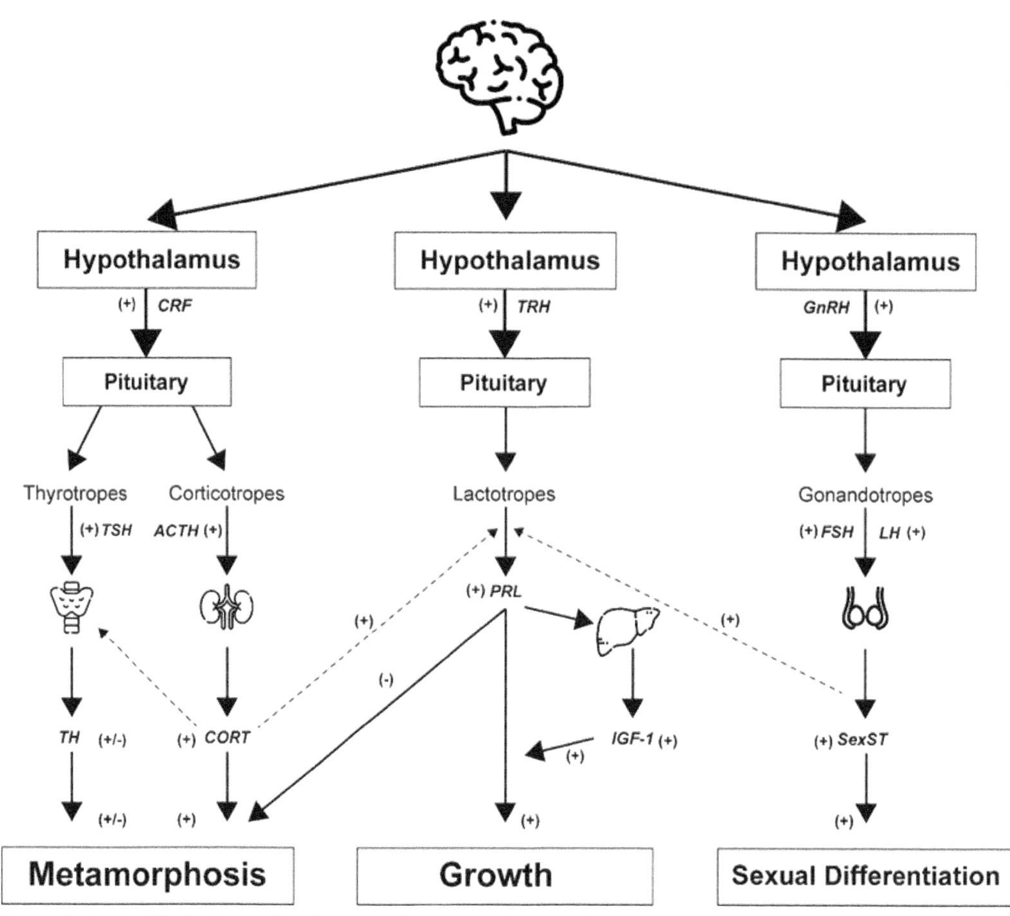

Figure 1. A simplified scheme showing the endocrine regulation and hormonal crosstalk of amphibian metamorphosis, growth, sexual differentiation, and stress in tadpoles. Dashed lines: Cross talks between endocrine axes. CRF: Corticotropin-releasing factor. TRH: thyrotropin-releasing hormone. GnRH: gonadotropin-releasing hormone. TSH: thyroid-stimulating hormone. ACTH: adrenocorticotropic hormone FSH: follicle-stimulating hormone. LH: luteinizing hormone. TH: thyroid hormones (i.e., T3 and T4). CORT: corticosteroids (i.e., corticosterone and aldosterone). PRL: Prolactin. IGF-1: insulin-like growth factor 1. SexST: sex steroids. Note that feedback regulation at the hypothalamus and pituitary by THs, CORT, and SexST exists, but is not depicted in the figure. For further details, refer to documented actions of endocrine disrupting chemicals in the main text (Denver 1997, Shi 2000, Tata 1998, 2006, Kloas et al. 2009a, Ishizuya-Oka et al. 2010, Trudeau et al. 2020).

system activates the hypothalamic–pituitary–interrenal axis, which is the neuroendocrine stress axis in amphibians (Sapolsky 2002, Dantzer et al. 2014, Rollins-Smith 2017). As a result, the CRF from the hypothalamus regulates the release of adrenocorticotropin from the pituitary (Denver 1998, 2013) and also stimulates TH production, since the CRF controls both the thyroid and interrenal axes. The latter produces corticosteroid stress hormones which can synergize with THs and promote metamorphosis (Glennemeier and Denver 2002, Kulkarni and Buchholz 2012).

In addition to these two main axes regulating amphibian metamorphosis, the hypothalamic-pituitary-gonadal (HPG) axis plays the major role in sexual differentiation during larval development (Kloas and Lutz 2006), which is driven by sex steroid hormones (Hayes 1998). However, it is possible that THs might also be involved in sexual differentiation and sex steroids could inhibit TH action (Hayes 1997, rev. in Hayes 1998). Further, liver and pancreatic hormones are also important for many aspects of growth, metabolism, and metamorphosis (Flood et al. 2013, Duarte-Gutermann et al. 2014).

Overall, amphibians appear to be particularly susceptible to disruptions in development, growth, and metabolism, because metamorphosis is dependent on THs, their complex feedback mechanisms, and crosstalk with other hormonal axes, all of which are under complex neuroendocrine control and susceptible to disruption by environmental factors.

3. Endocrine Disrupting Chemicals

Endocrine disruptors are exogenous compounds that alter functions of the endocrine system in animals at several levels (Kloas et al. 2009a). Since the endocrine system is critical to organisms as it plays a role in development, reproduction, metabolism, and immune system functioning, endocrine disruptors are considered as environmental hazards that cause adverse health effects (Vos et al. 2000). The most well-known endocrine disruptors are agrochemicals, industrial chemicals, pharmaceuticals, heavy metals, and personal care products in domestic sewage (rev. in Mann et al. 2009, rev. in Thambirajah et al. 2019, rev. in Trudeau et al. 2020). The main sink for endocrine disrupting pollutants are surface waters (Kime 1998, Kloas 2002). As a large proportion of the amphibian life cycle occurs in ponds, streams, and temporary pools that are often associated with agricultural or urban areas receiving endocrine disrupting pollutants, amphibians at all life stages are most endangered (Carr and Patiño 2011).

Tadpoles appear to be particularly susceptible to endocrine disruption because metamorphosis is the classical and unique example of endocrine regulation of development by the hypothalamic-pituitary-thyroid axis (HPT axis, Kikuyama et al. 1993, Shi 2000, Tata 2006). Beside the HPT axis, pollutants might also disrupt the endocrine axes regulating the reproductive system (i.e., the hypothalamic-pituitary-gonadal axis, Hayes et al. 2011, Li et al. 2017) and the brain-stress axis (i.e., the hypothalamic-pituitary-interrenal axis, e.g., Hayes et al. 2006, Burraco and Gomez-Mestre 2016) with consequences for multiple morphological and physiological endpoints (Tyler et al. 1998, Slaby et al. 2019, Fig. 1). Therefore, endocrine disruptors alter not only metamorphosis, but also growth (Glennemeier and Denver 2002, Sullivan and Spence 2003, Relyea 2004, Lanctôt et al. 2014), metabolism (Strong et al. 2017, Ruthsatz et al. 2019), behavior (Narayan et al. 2013, Polo-Cavia et al. 2016) and aspects of sex differentiation (Hayes et al. 2011, Orton et al. 2018). These effects are mainly driven by direct interferences with the hormone receptors or by indirect impacts on synthesis and bioavailability of hormones (rev. in Celino-Brady et al. 2021, Fig. 1) leading to inhibition or overstimulation of production pathways.

Endocrine disruptors of human origin are known to potentially contribute to the worldwide decline of amphibian populations (Carey and Bryant 1995, Hayes et al. 2010, Rohr et al. 2011). In addition, abiotic and biotic environmental stressors of which some are associated with anthropogenic global change affect endocrine processes in wildlife (desiccation: Gervasi and Foufopoulos 2008, UV-radiation: Belden et al. 2003, temperature variation: Narayan et al. 2012, disease: Gabor et al. 2015) and can therefore be considered as endocrine disruptors (Denver 1997b, 2009, 2021, Rollins-Smith 2017, Ruthsatz et al. 2018a, 2019, 2020a,b,c).

3.1 Pharmaceuticals

The extensive use of pharmaceuticals by humans and also in agriculture and aquaculture has led to environmental contamination. Following their use in humans and animals, some pharmaceuticals and their metabolites are excreted in urine in a biologically active form. Both human waste streams and animal effluent containing these active compounds can reach surface waters and have been widely detected in ground waters, rivers, and lakes (Jobling et al. 1998, Belfroid et al. 1999, Aris et al. 2014). These bioactive compounds can disrupt the endocrine systems of aquatic organisms such as tadpoles and affect their somatic and sexual development (Kloas 2002, Lintelmann et al. 2003, Armstrong et al. 2015). Most important pharmaceutical endocrine disrupting chemicals are (anti)thyroidal compounds that impair the hypothalamic-pituitary-thyroid axis (HPT axis) and affect tadpole development. In addition, (anti)estrogenic or (anti)androgenic substances can impair the functioning of the hypothalamic-pituitary-gonadal axis and in particular, sex differentiation and gonadal development (rev. in Aris et al. 2014, rev. in Säfholm et al. 2014, rev. in Ziková et al. 2017, rev. in Thambirajah et al. 2019).

3.1.1 Thyroid Hormones and Thyroid-Inhibiting Agents

Thyroid hormones are widely prescribed as drugs treating hypothyroidism and have been detected in biologically active concentrations in wastewater (Roterman et al. 2014, rev. in Thambirajah et al. 2019) with consequent effects on tadpole development and metamorphosis. A transcriptomic study by Maher et al. (2016) showed that environmentally relevant, albeit low concentrations of TH in water may be sufficient to affect metamorphosis in *Lithobates catesbeianus* tadpoles. Ruthsatz et al. (2018a,b, 2019a, 2020a,b,c) performed a series of experiments on *Rana temporaria* and *Xenopus laevis* showing direct effects of TH on anuran development. High TH levels in water lead to an acceleration of larval development and therefore an early metamorphosis, and to smaller and lighter tadpoles at metamorphosis (Coady et al. 2010, Miyata and Ose 2012, Ruthsatz et al. 2018a,b, 2019). These variables are known to affect individual fitness in tadpoles (Smith 1987, Berven 1990, Beck and Congdon 2000).

As expected, drugs for treating hyperthyroidism in humans, i.e., the excessive production of thyroid hormones, cause opposite effects on tadpole development. 6-Propylthiouracil (PTU) and ethylenethiourea (ETU) inhibit thyroid peroxidase, the enzyme that synthesizes TH, and are known to cause aberrant thyroid histology and delay or arrest of metamorphosis in a number of model tadpole species, such as *X. laevis*, *X. tropicalis* and *Rana rugosa* (Doerge and Takazawa 1990, Opitz et al. 2006, rev. in Thambirajah et al. 2019). Methimazole is another goitrogen and anti-thyroid drug that affects TH signaling by inhibiting thyroid peroxidase. Exposure of *X. laevis* tadpoles to methimazole caused decreased metamorphic rate at $\geq$ 20 mg/l (Coady et al. 2010) and $\geq$ 25 mg/l (Degitz et al. 2005), and also aberrant thyroid morphology at $\geq$ 20 mg/l (Coady et al. 2010) and $\geq$ 12.5 mg/l (Degitz et al. 2005).

Sodium perchlorate ($NaClO_4$) is a compound commonly used in fireworks and rocket fuels but has therapeutic effects by blocking iodine uptake before the administration of iodinated contrast agents in patients. $NaClO_4$ is an environmentally relevant and known endocrine disruptor in the TH system. As a goitrogen, $NaClO_4$ inhibits TH synthesis (Ortiz-Santaliestra and Sparling 2007) and consequently affects tadpole development and inhibits metamorphosis (Tietge et al. 2005). A number of experimental studies with different frog species (*L. sylvatica*, *R. temporaria*, *X. laevis*) have shown that $NaClO_4$ in water usually slows down tadpole development and delays metamorphosis, often associated with an increase in size and weight at metamorphosis (Bulaeva et al. 2015, Ruthsatz et al. 2018a,b, 2019). Sodium perchlorate also affects several pathways of growth and development. These include reducing heart rate (Ruthsatz et al. 2020c), reducing metabolic rate and improving body condition when compared to controls (Ruthsatz et al. 2018a,b, 2020a,c).

3.1.2 Pain Relievers and Anti-Depressants

Pain relieving pharmaceuticals are widely used by humans and are found in physiologically relevant concentrations in wastewater and waters where amphibians breed (Sui et al. 2015, Ebele et al. 2017). However, studies on the effect of pain relievers on the amphibian hormone system, and on amphibians in general show inconsistent results.

Ibuprofen is a widely used non-steroidal anti-inflammatory drug (NSAID) and acts in humans through the inhibition of the prostaglandin synthesis and other pathways (Rainsford 2009, rev. in Thambirajah et al. 2019). Analyses of transcriptomes of bullfrog tadpoles (*L. catesbeianus*) showed effects on TH-regulated processes, but direct fitness-relevant parameters were not investigated (Veldhoen et al. 2014). Chronic exposure to another NSAID, diclofenac, decreased TH-regulated larval development, growth, and body condition, and had further morphological and behavioral effects on tadpoles of two species from Argentina (*Trachycephalus typhonius* and *Physalaemus albonotatus*) (Peltzer et al. 2019). Acetaminophen ("Paracetamol"), one of the most widely used pain relievers, increases swimming activity in toad tadpoles (*Bufo americanus*). High concentrations of the drug also caused mortality, but with no effect on growth of tadpoles (Smith and Burgett 2005).

Carbamazepine is a pharmaceutical drug with anticonvulsant, analgesic, and mood-stabilizing properties. Tadpoles of *Rana dalmatina* and *Bufo bufo* exposed to 50 µg/l carbamazepine in water had impaired spleen development (i.e., decreased spleen size and increased spleen pigmentation). However, larval developmental time, post-metamorphic body mass, and sex ratios were not affected (Bókony et al. 2020). Similarly, no effect on growth and development was found in *Limnodynastes peronii* tadpoles (Melvin et al. 2014).

Sertraline ("Zoloft") and Fluoxetin are both selective serotonin reuptake inhibitory agents that are prescribed as antidepressant drugs. They have been detected in relevant concentrations in untreated sewage and wastewater (Metcalfe et al. 2010, Schultz et al. 2010, Styrishave et al. 2011), and are suspected of causing developmental abnormalities in some aquatic organisms (e.g., Brooks et al. 2003, Fong et al. 2005, Bringolf et al. 2010). In experimental studies, however, these drugs had only minor or inconclusive effects on tadpoles. They did not affect time to metamorphosis in *L. sylvatica* (Sertraline; Carfagno and Fong 2014) and *X. tropicalis* (Berg et al. 2013), but delayed development in *L. pipiens* tadpoles (fluoxetine, Foster et al. 2010) and accelerated *X. laevis* development (sertraline, Connors et al. 2009). Tadpoles exposed to sertraline and fluoxetine have shown reduced size at metamorphosis in *X. laevis* (Conners et al. 2009) and *L. pipiens* (Foster et al. 2010), but other hormone-related variables such as thyroid histology or gonadal sex determination were not affected (*X. tropicalis*, Berg et al. 2013). These contrasting results illustrate the uncertainty in understanding how these drugs may impact hormonally controlled aspects of tadpole development and in particular how they might affect natural populations of frogs.

3.1.3 Oral Contraceptive Agents

Humans consume and subsequently excrete a considerable amount of sex hormones (Vethaak et al. 2005, Pauwels et al. 2008), mainly for oral contraception and different hormonal therapies. These hormones are then incompletely removed in sewage plants and end up in surface and ground water, and in sediments particularly where sewage and manure are used as fertilizers (Belfroid et al. 1999, Aris et al. 2014). Sex hormones are also used in considerable amounts in agri- and aquaculture (e.g., Liu et al. 2012) to improve productivity in livestock (Gadd et al. 2010) and fish (rev. by Aris et al. 2014). Due to their relatively high lipophilicity and persistence, many sex hormones can bioaccumulate in aquatic organisms (Mazotto et al. 2008). Accordingly, sex hormones are found in relevant concentrations in potential breeding sites of amphibians (Kolodziej et al. 2004, Viglino et al. 2008, Vulliet et al. 2008, Vulliet and Cren-Olive 2011).

These sex hormones are steroid hormones that include (anti-) estrogens, (anti-) androgens and (anti-) gestogens. Although the sex of amphibians is genetically determined (Eggert 2004), the gonads are still undifferentiated in the early larval phase and potentially can develop into either testis or ovaries (Piprek and Kubiak 2014). Sex hormones interact with steroid hormone receptors of the endocrine system and impair the sexual differentiation and gonadal development of vertebrates, including amphibians, and accordingly lead to skewed sex ratios at metamorphosis (rev. in Hayes 1998, rev. in Aris et al. 2014, rev. in Säfholm et al. 2014, rev. in Zikova et al. 2017).

17α-ethinylestradiol (EE2) is a synthetically stabilized estrogen and is the main agent in most modern oral contraceptives. Additionally, EE2 is prescribed in other hormonal therapies (e.g., for menopausal syndrome), cancer therapies and for osteoporosis, but is also used to increase productivity in livestock and in aquaculture (rev. in Ying et al. 2002). When amphibians are exposed to EE2 in their larval phase, it can lead to sex reversals and mixed sex, and to anatomical impairments of the gonads. Generally, EE2 has a feminizing effect, which can lead to skewed sex ratios in amphibian populations (rev. in Aris et al. 2014, rev. in Zikova et al. 2017). Sex reversal from genetic males to phenotypic females have been demonstrated in the model species *X. laevis*

(Hecker et al. 2004, Hogan et al. 2008, Tamschick et al. 2016a), *X. tropicalis* (Gyllenhammar et al. 2009) and *L. pipiens* (Hogan et al. 2008) as well as in a number of non-model species such as *Hyla arborea* and *Bufo viridis* (Tamschick et al. 2016a). Mixed sexes as a result of EE2 treatments were found in *X. laevis* (Hecker et al. 2004, Tamschick et al. 2016a), *H. arborea* and *B. viridis* (Tamschick et al. 2016a). EE2 leads to gonadal impairment and underdeveloped gonads (Cevasco et al. 2008, Gyllenhammar et al. 2009, Piprek et al. 2012, Tamschick et al. 2016b), often to a degree that both male and female animals have reduced fertility or are sterile. Apart from these effects on the sexual differentiation, some studies have shown that exposure to EE2 leads to delays of larval development and metamorphosis (in *L. pipiens* and *X. laevis*, Hogan et al. 2008, Tompsett et al. 2012, 2015).

Natural (pro)gestagens are steroid hormones that activate the progesterone receptor and are important in the maintenance of pregnancy in humans (Norris and Carr 2013) and gamete maturation (in amphibians: Ogawa et al. 2011, Arias Torresa et al. 2016). Synthetic progestogens are used in humans for contraception, menopausal hormone therapy, transgender hormone therapy and cancer treatment. In general, gestagens impair the reproductive functions in amphibians at environmentally relevant exposure concentrations (rev. in Säfholm et al. 2014, rev. in Ziková et al. 2017). Also, females seem to be more affected than males by exposure to gestagens.

The gestagen levonorgestrel (LNG) has been shown to negatively affect reproduction in amphibians, both after larval and adult exposure (Kloas et al. 2009a, Kvarnryd et al. 2011, Säfholm et al. 2012). Female *X. tropicalis*, for example, exposed to LNG during the tadpole stage became sterile. These females completely lacked oviducts and the number of maturing oocytes was considerably decreased (Kvarnryd et al. 2011). In the same study, no effects on testicular development, sperm number or sperm fertility were observed in males of *X. tropicalis*. Ziková et al. (2017) confirmed direct impact of gestagens, i.e., LNG and progesterone (P4), on the gonadal differentiation in larval *X. laevis*, again with a greater effect in females than in males. Also, in *X. laevis*, larval exposure to LNG did not change sex ratio but disrupted male gonadal development. LNG at higher concentrations not only interferes with the hypothalamic-pituitary-gonadal axis but also the hypothalamic-pituitary-thyroid axis. In *X. laevis*, accordingly, LNG delayed or arrested metamorphosis and revealed tailed juvenile frogs (Kloas et al. 2009a, Lorenz et al. 2011, 2016).

In conclusion, the active substances in oral contraceptives, estrogens and gestagens, occur ubiquitously in aquatic ecosystems and have the potential to act as endocrine disruptors and impair sexual and somatic development in larval amphibians. The effects of these endocrine disruptors, including skewed sex ratios, gonadal malformations, sterility, and delays of metamorphosis, might have severe negative effects on amphibian populations and cause imbalances in the ecosystem.

3.2 Agrochemicals

Agrochemicals encompass a large variety of different pesticides and fertilizers. Among all aquatic contaminants introduced to the environment, agrochemicals play the most important role in the Anthropocene due to the diversity of active ingredients, formulations, and mixtures in combination with their widespread use, persistence, transport, and bioaccumulation in wildlife (Hayes et al. 2017). Over the last 75 years, the use of agrochemicals has increased dramatically (Sharma et al. 2019). Furthermore, agrochemicals in the form of pesticides and fertilizers are being applied in greater varieties, combinations, and to greater extent than ever before, and thus, represent a significant suite of pollutants (Mann et al. 2009, EUROSTAT 2019).

There is accumulating but also controversial evidence that some agrochemicals act as endocrine disruptors affecting amphibian development, reproduction, disease transmission, sexual differentiation as well as growth and therefore pose a threat to overall health, recruitment, and survival. Agrochemicals are receiving increasing attention as a potential cause of amphibian declines, acting singly or in combination with other stressors (Hayes et al. 2010, Relyea and Mills 2001, Rohr et al. 2011, Trudeau et al. 2020).

Amphibians in general and tadpoles in particular are non-target organisms of pesticides but are exposed to all kinds of agrochemicals since their aquatic habitats are the major sinks of aquatic contaminants. Agrochemicals and their transformation products are present in surface waters worldwide (de Souza et al. 2020), especially temporary ponds in which amphibians breed are often associated with agricultural landscapes receiving respective chemicals. Pesticides and fertilizers are mostly supplied as formulations based on a variety of active ingredients. Consequently, this section focuses on commonly used compounds to describe endocrine disruptive effects associated with agrochemicals.

3.2.1 Pesticides

Pesticides are natural or synthetic chemicals employed in various agricultural practices to control pests, weeds and diseases in crops. In the course of time, pesticides get accumulated in plant parts, water, soil, air and biota (de Souza et al. 2020). On global scale, the pesticide groups that are most frequently applied are the herbicides, fungicides, and insecticides.

Herbicides. Herbicides are synthetic chemical substances that control growth and death of undesirable plants, such as weeds. These compounds act selectively on specific weed species or non-selectively on all plant material with which they come into contact. Acetochlor (Arregui et al. 2010), atrazine, and glyphosate are amongst the most widely used herbicides on global scale which also have been detected in surface waters and have been shown to disrupt endocrine function in tadpoles (rev. in Trudeau et al. 2020, Shepard et al. 2004, Mann et al. 2009, Freitas et al. 2016, Gupta 2017, de Souza et al. 2020).

Acetochlor modulates the HPT axis in amphibians by affecting the expression of thyroid hormone related genes (Helbing et al. 2006, Veldhoen and Helbing 2001). The sensitivity to this modulation is suggested to be tissue specific (Thambirajah et al. 2019). Acetochlor treatment resulted in accelerated developmental rates in tadpoles of *Lithobates pipiens* (Cheek et al. 1999) and *Xenopus laevis* (Crump et al. 2002). However, a recent study in *Boana pardalis* tadpoles could not detect a significant alteration in developmental rate related to acetochlor exposure (Moutinho et al. 2020).

In contrast to acetochlor, evidence on an endocrine disruptive effect of the widely used herbicide atrazine during amphibian development is still controversial and is thus, suggested to be species-specific. However, several studies have not found evidence that atrazine disrupts amphibian development (Carr et al. 2003, Coady et al. 2005, Freeman et al. 2005, Orton et al. 2006, Kloas et al. 2009b) even in species where effects were demonstrated in earlier work (Langlois et al. 2010). Other studies demonstrated either a delaying or accelerating effect of atrazine on developmental rate (e.g., *Xenopus laevis*: Sullivan and Spence 2003, Freeman and Rayburn 2005, *Ambystoma tigrinum*: Larson et al. 1998, Rana clamitans: Coady et al. 2004). McMahon et al. (2017) showed a dysregulating effect on corticosterone release in tadpoles of the Cuban treefrog (*Osteopilus septentrionalis*) as intermediate concentrations of atrazine (10.2 and 50.6 μg/L) tended to lower corticosterone, whereas the lowest (0.1 μg/L) and highest atrazine concentrations (102 μg/L) elevated whole-body corticosterone. Further investigations are needed to verify these effects. Depending on developmental stage and species, different herbicide formulations containing glyphosate such as Roundup® and VisionMax® lead to metamorphic alterations in tadpoles of several species (acceleration: *Rana cascadae*, Cauble and Wagner 2005, delay: *Lithobates pipiens*, Williams and Semlitsch 2010, *Anaxyrus americanus, Lithobates pipiens*, Howe et al. 2004, *Lithobates sylvaticus*, Navarro-Martin et al. 2014, but not Lanctot et al. 2013). Delayed time to metamorphosis can either be caused by disruption of the thyroid axis or energy consumptive detoxification processes (Wagner et al. 2013). A disruption of the HPT axis could be confirmed for glyphosate-exposed tadpoles of the northern leopard frog (Howe et al. 2004). However, effects of glyphosate-based herbicides depend on the specific formulations due to different effects of additional ingredients such as the widely used surfactant polyethoxylated tallowamine (POEA, Mann et al. 2009, Thambirajah et al. 2019).

Insecticides. Insecticides are chemical products used to repel or kill insects. These substances have mechanisms of action that vary according to the type of insecticide (De Souza et al. 2020). Insecticides are claimed to be a major factor behind the increase in the 20th-century's agricultural productivity (Van Emden and Peakall 1996). Nevertheless, the vast majority of insecticides have the potential to significantly alter ecosystems and affect wildlife (Relyea 2005, Gupta 2019).

The most commonly used insecticides are the carbamates, organophosphates, and neonicotinoids (Cressey 2017, EPA 2021, Jeschke et al. 2011). Carbaryl is a carbamate insecticide (Gunasekara et al. 2008) and can modify acetylcholinesterase, which affects neurotransmission (Rosman et al. 2009). Depending on the timing of exposure, an endocrine disruptive effect of carbaryl on amphibian metamorphosis could be detected. Boone et al. (2013) proved carbaryl as an endocrine disruptor in green frog (*Lithobates clamitans*) by delaying development since expression of TH-responsive thyroid hormone receptor genes were altered. However, effects on developmental rate were not consistent across various studies (acceleration: *L. clamitans*, Boone 2008, delay: *Lithobates spehocephalus*, Bridges et al. 2000, no effect: *L. clamitans*, Boone et al. 2013).

Dichlorodiphenyltrichloroethane (DDT) has been the most commonly used insecticide since its introduction in the early 1940s. However, DDT and its metabolites persist in the environment and was therefore banned in most countries by 2004. Wildlife (Guillette et al. 2006, Matsushima 2018) including amphibians at all developmental stages (e.g., Clark et al. 1998, Arukwe and Jenssen 2005, Mortensen and Arukwe 2006) have been affected by the disrupting effect of DDT on different endocrine pathways such as the thyroid axis. Across different taxa, DDT and its metabolites resulted in hypothyroidism by various modes of action (e.g., Goldner et al. 2013, Liu et al. 2014, Yaglova and Yaglov 2014). In larvae of the common frog (*Rana temporaria*), exposure to a DDT metabolite (dichlorodiphenyldichloroethylene, DDE) caused a down-regulation on thyroid hormone receptor expression (Mortensen and Arukwe 2006). However, data on endocrine disruptive effects of DDT on amphibian larvae are underrepresented.

Since their introduction in the early 1990s, neonicotinoids have become the most commonly used type of insecticides (Jeschke et al. 2011) and imidacloprid in particular. A recent study demonstrated a decrease in thyroid hormone receptor expression during exposure to imidacloprid, but no effects on larval survival, body weight, and body length in larvae of the Northwestern salamander (*Ambystoma gracile,* Danis and Marlatt 2021). In tadpoles of *L. sylvaticus*, developmental rate was delayed by imidacloprid exposure (Robinson et al. 2017) and corticosterone concentrations were decreased in thiamethoxam-exposed tadpoles (Gavel et al. 2019), indicating that the HTP and HPI axes might be altered by neonicotinoids.

Fungicides. Fungicides are chemical substances applied to plants or seeds to prevent fungal infections. In general, they damage the cellular membrane, affect specific metabolic processes, inactivate enzymes, and/or alter processes related to energy production and respiration in the target fungi (Gupta 2017). The use of fungicides has significantly increased over the past few years (EUROSTAT 2019) and results of their repeated applications in agricultural practices, fungicides are frequently detected in amphibian habitats and tissues (Wightwick et al. 2012, Knäbel et al. 2014, Smalling et al. 2015, Battaglin et al. 2016, Bernabó et al. 2017).

Several fungicides have been directly or indirectly demonstrated to disrupt thyroid, interrenal, and gonadal functions in tadpoles. For example, after exposure to prochloraz, metamorphosis was delayed in larval *R. temporaria* through changes to T3 concentrations and thyroid hyperplasia (Brande-Lavridsen et al. 2010). Chlorothalonil elevated corticosterone levels but not linearly in *Rana sphenocephala* (McMahon et al. 2011). In the Italian tree frog (*Hyla intermedia*), pyrimethanil induced underdevelopment of ovaries and interfered with normal sexual differentiation (Bernabó et al. 2017) and pyrimethanil and tebuconazole have been shown to affect metamorphosis success (Bernabó et al. 2016). Both these studies suggest the endocrine disruption potential of these fungicides.

3.2.2 Fertilizers

Fertilizers are chemical substances that are used to enrich soils with specific nutrients to supply plant growth. In recent years, a great deal of attention has been focused on the use and effects of fertilizers (Mann et al. 2009). Commonly used agricultural fertilizers are urea, ammonium nitrate, calcium nitrate, and sodium nitrate. All fertilizers break down into their constituent components.

Fertilizers reach ground and surface waters via subsequent runoff after rainfall and result in elevated concentrations of inorganic nitrogenous compounds (i.e., NH_3/NH_4^+, NO_2^-, NO_3^-). All these chemicals can contribute to impaired larval development, mortality, decreased feeding activity and disruption of physiological activity, amongst other adverse effects (rev. in Van Meter et al. 2019, Guilette and Edwards 2005, Ruthsatz et al. 2022b, 2023a, Zamora-Camacho et al. 2023).

Nitrogenous fertilizers, in particular, can interfere with the HTP axis of amphibians since exposure to nitrate alters developmental rate during metamorphosis (e.g., Oldham et al. 1997, Ortiz et al. 2004, Edwards et al. 2006, Ortiz-Santaliestra and Sparling 2007, Wang et al. 2015, Ruthsatz et al. 2022). However, little attention has been paid to the molecular mechanisms through which nitrate disrupts the HTP axis in amphibians (Carr and Patiño 2011). Nitrate has been suggested to affect the HTP axis by inhibiting TH synthesis through competing with iodine uptake in thyroid follicles (rev. in Trudeau et al. 2020). When tadpoles were exposed to nitrate, Wang et al. (2015) demonstrated that a delayed metamorphosis was related to reduced T3 and T4 levels in tadpoles of the Chinese toad (*Bufo gargarizans*), whereas Xie et al. (2019) found an increase in developmental rate. Edwards et al. (2006), however, could not find any effect of nitrate exposure on T4 levels in tadpoles of *Bufo terrestris*. Consequently, the effects and sensitivity of tadpoles to nitrate seem to be species-specific (Lenuweit 2009). Ruthsatz et al. (2023a) could demonstrate an increase in glucocorticoid hormone levels (CORT) in response to nitrate exposure along with decreasing effects on developmental rate in larvae of the European common frog (*R. temporaria*). Recent studies suggest that nitrate does not interact with T4 but disrupts essential proteins and affects transcription and mRNA stability (Poulsen et al. 2018). In contrast to studies mainly carried out under laboratory or mesocosm conditions, studies under more natural conditions have disputed the notion that nitrogenous fertilizers persist in surface waters where amphibians breed. If so, this suggests that nitrate does not act as an endocrine disruptor in tadpoles in nature (rev. in Trudeau et al. 2020).

3.3 Industrial Chemicals

A multitude of substances that are the product or the by-product of industrial processes contaminate air, soil and water. Many of these substances are known as endocrine disruptors in amphibians and other taxa, mainly by influencing the TH-driven amphibian metamorphosis and subsequently affecting individual fitness and population dynamics (reviewed in Croteau et al. 2008, rev. in Thambirajah et al. 2019). Here, we mainly focus on heavy metals, flame retardants and plasticizers.

3.3.1 Heavy metals

Heavy metals, i.e., metals with high densities and high atomic weights, are natural elements occurring in a variety of chemical compounds in the environment. They are used in a multitude of industrial processes and are emitted into the environment in considerable concentrations (Nriagu and Pacyna 1988, Nriagu 1989). When heavy metals contaminate amphibian breeding waters, either directly via industrial run-off or by indirect deposition via air or soil (e.g., Garnaud et al. 1999), they can have a range of effects on amphibian larvae, from acute toxicity in high concentrations to sublethal effects on processes that are regulated by the endocrine axes, subsequently impairing development and metamorphosis (Thambirajah et al. 2019).

Cadmium (Cd)—Cadmium is found in environmentally relevant concentrations in terrestrial and aquatic ecosystems due to human activities, such as urban air pollution from medicinal and other waste incinerators, car exhausts, tire wear on asphalt, mining, industrial dumping, and the use of industrial sludge and fertilizers to landfills and agricultural fields (Eisler 1985, Garnaud et al. 1999). It has no known biological function in organisms and is known to be very toxic to aquatic and terrestrial wildlife (Eisler 1985), including tadpoles. Accordingly, environmentally relevant concentrations of Cd increase mortality in tadpoles of a number of amphibians such as *Bufo americanus* (James and Little 2003), *B. gargarizans* (Sun et al. 2018), *Microhyla fissipes* (Hu et al. 2019), *Pelophylax ridibunda* (Loumbardis et al. 1999) and *X. laevis* (Sharma and Patiño 2009). Also, Cd appears to be toxic at lower doses when exposed together with other metals (Lead, Zinc, in *Rana luteiventris*; Lefcort et al. 1998) or at high pH (Lu et al. 2021).

A number of studies showed sublethal effects of Cd in lower concentrations on tadpole development, presumably by disrupting the HPT axis. The magnitude of these effects generally depends on Cd-concentration, exposure time and the developmental stage to which the tadpole was exposed. In transcriptomic analyses, Sun et al. (2018) showed that tadpoles of *B. gargarizans* exposed to high-dose Cd exhibited a significant decrease in deiodinase (Dio2) and thyroid hormone receptor (TRβ) mRNA levels in late larval development. Accordingly, less tadpoles reached metamorphosis under high Cd in *B. americanus* (James and Little 2003), *X. laevis* (Sharma and Patiño 2009), *B. gargarizans* (Sun et al. 2018), *M. fissipes* (Hu et al. 2019), and metamorphosis was delayed in some species (*Rana zhenhaiensis*, Lu et al. 2021; *R. luteiventris*, Lefcort et al. 1998). These effects of Cd can be exacerbated by interactive effects with, e.g., estradiol (E2) (Sharma and Patiño 2009) or pH (Lu et al. 2021). Also, in *B. gargarizans*, histopathological changes of the thyroid gland showed that follicular cell hyperplasia and malformation were induced by high Cd concentrations (100 and 500 µg/l; Sun et al. 2018), and skeletal development was delayed. In contrast to these effects on the HPT axis, Cd was not strongly estrogenic, as testicular histology and population sex ratios were unaffected by Cd-exposure in *X. laevis* tadpoles (Sharma and Patiño 2009).

Copper (Cu)—Cu naturally and ubiquitously occurs in the environment and is an element that is essential for many biochemical and physiological processes in aerobic organisms (O'Dell 1990, Grosell 2011). Cu accumulates in aquatic systems due to numerous anthropogenic activities including industrial discharges, cement production, soil disturbances, urban and agricultural runoff, mining operations, erosion of automobile brake linings, water pipes and plant and algae control (Eisler 1998). Cu levels can be from 0.2 to 30 µg/L in natural waters, and up to 200 µg/L in heavily contaminated sites (Legret and Pagotto 1999, Grosell 2011), and it can be toxic to some animals even at concentrations as low as 5 µg/L (Brix et al. 2011). In larval amphibians, studies on Cu effects yielded somewhat ambiguous results, depending on Cu concentration, amphibian species, population, and exposure time. In general, it is known to cause mortality and retarded growth, and is presumed to affect the HPT axis leading to a delay of metamorphosis and morphological abnormalities in the thyroid system (Croteau et al. 2008, Thambarajah et al. 2019).

Cu at relevant environmental concentrations reduces tadpole survival and impairs growth rate in a wide variety of species from different phylogenetic lineages and geographic origin, but not in all species or populations (rev. in Croteau et al. 2008, rev. in Thambarajah et al. 2019, rev. in Azizishirazi et al. 2021). Disrupting effects on the HPT axis and subsequent sublethal effects of Cu on tadpole development and metamorphosis are also known from a multitude of species. As a consequence of Cu exposure, metamorphosis was delayed in tadpoles for example of *Lithobates sylvatica* (Peles 2013), *L. pipiens* (Chen et al. 2007), *B. gargarizans* (Chai et al. 2014), *Epidalea calamita* (García-Muñoz et al. 2008), and *Hyla chrysocelis* (Parris and Baud 2004). Also, Cu even in very low concentrations can damage (the liver and) the thyroid gland by reducing thyroid gland size, and the size and number of follicles (*B. gargarizans*; Chai et al. 2014), indicating a disrupting effect of Cu on TH homeostasis. This was also indicated by an effect on TH-related transcription upon Cu exposure in the same species (Wang et al. 2016). Furthermore, Cu caused morphological deformities and impaired behavioral measures of performance in several species (escape behavior in tadpoles of *Epidalea calamita*; García-Muñoz et al. 2008, swimming performance in tadpoles of *L. pipiens*; Chen et al. 2007, carry-over effects on jumping speed and endurance of juveniles in *Anaxyrus terrestris*; Rumrill et al. 2018).

Lead (Pb)—Pb was and is still released into the atmosphere and subsequently contaminates soils and wetlands as a result of the combustion of leaded fuels in all sorts of vehicles (Hares and Ward 1999, Legret and Pagotto 1999) and in industrial facilities. Further anthropogenic sources of environmental Pb include coal and oil combustion, as well as cement and metallurgic industries (Wang et al. 2000) or shooting ranges (Stansley and Roscoe 1996).

When Pb in contaminated sites is taken up by larval amphibians in high doses it can be highly toxic. A variety of sublethal effects can also be caused when Pb concentrations are lower. Accordingly, mortality and growth rate were concentration dependent in tadpoles of *Bufo arenarum* (Herkovits and Perez-Coll 1991), *B. gargarizans* (Yang et al. 2019), *Pelophylax nigromaculata* (Huang et al. 2014) and *Rana sphenocephala* (Sparling et al. 2006).

Studies on *B. gargarizans* suggest that Pb disrupts TH homeostasis in tadpoles by histological alterations of the thyroid gland, i.e., by follicular cell hyperplasia and colloid depletion, and disturbs the transcription of TH-related genes. This led to altered growth and tadpole development, as well as skeletal malformations (Yang et al. 2019), with a respective increase at low and a decrease at very high Pb concentrations. A developmental delay was also observed in *L. pipiens* (Cheng et al. 2007) and *L. sphenocephala* (Sparling et al. 2006). Further studies showed that Pb can cause skeletal malformations (*L. sphenocephala*; Sparling et al. 2006) and influence tadpole behavior such as swimming and jumping speed (*P. nigromaculata*; Huang et al. 2014) and surfacing (*L. catesbeianus*; Rice et al. 1999). Sex ratio and gonadal histology were not affected by Pb in *L. nigromaculata*, suggesting that Pb is not strongly estrogenic (Huang et al. 2014).

Other heavy metals—Zinc (Zn), iron (Fe), mercury (Hg), manganese (Mn), chromium (Cr), nickel (Ni) and uranium (U) are further heavy metals that are emitted from anthropogenic sources such as industrial processes, traffic, combustion of fuels and of other substances. These elements potentially accumulate in aquatic ecosystems in biologically relevant concentrations (e.g., Hares and Ward 1999, Legret and Pagotto 1999).

The acute exposure to Zn caused significant mortality in *B. arenarum* tadpoles (Herkovitz and Pérez-Coll 1991), as did exposure to Ni in *L. pipiens* tadpoles (Leduc et al. 2016), however, the Zn and Ni concentrations tested in these experiments were high compared to naturally occurring concentrations. Uranium (i.e., depleted uranium ^{234}U and ^{235}U), in contrast, did not affect mortality in *X. laevis*, although the tested concentrations were much higher than found in the environment near U-contaminated sites (i.e., breeding ponds at firing sites near the Los Alamos National Laboratory; Mitchell et al. 2005). Growth rate in tadpoles was decreased by Hg (*B. gargazensis*; Shi et al. 2018), Cr (*Duttaphrynus melanostictus*; Fernando et al. 2016), Ni (*L. pipiens*; Leduc et al. 2015) and U (*Rana perezi*; Marques et al. 2008), but not affected by U in *X. laevis* (Mitchell et al. 2005). Also, in *R. perezi*, U in high but relevant concentrations (i.e., testing the effluents of an Uranium mine) increased tail deformities (Marques et al. 2008). No such malformations were detected in *X. laevis* upon depleted U exposure (Mitchell et al. 2005). Disrupting effects on the HPT axis and subsequent sublethal retarding effects on tadpole development and metamorphosis were observed in *L. catesbeianus* by Fe and Mn in environmentally relevant concentrations (near an iron ore mine; da Silva Veronez et al. 2016), in *X. laevis* by depleted U in high concentrations (Mitchell et al. 2005) and in *D. melanostictus* by Cr in field concentrations (Fernando et al. 2016). High dose Hg disrupted the transcription of thyroid hormone-dependent genes in *B. gargazensis*, induced histopathological alterations in the thyroid, and subsequently decelerated metamorphosis rate and inhibited body size of *B. gargarizans* larvae (Shi et al. 2018). In conclusion, the known effects of environmentally relevant concentrations of these heavy metals on organisms, in particular on amphibians, are ambiguous and fragmentary.

Interactive effects—Most studies on the effects of environmental contaminants on amphibians, such as heavy metals here, are laboratory assays highlighting the effect of one stressor independent of other confounding stressors. In natural systems, however, tadpoles are rarely facing one stressor alone, but a combination of many. These concurrent stressors might act as endocrine disruptors on tadpoles independently, or they interact resulting in either stronger or weaker reactions compared to single stressors. Often, the exposure to two or more metals together is more toxic and/or delays development more strongly compared to single metals (e.g., Ld, Zn and Cd in *R. luteiventris*, Lefcort et al. 1998 or Cu and Ni in *L. pipiens*, Leduc et al. 2015).

3.3.2 Plasticizers

Plasticizers are the most common additives to a multitude of products made from plastic and are added to increase the flexibility of plastics. However, these substances are known to leach out of plastics and successively contaminate the environment, including amphibian breeding sites. Some plasticizers are known to bioaccumulate and are identified in high concentrations in animal tissue (Horn et al. 2004). The most important and most frequently used plasticizers are chemicals belonging to the polychlorinated bisphenols (PCBs), phthalates, and bisphenol A (BPA). These substances are known to cause a variety of adverse health effects in vertebrates, including disturbing their development by impairing the thyroid hormone axis (rev. in Heimeier and Shi 2010, rev. in Mathieu-Denoncourt al. 2015, rev. in Thambarajah et al. 2019).

Transcriptomic studies showed that all common plasticizers disrupt the HPT axis by impairing TH-mediated gene expression in tadpoles of the model species *X. laevis* (PCB: Gutleb et al. 2000, Lehigh Shirey et al. 2006, phthalates: Shen et al. 2011, BPA: Iwamuro et al. 2003, Heimeier et al. 2009). Accordingly, and not surprisingly, these plasticizers delay metamorphosis and/or decrease metamorphic rate in tadpoles (PCB: Gutleb et al. 2000 in *Rana temporaria* and *X. laevis*, Phthalate (Bis-(2-ethylhexyl)-phthalat): Dumpert and Zietz 1984, Shen et al. 2011, BPA: Iwamuro et al. 2003, Heimeier et al. 2009). In contrast, chronic exposure to a different phthalate, i.e., monomethyl phthalate, increased the metamorphic rate in *X. tropicalis* (Mathieu-Denoncourt et al. 2015) indicating that the TH-response mechanism to phthalates is substance specific.

Some plasticizers are additionally known as xenoestrogenics, i.e., they can function as natural estrogens and disrupt the hypothalamic-pituitary-gonadal axis affecting the reproductive system, in particular sex differentiation and gonadal development. Accordingly, plasticizer exposure in tadpoles led to a feminization of male gonads, to impaired gonadal development and to skewed sex ratios. *Xenopus laevis*, for example, developed intersexes upon BPA exposure (Levy et al. 2004) and some genetically male *Rana rugosa* developed ovaries upon dibutyl phthalate exposure. (Phenotypically) Female biased sex ratios upon BPA exposure were found in *X. laevis* (at a concentration of 10^{-7} M in Kloas et al. 1999, at 10^{-7} M and 10^{-8} M in Levy et al. 2004) and *X. tropicalis* (Mathieu-Denoncourt et al. 2015a), but not in Pickford et al. (2003) with *X. laevis* tadpoles exposed to BPA concentrations ranging from 0.83 to 497 μg/l.

3.3.3 Flame Retardants

Flame retardants are ubiquitously and extensively integrated into a multitude of materials to increase their fire resistance, including plastics, polyurethane foams, epoxy resins, wood, paper, and textiles. There is a wide variety of substances used as flame retardants, with brominated or chlorinated organic flame retardants as the most used chemical classes (e.g., Alaee et al. 2003, Letcher and Behnisch 2003, Birnbaum and Staskal 2004). Many flame retardants are not bound strongly to the material and can leach out easily, accordingly contaminating soil, sediments, and water. Many of these substances bioaccumulate, and increasing concentrations were found in animals of higher trophic positions, including humans (Alaee et al. 2002, Boon et al. 2002, de Wit 2002, Birnbaum and Staskal 2004).

Perchlorates—Perchlorates are commonly used in rockets and fireworks and in medical treatments for humans (sodium perchlorate $NaClO_4$, see subchapter "pharmaceuticals"), but also as flame retardants. Perchlorates are persistent contaminants that occur in environmentally relevant concentration even in aquatic systems (Dasgupta et al. 2006). Perchlorates are environmentally relevant endocrine disruptors on the HPT axis. They inhibit TH synthesis (Ortiz-Santaliestra and Sparling 2007) and consequently affect tadpole development and inhibit metamorphosis (Tietge et al. 2005, Carr and Theodorakis 2006, Thambarajah et al. 2019). A number of experimental studies on different frog species (*L. sylvatica, R. temporaria, X. laevis*) have shown that perchlorates in environmentally relevant concentrations retard tadpole development and decrease metamorphic rate (Goleman et al. 2002a,b, Opitz et al. 2009, Bulaeva et al. 2015, Ruthsatz et al. 2018a,b, 2019, 2020a,c). When tadpoles were exposed to perchlorates in the water, genes associated with TH-regulation and TH levels were decreased (Opitz et al. 2009, Titge et al. 2010, Flood and Langlois 2014), the thyroid gland was enlarged, and thyroid follicles were impaired (e.g., Tietge et al. 2010). Also, $NaClO_4$ affects a series of variables that are often associated with growth and development, such as reducing heart rate (Ruthsatz et al. 2020c), reducing metabolic rate, and increasing size at metamorphosis and body condition when compared to similar developed tadpoles from control treatments (Ruthsatz et al. 2018a; 2020a,c).

Brominated Flame Retardants (BFRs)—BFRs, mainly Polybrominated Diphenyl Ethers (PBDE) and Tetrabromobisphenol A (TBBPA), are the most extensively used flame retardant substances (Alaee et al. 2003). BFRs can leach out of the material and contaminate the environment, including amphibian breeding sites, and finally bioaccumulate in animals (de Wit 2002, Birnbaum and Staskal 2004). PBDEs and TBBPA have structural similarities to natural THs and may either enhance the excretion of TH and/or competitively replace THs in binding to thyroid transport proteins (Meerts et al. 2000, Balch et al. 2006, rev. in Alaee and Wenning 2002, rev. in Birnbaum and Staskal 2004, rev. in Thambarajah et al. 2019). Consequently, several studies indicated that these substances perturb TH homeostasis in animal model systems (rats and mice) as well as in *in vitro* systems (Meerts et al. 2000, Hallgren et al. 2001, Zhou et al. 2002). Also, in amphibians, the HPT axis is disturbed resulting in negative effects on tadpole development and metamorphosis. Transcriptomic changes of TH-regulating genes were identified in *X. laevis* upon PBDE and TBBPA exposure (Zhang et al. 2014, Yost et al. 2016, Mengeling et al. 2017, Wang et al. 2017). Accordingly, several studies found retarded development and a delayed or inhibited metamorphosis as a reaction to high concentrations of PBDE (in *X. laevis*, Balch et al. 2006; in *X. tropicalis*, Carlsson et al. 2007; in *L. pipiens*, Coyle et al. 2010). TBBPA, however, lead to either an increase or a retard of development depending on the life stages of the exposed tadpoles (Zhang et al. 2014), indicating different stage-dependent sensitivities of tadpoles to TBBPA.

3.4 Road Salt Exposure

Natural processes such as rainfall, rock weathering, and seawater intrusion are known to alter the salinity of freshwater ecosystems (rev. in Cañedo-Argüelles et al. 2018). However, human activities might accelerate these processes indirectly via increased desiccation rates due to global warming or directly through the applications of salts during winter to de-ice roads (Kaushal et al. 2018).

The most common road salts are inorganic chloride-based salts such as sodium chloride (NaCl), magnesium chloride ($MgCl_2$), and calcium chloride ($CaCl_2$, rev. in Tiwari and Rachlin 2018). After application, road salts dissolve into their constitutive ions and enter freshwater systems as saline flow generated by snowmelt or rain (Evans and Frick 2001, rev. in Hintz and Relyea 2018). The constitutive ions are persistent pollutants that expose freshwater organisms in rivers, lakes, and wetlands to osmotic stress for long periods (Findlay and Kelly 2011, Van Meter et al. 2011). Freshwater organisms need to maintain an osmotic balance between the ion concentration within their cells and their body fluids. Amphibians, however, are poor osmoregulators and tend to be particularly sensitive to salinity due to their permeable skin which is in turn used for osmoregulation (Wu et al. 2012).

During embryonic and larval development, amphibian species differ in their sensitivity to road salt pollution at environmentally realistic concentrations (Dougherty and Smith 2006, Collins and Russell 2009, Hua and Pierce 2013). In tadpoles, osmotic stress generally reduces survival to metamorphosis (Sanzo and Hecnar 2006, Van Meter et al. 2011) and during post-metamorphosis (Dananay et al. 2015) but also leads to sub-lethal effects on growth and developmental rate (Gomez-Mestre and Tejedo 2003, Wu and Kam 2009, Nakkrasae et al. 2016, Welch et al. 2019), immunocompetence (Milotic et al. 2017), metabolism (Gomez-Mestre et al. 2004), behavior (Wood and Welch 2015), and affects sexual differentiation (Leggett et al. 2021). These sub-lethal effects are attributable to physiological mechanisms involved in osmoregulation which also affect endocrine pathways controlling amphibian metamorphosis.

Prolactin and aldosterone are the major hormones regulating amphibian osmoregulation (Brown et al. 1991, Duellman and Trueb 1994, Lukens and Wilcoxon 2020) but are also known to interact with the thyroid hormone system leading to decreased thyroid hormone concentrations (Gomez-Mestre et al. 2004, Nakkrasae et al. 2016). Prolactin directly antagonizes the effect of thyroid hormones on growth and metamorphosis (Shi 2000, Denver 2013). Aldosterone is a mineralocorticoid which is under the same control as corticosterone, the endocrine stress biomarker in amphibians (Glennemeier and Denver 2001, Romero 2004, Burraco et al. 2013), and is also upregulated under osmotic stress (Chambers et al. 2011, Burraco and Gomez-Mestre 2016). Such biochemical changes are energetically demanding with ramifications on growth, development, and energy budgets available for metamorphosis (Gomez-Mestre et al. 2004, Peckett et al. 2011, Kirschman et al. 2017, Ruthsatz et al. 2019). Osmotic stress through road salt pollution is consequently an environmental stressor with the capability to disrupt the endocrine system of tadpoles.

3.5 Acidic Deposition

In 1852 Scottish chemist Robert Angus Smith linked the emission of sulfur dioxide (SO_2) by coal-fired factories in Manchester to the increasingly acidic precipitation (Smith 1872). Although there were early reports on the damaging effects of acidic rain on organisms in Europe and North America at the end of the 19th century, there was little scientific or public interest until the 1970's (Cowling 1982). The mechanisms underlying the acidification of soil and waters is well known by now. During the combustion of fossil fuels sulphur dioxide (SO_2), nitrogen oxides (NO_x) and volatile hydrocarbons (VOC) are released and (together with ammonium NH_4^+ from agricultural sources) are transformed in the atmosphere into sulphuric and nitrogen containing acids. This results in a decrease in soil and water pH (Flower und Battarbee 1983, Battarbee et al. 1985, Morling et al. 1985). The degree of acidification of water bodies depends on the natural mineral substrate which largely determines the buffering capacity of the water against H^+ ions.

Fish and amphibians are the vertebrates most sensitive to acidification of their habitats. Field studies on amphibian abundance and species diversity have shown clear correlations between the acidification of breeding ponds and the decline of amphibian populations (e.g., Beebee 1987). A multitude of experimental studies revealed negative effects of acidic water on embryos and larvae of different amphibian species (e.g., Andren et al. 1988, Rowe et al. 1992, Freda and McDonald 1993). During development, the embryonic stage appears to be most sensitive. Low pH leads to a denaturation of the hatching enzyme (Urch and Hedrick 1981) with subsequent deformations of the embryo and high embryonic mortality (Pahkala et al. 2001). Time to hatching (Pahkala et al. 2001) is prolonged and hatchling size is reduced in water of low pH (pH 4.5; Griffiths and DeWijer 1994). For the larval stage, low pH (mostly $<$ pH 4) also lead to high mortality rates (Grant and Licht 1993, Rosenberg and Pierce 1995). Low water pH also causes subtle negative effects, in particular on larval time and size at metamorphosis. In most species and populations, but not all (e.g., Ling et al. 1986, Kiesecker 1996), high acidity of water slows development and/or decreases size at metamorphosis (e.g., Beebee 1986, Rosenberg and Pierce 1995, Glos et al. 2003, Merilä et al. 2004, Skei and Dolmen 2006, Farquharson et al. 2016). Development is triggered by TH, which in turn is regulated via the HPT axis. The effects of low pH vary between individuals within one population, between populations within species (e.g., Glos et al. 2003, Teplitsky et al. 2007), and between species (e.g., Rowe et al. 1992, Griffiths and DeWijer 1994). Accordingly, populations from naturally acidic habitats seem to be more tolerant to acid stress (Freda 1986, Glos et al. 2003). High acidity of water can cause further sublethal effects such as deformities (Farquharson et al. 2016), alter predator-prey interactions (Brodman 1993, Kiesecker 1996), impact swimming performance (Freda and Dunson 1985, Rowe et al. 1992), alter morphology (Wijethunga et al. 2015), reduce splenic white blood cells (Simon et al. 2002), and enhance bacterial infections (Brodkin et al. 2003).

The proximate mechanism for these effects has been attributed to a disturbance of ion balance of the larvae (Freda and Dunson, 1984, 1985, McDonald et al. 1984, Freda 1986). Accordingly, sub-lethal pH effects might be driven by physiological mechanisms involved in osmoregulation which also affect endocrine pathways controlling amphibian metamorphosis. The regulation of CORT, for example, is a response to several kinds of stress in tadpoles including high acidity (Glennemeier and Denver 2001, Chambers 2009, Burraco et al. 2013, Chambers et al. 2013, Middlemis Maher et al. 2013). Burraco and Gomez-Mestre (2016) suggested that low pH might lead to an increase in CRF that subsequently elevates TH and CORT levels,further causing an increase in metabolic rate and ultimately energy expenditure (Denver et al. 2002, Wack et al. 2012). Moreover, the activation of CRF increases the expression of mineralocorticoid receptors (Gesing et al. 2001). These receptors are involved in the regulation of body fluid osmolality and ion balance (Terker and Ellison 2015), which is essential for amphibian osmoregulation, especially under osmotic stress from low pH (see Burraco and Gomez-Mestre 2016). Indeed, an effect of acid water on CORT has been observed in tadpoles of *Lithobathes sylvatica* and *Hyla versicolor* both in the field and in a mesocosm (Chambers 2009, Chambers et al. 2013). Likewise, low pH increased CORT concentrations in tadpoles of the European spadefoot toad *Pelobates cultripes* (Florencio et al. 2020). However, a further study on spadefoot toads (Burraco and Gomez-Mestre 2016) did not find any effect of water pH on CORT and suggested that factors such as local adaptations of populations might be important.

The negative effects of low pH on amphibians are relevant to the decline in amphibian populations in the last century, particularly in the northern hemisphere (Blaustein and Wake 1990). Even though the magnitude of acid deposition has decreased in the last decades, other stressors have increased (see this chapter). Many of these stressors interact synergistically with low pH causing negative effects on mortality, growth, and development of tadpoles. These additional stressors include aluminum (Bradford et al. 1992, Skei and Dolmen 2006), cadmium (Lu et al. 2021), nitrate and UVB light (Hatch and Blaustein 2000). The combination of stressors with low pH levels might endanger amphibian populations from stable habitats that are not locally adapted to low pH conditions.

3.6 Microplastics

Microplastics have become a pollutant of concern since the use of plastics increased in the 1950s (PlasticsEurope 2020). Today, microplastics are one of the fastest-growing sources of pollution (Akdogan and Guven 2019, Xu et al. 2020). Microplastics are defined as synthetic polymer particles below 5 mm in diameter (Horton et al. 2017) which originate from primary (i.e., textiles, medicines, personal care products, and pellets for plastic production, rev. in Rochman et al. 2015) or secondary sources (i.e., derived from the debris of plastic items, fishing nets, films, and tires) (Thompson 2015, Eriksen et al. 2014). Once plastic is disposed into the environment, weathering and aging processes such as UV radiation, wave and wind abrasion, hydrolysis, and microbial degradation lead to breakdown resulting in microplastic particles (rev. in Chen et al. 2019). Microplastics are usually composed of polyethylene, polypropylene, polystyrene, nylon, thermoplastic polyester, poly vinyl chloride (i.e., PVC), cellulose acetate, and polyethylene terephthalate (rev. in Bhattacharya and Khare 2020).

Bioaccumulation of microplastics via ingestion has been demonstrated in a variety of aquatic organisms from numerous taxa (rev. in Ribeiro et al. 2019), but evidence for consequent detrimental effects is much more limited in freshwater organisms (Horton et al. 2017). In particular, amphibians are largely under-represented in current research on microplastics, probably because they are not obligately aquatic. However, recent studies have confirmed that tadpoles ingest microplastics under laboratory (Hu et al. 2016, da Costa Araújo and Malafaia 2020, da Costa Araújo et al. 2020a, Ruthsatz et al. 2022b) and field conditions (Karaoglo and Gül 2020, Kolenda et al. 2020).

Ingested microplastics are distributed to different tissues and organs and lead to harmful, if not lethal, effects on growth, reproduction, energy metabolism, and survival together with oxidative stress and neurotoxicity (rev. in Prokić et al. 2019). In tadpoles of *Physalaemus cuvieri*, ingestion of microplastics affected anti-predatory behavior through compromising locomotion (da Costa Araújo and Malafaia 2020), and caused histopathological transformations (da Costa Araújo et al. 2020b) which led to morphological changes through cytotoxicity (da Costa Araújo et al. 2020b). Behavior, growth, and body condition was also impaired in tadpoles of *Alytes obstetricans* (Boyero et al. 2020). In contrast, growth (Ruthsatz et al. 2022b) and swimming activity (de Felice et al. 2018) were not impaired by ingestion of microplastics in *Xenopus laevis* tadpoles.

Although microplastics have been shown to disrupt endocrine systems in echinoderms (Trifuoggi et al. 2019), mollusks (Sussarellu et al. 2016), and fish (Rochman et al. 2014), similar effects have not been demonstrated in tadpoles and warrant further research. A recent study, however, found that tadpoles of *X. laevis* show increased corticosterone levels in response to polyethylene microplastics ingestion (Ruthsatz et al. 2023b). Nevertheless, microplastics contain additives such as phthalates and bisphenol A (rev. in Zhang et al. 2019) and pose a special threat to tadpoles through their disruptive effect on estrogenic activity (Ohtani et al. 2000, Xu and Gye 2018, Levy et al. 2004, rev. in Wu and Seebacher 2020). Moreover, the surface area of microplastics facilitates the reabsorption of other pollutants from the surrounding environment and might serve as vectors of other potentially endocrine disruptive pollutants (Endo et al. 2005, Chua et al. 2014, Rochman et al. 2014, rev. in Franzellitti et al. 2019). Overall, microplastic exposure is a source of stress for tadpoles, especially in addition to other pollutants, and might activate the stress hormone axis with ramifications for energy metabolism, growth and development.

4. The Interactions Between Global Change and Multiple Stressors in Larval Habitats

A major challenge for amphibian conservation is to identify the diverse and complex causes of population declines and species extinctions. Tadpoles live in a complex environment and are exposed to many stressors of natural and anthropogenic origin (Fig. 2). Consequently, a growing body of literature has examined possible interactions of pollutants and other stressors such as global warming (Rohr et al. 2011, Rohr and Palmer 2013, Baier et al. 2016, Quiroga et al. 2019), pathogen exposure (Rollins-Smith et al. 2006, Kerby and Storfer 2009), UV-radiation (Bridges and Boone 2003, Hatch and Blaustein 2003, rev. in Alton et al. 2017), and desiccation (Rohr and Palmer 2005). Several studies suggest that environmental stressors associated with global change including desiccation and temperature variation could increase the effects of aquatic pollutants on amphibians. Alternatively, pollutants might reduce the tolerance of amphibians to rising temperatures, increase the risk for desiccation or affect the immune system of larval amphibians, making them more susceptible to infections (rev. in Buck et al. 2011, rev. in Rohr et al. 2011). However, interactive effects of multiple stressors remain difficult to predict and existing ecotoxicological studies often oversimplify possible interactions between pollution and other stressors. To fully understand the combined effects of multiple stressors on tadpoles, researchers will undoubtedly have to integrate the effects of environmental stressors more thoroughly on both susceptibility and exposure to pollutants.

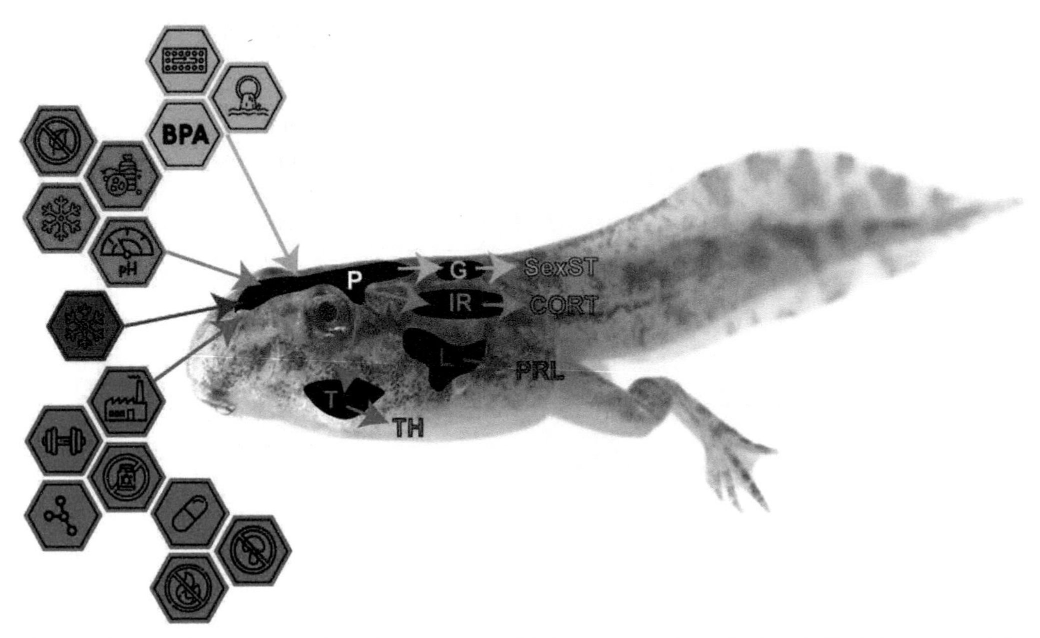

Figure 2. Effects of pollutants on the endocrine system of tadpoles. P: Pituitary. G: Gonads. IR: Interrenal (amphibian equivalent to mammalian adrenal cortex embedded within the kidney complex). L: Liver. T: Thyroid. TH: thyroid hormones (i.e., T3 and T4). CORT: corticosteroids (i.e., corticosterone and aldosterone). PRL: Prolactin. SexST: sex steroids. Yellow lines and symbols: hypothalamus-pituitary-thyroid-axis (HPT axis) and pollutants acting as endocrine disruptors on the HPT axis by affecting TH production pathways resulting in up- or downregulated endogenous TH levels and thus, impacting developmental rate. Blue lines and symbols: hypothalamus-pituitary-liver-axis (HPL axis) and pollutants acting as endocrine disruptors on the HPL axis by affecting PRL production pathways resulting in up- or downregulated endogenous PRL levels and thus, impacting growth rate. Red lines and symbols: hypothalamus-pituitary-interrenal-axis (HPI axis, i.e., stress axis) and pollutants acting as endocrine disruptors on the HPI axis by activating CORT production pathways resulting in upregulated endogenous CORT levels and thus, an endocrine stress response. Purple lines and symbols: hypothalamus-pituitary-gonad-axis (HPG axis) and pollutants acting as endocrine disruptors on the HPG axis by affecting SexST production pathways resulting in up- or downregulated endogenous SexST levels and thus, impacting sexual differentiation. In addition to pollutants which act as endocrine disruptors, several abiotic and biotic environmental factors might interact as stressors impacting the HPI axis and resulting in increased CORT levels (green symbols). Note that feedback regulation at the hypothalamus and pituitary by THs, CORT, and SexST exists, but is not depicted in the figure. For further details, refer to documented actions of endocrine disrupting chemicals in the main text.

Acknowledgements

The authors appreciatively acknowledge the constructive edits and comments provided by William J. Foley. We also thank Maya Riepe for conducting the literature review on which this chapter was based.

References Cited

Akdogan, Z. and B. Guven. 2019. Microplastics in the environment: A critical review of current understanding and identification of future research needs. Environ. Poll. 254: 113011.

Alaee, M., J.M. Luross, D.M. Whittle and D.B. Sergeant. 2002. Bioaccumulation of polybrominated diphenyl ethers in the Lake Ontario pelagic food web. Organohalog. Compounds 57: 427–430.

Alaee, M. and R.J. Wenning. 2002. The significance of brominated flame-retardants in the environment. Chemosphere 46: 579–796.

Alaee, M., P. Arias, A. Sjödin and A. Bergman. 2003. An overview of commercially used brominated flame retardants, their applications, their use patterns in different countries/regions and possible modes of release. Environ. Int. 29: 683–689.

Alton, L.A. and C.E. Franklin. 2017. Drivers of amphibian declines: Effects of ultraviolet radiation and interactions with other environmental factors. Clim. Change Resp. 4: 1–26.

Andren, C., L. Henrikson, M. Olsson and G. Nilson. 1988. Effects of pH and aluminum on embryonic and early larval stages of Swedish brown frogs *Rana arvalis*, *R. temporaria* and *R. dalmatina*. Holarctic Ecol. 11: 127–135.

Arias Torresa, A.J., M.J. Bühlera and L.I. Zelarayán. 2016. *In vitro* steroid-induced meiosis in *Rhinella arenarum* oocytes: Role of pre-MPF activation. Zygote 24: 252–258.

Aris, A.Z., A.S. Shamsuddin and S.M. Praveena. 2014. Occurrence of 17α-ethinylestradiol (EE2) in the environment and effect on exposed biota: A review. Environ. Int. 69: 104–119.

Armstrong, B.M., J.M. Lazorchak, C.A. Murphy, H.J. Haring, K.M. Jensen and M.E. Smith. 2015. Determining the effects of a mixture of an endocrine disrupting compound, 17α-ethinylestradiol, and ammonia on fathead minnow (*Pimephales promelas*) reproduction. Chemosphere 120: 108–114.

Arregui, M.C., D. Sánchez, R. Althaus, R.R. Scotta and I. Bertolaccini. 2010. Assessing the risk of pesticide environmental impact in several Argentinian cropping systems with a fuzzy expert indicator. Pest Manag. Sci. 66: 736–740.

Arukwe, A. and B.M. Jenssen 2005. Differential organ expression patterns of thyroid hormone receptor isoform genes in p, p′-DDE-treated adult male common frog, *Rana temporaria*. Environ. Toxicol. Pharmacol. 20: 485–492.

Azizishirazi, A., J.L. Klemish and G.G. Pyle. 2021. Sensitivity of amphibians to copper. Environ. Toxicol. Chem. 40: 1808–1819.

Baier, F., M. Jedinger, E. Gruber and J.G. Zaller. 2016. Temperature-dependence of glyphosate-based herbicide's effects on egg and tadpole growth of common toads. Environ. Sci. 4: 51.

Balch, G.C., L.A. Vélez-Espino, C. Sweet, M. Alaee and C.D. Metcalfe. 2006. Inhibition of metamorphosis in tadpoles of *Xenopus laevis* exposed to polybrominated diphenyl ethers (PBDEs). Chemosphere 64: 328–38.

Battaglin, W.A., K.L. Smalling, C. Anderson, D. Calhoun, T. Chestnut and E. Muths. 2016. Potential interactions among disease, pesticides, water quality and adjacent land cover in amphibian habitats in the United States. Sci. Total Environ. 566: 320–332.

Battarbee, R.W., R.J. Flower, A.C. Stevenson and B. Rippey. 1985. Lake acidification in Galloway: A palaeoecological test of competing hypotheses. Nature 314: 350–352.

Beachy, C.K., T.H. Surges and M. Reyes 1999. Effects of developmental and growth history on metamorphosis in the gray treefrog, *Hyla versicolor* (Amphibia, Anura). J. Exp. Zool. 283: 522–530.

Beck, C.W. and J.D. Congdon. 2000. Effects of age and size at metamorphosis on performance and metabolic rates of Southern Toad, *Bufo terrestris*, metamorphs. Funct. Ecol. 14: 32–38.

Beebee, T.J.C. 1986. Acid tolerance of natterjack toad (*Bufo calamita*) development. Herpetol. J. 1: 78–81.

Beebee, T.J.C. 1987. Eutrophication of heathland ponds at a site in southern England: Causes and effects, with particular reference to the amphibia. Biol. Conserv. 42: 39–52.

Belden, L.K., I.T. Moore, R.T. Mason, J.C. Wingfield and A.R. Blaustein. 2003. Survival, the hormonal stress response and UV-B avoidance in Cascades Frog tadpoles (*Rana cascadae*) exposed to UV-B radiation. Funct. Ecol. 17: 409–416.

Belfroid, A.C., A. Van der Horst, A.D. Vethaak, A.J. Schäfer, G.B.J. Rijs, J. Wegener and W.P. Cofino. 1999. Analysis and occurrence of estrogenic hormones and their glucuronides in surface water and wastewater in The Netherlands. Sci. Total Environ. 225: 101–108.

Berg, C., T. Backström, S. Winberg, R. Lindberg and I. Brandt. 2013. Developmental exposure to fluoxetine modulates the serotonin system in hypothalamus. PLoS ONE 8: e55053.

Bernabò, I., A. Guardia, R. Macirella, S. Sesti, A. Crescente and E. Brunelli. 2016. Effects of long-term exposure to two fungicides, pyrimethanil and tebuconazole, on survival and life history traits of Italian tree frog (*Hyla intermedia*). Aquat. Toxicol. 172: 56–66.

Bernabò, I., A. Guardia, R. Macirella, S. Tripepi and E. Brunelli. 2017. Chronic exposures to fungicide pyrimethanil: Multi-organ effects on Italian tree frog (*Hyla intermedia*). Sci. Rep. 7: 1–16.

Berven, K.A. 1990. Factors affecting population fluctuations in larval and adult stages of the wood frog (*Rana sylvatica*). Ecology 71: 1599–1608.

Bhattacharya, A. and S.K. Khare. 2020. Ecological and toxicological manifestations of microplastics: current scenario, research gaps, and possible alleviation measures. J. Environ. Sci. Health C 38: 1–20.

Birnbaum, L.S. and D.F. Staskal. 2004. Brominated flame retardants: Cause for concern? Environ. Health Perspect. 112: 9–17.

Blaustein, A.R. and D.B. Wake. 1990. Declining amphibian populations: A global phenomenon? Trends Ecol. Evol. 5: 203–204.

Bókony, V., V. Verebélyi, N. Ujhegyi, Z. Mikó, E. Nemesházi, M. Szederkényi et al. 2020. Effects of two little-studied environmental pollutants on early development in anurans. Environ. Pollut. 260: 114078.

Boon, J.P., W.E. Lewis, M.R. Tjoen-A-Choy, C.R. Allchin, R.J. Law, J. De Boer et al. 2002. Levels of polybrominated diphenyl ether (PBDE) flame retardants in animals representing different trophic levels of the North Sea food web. Environ. Sci. Technol. 36: 4025–4032.

Boone, M.D. 2008. Examining the single and interactive effects of three insecticides on amphibian metamorphosis. Environ. Toxicol. Chem. 27: 1561–1568.

Boone, M.D., S.A. Hammond, N. Veldhoen, M. Youngquist and C.C. Helbing. 2013. Specific time of exposure during tadpole development influences biological effects of the insecticide carbaryl in green frogs (*Lithobates clamitans*). Aquat. Toxicol. 130: 139–148.

Boyero, L., N. López-Rojo, J. Bosch, A. Alonso, F. Correa-Araneda and J. Pérez. 2020. Microplastics impair amphibian survival, body condition and function. Chemosphere 244: 125500.

Bradford, D.F., C. Swanson and M.S. Gordon. 1992. Effects of low pH and aluminum on two declining species of amphibians in the Sierra Nevada, California. J. Herpetol. 26: 369–377.

Brande-Lavridsen, N., J. Christensen-Dalsgaard and B. Korsgaard. 2010. Effects of Ethinylestradiol and the fungicide Prochloraz on metamorphosis and thyroid gland morphology in *Rana temporaria*. Open Zool. 3: 7–16.

Bridges, C.M. 2000. Long-term effects of pesticide exposure at various life stages of the southern leopard frog (*Rana sphenocephala*). Arch. Environ. Contam. Toxicol. 39: 91–96.

Bridges, C.M. and M.D. Boone. 2003. The interactive effects of UV-B and insecticide exposure on tadpole survival, growth and development. Biol. Conserv. 113: 49–54.

Bringolf, R.B., R.M. Heltsley, T.J. Newton, C.B. Eads, S.J. Fraley, D. Shea et al. 2010. Environmental occurrence and reproductive effects of the pharmaceutical fluoxetine in native freshwater mussels. Environ. Toxicol. Chem. 29: 1311–1318.

Brix, K.V., A.J. Esbaugh and M. Grosell. 2011. The toxicity and physiological effects of copper on the freshwater pulmonate snail, *Lymnaea stagnalis*. Comp. Biochem. Physiol. C. 154: 261–267.

Brodkin, M., I. Vatnick, M. Simon, H. Hopey, K. Butler-Holston and M. Leonard. 2003. Effects of acid stress in adult *Rana pipiens*. J. Exp. Zool. A 298: 16–22.

Brodman, R. 1993. The effect of acidity on interactions of *Ambystoma* salamander larvae. J. Freshwater Ecol. 8: 209–214.

Brooks, B.W., C.M. Foran, S.M. Richards, J. Weston, P.K. Turner, J.K. Stanley et al. 2003. Aquatic ecotoxicology of fluoxetine. Toxicol. Lett. 142: 169–183.

Brown, D.D. and L. Cai. 2007. Amphibian metamorphosis. Dev. Biol. 306: 20–33.

Brown, S.C., Brown, P.S., Yamamoto, K., Matsuda, K. and Kikuyama, S. 1991. Amphibian prolactins: Activity in the eft skin transepithelial potential bioassay. Gen. Comp. Endocrinol. 82: 1–7.

Buck, J.C., E.A. Scheessele, R.A. Relyea and A.R. Blaustein. 2012. The effects of multiple stressors on wetland communities: pesticides, pathogens and competing amphibians. Freshw. Biol. 57: 61–73.

Bulaeva, E., C. Lanctôt, L. Reynolds, V.L. Trudeau and L. Navarro-Martín. 2015. Sodium perchlorate disrupts development and affects metamorphosis- and growth-related gene expression in tadpoles of the wood frog (*Lithobates sylvaticus*). Gen. Comp. Endocrinol. 222: 33–43.

Burraco, P., L.J. Duarte and I. Gomez-Mestre. 2013. Predator-induced physiological responses in tadpoles challenged with herbicide pollution. Curr. Zool. 59: 475–484.

Burraco, P. and I. Gomez-Mestre. 2016. Physiological stress responses in amphibian larvae to multiple stressors reveal marked anthropogenic effects even below lethal levels. Physiol. Biochem. Zool. 89: 462–472.

Cañedo-Argüelles, M., B. Kefford and R. Schäfer. 2019. Salt in freshwaters: Causes, effects and prospects-introduction to the theme issue. Philos. Trans. R. Soc. B. 374: 1–6.

Carey, C. and C.J. Bryant. 1995. Possible interrelations among environmental toxicants, amphibian development, and decline of amphibian populations. Environ. Health Perspect. 103: 13–17.

Carfagno, G.L. and P.P. Fong. 2014. Growth inhibition of tadpoles exposed to sertraline in the presence of conspecifics. J. Herpetol. 48: 571–576.

Carlsson, G., P. Kulkarni, P. Larsson and L. Norrgren. 2007. Distribution of BDE-99 and effects on metamorphosis of BDE-99 and -47 after oral exposure in *Xenopus tropicalis*. Aquat. Toxicol. 84: 71–9.

Carr, J.A., L.J. Urquidi, W.L. Goleman, F. Hu, P.N. Smith and C.W. Theodorakis. 2003. Ammonium perchlorate disruption of thyroid function in natural amphibian populations: Assessment and potential impact. pp. 130–142. *In*: Lindner, G., S. Krest, D. Sparling and E. Little [eds.]. Multiple Stressor Effects in Relation to Declining Amphibian Populations. ASTM International, West Conshohocken, PA, USA.

Carr, J.A. and C. Theodorakis. 2006. Effects of perchlorate in amphibians. pp. 125–53. *In*: Kendall, R.J. and P.N. Smith [eds.]. Perchlorate Ecotoxicology. SETAC Press, Pensacola, FL, USA.

Carr, J.A. and R. Patiño. 2011. The hypothalamus–pituitary–thyroid axis in teleosts and amphibians: Endocrine disruption and its consequences to natural populations. Gen. Comp. Endocrinol. 170: 299–312.

Cauble, K. and R.S. Wagner. 2005. Sublethal effects of the herbicide glyphosate on amphibian metamorphosis and development. Environ. Contam. Toxicol. 75: 429–435.

Celino-Brady, F.T., D.T. Lerner and A.P. Seale. 2021. Experimental approaches for characterizing the endocrine-disrupting effects of environmental chemicals in fish. Front. Endocrinol. 11: 619361.

Cevasco, A., R. Urbatzka, S. Bottero, A. Massari, F. Pedemonte, W. Kloas et al. 2008. Endocrine disrupting chemicals (EDC) with (anti)estrogenic and(anti)androgenic modes of action affecting reproductive biology of *Xenopus laevis*: II. Effects on gonad histomorphology. Comp. Biochem. Physiol. Part C 147: 241–251.

Chen, Q., A. Allgeier, D. Yin and H. Hollert. 2019. Leaching of endocrine disrupting chemicals from marine microplastics and mesoplastics under common life stress conditions. Environ. Internat. 130: 104938.

Chai, L., H. Wang, H. Deng, H. Zhao and W. Wang. 2014. Chronic exposure effects of copper on growth, metamorphosis and thyroid gland, liver health in Chinese toad, *Bufo gargarizans* tadpoles. Chem. Ecol. 30: 589–601.

Chambers, D.L. 2009. Abiotic factors underlying stress hormone level variation among larval amphibians. Doctoral dissertation, Virginia Tech, Blacksburg, USA.

Chambers, D.L., J.M. Wojdak, P. Du and L.K. Belden. 2013. Pond acidification may explain differences in corticosterone among salamander populations. Physiol. Biochem. Zool. 86: 224–232.

Chambers, L.G., K.R. Reddy and T.Z. Osborne. 2011. Short-term response of carbon cycling to salinity pulses in a freshwater wetland. Soil Sci. Soc. Am. J. 75: 2000–2007.

Cheek, A.O., C.F. Ide, J.E. Bollinger, C.V. Rider and J.A. McLachlan. 1999. Alteration of leopard frog (*Rana pipiens*) metamorphosis by the herbicide acetochlor. Arch. Environ. Contam. Toxicol. 37: 70–77.

Chen, T.H., J.A. Gross and W.H. Karasov. 2007. Adverse effects of chronic copper exposure in larval northern leopard frogs (*Rana pipiens*). Environ. Toxicol. Chem. 26: 1470–1475.

Chua, E.M., J. Shimeta, D. Nugegoda, P.D. Morrison and B.O. Clarke. 2014. Assimilation of polybrominated diphenyl ethers from microplastics by the marine amphipod, *Allorchestes compressa*. Environ. Sci. Technol. 48: 8127–8134.

Clark, E.J., D.O. Norris and R.E. Jones. 1998. Interactions of gonadal steroids and pesticides (DDT, DDE) on gonaduct growth in larval tiger salamanders, *Ambystoma tigrinum*. Gen. Comp. Endocrinol. 109: 94–105.

Coady, K., M. Murphy, D. Villeneuve, M. Hecker, P. Jones, J. Carr et al. 2004. Effects of atrazine on metamorphosis, growth, and gonadal development in the green frog (*Rana clamitans*). J. Toxicol. Environ. Health. 67: 941–957.

Coady, K., T. Marino, J. Thomas, R. Currie, G. Hancock, J. Crofoot et al. 2010. Evaluation of the amphibian metamorphosis assay: Exposure to the goitrogen methimazole and the endogenous thyroid hormone L-thyroxine. Environ. Toxicol. Chem. 29: 869–80.

Coady, K.K., M.B. Murphy, D.L. Villeneuve, M. Hecker, P.D. Jones, J.A. Carr et al. 2005. Effects of atrazine on metamorphosis, growth, laryngeal and gonadal development, aromatase activity, and sex steroid concentrations in *Xenopus laevis*. Ecotoxicol. Environ. Saf. 62: 160–173.

Collins, S.J. and R.W. Russell. 2009. Toxicity of road salt to Nova Scotia amphibians. Environ. Pollut. 157: 320–324.

Colombo, L., L. Dalla Valle, C. Fiore, D. Armanini and P. Belvedere. 2006. Aldosterone and the conquest of land. J. Endocrinol. Investig. 29: 373–379.

Conners, D.E., E.D. Rogers, K.L. Armbrust, J.W. Kwon and M.C. Black. 2009. Growth and development of tadpoles (*Xenopus laevis*) exposed to selective serotonin reuptake inhibitors, fluoxetine and sertraline, throughout metamorphosis. Environ. Toxicol. Chem. 28: 2671–2676.

Cowling, E.B. 1982. Acid precipitation in historical perspective. Environ. Sci. Technol. 16: 110A–122A.

Coyle, T.L.C. and W.H. Karasov. 2010. Chronic, dietary polybrominated diphenyl ether exposure affects survival, growth, and development of *Rana pipiens* tadpoles. Environ. Toxicol. Chem. 29: 133–141.

Cressey, D. 2017. The bitter battle over the world's most popular insecticides. Nature News. 551: 156–158.

Croteau, M.C., N. Hogan, J.C. Gibson, D. Lean and V.L. Trudeau. 2008. Toxicological threats to amphibians and reptiles in urban environments. *In*: Mitchell, J.C., R.E. Jung Brown and B. Bartolomew [eds.]. Urban Herpetology. Society for the Study of Amphibians and Reptiles.

Crump, D., K. Werry, N. Veldhoen, G. Van Aggelen and C.C. Helbing. 2002. Exposure to the herbicide acetochlor alters thyroid hormone-dependent gene expression and metamorphosis in *Xenopus laevis*. Environ. Health Perspect. 110: 1199–1205.

da Costa Araújo, A.P. and G. Malafaia. 2020. Can short exposure to polyethylene microplastics change tadpoles' behavior? A study conducted with neotropical tadpole species belonging to order anura (*Physalaemus cuvieri*). J. Hazard. Mater 391: 122214.

da Costa Araújo, A.P., A.R. Gomes and G. Malafaia. 2020a. Hepatotoxicity of pristine polyethylene microplastics in neotropical *Physalaemus cuvieri* tadpoles (Fitzinger, 1826). J. Hazard. Mater 386: 121992.

da Costa Araújo, A.P., N.F.S. de Melo, A.G. de Oliveira Junior, F.P. Rodrigues, T. Fernandes, J.E. de Andrade Vieira et al. 2020b. How much are microplastics harmful to the health of amphibians? A study with pristine polyethylene microplastics and *Physalaemus cuvieri*. J. Hazard. Mater 382: 121066.

da Silva Veronez, A.C., R.V. Salla, V.D. Baroni, L.F. Barcarolli, A. Bianchini, C.B. dos Reis Martinez et al. 2016. Genetic and biochemical effects induced by iron ore, Fe and Mn exposure in tadpoles of the bullfrog *Lithobates catesbeianus*. Aquat. Toxicol. 174: 101–108.

Dananay, K.L., K.L. Krynak, T.J. Krynak and M.F. Benard. 2015. Legacy of road salt: Apparent positive larval effects counteracted by negative postmetamorphic effects in wood frogs. Environ. Toxicol. Chem. 34: 2417–2424.

Danis, B.E. and V.L. Marlatt. 2021. Investigating acute and subchronic effects of neonicotinoids on Northwestern salamander larvae. Arch. Envrion, Contam. Toxicol. 80: 691–707.

Dantzer, B., Q.E. Fletcher, R. Boonstra and M.J. Sheriff. 2014. Measures of physiological stress: A transparent or opaque window into the status, management and conservation of species? Conserv. Physiol. 2: 1–18.

Dasgupta, P.K., J.V. Dyke, A.B. Kirk and W.A. Jackson. 2006. Perchlorate in the United States. Analysis of relative source contributions to the food chain. Environ. Sci. Technol. 40: 6608–6614.

de Felice, B., R. Bacchetta, N. Santo, P. Tremolada and M. Parolini. 2018. Polystyrene microplastics did not affect body growth and swimming activity in *Xenopus laevis* tadpoles. Environ. Sci. Poll. Res. 25: 34644–34651.

de Souza, R.M., D. Seibert, H.B. Quesada, F. de Jesus Bassetti, M.R. Fagundes-Klen and R. Bergamasco. 2020. Occurrence, impacts and general aspects of pesticides in surface water: A review. Process Saf. Environ Prot. 135: 22–37.

de Wit, C.A. 2002. An overview of brominated flame retardants in the environment. Chemosphere 46: 583–624.

Degitz, S.J., G.W. Holcombe, K.M. Flynn, P.A. Kosian, J.J. Korte and J.E. Tietge. 2005. Progress towards development of an amphibian-based thyroid screening assay using *Xenopus laevis*. Organismal and thyroidal responses to the model compounds 6-propylthiouracil, methimazole, and thyroxine. Toxicol. Sci. 87: 353–364.

Denver, R.J. 1997a. Proximate mechanisms of phenotypic plasticity in amphibian metamorphosis. Am. Zool. 37: 172–184.

Denver, R.J. 1997b. Environmental stress as a developmental cue: Corticotropin-releasing hormone is a proximate mediator of adaptive phenotypic plasticity in amphibian metamorphosis. Horm. Behav. 31: 169–179.

Denver, R.J., N. Mirhadi and M. Phillips. 1998. Adaptive plasticity in amphibian metamorphosis: Response of *Scaphiopus hammondii* tadpoles to habitat desiccation. Ecology 79: 1859–1872.

Denver, R.J., K.A. Glennemeier and G.C. Boorse. 2002. Endocrinology of complex life cycles: Amphibians. Hormones, Brain and Behavior 2: 469–513.

Denver, R.J. 2009. Stress hormones mediate environment-genotype interactions during amphibian development. Gen. Comp. Endocrinol. 164: 20–31.

Denver, R.J. 2013. Neuroendocrinology of amphibian metamorphosis. Curr. Top. Dev. Biol. 103: 195–227.

Denver, R.J. 2021. Stress hormones mediate developmental plasticity in vertebrates with complex life cycles. Neurobiol. Stress. 14: 100301.

Dodd, C.K. 2010. Amphibian ecology and conservation: a handbook of techniques. Oxford University Press, UK.

Dodd, M.H.I. and J.M. Dodd. 1976. The biology of metamorphosis. Physiol. Amphib. 3: 467–599.

Doerge, D.R. and R.S. Takazawa. 1990. Mechanism of thyroid peroxidase inhibition by ethylenethiourea. Chem. Res. Toxicol. 3: 98–101.

Dougherty, C. and G. Smith. 2006. Acute effects of road de-icers on the tadpoles of three anurans. Appl. Herpetol. 3: 87–93.

Duarte-Guterman, P., L. Navarro-Martín and V.L. Trudeau. 2014. Mechanisms of crosstalk between endocrine systems: Regulation of sex steroid hormone synthesis and action by thyroid hormones. Gen. Comp. Endocrinol. 203: 69–85.

Duellman, W.E. and L. Trueb. 1994. Biology of Amphibians. JHU press, Baltimore, USA.

Dumpert, K. and E. Zietz. 1984. Platanna (*Xenopus laevis*) as a test organism for determining the embryotoxic effects of environmental chemicals. Ecotox. Environ. Safe. 8: 55–74.

Ebele, A.J., M.A.E. Abdallah and S. Harrad. 2017. Pharmaceuticals and personal care products (PPCPs) in the freshwater aquatic environment. Emerg. Contam. 3: 1–16.

Edwards, T.M., K.A. McCoy, T. Barbeau, M.W. McCoy, J.M. Thro and L.J. Guillette Jr. 2006. Environmental context determines nitrate toxicity in Southern toad (*Bufo terrestris*) tadpoles. Aquat. Toxicol. 78: 50–58.

Endo, S., R. Takizawa, K. Okuda, H. Takada, K. Chiba, H. Kanehiro et al. 2005. Concentration of polychlorinated biphenyls (PCBs) in beached resin pellets: variability among individual particles and regional differences. Mar. Pollut. Bull. 50: 1103–1114.

Eggert, C. 2004. Sex determination: The amphibian models. Reprod. Nutr. Dev. 44: 539–549.

Eisler, R. 1985. Cadmium hazards to fish, wildlife, and invertebrates: A synoptic review. USFWS Biological Report 85 (1.2). Final/Technical Report. U.S. Fish and Wildlife Service, Laurel, MD, USA.

Elinson, R.P. and E.M. del Pino. 2012. Developmental diversity of amphibians. J. Dev. Biol. 1: 345–369.

EPA. 2021. https://www.epa.gov/; downloaded on 30 June 2021.

Eriksen, M., L.C. Lebreton, H.S. Carson, M. Thiel, C.J. Moore, J.C. Borerro et al. 2014. Plastic pollution in the world's oceans: more than 5 trillion plastic pieces weighing over 250,000 tons afloat at sea. *PloS one* 9: e111913.

Etkin, W. 1968. Hormonal control of amphibian metamorphosis. pp. 313–348. *In*: Gilbert, W. and L.I. Gilbert. [eds.]. Metamorphosis: A Problem in Developmental Biology. Appleton-Century Crofts, New York, USA.

EUROSTAT. 2019. https://ec.europa.eu/eurostat/statistics-explained/index.php?title=Agri-environmental_indicator_-_consumption_of_pesticides; downloaded on 30 June 2021.

Evans, M. and C. Frick. 2001. The effects of road salts on aquatic ecosystems. Enironment Canada, NWRI Contributions Series.

Farquharson, C., V. Wepener and N.J. Smit. 2016. Acute and chronic effects of acidic pH on four subtropical frog species. Water SA 42: 52–62.

Fernando, V.A., J. Weerasena, G.P. Lakraj, I.C. Perera, C.D. Dangalle, S. Handunnetti et al. 2016. Lethal and sub-lethal effects on the Asian common toad *Duttaphrynus melanostictus* from exposure to hexavalent chromium. Aquat. Toxicol. 177: 98–105.

Findlay, S.E. and V.R. Kelly. 2011. Emerging indirect and long-term road salt effects on ecosystems. Ann. N. Y. Acad. Sci. 1223: 58–68.

Flood, D.E., J.I. Fernandino and V.S. Langlois. 2013. Thyroid hormones in male reproductive development: Evidence for direct crosstalk between the androgen and thyroid hormone axes. Gen. Comp. Endocrinol. 192: 2–14.

Flood, D.E.K and V.S. Langlois. 2014. Crosstalk between the thyroid hormone and androgen axes during reproductive development in *Silurana tropicalis*. Gen. Comp. Endocrinol. 203: 232-40.

Florencio, M., P. Burraco, M.Á. Rendón, C. Díaz-Paniagua and I. Gomez-Mestre. 2020. Opposite and synergistic physiological responses to water acidity and predator cues in spadefoot toad tadpoles. Comp. Biochem. Physiol. A 242: 110654.

Flower, R.J. and R.W. Battarbee. 1983. Diatom evidence for recent acidification of two Scottish lochs. Nature 305: 130-133.

Fong, P.P., A.L. Olex, J.E. Farrell, R.M. Majchrzak and J.W. Muschamp. 2005. Induction of preputium eversion by peptides, serotonin receptor antagonists, and selective serotonin reuptake inhibitors in *Biomphalaria glabrata*. Invertebr. Biol. 124: 296–302.

Foster, H.R., G.A. Burton, N. Basu and E.E. Werner. 2010. Chronic exposure to fluoxetine (Prozac) causes developmental delays in *Rana pipiens* larvae. Environ. Toxicol. Chem. 29: 2845–2850.

Franzellitti, S., L. Canesi, M. Auguste, R.H. Wathsala and E. Fabbri. 2019. Microplastic exposure and effects in aquatic organisms: A physiological perspective. Environ. Toxicol. Pharmacol. 68: 37–51.

Freda, J. and W.D. Dunson. 1984. Sodium balance of amphibian larvae exposed to low environmental pH. Physiol. Zool. 57: 435–443.

Freda, J. and W.D. Dunson. 1985. Field and laboratory studies of ion balance and growth rates of ranid tadpoles chronically exposed to low pH. Copeia 1985: 415–423.

Freda, J. 1986. The influence of acidic pond water on amphibians: A review. Water Air Soil Pollut. 30: 439–450.

Freda, J. and D.G. McDonald. 1993. Toxicity of amphibian breeding ponds in the Sudbury region. Can. J. Fish. Aquat. Sci. 50: 1497v1503.

Freeman, J.L. and A.L. Rayburn. 2005. Developmental impact of atrazine on metamorphing *Xenopus laevis* as revealed by nuclear analysis and morphology. Environ. Toxicol. Chem. 24: 1648–1653.

Freeman, J.L., N. Beccue and A.L. Rayburn. 2005. Differential metamorphosis alters the endocrine response in anuran larvae exposed to T3 and atrazine. Aquat. Toxicol. 75: 263–276.

Freitas, J.S., A. Kupsco, G. Diamante, A.A. Felicio, E.A. Almeida and D. Schlenk. 2016. Influence of temperature on the thyroidogenic effects of diuron and its metabolite 3,4-DCA in tadpoles of the American bullfrog (*Lithobates catesbeianus*). Environ. Sci. Technol. 50: 13095–13104.

Frieden, E. 1981. The dual role of thyroid hormones in vertebrate development and calorigenesis. pp. 545–563. *In*: Gilbert, L.I. and E. Frieden [eds.]. Metamorphosis. Springer, Boston, MA, USA.

Gabor, C.R., M.C. Fisher and J. Bosch. 2015. Elevated corticosterone levels and changes in amphibian behavior are associated with *Batrachochytrium dendrobatidis* (Bd) infection and Bd lineage. PLoS One 10: e0122685.

Gadd, J.B., L.A. Tremblay and G.L. Northcott. 2010. Steroid estrogens, conjugated estrogens and estrogenic activity in farm dairy shed effluents. Environ. Poll. 158: 730–736.

García-Muñoz, E., F. Guerrero and G. Parra. 2008. Effects of copper sulfate on growth, development, and escape behavior in *Epidalea calamita* embryos and larvae. Arch. Environ. Contam. Toxicol. 56: 557–565.

Garnaud, S., J.-M. Mouchel, G. Chebbo and D.R. Thevenot. 1999. Heavy metal concentrations in dry and wet atmospheric deposits in Paris district: Comparison with urban runoff. Sci. Total Environ. 235: 235–245.

Gavel, M.J., S.D. Richardson, R.L. Dalton, C. Soos, B. Ashby, L. McPhee et al. 2019. Effects of 2 neonicotinoid insecticides on blood cell profiles and corticosterone concentrations of wood frogs (*Lithobates sylvaticus*). Environ. Toxicol. Chem. 38: 1273–1284.

Gervasi, S.S. and J. Foufopoulos 2008. Costs of plasticity: Responses to desiccation decrease post-metamorphic immune function in a pond-breeding amphibian. Funct. Ecol. 22: 100–108.

Gesing, A., A. Bilang-Bleuel, S.K. Droste, A.C. Linthorst, F. Holsboer and J.M. Reul. 2001. Psychological stress increases hippocampal mineralocorticoid receptor levels: Involvement of corticotropin-releasing hormone. J. Neurosci. 21: 4822v4829.

Glennemeier, K.A. and R.J. Denver. 2001. Sublethal effects of chronic exposure to an organochlorine compound on northern leopard frog (*Rana pipiens*) tadpoles. Environ. Toxicol. 16: 287–297.

Glennemeier, K.A. and R.J. Denver. 2002. Small changes in whole-body corticosterone content affect larval *Rana pipiens* fitness components. Gen. Comp. Endocrinol. 127: 16–25.

Glos, J., T.U. Grafe, M.-O. Rödel and K.E. Linsenmair. 2003. Geographic variation in pH tolerance of two populations of the European common frog, *Rana temporaria*. Copeia 2003: 650–656.

Goldner, W.S., D.P. Sandler, F. Yu, V. Shostrom, J.A. Hoppin, F. Kamel and T.D. LeVan. 2013. Hypothyroidism and pesticide use among male private pesticide applicators in the agricultural health study. Int. J. Occup. Environ. Med. 55: 1171.

Goleman, W.L., J.A. Carr and T.A. Anderson. 2002a. Environmentally relevant concentrations of ammonium perchlorate inhibit thyroid function and alter sex ratios in developing *Xenopus laevis*. Environ. Toxicol. Chem. 21: 590–597.

Goleman, W.L., L.J. Urquidi, T.A. Anderson, E.E. Smit, R.J. Kendall and J.A. Carr. 2002b. Environmentally relevant concentrations of ammonium perchlorate inhibit development and metamorphosis in *Xenopus laevis*. Environ. Toxicol. Chem. 21: 424–430.

Gomez-Mestre, I. and M. Tejedo. 2003. Local adaptation of an anuran amphibian to osmotically stressful environments. Evolution 57: 1889–1899.

Gomez-Mestre, I., M. Tejedo, E. Ramayo and J. Estepa. 2004. Developmental alterations and osmoregulatory physiology of a larval anuran under osmotic stress. Physiol. Chem. Zool. 77: 267–274.

Grant, K.P. and L.E. Licht. 1993. Acid tolerance of anuran embryos and larvae from central Ontario. J. Herpetol. 27: 1–6.

Gray, K.M. and P.A. Janssens. 1990. Gonadal hormones inhibit the induction of metamorphosis by thyroid hormones in *Xenopus laevis* tadpoles *in vivo*, but not *in vitro*. Gen. Comp. Endocrinol. 77: 202–211.

Griffiths, R.A. and P.D. Wijer. 1994. Differential effects of pH and temperature on embryonic development in the British newts (*Triturus*). J. Zool. 234: 613–622.

Grosell, M. 2011. Copper. pp. 54–110. *In*: Wood, C., A. Farrel and C. Brauner [eds.]. Homeostasis and Toxicology of Essential Metals. Vol. 31—Fish Physiology. Academic, San Diego, CA, USA.

Guillette Jr, L.J. and T.M. Edwards. 2005. Is nitrate an ecologically relevant endocrine disruptor in vertebrates? Interg. Comp. Biol. 45: 19–27.

Guillette Jr., L.J., S. Kools, M.P. Gunderson and D.S. Bermudez. 2006. DDT and its analogues: New insights into their endocrine disruptive effects on wildlife. pp. 332–355. *In*: Norris, D.O. and J.A. Carr [eds.]. Endocrine Disruption: Biological Bases for Health Effects in Wildlife and Humans. Oxford University Press, UK.

Gunasekara, A.S., A.L. Rubin, K.S. Goh, F.C. Spurlock and R.S. Tjeerdema. 2008. Environmental fate and toxicology of carbaryl. Environ. Contam. Toxicol. 196: 95–121.

Gupta, P.K. 2017. Herbicides and fungicides. pp. 657–679. *In*: Gupta, R.C. [ed.]. Reproductive and Developmental Toxicology. Academic Press, Cambridge, MA, USA.

Gupta, R.C., I.R.M. Mukherjee, J.K. Malik, R.B. Doss, W.D. Dettbarn and D. Milatovic. 2019. Insecticides. pp. 455–475. *In*: Gupta, R.C. [ed.]. Biomarkers in Toxicology. Academic Press, Cambridge, MA, USA.

Gutleb, A.C., J. Appelman, M. Bronkhorst, J.H.J. van den Berg and A.J. Murk. 2000. Effects of oral exposure to polychlorinated biphenyls (PCBs) on the development and metamorphosis of two amphibian species (*Xenopus laevis* and *Rana temporaria*). Sci. Total Environ. 262: 147–57.

Gyllenhammar, I., I. Holm, R. Eklund and C. Berg. 2009 Reproductive toxicity in *Xenopus tropicalis* after developmental exposure to environmental concentrations of ethynylestradiol. Aquat. Toxicol. 91: 171–178.

Hallgren, S., T. Sinjari, H. Hakansson and P.O. Darnerud. 2001. Effects of polybrominated diphenyl ethers (PBDEs) and polychlorinated biphenyls (PCBs) on thyroid hormone and vitamin A levels in rats and mice. Arch. Toxicol. 75: 200–208.

Hares, R.J. and N.I. Ward. 1999. Comparison of the heavy metal content of motorway stormwater following discharge into wet biofiltration and dry detention ponds along the London Orbital (M25) motorway. Sci. Total Environ. 235: 169–178.

Hatch, A.C. and A.R. Blaustein. 2000. Combined effects of UV-B, nitrate, and low pH reduce the survival and activity level of larval cascades frogs (*Rana cascadae*). Arch. Environ. Con. Tox. 39: 494–499.

Hatch, A.C. and A.R. Blaustein. 2003. Combined effects of UV-B radiation and nitrate fertilizer on larval amphibians. Ecol. Appl. 13: 1083–1093.

Hayes, T.B. 1997. Hormonal mechanisms as potential constraints on evolution: Examples from the Anura. Am. Zool. 37: 482–490.

Hayes, T.B. 1998. Sex determination and primary sex differentiation in amphibians: Genetic and developmental mechanisms. J. Exp. Zool. 281: 373–399.

Hayes, T.B., P. Case, S. Chui, D. Chung, C. Haeffele, K. Haston et al. 2006. Pesticide mixtures, endocrine disruption, and amphibian declines: Are we underestimating the impact? Environ. Health Perspect. 114(Suppl 1): 40–50.

Hayes, T.B., P. Falso, S. Gallipeau and M. Stice. 2010. The cause of global amphibian declines: A developmental endocrinologist's perspective. J. Exp. Zool. 213: 921–933.

Hayes, T.B., L.L. Anderson, V.R. Beasley, S.R. De Solla, T. Iguchi, H. Ingraham et al. 2011. Demasculinization and feminization of male gonads by atrazine: Consistent effects across vertebrate classes. J. Steroid Biochem. Mol. Bio. 127: 64–73.

Hayes, T.B., M. Hansen, A.R. Kapuscinski, K.A. Locke and A. Barnosky. 2017. From silent spring to silent night: Agrochemicals and the anthropocene. Sci. Anthrop. 5: 57.

Hecker, M., J.P. Giesy, P.D. Jones, A.M. Jooste, J.A. Carr, K.R. Solomon et al. 2004. Plasma sex steroid concentrations and gonadal aromatase activities in African clawed frogs (*Xenopus laevis*) from South Africa. Environ. Toxicol. Chem. 23: 1996–2007.

Heimeier, R.A., B. Das, D.R. Buchholz and Y.-B. Shi. 2009. The xenoestrogen bisphenol A inhibits postembryonic vertebrate development by antagonizing gene regulation by thyroid hormone. Endocrinology 150: 2964–2973.

Heimeier, R.A. and Y.-B. Shi. 2010. Amphibian metamorphosis as a model for studying endocrine disruption on vertebrate development: Effect of bisphenol A on thyroid hormone action. Gen. Comp. Endocrinol. 168: 181–189.

Helbing, C.C., K. Ovaska and L. Ji. 2006. Evaluation of the effect of acetochlor on thyroid hormone receptor gene expression in the brain and behavior of *Rana catesbeiana* tadpoles. Aquat. Toxicol. 80: 42–51.

Herkovitz, J. and C.S. Pérez-Coll. 1991. Antagonism and synergism between lead and zinc in amphibian larvae. Eviron. Poll. 69: 217–221.

Hintz, W.D. and R.A. Relyea. 2019. A review of the species, community, and ecosystem impacts of road salt salinisation in fresh waters. Freschw. Biol. 64: 1081–1097.

Hogan, N.S., P. Duarte, M.G. Wade, D.R.S. Lean and V.L. Trudeau. 2008. Estrogenic exposure affects metamorphosis and alters sex ratios in the northern leopard frog (*Rana pipiens*): Identifying critically vulnerable periods of developments. Gen. Comp. Endocrinol. 156: 515–523.

Hopkins, W.A. 2007. Amphibians as models for studying environmental change. ILAR J. 48: 270–277.

Horn, O., S. Nalli, D. Cooper and J. Nicell. 2004. Plasticizer metabolites in the environment. Water. Res. 38: 3693–3698.

Horton, A.A., A. Walton, D.J. Spurgeon, E. Lahive and C. Svendsen. 2017. Microplastics in freshwater and terrestrial environments: Evaluating the current understanding to identify the knowledge gaps and future research priorities. Sci. Total Environ. 586: 127–141.

Howe, C.M., M. Berrill, B.D. Pauli, C.C. Helbing, K. Werry and N. Veldhoen. 2004. Toxicity of glyphosate-based pesticides to four North American frog species. Environ. Toxicol. Chem. 23: 1928–1938.

Hu, L., L. Su, Y. Xue, J. Mu, J. Zhu, J. Xu, J. and H. Shi 2016. Uptake, accumulation and elimination of polystyrene microspheres in tadpoles of *Xenopus tropicalis*. Chemosphere 164: 611–617.

Hu, Y.C., Y. Tang, Z.Q. Chen, J.Y. Chen and G.H, Ding. 2019. Evaluation of the sensitivity of *Microhyla fissipes* tadpoles to aqueous cadmium. Ecotoxicology 28: 1150–1159.

Hua, J. and B.A. Pierce. 2013. Lethal and sublethal effects of salinity on three common Texas amphibians. Copeia. 2013: 562–566.

Huang, H. and D.D. Brown. 2000. Prolactin is not a juvenile hormone in *Xenopus laevis* metamorphosis. PNAS 97: 195–199.

Huang, M.-Y., R.-Y. Duan and X. Ji. 2014. Chronic effects of environmentally-relevant concentrations of lead in *Pelophylax nigromaculata* tadpoles: Threshold dose and adverse effects. Ecotox. Environ. Safe. 104: 310–316.

Ishizuya-Oka, A., T. Hasebe and Y.B. Shi. 2010. Apoptosis in amphibian organs during metamorphosis. Apoptosis 15: 350–364.

Iwamuro, S., M. Sakakibara, M. Terao, A. Ozawa, C. Kurobe, T. Shigeura et al. 2003. Teratogenic and anti-metamorphic effects of bisphenol A on embryonic and larval *Xenopus laevis*. Gen. Comp. Endocrinol. 133: 189–198.

James, S.M. and E.E. Little. 2003. The effects of chronic cadmium exposure on American toad (*Bufo americanus*) tadpoles. Environ. Toxicol. Chem. 22: 377–380.

Jeschke, P., R. Nauen, M. Schindler and A. Elbert. 2011. Overview of the status and global strategy for neonicotinoids. J. Agric. Food Chem. 59: 2897–2908.

Jobling, S., M. Nolan, C.R. Tyler, G. Brighty and J.P. Sumpter. 1998. Widespread sexual disruption in wild fish. Environ. Sci. Technol. 32: 2498–2506.

Karaoğlu, K. and S. Gül. 2020. Characterization of microplastic pollution in tadpoles living in small water-bodies from Rize, the northeast of Turkey. Chemosphere 255: 126915.

Kaushal, S.S., G.E. Likens, M.L. Pace, R.M. Utz, S. Haq, J. Gorman and M. Grese. 2018. Freshwater salinization syndrome on a continental scale. Proceedings of the National Academy of Sciences 115(4): E574–E583.

Kiesecker, J. 1996. pH-mediated predator-prey interactions between *Ambystoma tigrinum* and *Pseudacris triseriata*. Ecol. Appl. 6: 1325–1331.

Kikuyama, S., K. Kawamura, S. Tanaka and K. Yamamoto. 1993. Aspects of amphibian metamorphosis: Hormonal control. Int. Rev. Cytol. 145: 105–148.

Kime, D.E. 1998. Endocrine Disruption in Fish. Springer Science & Business Media.

Kirschman, L.J., M.D. McCue, J.G. Boyles and R.W. Warne. 2017. Exogenous stress hormones alter energetic and nutrient costs of development and metamorphosis. J. Exp. Biol. 220: 3391–3397.

Kloas, W., I. Lutz and R. Einspanier. 1999. Amphibians as a model to study endocrine disruptors: II. Estrogenic activity of environmental chemicals *in vitro* and *in vivo*. Sci. Total Environ. 225: 59–68.

Kloas, W. 2002. Amphibians as a model for the study of endocrine disruptors. Int. Rev. Cytol. 216: 1–57.

Kloas, W. and I. Lutz. 2006. Amphibians as model to study endocrine disrupters. J. Chromatogr. A. 1130: 16–27.

Kloas, W., R. Urbatzka, R. Opitz, S. Würtz, T. Behrends, I. Lutz et al. 2009a. Endocrine disruption in aquatic vertebrates. Ann. N. Y. Acad. Sci. 1163: 187–200.

Kloas, W., I. Lutz, T. Springer, H. Krueger, J. Wolf, L. Holden et al. 2009b. Does atrazine influence larval development and sexual differentiation in *Xenopus laevis*? Toxicol. Sci. 107: 376–384.

Knäbel, A., K. Meyer, J. Rapp and R. Schulz. 2014. Fungicide field concentrations exceed FOCUS surface water predictions: Urgent need of model improvement. Environ. Sci. Technol. 48: 455–463.

Kolenda, K., N. Kuśmierek and K. Pstrowska. 2020. Microplastic ingestion by tadpoles of pond-breeding amphibians—first results from Central Europe (SW Poland). Environ. Sci. Pollut. Res. 27: 33380–33384.

Kolodziej, E.P., T. Harter and D.L. Sedlak. 2004. Dairy wastewater, aquaculture, and spawning fish as sources of steroid hormones in the aquatic environment. Environ. Sci. Technol. 38: 6377–6384.

Kulkarni, S.S. and D.R. Buchholz. 2012. Beyond synergy: Corticosterone and thyroid hormone have numerous interaction effects on gene regulation in *Xenopus tropicalis* tadpoles. Endocrinol. 153: 5309–5324.

Kvarnryd, M., R. Grabic, I. Brandt and C. Berg. 2011. Early life progestin exposure causes arrested oocyte development, oviductal agenesis and sterility in adult *Xenopus tropicalis* frogs. Aquat. Toxicol. 103: 18–24.

Lanctot, C., C. Robertson, L. Navarro-Martín, C. Edge, S.D. Melvin, V.L. Trudeau et al. 2013. Effects of the glyphosate-based herbicide Roundup WeatherMax® on metamorphosis of wood frogs (*Lithobates sylvaticus*) in natural wetlands. Aquat. Toxicol. 140: 48–57.

Lanctôt, C., L. Navarro-Martín, C. Robertson, B. Park, P. Jackman, V.L. Trudeau et al. 2014. Effects of glyphosate-based herbicides on survival, development, growth and sex ratios of wood frog (*Lithobates sylvaticus*) tadpoles. II: Agriculturally relevant exposures to Roundup WeatherMax® and Vision® under laboratory conditions. Aquat. Toxicol. 154: 291–303.

Langlois, V.S., A.C. Carew, B.D. Pauli, M.G. Wade, G.M. Cooke and V.L. Trudeau. 2010. Low levels of the herbicide atrazine alter sex ratios and reduce metamorphic success in *Rana pipiens* tadpoles raised in outdoor mesocosms. Environ. Health Perspect 118: 552–557.

Larson, D.L., S. McDonald, S.J. Hamilton, A.J. Fivizzani and W.E. Newton. 1998. Effects of the herbicide atrazine on *Ambystoma tigrinum* metamorphosis: duration, larval growth, and hormonal response. Physiol. Zool. 71: 671–679.

Leduc, J., P. Echaubard, V. Trudeau and D. Lesbarrères. 2015. Copper and nickel effects on survival and growth of northern leopard frog (*Lithobates pipiens*) tadpoles in field-collected smelting effluent water. Environ. Toxicol. Chem. 35: 687–694.

Lefcort, H., R.A. Meguire, L.H. Wilson and W.F. Ettinger. 1998. Heavy metals alter the survival, growth, metamorphosis, and antipredatory behavior of Columbia spotted frog (*Rana luteiventris*) tadpoles. Arch. Environ. Con. Tox. 35: 447–456.

Leggett, S., J. Borrelli, D.K. Jones and R. Relyea. 2021. The combined effects of road salt and biotic stressors on amphibian sex ratios. Environ. Toxicol. Chem. 40: 231–235.

Legret, M. and C. Pagotto. 1999. Evaluation of pollutant loadings in the runoff waters from a major rural highway. Sci. Total Environ. 235: 143–150.

Lehigh Shirey, E.A., A.J. Langerveld, D. Mihalko and C.F. Ide. 2006. Polychlorinated biphenyl exposure delays metamorphosis and alters thyroid hormone system gene expression in developing *Xenopus laevis*. Environ. Res. 102: 205–214.

Lenuweit, U. 2009. Beeinträchtigungen von Amphibien durch Düngemittel - ein Überblick. Rana 10: 14–25.

Letcher, R. and P. Behnisch. 2003. The state-of-science and trends of BFRs in the environment. Environ. Int. 29: 663–885.

Levy, G., I. Lutz, A. Krüger and W. Kloas. 2004. Bisphenol A induces feminization in *Xenopus laevis* tadpoles. Environ. Res. 94: 102–111.

Li, S., M. Li, W. Gui, Q. Wang and G. Zhu. 2017. Disrupting effects of azocyclotin to the hypothalamo-pituitary-gonadal axis and reproduction of *Xenopus laevis*. Aquat. Toxicol. 185: 121–128.

Ling, R.W., J.P. VanAmberg and J.K. Werner. 1986. Pond acidity and its relationship to larval development of *Ambystoma maculatum* and *Rana sylvatica* in upper Michigan. J. Herpetol. 20: 230–236.

Lintelmann, J., A. Katayama, N. Kurihara, L. Shore and A. Wenzel. 2003. Endocrine disruptors in the environment (IUPAC Technical Report). Pure Appl. Chem. 75: 631–681.

Liu, C., M. Ha, L. Li and K. Yang. 2014. PCB153 and p, p′-DDE disorder thyroid hormones via thyroglobulin, deiodinase 2, transthyretin, hepatic enzymes and receptors. Environ. Sci. Pollut. Res. 21: 11361–11369.

Liu, S., G.G. Ying, L.J. Zhou, R.Q. Zhang, Z.F. Chen and H.J. Lai. 2012. Steroids in a typical swine farm and their release into the environment. Water Res. 46: 3754–3768.

Lorenz, C., V. Contardo-Jara, S. Pflugmacher, C. Wiegand, G. Nützmann, I. Lutz and W. Kloas. 2011. The synthetic gestagen levonorgestrel impairs metamorphosis in *Xenopus laevis* by disruption of the thyroid system. Toxicol. Sci. 123: 94e102.

Lorenz, C., R. Opitz, A. Trubiroha, I. Lutz, A. Zikova and W. Kloas. 2016. The synthetic gestagen levonorgestrel is an endocrine disruptor of the amphibian thyroid system: Evidence for a direct effect on thyroidal gene expression. Aquat. Toxicol. 177: 63e73.

Loumbourdis, N.S., P. Kyriakopoulou-Sklavounou and G. Zachariadis. 1999. Effects of cadmium exposure on bioaccumulation and larval growth in the frog *Rana ridibunda*. Environ. Poll. 104: 429–433.

Lu, H., Y. Hu, C. Kang, Q. Meng and Z. Lin. 2021. Cadmium-induced toxicity to amphibian tadpoles might be exacerbated by alkaline not acidic pH level. Ecotox. Environ. Safe. 218: 112288.

Lukens, E. and T.E. Wilcoxen. 2020. Effects of elevated salinity on Cuban treefrog *Osteopilus septontrionalis* aldosterone levels, growth, and development. Mar. Freshwater Behav. Physiol. 53: 99–111.

Maher, S.K., P. Wojnarowicz, T.-A. Ichu, N. Veldhoen, L. Lu, M. Lesperance et al. 2016. Rethinking the biological relationships of the thyroid hormones, lL-thyroxine and 3,5,3′-triiodothyronine. Comp. Biochem. Physiol. D 18: 44–53.

Mann, R.M., R.V. Hyne, C.B. Choung and S.P. Wilson. 2009. Amphibians and agricultural chemicals: Review of the risks in a complex environment. Environ. Pollut. 157: 2903–2927.

Marques, S.M., F. Gonçalves and R. Pereira. 2008. Effects of a uranium mine effluent in the early-life stages of *Rana perezi* Seoane. Sci. Total Environ. 402: 29–35.

Mathieu-Denoncourt, J., S.J. Wallace, S.R. de Solla and V.S. Langlois. 2015. Plasticizer endocrine disruption: Highlighting developmental and reproductive effects in mammals and non-mammalian aquatic species. Gen. Comp. Endocrinol. 219: 74–88.

Matsushima, A. 2018. A novel action of endocrine-disrupting chemicals on wildlife; DDT and its derivatives have remained in the environment. Int. J. Mol. Sci. 19: 1377.

Mazotto, V., F. Gagne, M.G. Marin, F. Ricciardi and C. Blaise. 2008. Vitellogenin as a biomarker of exposure to estrogenic compounds in aquatic invertebrates: A review. Environ. Int. 34: 531–545.

McDonald, D.G., J.L. Ozog and B.P. Simons. 1984. The influence of low pH environments on ion regulation in the larval stages of the anuran amphibian, *Rana clamitans*. Can. J. Zool. 62: 2171–2177.

McMahon, T.A., N.T. Halstead, S. Johnson, T.R. Raffel, J.M. Romansic, P.W. Crumrine et al. 2011. Fungicide chlorothalonil is nonlinearly associated with corticosterone levels, immunity, and mortality in amphibians. Environ. Health Perspect. 119: 1098–1103.

McMahon, T.A., R.K. Boughton, L.B. Martin and J.R. Rohr. 2017. Exposure to the herbicide atrazine nonlinearly affects tadpole corticosterone levels. J. Herpetol. 51: 270–273.

Meerts, I.A., J.J. van Zanden, E.A. Luijks, I. van Leeuwen-Bol, G. Marsh, E. Jakobsson et al. 2000. Potent competitive interactions of some brominated flame retardants and related compounds with human transthyretin *in vitro*. Toxicol. Sci. 56: 95–104.

Melvin, S.D., M.C. Cameron and C.M. Lanctôt. 2014. Individual and mixture toxicity of pharmaceuticals naproxen, carbamazepine, and sulfamethoxazole to Australian striped marsh frog tadpoles (*Limnodynastes peronii*). J. Toxicol. Environ. Health A 77: 337–345.

Mengeling, B.J., Y. Wei, L.N. Dobrawa, M. Streekstra, J. Louisse, V. Singh et al. 2017. A multi-tiered, *in vivo*, quantitative assay suite for environmental disruptors of thyroid hormone signaling. Aquat. Toxicol. 190: 1–10.

Merilä, J., F. Söderman, R. O'Hara, K. Räsänen and A. Laurila. 2004. Local adaptation and genetics of acid-stress tolerance in the moor frog, *Rana arvalis*. Conserv. Gen. 5: 513–527.

Metcalfe, C.D., S. Chu, C. Judt, H. Li, K.D. Oakes, M.R. Servos et al. 2010. Antidepressants and their metabolites in municipal wastewater, and downstream exposure in an urban watershed. Environ. Toxicol. Chem. 29: 79–89.

Middlemis Maher, J., E.E. Werner and R.J. Denver. 2013. Stress hormones mediate predator-induced phenotypic plasticity in amphibian tadpoles. P. Roy. Soc. B-Biol. Sci. 280: 20123075.

Milotic, D., M. Milotic and J. Koprivnikar. 2017. Effects of road salt on larval amphibian susceptibility to parasitism through behavior and immunocompetence. Aquat. Toxicol. 189: 42–49.

Mitchell, S.E., C.A. Caldwell, G. Gonzales, W.R. Gould and R. Arimoto. 2005. Effects of depleted uranium on survival, growth, and metamorphosis in the African clawed frog (*Xenopus laevis*). J. Toxicol. Environ. Health A 68: 951–965.

Miyata, K. and K. Ose. 2012. Thyroid hormone-disrupting effects and the amphibian metamorphosis assay. J. Toxicol. Pathol. 25: 1–9.

Morling, G., C. Forsberg and R.G. Wetzel. 1985. Lake Anketjarn, a non-acidified lake in an acidified region. Oikos 44: 324–328.

Mortensen, A.S. and A. Arukwe. 2006. The persistent DDT metabolite, 1, 1-dichloro-2, 2-bis (p-chlorophenyl) ethylene, alters thyroid hormone-dependent genes, hepatic cytochrome P4503A, and pregnane× receptor gene expressions in atlantic salmon (*Salmo salar*) parr. Environ. Toxicol. Chem. 25: 1607–1615.

Moutinho, M.F., E.A. de Almeida, E.L. Espíndola, M.A. Daam and L. Schiesari. 2020. Herbicides employed in sugarcane plantations have lethal and sublethal effects to larval *Boana pardalis* (Amphibia, Hylidae). Ecotoxicol. 29: 1043–1051.

Nakkrasae, L.I., S. Phummisutthigoon and N. Charoenphandhu. 2016. Low salinity increases survival, body weight and development in tadpoles of the Chinese edible frog *Hoplobatrachus rugulosus*. Aquac Res. 47: 3109–3118.

Narayan, E.J., J.F. Cockrem and J.M. Hero. 2012. Effects of temperature on urinary corticosterone metabolite responses to short-term capture and handling stress in the cane toad (*Rhinella marina*). Gen. Comp. Endocrinol. 178: 301–305.

Narayan, E.J., J.F. Cockrem and J.M. Hero. 2013. Sight of a predator induces a corticosterone stress response and generates fear in an amphibian. PLoS one. 8: e73564.

Navarro-Martín, L., C. Lanctot, P. Jackman, B.J. Park, K. Doe, V.L. Trudeau et al. 2014. Effects of glyphosate-based herbicides on survival, development, growth and sex ratios of wood frogs (*Lithobates sylvaticus*) tadpoles. I: Chronic laboratory exposures to VisionMax®. Aquat. Toxicol. 154: 278–290.

Norris, D.O. and J.A. Carr. 2013. Vertebrate Endocrinology, fifth ed. Academic Press, San Diego, USA.

Nriagu, J.O. 1989. A global assessment of natural sources of atmospheric trace metals. Nature 338: 47–49.

Nriagu, J.O. and J.M. Pacyna. 1988. Quantitative assessment of worldwide contamination of air, water and soils by trace metals. Nature 333: 134–139.

O'Dell, L. 1990. Copper. pp. 261–267. *In*: Brown,M.L. [ed.]. Present Knowledge in Nutrition. International Life Sciences Institute, Washington, DC, USA.

Ogawa, A., J. Dake, Y.K. Iwashima and T. Tokumoto. 2011. Induction of ovulation in *Xenopus* without hCG injection: The effect of adding steroids into the aquatic environment. Reprod. Biol. Endocrinol. 9: 1e6.

Ohtani, H., I. Miura and Y. Ichikawa. 2000. Effects of dibutyl phthalate as an environmental endocrine disruptor on gonadal sex differentiation of genetic males of the frog *Rana rugosa*. Environ. Health Perspect. 108: 1189–1193.

Oldham, R.S., D.M. Latham, D. Hilton-Brown, M. Towns, A.S. Cooke, A. Burn et al. 1997. The effect of ammonium nitrate fertiliser on frog (*Rana temporaria*) survival. Agric. Ecosyst. Environ. 61: 69–74.

Opitz, R., S. Hartmann, T. Blank, T. Braunbeck, I. Lutz and W. Kloas. 2006. Evaluation of histological and molecular endpoints for enhanced detection of thyroid system disruption in *Xenopus laevis* tadpoles. Toxicol. Sci. 90: 337–348.

Opitz, R., F. Schmidt, T. Braunbeck, S. Wuertz and W. Kloas. 2009. Perchlorate and ethylenethiourea induce different histological and molecular alterations in a non-mammalian vertebrate model of thyroid goitrogenesis. Mol. Cell. Endocrinol. 298: 101–114.

Ortiz, M.E., A. Marco, N. Saiz and M. Lizana. 2004. Impact of ammonium nitrate on growth and survival of six European amphibians. Arch. Environ. Contam. Toxicol. 47: 234–239.

Ortiz-Santaliestra, M.E. and D.W. Sparling. 2007. Alteration of larval development and metamorphosis by nitrate and perchlorate in southern leopard frogs (*Rana sphenocephala*). Arch. Environ. Contam. Toxicol. 53: 639–646.

Orton, F., J.A. Carr and R.D. Handy. 2006. Effects of nitrate and atrazine on larval development and sexual differentiation in the northern leopard frog *Rana pipiens*. Environ. Toxicol. Chem. 25: 65–71.

Orton, F., M. Säfholm, E. Jansson, Y. Carlsson, A. Eriksson, J. Fick et al. 2018. Exposure to an anti-androgenic herbicide negatively impacts reproductive physiology and fertility in *Xenopus tropicalis*. Sci. Rep. 8: 1–15.

Pahkala, M., A. Laurila, L.O. Björn and J. Merilä. 2001. Effects of ultraviolet-B radiation and pH on early development of the moor frog *Rana arvalis*. J. Appl. Ecol. 38: 628–636.

Paris, M. and V. Laudet. 2008. The history of a developmental stage: Metamorphosis in chordates. Genesis 46: 657–672.

Parris, M.J. and D.R. Baud. 2004. Interactive effects of a heavy metal and chytridiomycosis on gray treefrog larvae (*Hyla chrysoscelis*). Copeia 2004: 344–350.

Pauwels, B., K. Wille, H. Noppe, H. De Brabander, T. Van de Wiele, W. Verstraete et al. 2008. 17α-ethinylestradiol cometabolism by bacteria degrading estrone, 17β-estradiol and estriol. Biodegradation 19: 683–693.

Peckett, A.J., D.C. Wright and M.C. Riddell. 2011. The effects of glucocorticoids on adipose tissue lipid metabolism. Metabolism. 60: 1500–1510.

Peles, J.D. 2013. Effects of chronic aluminum and copper exposure on growth and development of wood frog (*Rana sylvatica*) larvae. Aquat. Toxicol. 140: 242–248.

Peltzer, P.M., R.C. Lajmanovich, C. Martinuzzi, A.M. Attademo, L.M. Curi and M.T. Sandoval. 2019. Biotoxicity of diclofenac on two larval amphibians: Assessment of development, growth, cardiac function and rhythm, behavior and antioxidant system. Sci. Total Environ. 683: 624–637.

Pickford, D.B., M.J. Hetheridge, J.E. Caunter, A.T. Hall and T.H. Hutchinson. 2003. Assessing chronic toxicity of bisphenol A to larvae of the African clawed frog (*Xenopus laevis*) in a flow-through exposure system. Chemosphere 53: 223–235.

Piprek, R.P. and J.Z. Kubiak. 2014. Development of gonads, sex determination, and sex reversal in *Xenopus*. pp. 199–214. *In*: Kloc, M. and J.Z. Kubiak [eds.]. Xenopus Development. Wiley-Blackwell, Hoboken, NJ, USA.

Piprek, R.P., A. Pecio, J.Z. Kubiak and J.M. Szymura. 2012. Differential effects of testosterone and 17β-estradiol on gonadal development in five anuran species. Reproduction 144: 257–267.

Plastics Europe. 2020. Plastics-the Facts 2020. An analysis of European plastics production, demand and waste data.

Polo-Cavia, N., P. Burraco and I. Gomez-Mestre. 2016. Low levels of chemical anthropogenic pollution may threaten amphibians by impairing predator recognition. Aquat. Toxicol. 172: 30–35.

Poulsen, R., N. Cedergreen, T. Hayes and M. Hansen. 2018. Nitrate: An environmental endocrine disruptor? A review of evidence and research needs. Environ. Sci. Technol. 52: 3869–3887.

Prokić, M.D., T.B. Radovanović, J.P. Gavrić and C. Faggio. 2019. Ecotoxicological effects of microplastics: Examination of biomarkers, current state and future perspectives. Trends Analyt. Chem. 111: 37–46.

Quaranta, A., V. Bellantuono, G. Cassano and C. Lippe. 2009. Why amphibians are more sensitive than mammals to xenobiotics. PLoS One 4: 1–4.

Quiroga, L.B., E.A. Sanabria, M.W. Fornés, D.A. Bustos and M. Tejedo. 2019. Sublethal concentrations of chlorpyrifos induce changes in the thermal sensitivity and tolerance of anuran tadpoles in the toad *Rhinella arenarum*? Chemosphere 219: 671–677.

Rainsford, K.D. 2009. Ibuprofen: Pharmacology, efficacy and safety. Inflammopharmacology 17: 275–342.

Relyea, R.A. and N. Mills. 2001. Predator-induced stress makes the pesticide carbaryl more deadly to gray treefrog tadpoles (*Hyla versicolor*). PNAS 98: 2491–2496.

Relyea, R.A. 2004. Growth and survival of five amphibian species exposed to combinations of pesticides. Environ. Toxicol. Chem. 23: 1737–1742.

Relyea, R.A. 2005. The impact of insecticides and herbicides on the biodiversity and productivity of aquatic communities. Ecol. Appl. 15: 618–627.

Ribeiro, F., J.W. O'Brien, T. Galloway and K.V. Thomas. 2019. Accumulation and fate of nano-and micro-plastics and associated contaminants in organisms. Trends Analyt. Chem. 111: 139–147.

Rice, T.M., B.J. Blackstone, W.L. Nixdorf and D.H. Taylor. 1999. Exposure to lead induces hypoxia-like responses in bullfrog larvae (*Rana catesbeiana*). Environ. Toxicol. Chem. 18: 2283–2288.

Robinson, S.A., S.D. Richardson, R.L. Dalton, F. Maisonneuve, V.L. Trudeau, B.D. Pauli et al. 2017. Sublethal effects on wood frogs chronically exposed to environmentally relevant concentrations of two neonicotinoid insecticides. Environ. Toxicol. Chem. 36: 1101–1109.

Rochman, C.M., T. Kurobe, I. Flores and S.J. Teh. 2014. Early warning signs of endocrine disruption in adult fish from the ingestion of polyethylene with and without sorbed chemical pollutants from the marine environment. Sci. Total Environ. 493: 656–661.

Rochman, C.M. 2015. The complex mixture, fate and toxicity of chemicals associated with plastic debris in the marine environment. pp. 117–140. *In*: *Marine Anthropogenic Litter*. Springer, Cham.

Rohr, J.R. and B.D. Palmer. 2005. Aquatic herbicide exposure increases salamander desiccation risk eight months later in a terrestrial environment. Environ. Toxicol. Chem. 24: 1253–1258.

Rohr, J.R., T.M. Sesterhenn and C. Stieha. 2011. Will climate change reduce the effects of a pesticide on amphibians? Partitioning effects on exposure and susceptibility to contaminants. Glob. Change Biol. 17: 657–666.

Rohr, J.R. and B.D. Palmer. 2013. Climate change, multiple stressors, and the decline of ectotherms. Conservation Biology 27(4): 741–751.

Rollins-Smith, L.A. 2017. Amphibian immunity–stress, disease, and climate change. Dev. Comp. Immunol. 66: 111–119.

Romero, L.M. 2004. Physiological stress in ecology: Lessons from biomedical research. Trends Ecol. Evol. 19: 249–255.
Rosenberg, E.A. and B.A. Pierce. 1995. Effect of initial mass on growth and mortality at low pH in tadpoles of *Pseudacris clarkii* and *Bufo valliceps*. J. Herpetol. 29: 181–185.
Roser, M. 2019. Pesticides. Published online at OurWorldInData.org. Retrieved from: 'https://ourworldindata.org/pesticides' downloaded on 30 June 2021.
Rosman, Y., I. Makarovsky, Y. Bentur, S. Shrot, T. Dushnistky and A. Krivoy. 2009. Carbamate poisoning: Treatment recommendations in the setting of a mass casualties event. Am. J. Emerg. Med. 27: 1117–1124.
Rotermann, M., C. Sanmartin, D. Hennessy and M. Arthur. 2014. Prescription medication use by Canadians aged 6 to 79. Health Rep. 25: 3–9.
Rowe, C.L., W.J. Sadinski and W.A. Dunson. 1992. Effects of acute and chronic acidification on three larval amphibians that breed in temporary ponds. Arch. Environ. Con. Tox. 23: 339–350.
Rumrill, C.T., D.E. Scott and S.L. Lance. 2018. Delayed effects and complex life cycles: How the larval aquatic environment influences terrestrial performance and survival. Environ. Toxicol. Chem. 37: 2660–2669.
Ruthsatz, K., K.H. Dausmann, C. Drees, L.I. Becker, L. Hartmann, J. Reese et al. 2018a. Altered thyroid hormone levels affect body condition at metamorphosis in larvae of *Xenopus laevis*. J. Appl. Toxicol. 38: 1416–1425.
Ruthsatz, K., K.H. Dausmann, M.A. Peck, C. Drees, N.M. Sabatino, L.I. Becker et al. 2018b. Thyroid hormone levels and temperature during development alter thermal tolerance and energetics of *Xenopus laevis* larvae. Conserv. Phys. 6(1): 1–15.
Ruthsatz, K., K.H. Dausmann, S. Reinhardt, T. Robinson, N.M. Sabatino, M.A. Peck et al. 2019. Endocrine disruption alters developmental energy allocation and performance in *Rana temporaria*. Integ. Comp. Biol. 59(1): 70–88.
Ruthsatz, K., K.H. Dausmann, S. Reinhardt, T. Robinson, N.M. Sabatino, M.A. Peck et al. 2020a. Post-metamorphic carry-over effects of altered thyroid hormone level and developmental temperature: Physiological plasticity and body condition at two life stages in *Rana temporaria*. J. Comp. Physiol. B. 190(3): 297–315.
Ruthsatz, K., K.H. Dausmann, C. Drees, L.I. Becker, L. Hartmann, J. Reese et al. 2020b. Altered thyroid hormone levels affect the capacity for temperature-induced developmental plasticity in larvae of *Rana temporaria* and *Xenopus laevis*. J. Therm. Biol. 90: 102599.
Ruthsatz, K., K.H. Dausmann, K. Paesler, P. Babos, N.M. Sabatino, M.A. Peck et al. 2020c. Shifts in sensitivity of amphibian metamorphosis to endocrine disruption: The common frog (*Rana temporaria*) as a case study. Conserv. Physiol. 8: coaa100.
Ruthsatz, K., F. Bartels, D. Stützer and P.C. Eterovick. 2022a. Timing of parental breeding shapes sensitivity to nitrate pollution in the common frog *Rana temporaria*. J. Therm. Biol. 108: 103296.
Ruthsatz, K., M. Domscheit, K. Engelkes and M. Vences. 2022b. Microplastics ingestion induces plasticity in digestive morphology in larvae of *Xenopus laevis*. Com. Biochem. Physiol. A. Mol. Integr. Physiol., 111210.
Ruthsatz, K., P.C. Eterovick, F. Bartels and J. Mausbach. 2023a. Contributions of water-borne corticosterone as one non-invasive biomarker in assessing nitrate pollution stress in tadpoles of *Rana temporaria*. Gen. Comp. Endocrinol. 331: 114164.
Ruthsatz, K., A. Schwarz, I. Gomez-Mestre, R. Meyer, M. Domscheit, F. Bartels et al. 2023b. Life in plastic, it's not fantastic: Sublethal effects of polyethylene microplastics ingestion throughout amphibian metamorphosis. Sci. Total Environ. 885: 163779.
Säfholm, M., A. Norder, J. Fick and C. Berg. 2012. Disrupted oogenesis in the frog *Xenopus tropicalis* after exposure to environmental progestin concentrations. Biol. Reprod. 86: 1–7.
Säfholm, M., A. Ribbenstedt, J. Fick and C. Berg. 2014. Risks of hormonally active pharmaceuticals to amphibians: A growing concern regarding progestagens. Philos. T. Roy. Soc. B 369: 20130577.
Sanzo, D. and S.J. Hecnar. 2006. Effects of road de-icing salt (NaCl) on larval wood frogs (*Rana sylvatica*). Environ. Pollut. 140: 247–256.
Sapolsky, R.M. 2002. Endocrinology of the stress-response. pp. 409–450. *In*: Becker, J.B., S.M. Breedlove, D. Crews and M.M. McCarthy [eds.]. Behavioral Endocrinology. MIT Press, Cambridge, MA, USA.
Schultz, M.M., E.T. Furlong, D.W. Kolpin, S.L. Werner, H.L. Schoenfuss, L.B. Barber et al. 2010. Antidepressant pharmaceuticals in two US effluent-impacted streams: Occurrence and fate in water and sediment, and selective uptake in fish neural tissue. Environ. Sci. Technol. 44: 1918–1925.
Sharma, A., V. Kumar, B. Shahzad, M. Tanveer, G.P.S. Sidhu, N. Handa et al. 2019. Worldwide pesticide usage and its impacts on ecosystem. SN Appl. Sci. 1: 1–16.
Sharma, B. and R. Patiño. 2009. Effects of cadmium on growth, metamorphosis and gonadal sex differentiation in tadpoles of the African clawed frog, *Xenopus laevis*. Chemosphere 76: 1048–1055.
Shen, O., W. Wu, G. Du, R. Liu, L. Yu, H. Sun et al. 2011 Thyroid disruption by di-n-butyl phthalate (DBP) and mono-n-butyl phthalate (MBP) in *Xenopus laevis*. PLoS ONE 6: 0019159.

Shepard, J.P., J. Creighton and H. Duzan. 2004. Forestry herbicides in the United States: An overview. Wildl. Soc. Bull. 32: 1020–1027.

Shi, Q., N. Sun, H. Kou, H. Wang and H. Zhao. 2018. Chronic effects of mercury on *Bufo gargarizans* larvae: Thyroid disruption, liver damage, oxidative stress and lipid metabolism disorder. Ecotox. Environ. Safe. 164: 500–509.

Shi, Y.B. 2000. *Amphibian metamorphosis*. Wiley-Liss, New York City, USA.

Simon, M.P., I. Vatnick, H.A. Hopey, K. Butler, C. Korver, C. Hilton et al. 2002. Effects of acid exposure on natural resistance and mortality of adult *Rana pipiens*. J. Herpetol. 36: 697–699.

Skei, J.K. and D. Dolmen. 2006. Effects of pH, aluminium, and soft water on larvae of the amphibians *Bufo bufo* and *Triturus vulgaris*. Can. J. Zool. 84: 1668–1677.

Slaby, S., M. Marin, G. Marchand and S. Lemiere. 2019. Exposures to chemical contaminants: What can we learn from reproduction and development endpoints in the amphibian toxicology literature? Environ. Pollut. 248: 478–495.

Smalling, K.L., R. Reeves, E. Muths, M. Vandever, W.A. Battaglin, M.L. Hladik et al. 2015. Pesticide concentrations in frog tissue and wetland habitats in a landscape dominated by agriculture. Sci. Total Environ. 502: 80–90.

Smith, D.C. 1987. Adult recruitment in chorus frogs: effects of size and date at metamorphosis. Ecology 68: 344–350.

Smith, G.R. and A.A. Burgett. 2005. Effects of three organic wastewater contaminants on American toad, *Bufo americanus*, tadpoles. Ecotoxicology 14: 477–482.

Smith, R.A. 1872. Air and Rain: The Beginnings of a Chemical Climatology. Longmans, Green & Co, London, UK.

Sparling, D.W., S. Krest and M. Ortiz-Santaliestra. 2006. Effects of lead-contaminated sediment on *Rana sphenocephala* tadpoles. Arch. Environ. Con. Tox. 51: 458–466.

Stansley, W. and D.E. Roscoe. 1996. The uptake and effects of lead in small mammals and frogs at a trap and skeet range. Arch. Environ. Con. Tox. 30: 220–226.

Strong, R., F.L. Martin, K.C. Jones, R.F. Shore and C.J. Halsall. 2017. Subtle effects of environmental stress observed in the early life stages of the Common frog, *Rana temporaria*. Sci Rep. 7: 1–13.

Styrishave, B., B. Halling-Sørensen and F. Ingerslev. 2011. Environmental risk assessment of three selective serotonin reuptake inhibitors in the aquatic environment: A case study including a cocktail scenario. Environ. Toxicol. Chem. 30: 254–261.

Sui, Q., X. Cao, S. Lu, W. Zhao, Z. Qiu and G. Yu. 2015. Occurrence, sources and fate of pharmaceuticals and personal care products in the groundwater: A review. Emerg. Contam. 1: 14–24.

Sullivan, K.B. and K.M. Spence. 2003. Effects of sublethal concentrations of atrazine and nitrate on metamorphosis of the African clawed frog. Environ. Toxicol. Chem. 22: 627–635.

Sun, N., H. Wang, Z. Ju and H. Zhao. 2018. Effects of chronic cadmium exposure on metamorphosis, skeletal development, and thyroid endocrine disruption in Chinese toad *Bufo gargarizans* tadpoles. Environ. Toxicol. Chem. 37: 213–223.

Sussarellu, R., M. Suquet, Y. Thomas, C. Lambert, C. Fabioux, M.E.J. Pernet et al. 2016. Oyster reproduction is affected by exposure to polystyrene microplastics. Proc. Natl. Acad. Sci. 113: 2430–2435.

Tamschick, S., B. Rozenblut-Kościsty, M. Ogielska, A. Lehmann, P. Lymberakis, F. Hoffmann et al. 2016a. Sex reversal assessments reveal different vulnerability to endocrine disruption between deeply diverged anuran lineages. Sci. Rep. 6: 1–8.

Tamschick, S., B. Rozenblut-Kościsty, M. Ogielska, A. Lehmann, P. Lymberakis, F. Hoffmann et al. 2016b. Impaired gonadal and somatic development corroborate vulnerability differences to the synthetic estrogen ethinylestradiol among deeply diverged anuran lineages. Aquat. Toxicol. 177: 503–514.

Tata, J.R. 1998. Amphibian metamorphosis as a model for studying the developmental actions of thyroid hormone. Cell Res. 8: 259–272.

Tata, J.R. 2006. Amphibian metamorphosis as a model for the developmental actions of thyroid hormone. Mol. Cell. Endocrinol. 246: 10–20.

Teplitsky, C., K. Räsänen and A. Laurila. 2007. Adaptive plasticity in stressful environments: Acidity constrains inducible defences in *Rana arvalis*. Evol. Ecol. Res. 9: 447–458.

Terker, A.S. and D.H. Ellison. 2015. Renal mineralocorticoid receptor and electrolyte homeostasis. Am. J. Physiol.-Reg. I. 309: 1068–1070.

Thambirajah, A.A., E.M. Koide, J.J. Imbery and C.C. Helbing. 2019. Contaminant and environmental influences on thyroid hormone action in amphibian metamorphosis. Front. Endocrinol. 10: 276.

Thompson, R.C. 2015. Microplastics in the marine environment: sources, consequences and solutions. pp. 185–200. *In*: *Marine Anthropogenic Litter*. Springer, Cham.

Tietge, J.E., G.W. Holcombe, K.M. Flynn, P.A. Kosian, J.J. Korte, L.E. Anderson et al. 2005. Metamorphic inhibition of *Xenopus laevis* by sodium perchlorate: Effects on development and thyroid histology. Environ. Toxicol. Chem. 24: 926–933.

Tietge, J.E., B.C. Butterworth, J.T. Haselman, G.W. Holcombe, M.W. Hornung, J.J. Korte et al. 2010. Early temporal effects of three thyroid hormone synthesis inhibitors in *Xenopus laevis*. Aquat. Toxicol. 98: 44–50.

Tiwari, A. and J.W. Rachlin. 2018. A review of road salt ecological impacts. Northeast. Nat. 25: 123–142.

Tompsett, A.R., S. Wiseman, E. Higley, S. Pryce, H. Chang, J.P. Giesy et al. 2012. Effects of 17α-ethynylestradiol on sexual differentiation and development of the African clawed frog (*Xenopus laevis*). Comp. Biochem. Physiol. C 156: 202–210.

Tompsett, A.R., E. Higley, S. Pryce, J.P. Giesy, M. Hecker and S. Wiseman. 2015. Transcriptional changes in African clawed frogs (*Xenopus laevis*) exposed to17α-ethinylestradiol during early development. Ecotoxicology 24: 321–329.

Tribondeau, A., L.M. Sachs and N. Buisine. 2021. Are paedomorphs actual larvae? Dev. Dyn. 250: 779–787.

Trifuoggi, M., G. Pagano, R. Oral, D. Pavičić-Hamer, P. Burić, I. Kovačić et al. 2019. Microplastic-induced damage in early embryonal development of sea urchin *Sphaerechinus granularis*. Environ. Res. 179: 108815.

Trudeau, V.L., P. Thomson, W.S. Zhang, S. Reyaud, L. Navarro-Martin and V.S. Langlois. 2020. Agrochemicals disrupt multiple endocrine axes in amphibians. Mol. Cell. Endocrinol. 513: 110861.

Tyler, C., S. Jobling and J.P. Sumpter. 1998. Endocrine disruption in wildlife: A critical review of the evidence. Crit. Rev. Toxicol. 28: 319–361.

Urch, U.A. and J.L. Hedrick. 1981. Isolation and characterisation of the hatching enzyme from the amphibian, *Xenopus laevis*. Arch. Biochem. Biophys. 206: 424–431.

Van Emden, H.F. and D.B. Peakall 1996. Beyond silent spring: Integrated pest management and chemical safety. Chapman & Hall Ltd.

Van Meter, R.J., C.M. Swan, J. Leips and J.W. Snodgrass. 2011. Road salt stress induces novel food web structure and interactions. Wetlands. 31: 843–851.

Van Meter, R.J., D.A. Glinski, T. Hong, M. Cyterski, W.M. Henderson and S.T. Purucker. 2014. Estimating terrestrial amphibian pesticide body burden through dermal exposure. Environ. Pollut. 193: 262–268.

Van Meter, R.J., D.A. Glinski, W.M. Henderson, A.W. Garrison, M. Cyterski and S.T. Purucker. 2015. Pesticide uptake across the amphibian dermis through soil and overspray exposures. Arch. Environ. Contam. Toxicol. 69: 545–556.

Van Meter, R.J., R. Adelizzi, D.A. Glinski and W.M. Henderson. 2019. Agrochemical mixtures and amphibians: The combined effects of pesticides and fertilizer on stress, acetylcholinesterase activity, and bioaccumulation in a terrestrial environment. Environ. Toxicol. Chem. 38: 1052–1061.

Veldhoen, N. and C.C. Helbing. 2001. Detection of environmental endocrine-disruptor effects on gene expression in live *Rana catesbeiana* tadpoles using a tail fin biopsy technique. Environ. Toxicol. Chem. 20: 2704–2708.

Veldhoen, N., R.C. Skirrow, L.L.Y. Brown, G. van Aggelen and C.C. Helbing. 2014. Effects of acute exposure to the non-steroidal anti-inflammatory drug ibuprofen on the developing North American bullfrog (*Rana catesbeiana*) tadpole. Environ. Sci. Technol. 48: 10439–10447.

Vethaak, A.D., J. Lahr, S.M. Schrap, A.C. Belfroid, G.B. Rijs, A. Gerritsen et al. 2005. An integrated assessment of estrogenic contamination and biological effects in the aquatic environment of The Netherlands. Chemosphere 59: 511–524.

Viglino, L., K. Aboulfadl, M. Prévost and S. Sauvé. 2008. Analysis of natural and synthetic estrogenic endocrine disruptors in environmental waters using online preconcentration coupled with LC-APPI-MS/MS. Talanta 76: 1088–1096.

Vos, J.G., E. Dybing, H.A. Greim, O. Ladefoged, C. Lambré, J.V. Tarazona, I. Brandt and A.D. Vethaak. 2000. Health effects of endocrine-disrupting chemicals on wildlife, with special reference to the European situation. Crit. Rev. Toxicol. 30: 71–133.

Vulliet, E., L. Wiest, R. Baudot and M.F. Grenier-Loustalot. 2008. Multi-residue analysis of steroids at sub-ng/L levels in surface and ground-waters using liquid chromatography coupled to tandem mass spectrometry. J. Chromatogr. A. 1210: 84–91.

Vulliet, E. and C. Cren-Olive. 2011. Screening of pharmaceuticals and hormones at the regional scale, in surface and groundwaters intended to human consumption. Environ. Pollut. 159: 2929–2934.

Wack, C.L., S.E. DuRant, W.A. Hopkins, M.B. Lovern, R.C. Feldhoff and S.K. Woodley. 2012. Elevated plasma corticosterone increases metabolic rate in a terrestrial salamander. Comp. Biochem. Physiol. A 161: 153–158.

Wagner, N., W. Reichenbecher, H. Teichmann, B. Tappeser and S. Lötters. 2013. Questions concerning the potential impact of glyphosate-based herbicides on amphibians. Environ. Toxicol. Chem. 32: 1688–1700.

Wake, D.B. and V.T. Vredenburg. 2008. Are we in the midst of the sixth mass extinction? A view from the world of amphibians. PNAS 105: 11466–11473.

Wang, C., G. Liang, L. Chai and H. Wang. 2016. Effects of copper on growth, metamorphosis and endocrine disruption of *Bufo gargarizans* larvae. Aquat. Toxicol. 170: 24–30.

Wang, J., P. Guo, X. Li, J. Zhu, T. Reinert, J. Heitmann et al. 2000. Source identification of lead pollution in the atmosphere of Shanghai City by analyzing single aerosol particles (SAP). Environ. Sci. Technol. 34: 1900–1905.

Wang, M., L. Chai, H. Zhao, M. Wu and H. Wang. 2015. Effects of nitrate on metamorphosis, thyroid and iodothyronine deiodinases expression in *Bufo gargarizans* larvae. Chemosphere 139: 402–409.

Wang, Y., Y. Li, Z. Qin and W. Wei. 2017. Re-evaluation of thyroid hormone signaling antagonism of tetrabromobisphenol A for validating the T3-induced *Xenopus* metamorphosis assay. J. Environ. Sci. China. 52: 325–332.

Welch, A.M., J.P. Bralley, A.Q. Reining and A.M. Infante. 2019. Developmental stage affects the consequences of transient salinity exposure in toad tadpoles. Integ. Comp. Biol. 59: 1114–1127.

White, B.A. and C.S. Nicoll. 1981. Hormonal control of amphibian metamorphosis. pp. 363–396. *In*: Gilbert, L.I. and E. Frieden [eds.]. Metamorphosis. Springer, Boston, MA.

Wightwick, A.W., A.D. Bui, P. Zhang and G. Rose. 2012. Environmental fate of fungicides in surface waters of a horticultural-production catchment in southeastern Australia. Arch. Environ. Contam. Toxicol. 62: 380–390.

Wijethunga, U., M. Greenlees and R. Shine. 2015. The acid test: pH tolerance of the eggs and larvae of the invasive cane toad (*Rhinella marina*) in southeastern Australia. Physiol. Biochem. Zool. 88: 433–443.

Willens, S., M.K. Stoskopf, R.E. Baynes, G.A. Lewbart, S.K. Taylor and S. Kennedy-Stoskopf. 2006. Percutaneous malathion absorption by anuran skin in flow-through diffusion cells. Environ. Toxicol. Pharmacol. 22: 255–262.

Williams B.K. and R.D. Semlitsch. 2010. Larval responses of three midwestern anurans to chronic, low-dose exposures of four herbicides. Arch. Environ. Contam. Toxicol. 58: 819–827.

Wood, L. and A.M. Welch. 2015. Assessment of interactive effects of elevated salinity and three pesticides on life history and behavior of southern toad (*Anaxyrus terrestris*) tadpoles. Environ. Toxicol. Chem. 34: 667–676.

Wu, C.S. and Y.C. Kam. 2009. Effects of salinity on the survival, growth, development, and metamorphosis of *Fejervarya limnocharis* tadpoles living in brackish water. Zool. Sci. 26: 476–482.

Wu, C.S., I. Gomez-Mestre and Y.C. Kam. 2012. Irreversibility of a bad start: Early exposure to osmotic stress limits growth and adaptive developmental plasticity. Oecologia 169: 15–22.

Wu, N.C. and F. Seebacher. 2020. Effect of the plastic pollutant bisphenol A on the biology of aquatic organisms: A meta-analysis. Glob. Chang Biol. 26: 3821–3833.

Xie, L., Y. Zhang, X. Li, L. Chai and H. Wang. 2019. Exposure to nitrate alters the histopathology and gene expression in the liver of *Bufo gargarizans* tadpoles. Chemosphere 217: 308–319.

Xu, S., J. Ma, R. Ji, K. Pan and A.J. Miao. 2020. Microplastics in aquatic environments: Occurrence, accumulation, and biological effects. Sci. Total Environ. 703: 134699.

Xu, Y. and M.C. Gye. 2018. Developmental toxicity of dibutyl phthalate and citrate ester plasticizers in *Xenopus laevis* embryos. Chemosphere 204: 523–534.

Yaglova, N.V. and V.V. Yaglov. 2014. Changes in thyroid status of rats after prolonged exposure to low dose dichlorodiphenyltrichloroethane. Bull. Exp. Biol. Med. 156: 760–762.

Yang, H., R. Liu, Z. Liang, R. Zheng, Y. Yang, L. Chai et al. 2019. Chronic effects of lead on metamorphosis, development of thyroid gland, and skeletal ossification in *Bufo gargarizans*. Chemosphere 236: 124251.

Yaoita, Y. and D.D. Brown. 1990. A correlation of thyroid hormone receptor gene expression with amphibian metamorphosis. Genes Dev. 4: 1917–1924.

Ying, G.G., R.S. Kookana and Y.-J. Ru. 2002. Occurrence and fate of hormone steroids in the environment. Environ. Int. 28: 545–551.

Yu, S., M.R. Wages, G.P. Cobb and J.D. Maul. 2013. Effects of chlorothalonil on development and growth of amphibian embryos and larvae. Environ. Pollut. 181: 329–334.

Zamora-Camacho, F.J., P. Burraco, S. Zambrano-Fernández and P. Aragón. 2023. Ammonium effects on oxidative stress, telomere length, and locomotion across life stages of an anuran from habitats with contrasting land-use histories. Sci. Total. Environ. 862: 160924.

Zhang, S., J. Wang, X. Liu, F. Qu, X. Wang, X. Wang et al. 2019. Microplastics in the environment: a review of analytical methods, distribution, and biological effects. Trends Analyt. Chem. 11: 62–72.

Zhang, Y.-F., W. Xu, Q.-Q. Lou, Y.-Y. Li, Y.-X. Zhao, W.-J. Wie et al. 2014. Tetrabromobisphenol A disrupts vertebrate development via thyroid hormone signaling pathway in a developmental stage-dependent manner. Environ. Sci. Technol. 48: 8227–8234.

Zhou, T., M.M. Taylor, M.J. DeVito and K.M. Crofton. 2002. Developmental exposure to brominated diphenyl ethers results in thyroid hormone disruption. Toxicol. Sci. 66: 105–116.

Ziková, A., C. Lorenz, F. Hoffmann, W. Kleiner, I. Lutz, M. Stöck et al. 2017. Endocrine disruption by environmental gestagens in amphibians—A short review supported by new *in vitro* data using gonads of *Xenopus laevis*. Chemosphere 181: 74–82.

6

Pollutants and Oxidative Stress in Tadpoles

Juliane Silberschmidt Freitas

1. A Brief Introduction to Reactive Oxygen Species

Aerobic organisms naturally produce reactive oxygen species (ROS), which are formed by the partial reduction of molecular oxygen (Winston and Di Giulio 1991). Among the different types of ROS, the most common are the superoxide anion ($O_2^{\bullet-}$), hydrogen peroxide (H_2O_2) and hydroxyl radical (•OH) (Winston and Di Giulio 1991). Even though H_2O_2 is not a radical, it is a source of oxygen-derived free radical. When present at high concentrations, ROS can quickly react with available biomolecules, potentially damaging biological systems. As an evolutionary process, aerobic organisms have developed appropriate mechanisms to keep ROS concentrations in balance, preventing further damage such as oxidative stress, which will be discussed later in this chapter.

ROS are continuously produced in the mitochondria during cellular respiration as by-products of partial reduction of molecular oxygen. More than 90% of the oxygen consumed is used for energy production in the electron-transport chains and up to 10% is partially reduced by one electron in other enzymatic oxidative reactions (Papa and Skulachev 1997). Some authors have suggested that about 0.2% of the oxygen consumed by aerobic cells is converted into ROS during normal cell respiration (Fridovich 2004). Regarding cellular respiration process, oxygen is usually reduced to water in four-electron oxygen reduction steps; however, oxygen reduction by one electron at time produces $O_2^{\bullet-}$, a relatively stable intermediate that is a precursor of most other ROS. The path of electrons through the respiratory complexes in the electron transport chain of the inner mitochondrial membrane, with potential $O_2^{\bullet-}$ formation, is shown in Fig. 1.

In the mitochondrial electron-transport chain, sequential addition of a single electron to O_2 forms $O_2^{\bullet-}$, which is further reduced to H_2O_2 and finally to hydroxyl radical (•OH) and hydroxyl anion (OH^-) (Lushchak 2011) (Fig. 2). •OH has potent oxidizing ability and can react with important cellular macromolecules; if not neutralized by the antioxidant system, its excess can cause lipid peroxidation, DNA damage and even cell death (Winston and Di Giulio 1991). The main ROS sources in eucaryotic cells during mitochondrial electron-transport chain are Coenzyme Q and Complex III. In these places, $O_2^{\bullet-}$ is formed by the interaction between molecular oxygen and

Department of Agrarian and Natural Sciences, Minas Gerais State University (UEMG), R. Ver. Geraldo Moisés da Silva, s/n - Universitário, 38302-192, Ituiutaba, MG, Brazil. Email: juliane.freitas@uemg.br

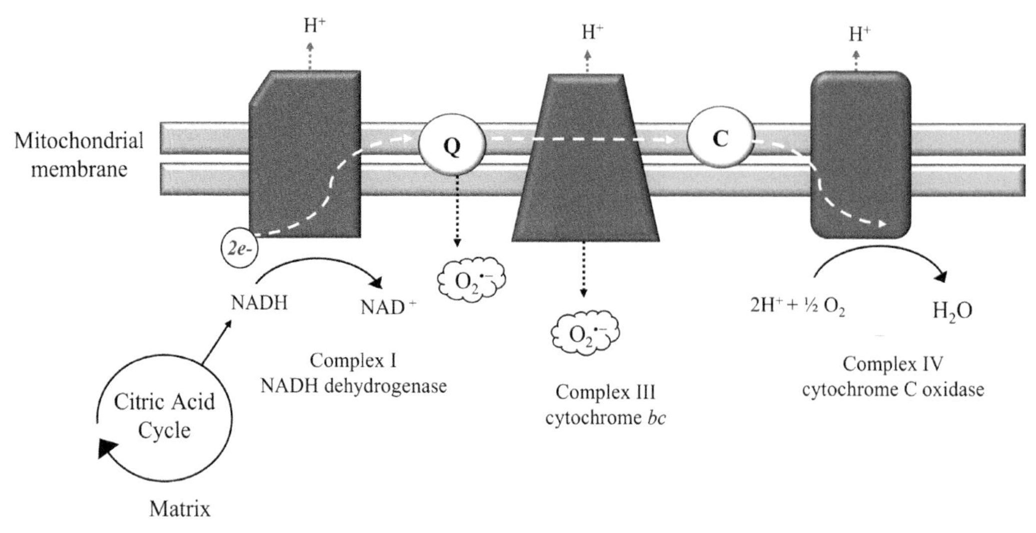

Figure 1. Scheme regarding the transport of electrons through the respiratory complexes (I, III and IV) along the electron transport chain in the inner mitochondrial membrane. Succinate Dehydrogenase (Complex II) is not represented in the figure. Complex I, NADH-dehydrogenase, is the electron acceptor of NADH, which was previously produced by the citric acid cycle reactions in the mitochondrial matrix. Thus, electron transport begins when a hydride ion is removed from NADH, producing $NADP^+$ and two high-energy electrons (*2e-*). The electrons are then transferred along the chain (dashed white arrow) to each of the other respiratory complexes, using mobile electron carriers (Q: ubiquinone or Coenzyme Q; C: cytochrome C). In Complex IV (cytochrome C oxidase), electrons (now, with less energy than at the beginning of the chain) are combined with oxygen, producing H_2O. During the described steps, $O_2^{\bullet-}$ can be produced by the interaction between molecular oxygen and escaping electrons.

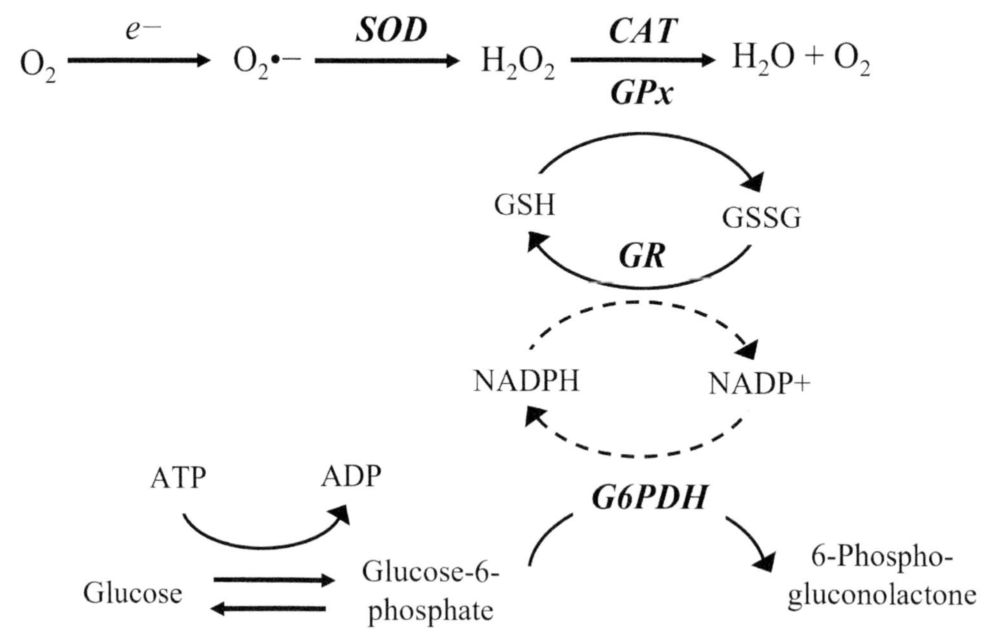

Figure 2. Summary scheme of the antioxidant enzymes activities. Superoxide dismutase (SOD) converts $O_2^{\bullet-}$ into H_2O_2, which is further decomposed by catalase (CAT) into H_2O and O_2. Glutathione peroxidase (GPx) also catalyzes the conversion of H_2O_2 into H_2O, but its activity depends on the electrons from the reduced glutathione (GSH), which is converted to glutathione disulfide (GSSG). GSH levels are maintained by glutathione reductase (GR), with the concomitant oxidation of NADPH to $NADP^+$. Finally, glucose-6- phosphate dehydrogenase (G6PDH) recycles NADPH levels in a $NADP^+$ dependent reaction, which includes the dehydrogenation of glucose-6-phosphate (G6PDH) to form 6-phosphogluconolactone.

escaping electrons (Demin et al. 1998) (Fig. 1). Usually, ROS do not exceed the concentration of 10^{-8} M (Halliwell and Gutteridge 1989), but when their production and elimination are unbalanced and ROS concentrations increase, a disturbance in the redox state known as oxidative stress is established. As described by Lushchak (2011), oxidative stress occurs when the steady-state ROS concentration is transiently or chronically enhanced, disrupting cellular metabolism and causing cellular damage. Although ROS are commonly associated with oxidative stress, they play several roles in cellular signaling processes in the animals, which include cell growth, metabolism, immune responses to xenobiotics and bacteria, cell communication, proliferation, and apoptosis (Ray et al. 2012, Costantini 2014). This also means that any disturbance on ROS production can have important cellular signaling consequences. Therefore, ROS levels are under critical control in cells by a complex and integrated antioxidant system.

2. Antioxidant Mechanisms

To avoid damage caused by oxidative stress, which is induced under different stressful circumstances, such as exposure to environmental contaminants, organisms have an effective antioxidant defense system, which includes enzymatic and non-enzymatic components. Antioxidant enzymes, such as superoxide dismutase (SOD), catalase (CAT), glutathione peroxidase (GPx) and glutathione reductase (GR), have critical role in detoxifying radicals to non-reactive forms (Van der Oost et al. 2003). SODs are a group of metalloenzymes that play essential antioxidant role in all aerobic organisms (Stegeman et al. 1992). SOD activity neutralizes the harmful effects of $O_2^{\bullet-}$ by converting it into H_2O_2. Since H_2O_2 is also reactive, it is subsequently detoxified by CAT and GPx. CAT catalyzes the decomposition of the H_2O_2 into H_2O and O_2. These heme-containing enzymes are located in the peroxisomes and are highly specific on H_2O_2 reduction (Stegeman et al. 1992, Filho 1996). GPx catalyzes the conversion of H_2O_2 into H_2O and its activity directly depends on the electrons from the reduced form of the tripeptide glutathione (GSH), which is converted to glutathione disulfide (GSSG). As endogenous GSH is an important antioxidant in cells, the balance on GSH/GSSG levels are supplied by GR activity (Winston and Di Giulio 1991), which provides GSH from GSSG by the concomitant oxidation of NADPH to $NADP^+$ (Worthington and Rosemeyer 1974) (Fig. 2).

Activity of antioxidant enzymes in aquatic vertebrates has been widely used in environmental monitoring studies, given their sensitivity and importance on the first-line defense response against ROS. Later, it will be discussed how antioxidant enzymes can be modulated and impaired by environmental contaminants in tadpoles. In aquatic animals, activity of these enzymes can be easily measured in laboratory by biochemical assays, mostly by spectrometric analysis.

Non-enzymatic antioxidants include low-molecular molecules, such as GSH, β-carotene (a precursor of vitamin A), ascorbic acid (vitamin C), α-tocopherol (vitamin E) and $ubiquinol_{10}$ (Stegeman et al. 1992, Lopez-Torres et al. 1993), that often operate as free radical scavengers. Besides being an essential substrate for GPx, GSH was previously mentioned as an important tripeptide protecting cells from oxidative damage (Almeida et al. 2011). GSH may also act as a cofactor for glutathione-*S*-transferases (GST), a superfamily of dimeric and multifunctional enzymes involved in the phase II of biotransformation that catalyzes the conjugation of GSH with reactive xenobiotics, making them more soluble and consequently more excretable (Van der Ost et al. 2003). GSH is found in all organisms and its content can be measured (along with other markers) to indicate the antioxidant equilibrium of the organism. Ascorbic acid (also known as vitamin C) is a water-soluble antioxidant (as well as GSH) that scavenges and reduces H_2O_2, $O_2^{\bullet-}$, and •OH (Lesser 2006). Vitamin C is also mentioned as a cofactor for several enzymes involved in collagen biosynthesis or neurotransmitter conversions (Stegeman et al. 1992, Lopez-Torres et al. 1993). β-carotene, the precursor of retinol (vitamin A), and α -tocopherol (vitamin E) are lipid-soluble antioxidants. Vitamin E is required in animal diets and is reported to play an important role in preventing lipid peroxidation in cell membranes (Stegeman et al. 1992, Van der Oost et al. 2003). The main mechanisms of action of enzymatic and non-enzymatic antioxidants are summarized in Table 1.

Table 1. A summary of the main antioxidant enzymes and their functions.

Enzyme	Function
Superoxide dismutase (SOD)	Metalloenzymes that converts superoxide ($O_2^{\cdot -}$) to hydrogen peroxide (H_2O_2)
Catalase (CAT)	Catalyzes the decomposition of the H_2O_2 into H_2O and O_2
Glutathione peroxidase (GPx)	Catalyzes the conversion of H_2O_2 into H_2O using GSH as electron donor
Glutathione reductase (GR)	Converts GSSG into GSH using electrons from NADPH
Glucose-6-phosphate dehydrogenase (G6PDH)	Action of G6PDH is important to maintain intracellular levels of NADPH
Glutathione S-transferases (GSTs)	Multifunctional enzymes with important role on phase II of xenobiotic biotransformation. Some isoforms have peroxidase activity and can also conjugate lipid peroxidation products with GSH

Some associated enzymes are also important in the redox status of aquatic animals providing cofactors. For example, glucose-6-phosphate dehydrogenase (G6PDH) participates in the pentose phosphate pathway by recycling NADPH, which is an important reducing agent helping to maintain GSH levels in the cells through the action of GR (Almeida and Mascio 2011, Freitas et al. 2017b). G6PDH performs the enzymatic dehydrogenation of glucose-6-phosphate to form 6-phosphogluconolactone, a NADP+ dependent reaction. Since NADP+ is the electron acceptor, G6PDH maintains NADPH levels, which are crucial for the reduction of GSSG to GSH (Fig. 1). Decreases in G6PDH activity may indirectly indicate reductions on GSH levels. Alternatively, G6PDH can be increased to supply the demands for GSH, thus reducing chances of oxidative stress. Associated pathways also include GSTs. For the most, several GSTs isoforms are not directly involved in antioxidant response, but some of them can help to prevent lipid peroxidation. For example, some GSTs can catalyze the reduction of cellular peroxides, especially fatty acid hydroperoxides (Prohaska and Ganther 1977, Sandeep Prabhu et al. 2004), while others (GSTs) may determine intracellular concentrations of 4-hydroxynonenal (by conjugation reaction), a prototypical α,β–unsaturated aldehyde that triggers cellular events associated with oxidative stress (Balogh and Atkins 2011).

3. Oxidative Damage Detection

Oxidative damage in aquatic vertebrates, including tadpoles, has been commonly assessed by measurements of lipid peroxidation and carbonyl proteins. These biomarkers are indicative of injuries caused by ROS in carbohydrates, lipids, and proteins (Miyata et al. 1993). For lipid peroxidation biomarkers, it is the usual application of methods using thiobarbituric-acid (TBARs) that analyze the product formed between malondialdehyde (MDA) and TBA. Increases on MDA levels may be indicative of oxidative stress in animals, as it is one of the final products of polyunsaturated fatty acids peroxidation in the cells. In addition, damage to proteins or chemical modification of amino acids in proteins during oxidative stress can produce protein carbonyls (Stadtman and Berlett 1998, Zusterzeel et al. 2001, Parvez and Raisuddin 2005). Although protein carbonyl is well established as a biomarker of oxidative stress in mammalian, including humans, less data is available yet for cold-blooded aquatic vertebrates, especially tadpoles.

4. ROS Production and Oxidative Stress During Tadpole Metamorphosis

Although this chapter focuses on the effects of environmental contaminants on antioxidant defense and oxidative lesions in tadpoles, it is important to emphasize that oxidative stress is a naturally occurring process during metamorphic transformation in amphibians. As mentioned before, ROS production is a normal consequence of the aerobic metabolism. However, some organisms are known to be more vulnerable to oxidative damage at certain stages of life or phases of development (Smith et al. 2016). Even if healthy organisms have an efficient antioxidant system to deal with ROS excess, this natural self-defense mechanism has energy costs (Prokić et al. 2019); thus, managing oxidative stress is an important strategy for an organism that depends on its macromolecules for growth and development.

First, it is easy to imagine that cellular activity is readily increased as morphological changes associated with metamorphosis occur in an amphibian. The transition from a herbivorous and aquatic tadpole to a terrestrial and carnivorous adult frog requires major changes in metabolic rates, which involves cell proliferation and apoptosis. With increased cell activity, metabolic rates are consequently improved, and ROS can be generated at higher levels as by-product. Thus, for developing tadpoles, oxidative stress could be a "price" for the animal growth (Smith et al. 2016), depending on how ROS are managed. If ROS levels are controlled by enzymatic and non-enzymatic antioxidant mechanisms, as mentioned earlier in this chapter, organisms follow their natural development process without any harm on the processes of tissues and organs remodeling (Allen and Tresini 2000). However, if oxidative stress is established, important molecules used for critical metabolic functions in the body (e.g., lipids, proteins, and DNA) can be damaged, resulting in altered development of the animal, with consequent malformed larvae or adults (Aliko et al. 2018, Burgos-Aceves et al. 2018, Gobi et al. 2018, Hodkovicova et al. 2019, Stara et al. 2019). In this scenario, amphibians are very interesting models of study to investigate changes in oxidative stress during larval development, since most species have a unique life cycle that includes drastic morphological and physiological alterations during metamorphosis (Burraco et al. 2018, Prokić et al. 2018a, Prokić et al. 2018b, Turani et al. 2018).

Many authors have suggested that the antioxidant response plays a crucial role in the management of development in amphibian larvae (Melvin 2016, Pinya et al. 2016, Prokić et al. 2017, Prokić et al. 2019). For example, SOD and CAT seems to remove ROS excess in tadpoles at earlier stages (Rizzo et al. 2007). These strategies are likely to be altered in an adult frog, but it is still unclear how the mechanisms involved in combating ROS evolve in amphibians (Samanta and Paital 2016, Prokić et al. 2018b). In a study with *Xenopus laevis*, authors observed that depletion in CAT activity and GSH levels with consequent increase of lipid peroxidation was associated with intestinal remodeling and tail regression in tadpoles (Menon and Rozman 2007). They also showed that gene expression for SOD and CAT in the intestine was downregulated at stage 61/62 (Nieuwkoop and Faber 1967); in tail, both enzymes were downregulated at stage 63, accompanying the beginning of tail regression (Menon and Rozman 2007). Finally, Menon and Rozman (2007) also found a drastic increase in ascorbic acid in both organs during metamorphic climax. In this way, intestinal morphological changes and tail regression seems to be followed by increased production of ROS and oxidative stress in tadpoles (Menon and Rozman 2007, Johnson et al. 2013).

Oxidative stress parameters were also investigated in the amphibian *Pelophylax esculentus* during premetamorphosis, prometamorphosis, metamorphic climax and juvenile stages (Prokić et al. 2019). The authors aimed to understand the functioning of the antioxidant system during the transition from an aquatic larva to a terrestrial adult. The results showed that different biochemical pathways were accessed depending on the developmental stage of the tadpole, with distinct levels of oxidative stress accompanying tissue modifications. Responses varied from: (i) increased enzymatic activities (e.g., SOD, GST, GPx and GR) during premetamorphic stages, to (ii) modification in non-enzymatic antioxidants (GSH) during the metamorphic climax and (iii) changes in both the enzymatic and non-enzymatic components in juveniles (Prokić et al. 2019). Biochemical alterations were also associated with alterations in tadpoles' life cycle, such as swimming and alimentary behavior. For example, premetamorphic tadpoles are exposed to the aquatic environment for the first time, now showing free swimming, intense feeding (Rizzo et al. 2007), and consequently increased metabolic rates (Kirschman et al. 2017). At this stage, higher levels of lipid peroxidation were reported, indicating susceptibility to oxidative damage (Prokić et al. 2019).

H_2O_2 is also mentioned to participate in cell cycle progression early in anuran development (Han et al. 2018) and during organ remodeling throughout the metamorphosis (Kashiwagi et al. 1999, Mahapatra et al. 2001, Menon and Rozman 2007, Johnson et al. 2013). Metamorphic individuals (*P. esculentus*) also showed higher MDA levels as consequence of the rapid and intense modulation of organs, which enhanced ROS concentrations. Increased lipid peroxidation in metamorphic amphibians were also observed in *X. laevis* (Menon and Rozman 2007). It is noteworthy that the changes in respiration and oxygen consumption that occur during the transformation from aquatic larvae to terrestrial adults are accompanied by several physiological adaptions (Burggren and Doyle 1986, Crowder et al. 1998), including higher metabolic rates in juvenile anurans (Kirschman et al. 2017). It can explain why enzymatic and non-enzymatic antioxidants are both activated in juveniles.

Summarizing, tadpoles have evolved mechanisms to counteract the effects of ROS by using different antioxidant components during their metamorphosis process (Metcalfe and Alonso-Alvarez 2010). In addition, ROS production and oxidative stress have important influence and pressure on tadpole development, making some specific stages more vulnerable than others when they are exposed to environmental contaminants. Despite this, it is worth noting that species distributed in different regions of the planet (with varied external environmental inputs) have developed physiological strategies to deal with the ecosystem pressure. For example, acceleration of metamorphosis is a common response in several species of tadpoles that experience puddle drying, and it is associated with many physiological alterations, such as increased production of thyroid hormones and corticosterone (Buchholz and Hayes 2005, Kulkarni et al. 2012, Gomez-Mestre et al. 2013, Kulkarni et al. 2017), as well as higher measures of oxidative stress (Gomez-Mestre et al., 2013). In western spadefoot toad tadpoles, acceleration on *Pelobates cultripes* development in response to decreased water levels was accompanied by increased activity of the antioxidant enzymes CAT, SOD and GPx (Gomez-Mestre et al. 2013). Authors reported that *P. cultripes* tadpoles shortened their larval period by around 30% in response to puddle drying, but with some physiological costs that included oxidative stress, possibly due to increased lipid catabolism in the animals (Gomez-Mestre et al. 2013).

Physiological aspects of antioxidant functioning in tadpoles may also, and probably will, differ in model (e.g., *Xenopus laevis*) and wild species, as well as in animals examined under laboratory and field conditions. Thus, it is important that these studies are carried out in parallel to obtain a more complete picture of the occurrence of oxidative stress in amphibians during metamorphosis. In the next sessions, we will discuss how the antioxidant status in tadpoles can be altered by the exposure to environmental contaminants, and how tadpoles have modulated their antioxidant mechanisms to coexist in anthropized areas. The influence of environmental factors on the antioxidant system of tadpoles exposed to environmental contaminants will also be explored.

5. Pollutants, Antioxidant Responses and Oxidative Stress in Tadpoles

As mentioned before, ROS are produced in all aerobic organisms and, under normal circumstances, the production and elimination of ROS is usually in balance by the action of an efficient antioxidant system. However, the rate of ROS generation is dynamic and can be increased by stressful conditions. In addition, oxidative stress is pointed out as a frequent consequence of physiological adjustments of animals to a wide range of environmental stressors.

Several types of environmental contaminants are known to enhance ROS production in tadpoles and other aquatic animals, potentially inducing oxidative stress. Xenobiotics can increase intracellular ROS by different pathways, including alterations on redox cycling, increases in basal metabolism with consequent rise of mitochondrial activity, as a by-product of reactions mediated by cytochrome P450 enzymes (CYPs), or by increases in Fenton and Haber-Weiss reactions due to excessive iron and copper ions. Oxidative stress biomarkers are first-line responses in animals and are very sensitive to contaminants even at relatively low concentrations. These biomarkers are also interesting because they combine a common response (not necessarily specific) for several types of environmental contaminants. In recent years, researchers have standardized laboratory protocols for the evaluation of biochemical biomarkers in different tadpole species, enabling their application in environmental biomonitoring studies. In the next section, it will be discussed how different classes of pollutants, such as pesticides, metals, drugs, nanoparticles, and oil derivatives, can modify the antioxidant responses of amphibians during their larval stages, potentially compromising their survival through metamorphosis (Fig. 3).

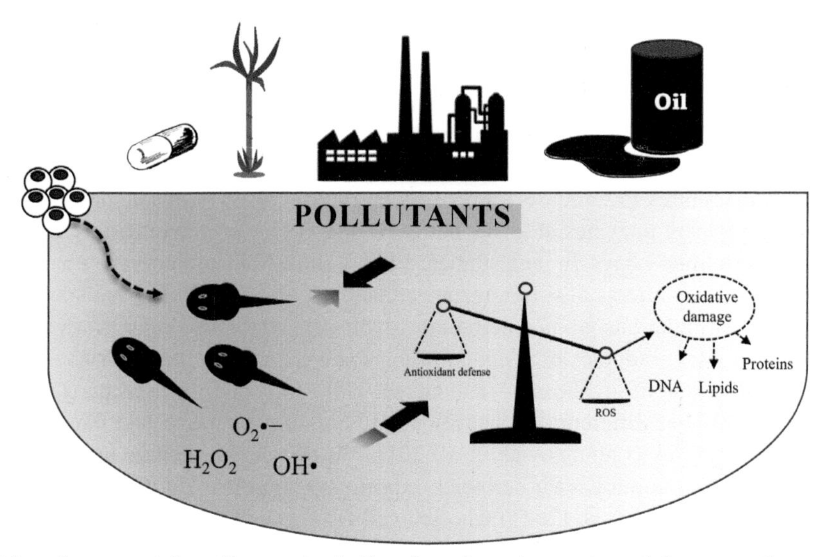

Figure 3. Schematic representation of how contamination of aquatic environments can influence reactive oxygen species (ROS) production in tadpoles, inducing oxidative stress.

5.1 Pesticides

Amphibians are known to be more sensitive to pesticide contamination during their larval stage because tadpoles are strictly aquatic and have higher skin permeability (Yan et al. 2008). In addition, several tadpole species inhabit ephemeral ponds formed in agriculture flooded areas, where the concentrations of pesticides is usually high due to the low volume of water. In tropical regions of the planet, application of pesticides is also intensified during the rainy season in the crops, since chemical management is more effective in humid soils. Thus, it is expected that many tadpole species are being exposed to several classes of pesticides, including their mixtures.

There is increasing evidence that pesticides may alter the antioxidant defense response of several tadpole species, inducing oxidative stress. Atrazine, glyphosate and quinclorac, the most widely used herbicides worldwide, increased lipid peroxidation in tadpoles of *Lithobates catesbeianus* after 7 or 14 days, with more prominent effects after the longest exposure (Dornelles and Oliveira 2014). In this study, concentrations were 5–20 µg/L of atrazine, 36–144 µg/L of glyphosate and 0.05–0.20 µg/L of quinclorac, the last herbicide being the most harmful of the three products. Dornelles and Oliveira (2014) also observed that these decreased glycogen and protein levels and total lipids in liver and muscle of tadpoles, suggesting their negative influence on energy metabolism of tadpoles. The effects of the same three herbicides were also tested on oxidative balance of *Rhinella icterica* tadpoles at concentrations of 10, 20 and 40 µg/L of atrazine and quinclorac, and 100, 250 and 500 µg/L of glyphosate; however, authors found no significant variations for any of the studied oxidative markers (SOD, CAT, lipid peroxidation, and carbonylated proteins) (Reichert et al. 2022). Despite this, atrazine and glyphosate accelerated *R. icterica* development, and quinclorac retarded this process.

Although herbicides target plants, there is ample evidence that they can interfere with tadpole physiology associated with antioxidant responses. For example, the herbicide sulfentrazone induced biochemical changes in *Physalaemus nattereri* and *Rhinella schneideri* tadpoles after 3 or 8 days of exposure. In this study, authors reported that sulfentrazone had temporal- and temperature-dependent responses in antioxidant enzymes and MDA levels for both species (Freitas et al. 2017a). In *P. nattereri*, those changes included increases in CAT, G6PDH and GST activities, in addition to decreases in SOD, with pronounced effects after 8-days exposure (Freitas et al. 2017a). For *R. schneideri*, most of the antioxidant enzymes had their activities altered after 3 days of exposure, with more prominent responses at higher temperatures (32 and 36°C). MDA levels was unaltered by sulfentrazone in this species. The differential responses of these two species to the same herbicide (sulfentrazone) were attributed to a differential stimulation of the antioxidant system in the tadpoles, which were mentioned to be a species-specific strategy of the animals to adapt to the contaminated condition. Another study with the same tadpole's species (*P. nattereri* and *R. schneideri*) showed that a clomazone-based herbicide (Gamit®) increased CAT, SOD, G6PDH and GST in *P. nattereri* under higher temperatures (32 and 36°C) (Freitas et al. 2017b). In *R. schneideri*, clomazone had no effects on antioxidant enzymes at 28°C, but SOD and GST were increased by the herbicide at higher temperatures after 3 days. In both studies, authors showed that temperature had an important influence inducing toxic effects of sulfentrazone and clomazone in tadpoles.

Commercial formulations of the herbicide glyphosate are also well-known for disturbing redox status in tadpoles. Tadpoles of *Rhinella arenarum* were short-term (6–48 h) exposed to four commercial formulations of glyphosate—Roundup® Ultra-Max (ULT), Infosato (INF), Glifoglex, and C-K Yuyos FAV—at different concentrations (1.85–240 mg/L) and had their GST activities decreased in all scenarios (Lajmanovich et al. 2011). In this study, authors indicated GST as an important and sensitive biomarker for herbicide exposure in tadpoles. This is in fact consistent with other authors, such as Silva et al. (2021) who pointed GST as a crucial enzyme in the metabolism of the herbicide Roundup® in the endemic tadpole from Atlantic Forest (Brazil) *Melanophryniscus admirabilis*. Silva et al. (2021) also observed decreases in SOD and CAT activities in *M. admirabilis* exposed to commercial formulation of the sulfentrazone-based herbicide Boral® 500 SC (at concentrations of 130 and 980 μg/L) and glyphosate-based herbicide Roundup® Original (at 234 and 2340 μg/L). They suggested that the high antioxidant capacity of this tadpole species could have contributed to its survival in areas contaminated by both herbicides. In *L. catebeianus* tadpoles, the herbicide Roundup® Original (glyphosate 41%) increased SOD and CAT activities in liver and decreased both enzymes in muscle after 48 h exposure at 1 mg/L (Costa et al. 2008). In both tissues, Roundup® also raised lipid peroxidation, suggesting that production of ROS and oxidative stress are involved in its toxicity. Wilkens et al. (2019) found that isolated exposure to the herbicides glyphosate (234 μg/L) and sulfentrazone (130 μg/L), as well as their mixture, changed oxidative stress markers with tissue-dependent responses in tadpoles of *L. castebeianus* after 7 days. In the liver, tadpoles from mixture treatment (glyphosate + sulfentrazone) had SOD and CAT decreases, in addition to increased GST. In the caudal muscle, declined levels of GST and TBARS were observed only in the blend group (Wilkens et al. 2019).

The herbicide butachlor in a range of 7–11 μg/L decreased GPx, CAT and SOD activities in *Bufo regularis* tadpoles, in addition to increasing GR, GST and GSH after 96 h (Ejilibe et al. 2018). Tadpoles of *Leptodactylus fuscus* had their GST activity increased after 7-days exposure to environmentally relevant concentrations of 2,4-D-herbicide in mesocosms under tropical conditions in Brazil (Freitas et al. 2022). In the same experiment, tadpoles of *L. catesbeianus* also had their CAT increased and G6PDH decreased by 2,4-D. An interesting result of this study was the prominent increase in GST in both species when the tadpoles were exposed to 2,4-D mixed with the insecticide fipronil. Those results, combined with previous results from literature (as mentioned before by Lajmanovich et al. 2011 and Silva et al. 2021) indicate the importance of GST on detoxification processes of tadpoles inhabiting contaminated agricultural areas. As mentioned earlier, GST acts on the detoxification pathways of pesticides, making the compounds more soluble and easily excreted by animals.

A study with paraquat, an herbicide which is currently banned in several countries, showed that tadpoles of the American Bullfrog (*L. catesbeianus*) exposed to concentrations between 0.1–2.0 mg/L for 24 h had increases in SOD, peroxidases, and GR activities in the hepatic tissue. The study suggests that the relative tolerance of bullfrog tadpoles to paraquat can be due to stress-induced raises of antioxidant enzymes (Jones et al. 2010). Exposure of *X. laevis* to metamifop, an aryloxyphenoxy propionate herbicide, induced severe lipid oxidative damage in tadpoles after 35 days-exposure at concentration of 0.063 mg/L (Liu et al. 2021). In this study, SOD activity was also 35% higher in metamifop-exposed tadpoles than those from the control after 35 days, whereas CAT had increases of 135.82% (Liu et al. 2021).

Insecticides are also known to disrupt antioxidant responses and cause oxidative stress in tadpoles. Ferrari et al. (2009) discussed the effects of the carbamate carbaryl and the organophosphate (OP) azinphos-methyl on the antioxidant defense of developing embryos of *R. arenarum*. Although both insecticides are known to be anticholinesterase agents, OP and carbamates may also induce oxidative stress and modulate antioxidant responses in aquatic animals (Oruc and Üner 2000, Anguiano et al. 2001, Ferrari et al. 2007, Matos et al. 2007, Bianchini and Monserrat 2007, Isik and Celik 2008). Some carbamates are reported to induce cytochrome P450 1A (CYP1A) activity (Denison et al. 1998, Ferrari et al. 2007), which is an enzyme associated with ROS generation and, consequently, to oxidative stress (Livingstone 2001). Ferrari et al. (2009) found that GSH levels were raised in early tadpoles' stages by carbaryl, but its content was decreased by both insecticides (carbaryl and azinphos-methyl) at the end of the embryonic development. GST in *R. arenarum* was also increased during embryonic stage and at the end of tadpoles' development. GR, GPx and CAT showed oscillation in their activities suggesting induction and/or inactivation in response to pesticides' exposure in *R. arenarum* (Ferrari et al. 2009). Enzymatic inactivation, as reported by the authors, could be driven by an excess of ROS, which acted on the enzyme's active sites.

Fipronil, one of the most used insecticides in the world (although already banned in some countries), acts on γ-aminobutyric acid (GABA) chloride channels in insects, disrupting their neuronal signaling (Gunasekara and Troung 2007). GABA antagonists, such as the case of fipronil, can cause hyperactivity, convulsions and even death in fish (Beggel et al. 2012), as other aquatic animals. Despite this specific action, fipronil has also been reported to alter the antioxidant responses of tadpoles, producing oxidative stress in some species. Gripp et al. (2017) showed that tadpoles of *P. nattereri* exposed to fipronil in sediment (35–180 μg/kg) had their CAT and G6PDH enzymes decreased, and the MDA levels increased in tadpoles exposed to all concentrations of the insecticide. On the other hand, this study reported that fipronil metabolites—fipronil sulfone and fipronil sulfide—had less potential to disrupt the tadpole antioxidant system than the original product. In *Scinax fuscovarious* tadpoles, GST was inhibited in the treatments containing fipronil (Regent®800WG) and GR activity varied according to the exposure period (5, 14 and 30 days) and pesticide concentration (0.005, 0.02, 0.05 and 0.1 mg/L) (Margarido et al. 2013). CAT and G6PDH were unchanged following the insecticide exposure in *S. fuscovarious* (Margarido et al. 2013). In the study of Freitas et al. (2022), fipronil-contaminated mesocosms enhanced G6PDH activity in *L. fuscus* and decreased its activity in *L. catesbeianus*, which also had an increase in CAT. Enhances on G6PDH activity are important to supply the demands for GSH, thus reducing chances of oxidative stress in the animals (Freitas et al. 2022). In this semi field experiment, CAT was also increased in *L. fuscus* by the mixture of fipronil with 2,4-D under tropical conditions (Freitas et al. 2022).

Other insecticides also show evidence of interfering with the tadpole antioxidant system of various species. A study with *B. regularis* tadpoles and the insecticide Termex® (imidaclorpid 30.5%) at concentrations ranging between 15–25 μg/L indicated that the antioxidant enzymes GPX, CAT and SOD were inhibited after 24 and 96 h of exposure (Ejilibe et al. 2018). In *Rana nigromaculata* tadpoles, the insecticides acetamiprid and α-cypermethrin caused toxicity by changing antioxidant enzymes or inducing lipid peroxidation (Xu and Huang 2017, Guo et al. 2022). In the first study, oxidative stress was evidenced in *R. nigromaculata* tadpoles exposed to α-cypermethrin (0.5 μg/L) by the increases in SOD, CAT, GST, and MDA levels (Xu and Huang 2017). For the second, acetamiprid increased GR and SOD activities of *R. nigromaculata* tadpoles, as well as the lipid peroxidation, at concentrations of 0.185 and 1.85 mg/L for chronic exposure (7–28 days) (Guo et al. 2022). Cypermethrin also decreased CAT and GST, and increased levels of hydrogen peroxide, GSH and MDA in tadpoles of *Duttaphrynus melanostictus* at 1.11 µg/L for 2–6 days of exposure (Muniswamy et al. 2016), indicating the dangerous effects of this synthetic pyrethroid insecticide even at very low concentrations.

In the Chinese toad *Bufo bufo gargarizans*, the systemic insecticide spirotetramat (0.03–3.23 mg/L) increased SOD activity in all experimental groups after 4 days of exposure, except in the group exposed to the highest concentration (Yin et al. 2014). Yin et al. (2014) also indicated GPx as the most sensitive enzyme to spirotetramat in *B. gargarizans*, as it markedly increased on day 4 of exposure and inhibited after prolonged exposure of 15 and 30 days. Clothianidin also induced oxidative stress in tadpoles of *R. pipiens* at 0.23, 1.0 and 100 µg/L after 8 weeks of exposure; the effects of this insecticide were accessed by the increases in protein carbonyl carboxylation and levels of 4-hydroxylnonenal (HNE) protein adducts (a product of lipid peroxidation) in liver of the tadpoles (Robinson et al. 2021). Effects (inhibition/activation) on the antioxidant enzymes GST, GR, GPx, and CAT were also stimulated by the insecticide thiacloprid in *X. laevis* tadpoles on early developmental stages, with synergistic effect when the insecticide was applied as a mixture with the fungicide trifloxystrobin (Uçkun and Özmen 2021).

Regarding the fungicides, investigation on the effects of vinclozolin (17.5, 174.8 and 1748.6 nM) and propiconazole (5.8, 58.4 and 584.4 nM) on tadpoles of striped marsh frog (*Limnodynastes peronii*) for 96h showed a broad downregulation of the tricarboxylic acid (TCA) cycle, an indicative of oxidative stress for both contaminants (Melvin et al. 2018). Cyproconazole at 1 mg/L increased SOD, CAT and GSH levels in *R. nigromaculata* tadpoles after 90 days of exposure, but the activities of these enzymes decreased when animals were exposed to 10 mg/L (Zhang et al. 2019). The fungicides pyrimethanil and tebuconazole affected metamorphosis success of the Italian tree frog *Hyla intermedia* exposed for more than 10 weeks to 5 and 50 µg/L (Bernabo et al. 2016), but no evidence on antioxidant responses has been reported.

Currently, the volume of studies using biomarkers of oxidative stress in tadpoles exposed to agricultural settings is increasing, but there is still a great scarcity of information for many substances. It is also important to mention that most commercial formulations of different pesticides are applied in combination in the field, thus favoring tadpoles' exposure to mixed conditions. Nevertheless, the understanding of this interaction is complex and depends on the species, chemical concentrations, and exposure scenarios (e.g., Kretschmann et al. 2015, Gottardi et al. 2017, Freitas et al. 2022).

5.2 Metals

Metal pollution is a threat to the health of ecosystems and wild organisms, including amphibians (Brühl et al. 2013, Guo et al. 2015, Banday et al. 2019). With the rapid development of industrial activities and expansion of agriculture, metals have rapidly spread in natural systems. It is also worth mentioning that mining activities and disasters involving the collapse of mining dams are currently potent sources of contamination by metals in the environment. Studies have described the negative effects caused by exposure to metals in amphibians during their larval stages (Prokić et al. 2016, 2017, Veronez et al. 2016, Jayawardena et al. 2017, Slaby et al. 2019). Chapter 13 addresses more accurately the effects of metal exposure in different biological responses of tadpoles. However, few studies have been directed yet to understand how such contaminants at environmental concentrations can affect the antioxidant defense of tadpoles.

Lu et al. (2021) showed that exposure to cadmium (Cd) decreased CAT, SOD and GPx activities as well as GSH contents in tadpoles of *R. zhenhaiensis*, especially under high pH (9.0), an effect attributed to the accumulative effects of ROS. Antioxidant enzymes in *L. catesbeianus* tadpoles were also very sensitive to zinc (Zn), copper (Cu), and cadmium (Cd) under isolated or mixed conditions (Carvalho et al. 2020). Authors observed that exposure to these three metals alone or co-exposed (1:1 and 1:1:1) at concentration of 1 μg/L increased SOD, CAT, GPx, GST activities in liver, kidney, and muscle after 2 days. After 16 days, a decrease in SOD, CAT and GST was observed in liver and kidney of tadpoles. Inhibition of these antioxidant enzymes was reported as a consequence of binding metals in their prosthetic, sulfhydryl, or other functional groups, and/or the removal of essential metals from the protein (Atli and Canli 2007, Sevcikova et al. 2011, Safari 2015, Carvalho et al. 2020). Additionally, increases on lipid peroxidation was observed in tadpoles exposed to the metals after both experimental periods (2 and 16 days), with prominent effects for co-exposure.

Another study with *L. catesbeianus* showed that iron ore (3.79 mg/L), Fe (0.51 mg/L) and Mn (5.23 mg/L) augmented erythrocyte DNA damage and micronuclei frequency, in addition to increased GST activity in tadpoles (Veronez et al. 2016). Iron ore also enhanced CAT in exposed tadpoles. In this study, authors indicated that tadpoles accumulated Fe and Mn in their bodies, which could be a biological factor inducing oxidative stress, as well as genotoxic and developmental effects. Fe is known to play a central role on oxidative stress, stimulating hydroxyl radical's production through Fenton and Haber-Weiss reactions (Kanti Das et al. 2015). On the other hand, Mn act as a cofactor for Mn-dependent SOD, which could help on cellular protection against ROS damage. Despite that, Veronez et al. (2016) showed that after longer exposure (30 days), Mn disturbed antioxidant response in tadpoles, potentially contributing to oxidative stress. In this way, Fe and Mn are both considered redox-active metals involved in cellular oxidative status of the animals.

Mining activities produce large volume of metal contaminated effluents that are commonly discharged into aquatic environments. Flooded environments in areas with mining activity also make metals bioavailable to the water fraction, potentially impacting tadpoles. Considering this scenario, mineralogical composition of the area contributes significantly to the hazardousness of the effluent, as well as the pH of the medium, as this parameter interferes with metal bioavailability and toxicity (Lopes et al. 1999, Williams 2001). Previous studies with adults of the Iberian green frog *Pelophylax perezi* showed that populations inhabiting a deactivated uranium mine presented metal accumulation in various tissues, in addition to histological changes in liver, kidney and lungs (Marques et al. 2009, 2011). During larval phase, tadpoles of *P. perezi* collected in ponds containing mine effluents presented higher metal bioaccumulation and GPx activity than animals from the reference area. Tadpoles with the highest levels of lipid peroxidation were also from the pond containing a complex mixture of metals (Marques et al. 2009, 2011). A study with *L. catesbeianus* tadpoles exposed to mining tailings from the Fundão dam failure in Brazil reported that animals had reduced swimming activity and lower respiration rates (Girotto et al. 2021), which possibly may have interfered with their antioxidant status; however the biochemical parameters related to oxidative stress were not evaluated in the study.

Due to the scarcity of studies exploring how metals can induce oxidative stress in amphibians, comparative effects among different species are usually based on available data for fish. However, it is already clear in the literature that the action pathways of metals, as well as the adaptive mechanisms used by tadpoles and fish to survive in metal contaminated areas, can vary on a large scale. In addition, metal accumulation in tadpoles' tissues can have relevant ecological consequences since amphibians play an important role as prey or predators (depending on the life stage) in the trophic chains. Thus, amphibians are an important source of transfer of toxic substances (Murphy et al. 2000). Considering the data available so far, more research is required to better distinguish and understand physiological responses of tadpoles to metals, especially considering environmentally relevant concentrations. The effects of metals on tadpoles beyond oxidative stress are available in Chapter 13.

5.3 Pharmaceuticals

Pharmaceuticals from different sources are among the emerging contaminants that cause negative effects on aquatic organisms. These pollutants deserve special attention because they are not commonly monitored in the environment, but are entering continuously in large volumes in water bodies of different magnitudes. Discharge of treated municipal wastewater is the main source of drug introduction into the environment, but other polluting activities such as the discharge of unused medicines, aquaculture, and pharmaceutical production facilities are also significant (Santos et al. 2010). Although most pharmaceuticals occur at trace levels in water, which could erroneously lead to the thought that they are harmless contaminants, they are in fact designed to act at very low concentrations on target organisms (humans, or other animals in veterinary). How aquatic vertebrates are affected by such pollutants is still unclear, and this is especially true for amphibians. What we do know is that many physiological features of vertebrates are evolutionarily conserved and it is similar for amphibians and mammals, the latter for which drugs have been designated.

A study with the Australian striped-marsh frog (*Limnodynastes peronii*) exposed to a combination of common pharmaceutical contaminants (i.e., diclofenac, naproxen, atenolol, and gemfibrozil) showed that tadpoles had their peroxidase activities increased (at concentrations of 1000 μg/L), which was negatively correlated with levels of triglycerides (Melvin et al. 2016). These opposing but related responses from both biomarkers may indicate increased ROS production, which resulted in lipid peroxidation. Melvin et al. (2016) also reported that SOD and GR activities were unchanged following pharmaceutical exposure.

Nascimento et al. (2021) investigated the effects of short-time exposure (72 h) to polyethylene glycol (PEG) in oxidative stress and neurotoxicity of *P. cuvieri* tadpoles. PEG is commonly used in the pharmaceutical industry to produce drug capsules, suppositories, and ointment bases (Hutanu et al. 2014). The study observed that tadpoles exposed to PEG presented increased CAT activity compared to control, and there was a positive correlation between CAT activity and PEG accumulation in the animals, as well as with H_2O_2 production in the treated group.

Environmentally realistic concentrations (20, 90 and 460 ng/L) of the antibiotics sulfamethoxazole (SMX) and oxytetracyclyne (OTC) were analyzed on oxidative stress biomarkers of *L. catesbeianus* tadpoles by Rutkoski et al. (2022). In the study, tadpoles exposed for 16 days to the two highest concentrations (90 and 460 ng/L) of SMX and all concentrations of OTC (20, 90 and 460 ng/L) presented decreases in GST activity, which indicates a compromise in the detoxification activity of the exposed animals. In addition, OTC at concentrations of 90 and 460 ng/L reduced SOD and G6PDH activities, whereas GPx was inhibited and protein carbonyl levels were increased by 460 ng/L (Rutkoski et al. 2022). Associated to oxidative stress parameters, authors also identified increases in the number of lymphocytes, decreases in neutrophils, and histopathological changes in the liver (e.g., presence of inflammatory infiltrate, higher melanomacrophages, vascular congestion, and increased eosinophils) of tadpoles exposed to SMX and OTC. This study was pioneer demonstrating the possible effects caused by antibiotics in tadpoles from an integrated analysis of different biomarkers, including those of oxidative stress.

Exposure of *R. arenarum* to the mixture of the antibiotics ciprofloxacin and amoxicillin with the pesticides chlorpyrifos and glyphosate resulted in lower activity of GST in the tadpoles (Boccioni et al. 2021). The effects of the anti-inflammatory drug diclofenac (DCF) at concentrations ranging between 125 and 4000 μg/L in two amphibian species from Argentina (*Trachycephalus typhonius* and *Physalaemus albonotatus*) included biochemical imbalance between ROS production and antioxidant systems (Peltzer et al. 2019). Peltzer et al. (2019) also showed that GST activity was inhibited by diclofenac in *T. typhonius* tadpoles exposed to 125 and 2000 μg/L, as well as in *P. albonotatus* exposed to 125 μg/L of the same drug. Authors suggested these inhibitions as indicative of an overload of the antioxidant system, which could impair the drug detoxification (Peltzer et al. 2019). Another study also conducted with the species *R. arenarum* showed that the antibiotic ciprofloxacin at concentration of 1000 μg/L increased GST activity of tadpoles after 96 h-exposure, whereas larvae exposed to enrofloxacin had GST and CAT inhibition (Peltzer et al. 2017). GST activity also showed significant raises in *R. arenarum* tadpoles exposed for 48 h to the antiretrovirals lamivudine, stavudine, zidovudine and nevirapine at sublethal concentrations

(0.5–4 μg/mL) (Fernández et al. 2020). In all the studies cited, the great responsiveness of GST in treated tadpoles indicates the important role of this enzyme in the pharmaceutical detoxification process in amphibians.

5.4 Nanoparticles

Exposure to nanomaterials (NMs) has also received attention in recent decades due to its large and varied source of release by sectors of society with technological advances. They are notably products from the pharmaceutical and medical industries which may present high toxicity in aquatic animals due to their tiny size and relatively high surface area (Rozita et al. 2010, Svartz et al. 2020). In addition, NMs have a high potential for accumulation in sediment, which makes them bioavailable both in the water fraction for aquatic animals and in the sediment itself, impacting mainly benthic species (Rocha et al. 2015). For example, Svartz et al. (2020) investigated the effects of a nanoceramic catalyst Ni<gamma>-Al_2O_3 (NC) and the NMs involved in their synthesis, <gamma>-Al_2O_3 support (SPC) and NiO<gamma>-Al2O3 precursor (PC) on *R. arenarum* tadpoles. Alumina (Al) is one of the most widely used oxide ceramic material and Al-based NMs are frequently applied for medical prostheses production, as well as in several other commercial products (e.g., electronic substrates and wear components) (Rozita et al. 2010). In this study, authors exposed tadpoles at Gosner stage 25 to sublethal NMs concentrations (5 and 25 mg/L) for 96 h and accessed responses on CAT, SOD, GST, GSH and lipid peroxidation. Results showed that tadpoles had a CAT and GST inhibition after exposure to all Al-NMs types and SOD activity was also inhibited by both concentrations (5 and 25 mg/L) of PC and NC (Rozita et al. 2010). Animals from PC group (NiO<gamma>-Al2O3 precursor) exposed to the highest concentration also experienced lipid peroxidation, which was expressed as augmented levels of TBARS/mg protein. The effects of silicon dioxide (SiO_2) nanoparticles (SiO_2-NPs) in *R. arenarum* larvae for 48 h also showed that GST activity was increased in tadpoles exposed to 0.001 mg/L, a value considered as no-observed-effect concentration ($NOEC_{48\ h}$) in the study (Lajmanovich et al. 2018).

5.5 Oil Derivates

Increased environmental pollution by oil and petroleum is a modern problem generated by the growing demand for these products in the market. Especially for countries with large sources of oils (the oil-producing regions), the contamination of environments by these products has generated mild impacts on wildlife. Knowledge about the negative effects caused by exposure to oil derivatives in the environment is still very scarce for amphibians. Oil-exposure is known to cause high mortality, reduced growth, and several malformations in amphibians (Sparling 2010). Morphological deformities caused by petroleum derivates are also mentioned to be accompanied by changes in the functioning of the antioxidant system of amphibians (Amaeze et al. 2014). Despite this, only a small number of works for a select group of species are available reporting the effects of these contaminants on biochemical pathways associated with oxidative stress in tadpoles.

Sutuyeva et al. (2019) tested the effects of the water-soluble fraction of crude oil (WSFO) from the Zhanazhol oil field (Aktobe region, Kazakhstan) at different dilutions in the marsh frog (*Rana ridibunda*). They found that 0.5 and 1.5 mg/L of WSFO raised lipid hyperoxide (LHO) and MDA levels by approximately 80 and 50%, respectively, in the tadpoles after 60 days of exposure. In addition, SOD and CAT activities were decreased in animals from the same groups. In this same study, authors compared the effects observed in tadpoles exposed to WSFO with those exposed to o-xylene, a substance which was prevalent in the used oil. Although they observed damage to the tadpoles' antioxidant mechanisms, such as increases in LHO and MDA levels and decreases in SOD and CAT activities, the effects of o-xylene were reported as less pronounced due to the magnitude of the changes (Sutuyeva et al. 2019). It is worth mentioning that concentrations of o-xylene used in the study (0.5 mg/L) corresponded to the drinking water limit concentration, according to the World Health Organization (WHO) (Gorchev and Ozolins 2011).

Another study exposing tadpoles of the common toad *Amietophrynus regularis* to kerosene and unused engine oil for 96 h reported increases of MDA levels at concentrations of 1/100th $LC50_{96\ h}$. Authors also showed reductions in SOD activity in tadpoles exposed to all these products (Amaeze et al. 2014). Eriyamrem et al. (2008) observed sub-lethal effects of Bonny Light crude oil (WC) and its water soluble (WSF) and insoluble (WIF) fractions on oxidative parameters of *X. laevis* tadpoles following two- and 4-weeks exposure at different concentrations (10–30 mg/L). They found that WC, WSF, or WIF treatments increased SOD and GR activities at lower concentrations, but these enzymes were decreased by higher concentrations. Enhanced MDA levels and decrease in tadpoles' weight were also reported for tadpoles exposed to all treatments (WC, WSF or WIF). Authors suggested in this study that longer exposure to oils in soluble or insoluble aqueous fractions may have metabolic costs for tadpoles as a consequence of reduced antioxidant enzymes and membrane peroxidation.

Exposure of *Euphlyctis cyanophlyctis* to phenanthrene (PHE), a tricyclic polycyclic aromatic hydrocarbon (PAH) ubiquitously found in aquatic environments, increased SOD activity and GSH levels, as well as induced lipid peroxidation in tadpoles at environmentally relevant concentrations (Bhuyan et al. 2020). In this study, PHE $LC50_{96\ h}$ (12.28 mg/L) was significantly higher than reported in the environment (i.e., around 1460 µg/L, as reported by Anyakora et al. 2005), and oxidative stress parameters were analyzed in tadpoles exposed to 5 and 10% of $LC50_{96\ h}$ value. Oxidative stress induced by PEH was also correlated with the increased frequency of micronucleus and DNA strand break in *E. cyanophlyctis* (both biomarkers are indicative of genotoxicity. See more details in Chapter 7). Accessing the results, PEH exposure in early life stages was considered a potential threat to amphibians, especially because these compounds are persistent and continually released into the environment, in addition to being subject to environmental translocations (Jiao et al. 2017, Zhang and Tao 2009, Abdel-Shafy and Mansour 2016, Bhuyan et al. 2020).

6. Abiotic Factor Influencing Antioxidant Response in Tadpoles Exposed to Pollutants

Abiotic factors, such as changes in temperature and pH, can have important effects on physiology of aquatic animals. During larval stages, amphibians are particularly susceptible to environmental changes since several species occupy ephemeral aquatic habitats, where the volume of water is constantly changing due to evaporation and precipitation, so consequently the temperatures. In these water bodies, sunlight incidence (UV radiation) and biological reactions, such as alterations in the photosynthetic rate of plants also have important influence. Thus, tadpoles are expected to experience continuous regulation in their molecular, biochemical, and physiological processes associated with homeostasis (Freitas and Almeida 2016).

For amphibians, temperature is an important parameter regulating from biochemical to behavioral aspects, since they are not able to regulate their internal temperature (Fontenot and Lutterschmidt 2011, Freitas et al. 2016, Freitas et al. 2017a,b). Metabolism, digestion, vision, locomotion, growth, and development in amphibians are virtually affected by environmental temperature (Rome 2007, Freitas et al. 2017a,b, Grott et al. 2021). Biological reactions also become faster in warmer waters as respiration rates are improved, contributing to ROS production. In fact, changes in several environmental conditions can lead to oxidative stress in organisms by increased production of ROS (Ahmed 2005, Halliwell and Gutteridge 2007). In tadpoles of *P. nattereri* for example, GST and GR showed temperature-dependent activities and CAT was increased when tadpoles were maintained in water with acid (pH 5.0) or alkaline (pH 8.5) pH, compared to those kept under neutral pH (pH 7.0) (Freitas and Almeida 2016). In this study, data demonstrated that *P. nattereri* had an efficient antioxidant response for dealing with heat and acidity/alkalinity conditions in water.

In addition to the direct biochemical effects caused by changes of environmental factors in tadpoles, the indirect effects induced by the interaction between abiotic factors and pollutants have been also discussed for oxidative markers. For example, the combined effects of pesticides and temperature on antioxidant responses of two Brazilian tadpoles was shown by Freitas et al. (2017a,b). In the first study, biochemical effects of clomazone (Gamit®) in *P. nattereri* and *R. schneideri* tadpoles were modulated by the thermal gradient (28, 32 and 36°C). Activities of the enzymes CAT, SOD, G6PDH, and GST in *P. nattereri* were all increased in animals exposed to the herbicide at higher temperatures. In *R. schneideri*, clomazone failed to alter antioxidant enzymes at 28°C, but their activities were raised at higher temperatures similar to the observed in *P. nattereri*. Lipid peroxidation (accessed by MDA levels) was induced in both species exposed to clomazone at 32 and 36°C, but not at 28°C. An integrated analysis of biomarker responses also revealed a synergic effect of temperature (higher temperatures) and clomazone after shorter exposure periods (3 days) (Freitas et al. 2017a). In the second study (Freitas et al. 2017b), a similar experimental design was established to investigate the combined effects of the herbicide sulfentrazone (Boral® SC) at different temperatures in tadpoles of *P. nattereri* and *R. schneideri*. In this study, authors observed that the combined effects had species-specific toxicity pattern in tadpoles. The integrated analysis of all biomarkers indicated a synergic effect of temperature and sulfentrazone for tadpoles of *P. nattereri*, while antioxidant responses were mostly influenced by temperature alone in *R. schneideri*. Lipid peroxidation was particularly reported in *P. nattereri* tadpoles exposed to sulfentrazone (Freitas et al. 2017b). Both studies indicated that tadpole's species from tropical areas may have the ability to modulate their physiological mechanisms associated to antioxidant defense to deal with temperature fluctuations and pesticide presence in their habitats. Therefore, it should be noted that contaminants may enhance toxicological effects on oxidative response of tadpoles that inhabit areas with naturally higher temperatures.

A multibiomarker approach on the influence of temperature (25 and 32°C) in the toxicity of the herbicide tebuthiuron (10, 50, and 200 ng/L) was analyzed in bullfrog tadpoles (*L. catesbeianus*) after 16 days of exposure. Combined effects of the highest temperature and tebuthiuron increased GPx, G6PDH and CAT activities compared to control tadpoles at 25°C, but the treatment decreased SOD (Grott et al. 2021). In this study, MDA levels remained unchanged after exposure to tebuthiuron at any temperature, indicating that the experimental conditions studied did not cause oxidative stress. Authors suggested that this lack of MDA alterations were probably a consequence of the observed increases in antioxidant responses, which although indicating an increase in ROS production, prevented enhancement of lipid peroxidation in tadpoles (Grott et al. 2021).

Presence of natural predators in the aquatic environment are also known to induce changes in the antioxidant responses of tadpoles. For example, tadpoles of *Pelobates cultripes* exposed to predator cues (*Dytiscus circumflexus* larvae) presented decreased SOD and GR activities, a response that was associated with a general reduction in animal activity and a concomitant reduction in catabolic pathways (Burraco et al. 2013). Despite this, oxidative damage was not observed since the levels of TBARs did not vary in tadpoles in the presence/absence of predator cues. The same study investigated the potential interaction between the predator and the herbicide glyphosate on the oxidative stress parameters of tadpoles, but no combined effects were observed at the studied concentrations of glyphosate (0.5 and 1.0 mg/L) and experimental period (20 days) (Burraco et al. 2013). The low toxicity of glyphosate in this case can also be associated with the advanced stages to which the tadpoles were exposed to the substances (stages 35–36 Gosner), since younger stages are usually more susceptible to the toxicants.

Although not directly related to oxidative stress, a study with Northern leopard tadpoles *Lithobates pipiens* showed that toxicokinetic elimination of polychlorinated biphenyls (PCBs) and polybrominated diphenyl ethers (PBDEs) increased in warmer temperatures (Brown et al. 2021). However, the authors highlighted that in a climate change scenario, PCB and PBDE should not be interpreted as having minor effects on tadpoles, since understanding the interaction of temperature and toxicant in the field is quite challenging. For example, faster metabolic rates resulting in faster chemical depuration can enhance the production of metabolites, which can be retained longer in the animal's body, producing various disorders that can include oxidative stress and changes in redox balance (Brown et al. 2021).

As discussed, the combined effects of abiotic factors and contaminants are extremely complex to interpret. For tadpoles (or even other ectotherms), some hypothesis should be considered in this context, especially for a scenario of rising global temperature: (i) heat stress increases reaction rates and then the uptake of environmental contaminants, making them more toxic to tadpoles; (ii) warmer waters have a lower rate of dissolved oxygen, which consequently increases animal ventilation and the acquisition of aquatic contaminants; (iii) exposure to contaminants makes tadpoles more susceptible to the effects of heat stress; (iv) increased metabolic rates increase the elimination of contaminants from the animal's body; (v) or increased metabolic rates produce metabolites that can be even more toxic to tadpoles than the original chemical (Blaustein et al. 2010, Hooper et al. 2013, Moe et al. 2013, Freitas et al. 2016, Freitas et al. 2017a,b, Grott et al. 2021). Despite evidence of the interaction between temperature and pesticides in amphibians, to date there is a huge deficit of information on how metabolic homeostasis associated with antioxidant systems in tadpoles can be disturbed by environmental chemicals and thermal stress. The little we know so far is restricted to just a few classes of pesticides. This demonstrates the need to expand knowledge of these effects for different chemicals, especially regarding species from tropical regions, as they harbor the highest amphibian species richness in the world.

7. Conclusion and Perspectives

Biochemical biomarkers assessing antioxidant response or oxidative stress in tadpoles are sensitive and very useful tools for investigating environmental pollution in different amphibian species. Antioxidant markers are also important indicators of sublethal environmental stress in amphibians. Although these biomarkers do not reflect specific mechanisms of action of contaminants, they may indicate the physiological status of animals and, therefore, the health of organisms. In this chapter, it was showed how environmental contaminants are likely to occur in tadpoles' environments and can impact their antioxidant defense systems, inducing oxidative stress. We also briefly discussed how environmental variables can interfere with the biochemical responses of tadpoles, making them more susceptible to toxic effects. So far, it is clear the presence of several gaps that need to be answered for amphibians, such as (i) the effects of environmental pollutants on the antioxidant systems of species from tropical and temperate regions; (ii) the effects caused by the mixture of contaminants; and (iii) compared studies with field and laboratory conditions, assisting in environmental monitoring measures.

Acknowledgment

The Productivity Scholarship Program in Research – PQ/UEMG (06/2021) provided financial support to the author during the production of this Chapter.

References Cited

Abdel-Shafy, A.I. and M.S.M. Mansour. 2016. A review on polycyclic aromatic hydrocarbons: Source, environmental impact, effect on human health and remediation. Egyptian J. Petroleum. 25(1): 107–123.

Ahmed, R.G. 2005. Is there a balance between oxidative stress and antioxidant defense system during development? Medical J. of Islamic World Acad. of Sci. 15(2): 55–63.

Aliko, V., M. Qirjo, E. Sula, V. Morina and C. Faggio. 2018. Antioxidant defense system, immune response and erythron profile modulation in Goldfish, *Carassius auratus*, after acute manganese treatment. Fish Shellfish Immunol. 76: 101–109.

Allen, R.G. and M. Tresini. 2000. Oxidative stress and gene regulation. Free Radic. Biol. Med. 28: 463–499.

Almeida, E.A. and P.D. Mascio. 2011. Hypometabolism: Strategies of Survival in Vertebrates and Invertebrates, 39-55 ISBN: 978-81-308-0471-2 Editors: Anna Nowakowska and Michał Caputa.

Amaeze, N.H., A. Onadeko and C.C. Nwosu. 2014. Comparative acute toxicity and oxidative stress responses in tadpoles of *Amietophrynus regularis* exposed to refined petroleum products, unused and spent engine oils. Afr. J. Biotechnol. 13: 4251–58.

Anguiano, O.L., A. Caballero and A.M. Pechen de D'Angelo. 2001. The role of glutathione conjugation in the regulation of early toad embryo's tolerance to pesticides. Comp. Biochem. Physiol. C. 128: 35–43.

Anyakora, C., A. Ogbeche, P. Palmer P. and H. Coker. 2005. Determination of polynuclear aromatic hydrocarbons in marine samples of Siokolo fishing settlement. J. Chromatogr. A. 1073(1-2): 323–330.

Atli, G. and C. Canli. 2007. Enzymatic responses to metal exposures in a freshwater fish *Oreochromis niloticus.* Comp. Biochem. Physiol. (C) 145: 282–287.

Balogh, L.M. and W.M. Atkins. 2011. Interactions of glutathione transferases with 4-hydroxynonenal. Drug Metab Rev. 43(2): 165–78.

Banday, U.Z., S.B. Swaleh and N. Usmani. 2019. Insights into the heavy metal-induced immunotoxic and genotoxic alterations as health indicators of *Clarias gariepinus* inhabiting a rivulet. Ecotoxicol. Environ. Saf. 183: 109584.

Beggel, S., I. Werner, R.E. Connon and J.P. Geist. 2012. Impacts of the phenylpyrazole Insecticide fipronil on larval fish: time-series gene transcription responses in fathead minnow (*Pimephales promelas*) following short-term exposure. Sci. Total Environ. 426: 160–165.

Bernabo, I., A. Guardia, R. Macirella, S. Sesti, A. Crescente and E. Brunelli. 2016. Effects of long-term exposure to two fungicides, pyrimethanil and tebuconazole, on survival and life history traits of Italian tree frog (Hyla intermedia). Aquat. Toxicol. 172: 56–66.

Bhuyan, K., P. Arabina, U. Singha, S. Giri and A. Giri. 2020. Phenanthrene alters oxidative stress parameters in tadpoles of *Euphlyctis cyanophlyctis* (Anura, Dicroglossidae) and induces genotoxicity assessed by micronucleus and comet assay. Environ. Sci. Pollut. Res. 27: 20962–20971.

Bianchini, A. and J.M. Monserrat. 2007. Effects of methyl parathion on *Chasmagnathus granulatus* hepatopancreas: Protective role of sesamol. Ecotoxicol. Environ. Saf. 67: 100–108.

Blaustein, A.R., S.C. Walls, B.A. Bancroft, J.J. Lawler, C.I. Searle and S.S. Gervasi. 2010. Direct and indirect effects of climate change on amphibian populations. Diversity 2(2): 281–313, https://doi.org/10.3390/d2020281.

Boccioni, A.P.C., R.C. Lajmanovich, P.M. Peltzer, A.M. Attademo and A.S. Martinuzzi. 2021. Toxicity assessment at different experimental scenarios with glyphosate, chlorpyrifos and antibiotics in *Rhinella arenarum* (Anura: Bufonidae) tadpoles. Chemosphere 273: 128475.

Brown, C.T., J.M. Yahn and W. Karasov. 2021. Warmer temperature increases toxicokinetic elimination of PCBs and PBDEs in Northern leopard frog larvae (*Lithobates pipiens*). Aquat. Toxicol. 234: 105806.

Brühl, C.A., T. Schmidt, S. Pieper and A. Alscher. 2013. Terrestrial pesticide exposure of amphibians: An underestimated cause of global decline? Sci. Rep. 3: 1135.

Buchholz, D.R. and T.B. Hayes. 2005. Variation in thyroid hormone action and tissue content underlies species differences in the timing of metamorphosis in desert frogs. Evol. Dev. 7: 458–467.

Burggren, W. and M. Doyle. 1986. Ontogeny of regulation of gill and lung ventilation in the bullfrog, *Rana catesbeiana*. Respir. Physiol. 66: 279–291.

Burgos-Aceves, M.A., A. Cohen, Y. Smith and C. Faggio. 2018. MicroRNAs and their role on fish oxidative stress during xenobiotic environmental exposures. Ecotoxicol. Environ. Saf. 148: 995–1000.

Burraco, P., L.J. Duarte and I. Gomez-Mestre. 2013. Predator-induced physiological responses in tadpoles challenged with herbicide pollution. Curr. Zool. 59(4): 475–484.

Burraco, P., M. Iglesias-Carrasco, C. Cabido and I. Gomez-Mestre. 2018. Eucalypt leaf litter impairs growth and development of amphibian larvae, inhibits their antipredator responses and alters their physiology. Conserv. Physiol. 6, coy066.

Carvalho, A.S., H.S.M. Utsunomiya, T. Pasquoto-Stiglianic, M.J. Costa and M.N. Fernandes. 2020. Biomarkers of the oxidative stress and neurotoxicity in tissues of the bullfrog, *Lithobates catesbeianus* to assess exposure to metals. Ecotoxicol. Environm. Saf. 196: 110560.

Costantini, D. 2014. Oxidative stress and hormesis in evolutionary ecology and physiology: A marriage between mechanistic and evolutionary approaches, Berlin (Germany), Springer Science+Business Media.

Costa, M.J., D.A. Monteiro, A.I. Oliveira-Neto, F.T. Rantin and A.L. Kalinin. 2008. Oxidative stress biomarkers and heart function in bullfrog tadpoles exposed to Roundup Original. Ecotoxicology 17: 153–163.

Crowder, W.C., M. Nie and G.R. Ultsch. 1998. Oxygen uptake in bullfrog tadpoles (*Rana catesbeiana*). J. Exp. Zool. 280: 121–134.

Dai, Z.Y., J. Cheng, L.S. Bao, X. Zhu, H.H. Li, X. Chen et al. 2020. Exposure to waterborne cadmium induce oxidative stress, autophagy and mitochondrial dysfunction in the liver of *Procypris merus*. Ecotoxicol. Environ. Saf. 204: 111051.

Demin, O.V., B.N. Kholodenko and V.P. Skulachev. 1998. A model of O2.-generation in the complex III of the electron transport chain. Mol. Cell. Biochem. 184: 21–33.

Denison, M.S., D. Phelan, G.M. Winter and M.H. Ziccardi. 1998. Carbaryl, a carbamate insecticide, is a ligand for the hepatic Ah (dioxin) receptor. Toxicol. Appl. Pharmacol. 152: 406–414.

Dornelles, M.F. and G.T. Oliveira. 2014. Effect of Atrazine, Glyphosate and Quinclorac on biochemical parameters, lipid peroxidation and survival in bullfrog tadpoles (*Lithobates catesbeianus*). Arch. Environ. Contam. Toxicol. 66: 415–429.

Ejilibe, C.O., H.O. Nwamba, C.L. Ani, J. Madu, G.C. Onyishi and C.D. Nwani. 2018. Oxidative stress responses in *Bufo regularis* tadpole exposed to Butaforce® and Termex®. Fisheries Livest. Prod. 6: 2.

Eriyamremua, G.E., V.E. Osagie, S.E. Omoregiea and C.O. Omofomac. 2008. Alterations in glutathione reductase, superoxide dismutase, and lipid peroxidation of tadpoles (*Xenopus laevis*) exposed to Bonny Light crude oil and its fractions. Ecotoxicol. Environ. Saf. 71: 284–290.

Fernández, L.P., R. Brasca, A.M. Attademo, P.M. Peltzer, R.C. Lajmanovich and M.J. Culzoni. 2020. Bioaccumulation and glutathione S-transferase activity on *Rhinella arenarum* tadpoles after short-term exposure to antiretrovirals. Chemosphere 246: 125830.

Ferrari, A., A. Venturino and A.M. Pechen de D'Angelo. 2007. Effects of carbaryl and azinphos methyl on juvenile rainbow trout (*Oncorhynchus mykiss*) detoxifying enzymes. Pest. Biochem. Physiol. 88: 134–142.

Ferrari, A., C.I. Lascano, O.I. Anguiano, A.M.P. D´Angelo and A. Venturino. 2009. Antioxidant responses to azinphos methyl and carbaryl during the embryonic development of the toad *Rhinella (Bufo) arenarum* Hensel. Aquat. Toxicol. 93: 37–44.

Filho, D.W. 1996. Fish antioxidant defenses—a comparative approach. Braz. J. Med. Biol. Res. 29: 1735–1742.

Fontenot, C.L. and W.I. Lutterschmidt. 2011. Thermal selection and temperature preference of the aquatic salamander, *Amphiuma tridactylum*. Herpetol. Conserv. Biol. 6: 395–399.

Freitas, J. and E.A.A. Almeida. 2016. Antioxidant defense system of tadpoles (*Eupemphix nattereri*) exposed to changes in temperature and pH. Zool. Sci. 33: 186–194.

Freitas, J.S., A. Kupsco, G. Diamante, A.A. Felicio, E.A. Almeida and D. Schlenk. 2016. Influence of temperature on the thyroidogenic effects of diuron and its metabolite 3,4-DCA in tadpoles of the american bullfrog (*Lithobates catesbeianus*). Environ. Sci. Technol. 50(23): 13095–13104.

Freitas, J.S., F.T. Teresa and E.A. Almeida. 2017a. Influence of temperature on the antioxidant responses and lipid peroxidation of two species of tadpoles (*Rhinella schneideri* and *Physalaemus nattereri*) exposed to the herbicide sulfentrazone (Boral 500SC®). Comp. Biochem. Physiol., C. 197: 32–44.

Freitas, J.S., A.A. Felicio, F.T. Teresa and E.A. Almeida. 2017b. Combined effects of temperature and clomazone (Gamit®) on oxidative stress responses and B-esterase activity of *Physalaemus nattereri* (Leiuperidae) and *Rhinella schneideri* (Bufonidae) tadpoles. Chemosphere 185: 548–562.

Freitas, J.S., L. Girotto, B.V. Goulart, L.O.G. Alho, R.C. Gebara, C.C. Montagner et al. 2019. Effects of 2,4-D-based herbicide (DMA® 806) on sensitivity, respiration rates, energy reserves and behavior of tadpoles. Ecotoxicol. Environ. Saf. 82: 109446.

Freitas, J.S., T.J.S. Pinto, M.P.C. Yoshii, L.C.M. Silva, L.F.P. Lopes, A.P. Ogura et al. 2022. Realistic exposure to fipronil, 2,4-D, vinasse and their mixtures impair larval amphibian physiology. Environ. Pollut. 299: 118894.

Fridovich, I. 2004. Mitochondria: Are they the seat of senescence? Aging Cell. 3: 13–16.

Girotto, L., E.L.G. Espíndola, R.C. Gebara and J.S. Freitas. 2020. Acute and chronic effects on tadpoles (*Lithobates catesbeianus*) exposed to mining tailings from the dam rupture in Mariana, MG (Brazil). Water, Air, Soil Pollut. 231: 325.

Gobi, N., B. Vaseeharan, R. Rekha, S. Vijayakumar and C. Faggio. 2018. Bioaccumulation, cytotoxicity and oxidative stress of the acute exposure selenium in *Oreochromis mossambicus*. Ecotoxicol. Environ. Saf. 162: 147–159.

Gomez-Mestre, I., S. Kulkarni and D.R. Buchholz. 2013. Mechanisms and consequences of developmental acceleration in tadpoles responding to pond drying. Plos One 8: 84266.

Gorchev, H.G. and G. Ozolins. 2011. WHO Guidelines for drinking-water quality. WHO Chron. 38: 104–08.

Gottardi, M., M.R. Birch, K.R. Dalhoff and N. Cedergreen. 2017. The effects of epoxiconazole and α-cypermethrin on *Daphnia magna* growth, reproduction, and offspring size. Environ. Toxicol. Chem. 36: 2155–2166.

Gripp, H.S., J.S. Freitas, E.A. Almeida, M.C. Bisinoti and A.B. Moreira. 2017. Biochemical effects of fipronil and its metabolites on lipid peroxidation and enzymatic antioxidant defense in tadpoles (*Eupemphix nattereri*: Leiuperidae). Ecotoxicol. Environ. Saf. 136: 173–179. https://doi.org/10.1016/j.ecoenv.2016.10.027.

Grott, S.C., D. Bitschinski, N.G Israel, G. Abel, S.P. Silva, T.C. Alves et al. 2021. Influence of temperature on biomarker responses and histology of the liver of American bullfrog tadpoles (Lithobates catesbeianus, Shaw, 1802) exposed to the herbicide Tebuthiuron. Sci. Total Environ. 771: 144971.

Gunasekara, A. and T. Troung. 2007. Environmental Fate of Fipronil. Report. Environmental Monitoring Branch, Department of Pesticide Regulation, California Environmental Protection Agency, Sacramento, CA.

Guo, W., Y. Yang, R. Ming, D. Hu and P. Lu. 2022. Insight into the toxic effects, bioconcentration and oxidative stress of acetamiprid on *Rana nigromaculata* tadpoles. Chemosphere. 305: 135380.

Guo, W.L., J. Zhang, W.J. Li, M. Xu and S.J. Liu. 2015. Disruption of iron homeostasis and resultant health effects upon exposure to various environmental pollutants: a critical review. J. Environ. Sci. 34: 155–164.

Halliwell, B. and J.M.C. Gutteridge. 1989. Free Radicals in Biology and Medicine, second ed. Clarendon Press, Oxford, UK.

Halliwell, B. and J.M.C. Gutteridge. 2007. Reactive species can be poisonous. pp. 440–487. *In*: Halliwell, B. and J.M.C. Gutteridge [eds.]. Free Radicals in Biology and Medicine 4th ed. Oxford University Press, New York.

Han, Y., S. Ishibashi, J. Iglesias-Gonzalez, Y. Chen, N.R. Love and E. Amaya. 2018. Ca^{2+}-Induced mitochondrial ROS regulate the early embryonic cell cycle. Cell Rep. 22: 218–231.

Hodkovicova, N., L. Chmelova, P. Sehonova, J. Blahova, V. Doubkova, L. Plhalova, E. Fiorino et al. 2019. The effects of a therapeutic formalin bath on selected immunological and oxidative stress parameters in common carp (*Cyprinus carpio*). Sci. Total Environ. 653: 1120–1127.

Hooper, M.J., G.T. Ankley, D.A. Cristol, L.A. Maryoung, P.D. Noyes and K.E. Pinkerton. 2013. Interactions between chemical and climate stressors: A role for mechanistic toxicology in assessing climate change risks. Environ. Toxicol. Chem. 32: 32–48.

Hutanu, D., M.D. Frishberg, L. Guo and C.C. Darie. 2014. Recent applications of polyethylene glycols (PEGs) and PEG derivatives. Mod. Chem. Appl. 2(2): 1–6.

Isik, I. and I. Celik. 2008. Acute effects of methyl parathion and diazinon as inducers for oxidative stress on certain biomarkers in various tissues of rainbowtrout (*Oncorhynchus mykiss*). Pestic. Biochem. Physiol. 92: 38–42.

Jayawardena, U.A., P. Angunawela, D.D. Wickramasinghe, W.D. Ratnasooriya and P.V. Udagama. 2017. Heavy metal-induced toxicity in the Indian green frog: Biochemical and histopathological alterations. Environ. Toxicol. Chem. 36(10): 2855–2867.

Jiao, H., Q. Wang, N. Zhao, B. Jin, X. Zhuang and Z. Bai. 2017. Distributions and sources of polycyclic aromatic hydrocarbons (PAHs) in soils around a chemical plant in Shanxi, China. Int. J. Environ. Res. Public. Health. 14(10): 1198.

Johnson, J., W. Manzo, E. Gardner and J. Menon. 2013. Reactive oxygen species and antioxidant defenses in tail of tadpoles, *Xenopus laevis*. Comp. Biochem. Physiol. C. Toxicol. Pharmacol. 158: 101–108.

Jones, L., D.R. Gosset, S.W. Banks and M. MCCallum. 2010. Antioxidant defense system in tadpoles of the american bullfrog (*Lithobates catesbeianus*) exposed to paraquat. Journal of Herpetology 44 (2): 222–228.

Kanti Das, T., M.R. Wati and K. Fatima-Shad. 2015. Oxidative stress gated by Fentonand Haber Weiss reactions and its association with Alzheimer's disease. Arch. Neurosci. 2(2): e20078.

Kashiwagi, A., H. Hanada, M. Yabuki, T. Kanno, R. Ishisaka, J. Sasaki et al. 1999. Thyroxine enhancement and the role of reactive oxygen species in tadpole tail apoptosis. Free Radic. Biol. Med. 26: 1001–1009.

Kirschman, L.J., M.D. McCue, J.G. Boyles and R.W. Warne. 2017. Exogenous stress hormones alter energetic and nutrient costs of development and metamorphosis. J. Exp. Biol. 220: 3391–3397.

Kretschmann, A., M. Gottardi, K. Dalhoff and N. Cedergreen. 2015. The synergistic potential of the azole fungicides prochloraz and propiconazole toward a short alphacypermethrin pulse increases over time in *Daphnia magna*. Aquat. Toxicol. 162: 94–101.

Kulkarni, S.S. and D.R. Buchholz. 2012. Beyond synergy: Corticosterone and thyroid hormone have numerous interaction effects on gene regulation in *Xenopus tropicalis* tadpoles. Endocrinology 153: 5309–5324.

Kulkarni, S.S., R.J. Denver, I. Gomez-Mestre and D.R. Buchholz. 2017. Genetic accommodation via modified endocrine signalling explains phenotypic divergence among spadefoot toad species. Nat. Commun. 8: 993.

Lajmanovich, R.C., A.M. Attademo, P.M. Peltzer, C.M. Junges and M.C. Cabagna. 2011. Toxicity of four herbicide formulations with Glyphosate on *Rhinella arenarum* (Anura: Bufonidae) Tadpoles: B-esterases and Glutathione S-transferase Inhibitors. Arch. Environ. Contam. Toxicol. 60: 681–689.

Lajmanovich, R.C., P.M. Peltzer, C.S. Martinuzzi, A.M. Attademo, C.L. Colussi and A. Bassó. 2018. Acute toxicity of colloidal silicon dioxide nanoparticles on amphibian larvae: Emerging environmental concern. Int. J. Environ. Res. 12: 269–278.

Lesser, M.P. 2006. Oxidative stress in marine environments: Biochemistry and physiological ecology. Annu. Rev. Phys. 68: 253–278.

Liu, R., Y. Qin, J. Diao and H. Zhang. 2021. *Xenopus laevis* tadpoles exposed to metamifop: Changes in growth, behavioral endpoints, neurotransmitters, antioxidant system and thyroid development. Ecotoxicol. Environ. Saf. 220: 112417.

Livingstone, D.R. 2001. Contaminant-stimulated reactive oxygen species production and oxidative damage in aquatic organisms. Mar. Pollut. Bull. 42: 656– 666.

Lopes, I., F. Gonçalves, A.M.V.M. Soares and R. Ribeiro. 1999. Discriminating the ecotoxicity due to metals and to low pH in acid mine drainage. Ecotoxicol. Environ. Saf. 44: 207–14.

Lopez-Torres, M., R. Perez-Campo, S. Cadenas, C. Rojas and G. Barja. 1993. A comparative research of free radicals in vertebrates—Non-enzymatic antioxidants and oxidative stress. Comp. Biochem. Physiol. 105: 757–763.

Lu, H., Y. Hu, C. Kang, Q. Meng and Z. Lin. 2021. Cadmium-induced toxicity to amphibian tadpoles might be exacerbated by alkaline not acidic pH level. Ecotoxicol. Environm. Saf. 218: 112288.

Lushchak, V.I. 2011. Environmentally induced oxidative stress in aquatic animals. Aquat. Toxicol. 101: 13–30.

Mahapatra, P.K., P. Mohanty-Hejmadi and G.B. Chainy. 2001. Changes in oxidative stress parameters and acid phosphatase activity in the pre-regressing and regressing tail of Indian jumping frog *Polypedates maculatus* (Anura, Rhacophoridae). Comp. Biochem. Physiol. C Toxicol. Pharmacol. 130: 281–288.

Margarido, T.C.S., A.A. Felicio, D.C. Rossa-Feres and E.A. Almeida. 2013. Biochemical biomarkers in *Scinax fuscovarius* tadpoles exposed to a commercial formulation of the pesticide fipronil. Mar. Environ. Res. 91: 61–67.

Marques, S., S. Antunes, B. Nunes, F. Gonçalves and R. Pereira. 2011. Antioxidant response and metal accumulation in tissues of Iberian green frogs (*Pelophylax perezi*) inhabiting a deactivated uranium mine. Ecotoxicology 20: 1315–27.

Marques, S.M., S.C. Antunes, H. Pissarra, M.L. Pereira, F. Gonçalves and R. Pereira 2009. Histopathological changes and erythrocytic nuclear abnormalities in Iberian green frogs (*Rana perezi Seoane*) from a uranium mine pond. Aquat. Toxicol. 91: 187–95.

Matos, P., A. Fontainhas-Fernandes, F. Peixoto, J. Carrola and E. Rocha. 2007. Biochemical and histological hepatic changes of Nile tilapia *Oreochromis niloticus* exposed to carbaryl. Pestic. Biochem. Physiol. 89: 73–80.

Melvin, S.D. 2016. Oxidative stress, energy storage, and swimming performance of *Limnodynastes peronii* tadpoles exposed to a sub-lethal pharmaceutical mixture throughout development. Chemosphere (150): 790–797.

Melvin, S.V., F.D.L. Leusch and A.R. Carroll. 2018. Metabolite profiles of striped marsh frog (Limnodynastes peronii) larvae exposed to the anti-androgenic fungicides vinclozolin and propiconazole are consistent with altered steroidogenesis and oxidative stress. Aquat. Toxicol. 199: 232–239.

Menon, J. and R. Rozman. 2007. Oxidative stress, tissue remodeling and regression during amphibian metamorphosis. Comp. Biochem. Physiol. C Toxicol. Pharmacol. 145: 625–631.

Metcalfe, N.B. and C. Alonso-Alvarez. 2010. Oxidative stress as a life-history constraint: The role of reactive oxygen species in shaping phenotypes from conception to death. Funct. Ecol. 24: 984–996.

Miyata, H., O. Aozasa, S. Ohta, T. Chang and Y. Yasuda. 1993. Estimated daily intakes of PCDDs, PCDFs and non-ortho coplanar PCBs via drinking water in Japan. Chemosphere. 26: 1527.

Moe, S.J., K. De Schamphelaere, W.H. Clements, M.T. Sorensen, P.J. Van den Brink and M. Liess. 2013. Combined and interactive effects of global climate change and toxicants on populations and communities. Environ. Toxicol. Chem. 32: 49–61.

Muniswamy, D., S.S. Jadhav and K.R. Malowade. 2016. Biochemical Modulations in Duttaphrynus melanostictus tadpoles, following exposure to commercial formulations of cypermethrin: an overlooked impact of extensive cypermethrin use. J. App. Biol. Biotech. 4 (06): 032–037.

Murphy, J.E., C.A. Phillips and V.R. Beasley. 2000. Aspects of amphibian ecology. pp. 141–78. *In*: Sparling, D.W., G. Linder and C.A. Bishop [eds.]. Ecotoxicology of amphibians and reptiles. Columbia, USA: SETAC technical publication series.

Nascimento, I.F., A.T.B. Guimarães, F. Ribeiro, A.S.L. Rodrigues, F.N. Estrela, T.M. Luz and G. Malafaia. 2021. Polyethylene glycol acute and sub-lethal toxicity in neotropical *Physalaemus cuvieri* tadpoles (Anura, Leptodactylidae). Environm. Poll. 283: 117054.

Nieuwkoop, P.D. and J. Faber. 1967. Normal Table of *Xenopus laevis*. North Holland Publishing Company, Amsterdam.

Oruc, E.Ö. and N. Üner. 2000. Combined effects of 2,4-d and azinphosmethyl on antioxidant enzymes and lipid peroxidation in liver of *Oreochromis niloticus*. Comp. Biochem. Physiol. C. 127: 291–296.

Papa, S. and V.P. Skulachev. 1997. Reactive oxygen species, mitochondria, apoptosis and aging. Mol. Cell. Biochem. 174: 305–319.

Parvez, S. and S. Raisuddin. 2005. Protein carbonyls: Novel biomarkers of exposure to oxidative stress-inducing pesticides in freshwater fish Channa punctata (Bloch). Environ. Toxicol. Pharmacol. 20: 112–117.

Peltzer, P.M., R.C. Lajmanovich, A.M Attademo, C.M. Junges, C.M. Teglia, C. Martinuzzi et al. 2017. Ecotoxicity of veterinary enrofloxacin and ciprofloxacin antibiotics on anuran amphibian larvae. Environ. Toxicol. Pharmacol. 51: 114–123.

Peltzer, P.M., R.C. Lajmanovich, C. Martinuzzi, A.M. Attademo, L.M. Curi and M.T. Sandoval. 2019. Biotoxicity of diclofenac on two larval amphibians: Assessment of development, growth, cardiac function and rhythm, behavior and antioxidant system. Sci. Total Environ. 683: 624–637.

Pinya, S., S. Tejada, X. Capó and A. Sureda. 2016. Invasive predator snake induces oxidative stress responses in insular amphibian species. Sci. Total Environ. 566: 57–62.

Prohaska, J.R. and H.E. Ganther. 1977. Glutathione peroxidase activity of glutathione-s-transferases purified from rat liver. Biochem. Biophys. Res. Commun. 76(2): 437–445.

Prokić, M.D., S.S. Borković-Mitić, I.I. Krizmanić, J.J. Mutić, V. Vukojević, M. Nasia et al. 2016. Antioxidative responses of the tissues of two wild populations of Pelophylax kl. esculentus frogs to heavy metal pollution. Ecotoxicol. Environ. Saf. 128: 21–29.

Prokić, M.D., S.S. Borković-Mitić, I.I. Krizmanić, J.J. Mutić, J.P. Gavrić, S.G. Despotović et al. 2017. Oxidative stress parameters in two *Pelophylax esculentus* complex frogs during pre-and posthibernation: Arousal vs heavy metals. Comp. Biochem. Physiol. C Toxicol. Pharmacol. 202: 19–25.

Prokić, M.D., S.G. Despotović, T.Z. Vučić, T.G. Petrović, J.P. Gavrić, B.R. Gavrilović et al. 2018a. Oxidative cost of interspecific hybridization: A case study of two *Triturus* species and their hybrids. J. Exp. Biol. 221: jeb182055.

Prokić, M.D., T.G. Petrović, J.P. Gavrić, S.G. Despotović, B.R. Gavrilović, T.B. Radovanović et al. 2018b. Comparative assessment of the antioxidative defense system in subadult and adult anurans: A lesson from the *Bufotes viridis* toad. Zoology 130: 30–37.

Prokić, M.D., J.P. Gavrić, T.G. Petrović, S.G. Despotović, B.R. Gavrilović, T.B. Radovanović et al. 2019. Oxidative stress in *Pelophylax esculentus* complex frogs in the wild during transition from aquatic to terrestrial life. Comp. Biochem. Physiol., A. 234: 98–105.

Ray, P.D., B.W. Huang and Y. Tsuji. 2012. Reactive oxygen species (ROS) homeostasis and redox regulation in cellular signaling. Cell Signal. 24 (5): 981–990.

Reichert, L.M.M., D.R. Oliveira, J.L. Papaleo, A.A.N. Valgas and G.T. Oliveira. 2022. Biochemical and body condition markers in *Rhinella icterica* tadpoles exposed to atrazine, glyphosate, and quinclorac based herbicides in ecologically relevant concentrations. Environ Toxicol. Pharmacol. 93: 103884.

Rizzo, A.M., L. Adorni, G. Montorfano, F. Rossi and B. Berra. 2007. Antioxidant metabolism of *Xenopus laevis* embryos during the first days of development. Comp. Biochem. Physiol. B Biochem. Mol. Biol. 146: 94–100.

Robinson, S.A., R.J. Chlebak, S.D. Young, R.L. Dalton, M.J. Gavel, R.S. Prosser et al. 2021. Clothianidin alters leukocyte profiles and elevates measures of oxidative stress in tadpoles of the amphibian, Rana pipiens. Environ. Pollut. 284: 117149.

Rocha, T.L., T. Gomes, V.S. Sousa, N.C. Mestre and M.J. Bebianno. 2015. Ecotoxicological impact of engineered nanomaterials in bivalve molluscs: An overview. Mar. Environ. Res. 111: 74–88.

Rome, L.C. 2007. The effect of temperature and thermal acclimation on the sustainable performance of swimming scup. Philos. Trans. R. Soc. Lond. B. Biol. Sci. 362: 1995e2016.

Rozita, Y., R. Brydson, A.J. Scott. 2010. An investigation of commercial gamma-Al2O3 nanoparticles. J. Phys.: Conf. Ser. 241(1): 012096. IOP Publishing.

Rutkoski, C.F., S.C. Grott, N.G. Israel, F.E. Carneiro, F.C. Guerreiro, S. Santos et al. Almeida 2022. Hepatic and blood alterations in *Lithobates catesbeianus* tadpoles exposed to sulfamethoxazole and oxytetracycline. Chemosphere. 307, Part 4, 136215.

Safari, R. 2015. Toxic effects of cadmium on antioxidant defense systems and lipid peroxidation in *Acipenser persicus* (Borodin, 1897). Int. J. Aquat. Biol. 3(6): 425–432.

Samanta, L. and B. Paital, B. 2016. Effects of seasonal variation on oxidative stress physiology in natural population of toad *Bufo melanostictus*; clues for analysis of environmental pollution. Environ. Sci. Pollut. Res. 23: 22819–22831.

Sandeep Prabhu, K., P.V. Reddy, E.C. Jones, A.D. Liken and C.C. Reddy. 2004. Characterization of a class alpha glutathione-S-transferase with glutathione peroxidase activity in human liver microsomes. Arch Biochem. Biophys. 424(1): 72–80.

Santos, L.H.M.LM., A.N. Araújo, A. Fachini, A. Pena, C. Delerue-Matos and M.C.B.S.M. Montenegro. 2010. Ecotoxicological aspects related to the presence of pharmaceuticals in the aquatic environment. J. Hazard. Mater. 175: 45–95.

Sevcikova, M., H. Modra, A. Slaninova and Z. Svobodova. 2011. Metals as a cause of oxidative stress in fish. A review. Vet. Med. 56: 537–546.

Silva, P.R., M. Borges-Martins and G.T. Oliveira. 2021. *Melanophryniscus admirabilis* tadpoles' responses to sulfentrazone and glyphosate-based herbicides: An approach on metabolism and antioxidant defenses. Environm. Sci. Pollut. Res. 28: 4156–4172.

Slaby, S., M. Marin, G. Marchand and S. Lemiere. 2019. Exposures to chemical contaminants: what can we learn from reproduction and development endpoints in the amphibian toxicology literature? Environ. Poll. 248: 478–495.

Smith, S.M., R.G. Nager and D. Costantini. 2016. Meta-analysis indicates that oxidative stress is both a constraint on and a cost of growth. Ecol. Evol. 6: 2833–2842.

Sparling, D.W. 2010. Ecotoxicology of organic contaminants to amphibians. pp. 261–88. *In*: Sparling, D.W., G. Linder, C.A. Bishop and S.K. Krest [eds.]. Ecotoxicology of Amphibians and Reptiles. 2nd ed. Pensacola, FL: SETAC Press.

Stadtman, E.R. 1986. Oxidation of proteins by mixed function oxidation system: implication in protein turnover, aging and neutrophil function. Trends Biochem. 11: 11.

Stadtman, E.R. and B.S. Berlett. 1998. Reactive oxygen-mediated protein oxidation in aging and disease. Drug Metab. Rev. 30(2): 225–43. doi: 10.3109/03602539808996310.

Stara, A., R. Bellinvia, J. Velisek, A. Strouhova, A. Kouba and C. Faggio. 2019. Acute exposure of neonicotinoid pesticide on common yabby (*Cherax destructor*). Sci. Total Environ. 665: 718–723.

Stegeman, J.J., M. Brouwer, T.D.G. Richard, L. Forlin, B.A. Fowler, B.M. Sanders et al. 1992. Molecular responses to environmental contamination: Enzyme and protein systems as indicators of chemical exposure and effect. pp. 235–335. *In*: Huggett, R.J., R.A. Kimerly, P.M. Mehrle, Jr. and H.L. Bergman [eds.]. Biomarkers: Biochemical, Physiological and Histological markers of Anthropogenic Stress. Lewis Publishers, Chelsea, MI, USA.

Sutuyeva, L.R., V.L. Trudeau and T.M. Shalakhmetova. 2019. Mortality of embryos, developmental disorders and changes in biochemical parameters in marsh frog (*Rana ridibunda*) tadpoles exposed to the water-soluble fraction of Kazakhstan crude oil and O-Xylene. Journal of Toxicology and Environmental Health, Part A.

Svartz, G., C. Aronzon, S.P. Catãn, S. Soloneski and C.P. Coll. 2020. Oxidative stress and genotoxicity in *Rhinella arenarum* (Anura: Bufonidae) tadpoles after acute exposure to Ni-Al nanoceramics. Environ. Toxicol. Pharm. 80: 103508.

Turani, B., V. Aliko and C. Faggio. 2018. Allurin and egg jelly coat impact on *in-vitro* fertilization success of endangered albanian water frog, *Pelophylax shqipericus*. Nat. Prod. Res. https://doi.org/10.1080/14786419.2018.1508147.

Uçkun, M. and M. Özmen. 2021. Evaluating multiple biochemical markers in *Xenopus laevis* tadpoles exposed to the pesticides thiacloprid and trifloxystrobin in single and mixed forms. Environ. Toxicol. 40(10): 2846–2860.

van der Oost, R., J. Beyer and N.P.E. Vermeulen. 2003. Fish bioaccumulation and biomarkers in environmental risk assessment: A review. Environ. Toxicol. Pharmacol. 13: 57–149.

Veronez, A.C.S., R.V. Salla, V.D. Baroni, I.F. Barcarolli, A. Bianchini, C.B.R. Martinez et al. 2016. Genetic and biochemical effects induced by iron ore, Fe and Mn exposure in tadpoles of the bullfrog *Lithobates catesbeianus*. Aquat. Toxicol. 174: 101–108.

Wilkens, A.L., A.N. Valgas and G.T. Oliveira. 2019. Effects of ecologically relevant concentrations of Boral® 500 SC, Glifosato® Biocarb, and a blend of both herbicides on markers of metabolism, stress, and nutritional condition factors in bullfrog tadpoles. Environ. Sci. Pollut. Res. Int. 26(23): 23242–23256.

Williams, M. 2001. Arsenic in mine waters: an international study. Environ. Geol. 40: 267–78.

Winston, G.W. 1991. Oxidants and antioxidants in aquatic animals. Comp. Biochem. Physiol. Part. C. 100: 173–176.

Winston, G.W. and R.T. Di Giulio. 1991. Prooxidant and antioxidant mechanisms in aquatic organisms. Aquat. Toxicol. 19: 137–161.

Worthington, D.J. and M.A. Rosemeyer. 1974. Human glutathione reductase: purification of the crystalline enzym from erythrocytes. Eur. J. Biochem. 48: 167–177.

Xu, P. and L. Huang. 2017. Effects of α-cypermethrin enantiomers on the growth, biochemical parameters and bioaccumulation in *Rana nigromaculata* tadpoles of the anuran amphibians. Ecotoxicol. Environ. Saf. 139: 431–438.

Yan, D., X. Jiang, S. Xu, L. Wang, Y. Bian and G. Yu. 2008. Quantitative structure-toxicity relationship study of lethal concentration to tadpole (*Bufo vulgaris formosus*) for organophosphorous pesticides. Chemosphere 71: 1809–1815.

Yin, X., S. Jiang, J. Yu, G. Zhu, H. Wu and C. Mao. 2014. Effects of spirotetramat on the acute toxicity, oxidative stress, and lipid peroxidation in Chinese toad (*Bufo bufo gargarizans*) tadpoles. Environ. Toxicol. Pharmacol. 37: 1229–1235.

Zhang, W., L. Chen, Y. Xu, Y. Deng, L. Zhang, Y. Qin et al. 2019. Amphibian (*Rana nigromaculata*) exposed to cyproconazole: Changes in growth index, behavioral endpoints, antioxidant biomarkers, thyroid and gonad development. Aquat. Toxicol. 208: 62–70.

Zhang, Y. and S. Tao. 2009. Global atmospheric emission inventory of polycyclic aromatic hydrocarbons (PAHs) for 2004. Atmos. Environ. 43(4): 812–819.

Zusterzeel, P.L.M., H. Rutten, H.M.J. Roelefs, W.H.M. Peters and E.A.P. Steegers. 2001. Protein carbonyls in decidua and placenta of preeclamptic women as markers of oxidative stress. Placenta. 22: 213.

7

Genotoxicity and Mutagenicity in Anuran Amphibians

Daniela de Melo e Silva,[1] *Tiago Quaggio Vieira,*[2] *Marcelino Benvindo de Souza,*[1] *Flávia Regina de Queiroz Batista,*[2] *Hugo Freire Nunes,*[1] *Alessandro Ribeiro Morais*[3] and *Eduardo Alves de Almeida*[4,*]

1. Introduction

Genotoxicity studies analyze the effects of xenobiotic agents on genetic material, which can result in modifications of nucleotide structures of nucleic acids that eventually lead to mutations, single or double-strand breaks in DNA or impaired gene expression. It also includes indirect effects by which xenobiotics cause alterations in enzymes and proteins involved in the process of RNA transcription or DNA replication, resulting in alterations of nucleus' metabolism. Evaluation of alterations of gene products (RNAs, polypeptides, and proteins) resulting from negative effects on genetic material is also part of the study of genotoxicity of xenobiotics, since these alterations may compromise the physiological homeostasis of exposed organisms altering their capacity to respond to environmental stressors (Grisolia 2005). These injuries may result in tumors or be transmitted from generation to generation, affecting populations for a long-term period.

[1] Laboratório de Mutagênese, Universidade Federal de Goias-UFG, Av. Esperança, s/n - Campus Samambaia, Goiânia - GO, 74045-155. Brazil. Emails: danielamelosilva@ufg.br; marcelinobenvindo@gmail.com; hugofreire@ufg.br
[2] Centro Nacional de Pesquisa e Conservação de Répteis e Anfíbios – RAN/ICMBio, Rua 229, nº 95, Edifício IBAMA, 4o andar, Setor Leste Universitário, Goiânia/GO, 74.605.090. Brazil. Emails: tiagovieira@icmbio.gov.br; frqbatista@gmail.com
[3] Laboratório de Biologia Animal, Instituto Federal Goiano - IFGoiano, Rodovia Sul Goiana, Km 01, s/n - Zona Rural, Rio Verde - GO, 75901-970. Brazil. Email: alessandro.morais@ifgoiano.edu.br
[4] Centro de Estudos em Toxicologia Aquática, Universidade Regional de Blumenau, Rua São Paulo, 3366, Bloco Q, setor Q101, Blumenau, SC, 89030-000, Brazil.
* Corresponding author: eduardoalves@furb.br

Most part of xenobiotics that are released into the environment can exert genotoxic effects on exposed biota. Some of these are called eugenics, as they act by causing changes in the distribution of chromosomes during the process of cell division, giving rise to aneuploidies. In turn, clastogenic compounds induce breaks and produce changes in chromosome structure. DNA breaks occur in general due to the action of reactive free radicals that are often formed excessively during intoxication (See Chapter 6 for more details), due to the direct binding of genotoxic compounds or its metabolites to the nucleotides, or because of the enzymatic action of excision repair mechanisms (Speit and Hartmann 1999).

Through different genotoxicity and cytogenetic tests, it is possible to evaluate the genotoxic effects of a particular compound, which makes these types of tests relevant in risk assessments of environmental contaminants (Rabello-Gay et al. 1991). Assays to evaluate genotoxicity of specific compounds or complex mixtures can be performed under laboratory conditions using various types of organisms, and many studies have been done using amphibian tadpoles. Physiological conditions that involve substantial cell divisions, such as the process of metamorphosis of amphibians, may increase the susceptibility of the organism to alterations in the genetic material. During the development of tadpoles from the aquatic larvae to the terrestrial adult, the animals undergo intense morphological and physiological alterations that involves apoptotic processes and extensive cell proliferation and differentiation. Any error that eventually occur in the genetic material during this process can propagate to daughter cells resulting in malformations, which may affect adult's capacity to reproduce and survive. In this sense, the exposure of developing tadpoles to environmental stressors, especially anthropogenic xenobiotics, are supposed to increase the vulnerability of the organism to genotoxic effects.

Numerous contaminants are known to produce genotoxicity in tadpoles, such as pesticides, metals, petroleum derivatives and emerging contaminants, many of them being carcinogenic. A compilation of substances that were proven to cause genotoxic effects in different tadpole species is given in Table 1. In these studies, genotoxic evaluation was performed using three main genotoxicity tests: the comet assay, the evaluation of micronucleus frequencies and the analyses of nuclear abnormalities.

2. The Comet Assay

DNA lesions in organisms can be evaluated through the comet assay, also known as SCGE (Single Cell Gel Electrophoresis). This assay can detect breaks in the DNA structure, being one of the main tools in genotoxicity studies with *in vivo* approaches (Tice et al. 2000, Collins 2004, Kumaravel et al. 2006). The comet assay was first performed by Ostling and Johanson (1984) as a micro-electrophoretic technique for directly visualizing DNA damage in individualized cells. The electrophoretic run was performed under neutral pH conditions, which, according to the authors, allowed the detection of double-strand breaks in DNA. Subsequently, Singh et al. (1988) proposed the alkaline version of the comet assay (pH>13). Initially, it was believed that the alkaline version of the test enabled the detection of single-strand breaks (SSBs), while the neutral version provided double-strand breaks (DSBs). However, both versions of the comet assay are capable of detecting SSBs and DSBs, in addition to cross-links and alkali labile sites (Dhawan et al. 2009, Azqueta et al. 2013b).

The comet assay is based on the behavior of the DNA molecule when subjected to a constant and uniform electric field. Suppose a damage has occurred to the DNA molecule, in this case, during the electrophoresis run, DNA fragments will migrate from the nucleoid at different speeds towards the positive pole, forming the typical figure of a "comet" (Godschalk et al. 2013). If no DNA damage has occurred, the entire nucleoid will migrate uniformly, forming a circle. To read and analyze the results obtained by the comet assay, the nucleoid is divided into the head and tail (Ostling and Johanson 1984, Singh et al. 1988). Thus, nucleoids with little or no DNA damage

Table 1. Evidences of genotoxic effects of environmentally realistic concentrations of different classes of contaminants in amphibian tadpoles. MN = micronucleus; NA = nuclear abnormalities; CA = DNA damage through the comet assay; SB = strand breaks.

Xenobiotic	Effect	Tadpole Species	References
Pesticides: *Herbicides*			
Tadpoles collected in agricultural areas			
	↑MN	*Bokermannohyla oxente*	Silva et al. (2021)
	↑MN	*Leptodactylus latrans*	Silva et al. (2021)
	↑CA	*Physalaemus cuvieri*	Gonçalves et al. (2017)
	↑NA	*Scinax fuscovarius*	Borges et al. (2019)
	↑MN	*Rhinella arenarum*	Peluso et al. (2019)
Glyphosate	↑MN/NA	*Physalaemus cuvieri*	Herek et al. (2021)
	↑MN/NA	*Physalaemus gracilis*	Herek et al. (2021)
	↑MN	*Euflictis cyanophlyctis*	Yadav et al. (2013)
	↑CA	*Dendropsophus minutus*	Carvalho et al. (2018)
	↑MN	*Odontophrynus cordobae*	Bosch et al. (2011)
	↑MN	*Rhinella arenarum*	Bosch et al. (2011)
Atrazine	↑CA	*Aquarana catesbeiana*	Clements et al. (1997)
	↑MN/CA	*Dendropsophus minutus*	Gonçalves et al. (2017)
	↑MN	*Sclerophrys regularis*	Said et al. (2022)
Metalochlor	↑CA	*Aquarana catesbeiana*	Clements et al. (1997)
Metribuzin	↑CA	*Aquarana catesbeiana*	Clements et al. (1997)
Imazethapyr	↑MN/CA/SB	*Leptodactylus latinasus*	Pérez-Iglesias et al. (2020)
	↑MN/CA/SB	*Boana pulchella*	Pérez-Iglesias et al. (2018)
Fenoxaprop-ethyl	↑MN	*Aquarana catesbeiana*	Jing et al. (2017)
Dicamba+glyphosate	↑CA	*Rhinella arenarum*	Soloneski et al. (2016)
Flurochloridone	↑MN/CA	*Rhinella arenarum*	Nikoloff et al. (2014)
Glufosinate	↑MN/NA	*Rhinella arenarum*	Lajmanovich et al. (2014)
Acetochlor	↑CA	*Bufo bufo gargarizans*	Yin et al. (2008)
Paraquat	↑CA	*Bufo bufo gargarizans*	Yin et al. (2008)
Butachlor	↑CA	*Bufo bufo gargarizans*	Yin et al. (2008)
	↑CA	*Fejervarya limnocharis*	Liu et al. (2011)
Insecticides			
Abamectin	↑MN/NA	*Aquarana catesbeiana*	Montalvão and Malafaia (2017)
	↑MN/NA	*Aquarana catesbeiana*	Amaral et al. (2018)
Lambda-cyhalothrin	↑MN	*Aquarana catesbeiana*	Campana et al. (2003)
Imidacloprid	↑MN/CA	*Hypsiboas pulchellus*	Pérez-Iglesias et al. (2014)
	↑MN/NA	*Leptodactylus luctator*	Samojeden et al. (2022)
	↑MN/NA	*Physalaemus cuvieri*	Samojeden et al. (2022)
Fipronil	↑NA	*Aquarana catesbeiana*	Santos et al. (2021)
Pyretrum	↑NA/CA	*Aquarana catesbeiana*	Oliveira et al. (2019)
Trichlorfon	↑MN	*Rana chensinensis*	Ma et al. (2019)
Pirimicarb	↑MN/NA	*Boana pulchella*	Natale et al. (2018)
	↑MN	*Rhinella arenarum*	Candioti et al. (2010)
Malathion	↑MN	*Euflictis cyanophlyctis*	Giri et al. (2012)
Endosulfan	↑MN	*Hyla pulchella*	Lajmanovich et al. (2005)
Profenofos	↑MN/CA	*Rana spinosa*	Li et al. (2010)
Cypermethrin	↑MN	*Odontophrynus americanus*	Cabagna et al. (2006)

Table 1 contd. ...

...Table 1 contd.

Xenobiotic	Effect	Tadpole Species	References
Fungicides			
ARTEA 330 EC (250g/l propiconazole + 80g/l Cyproconazole)	↑MN	*Rana saharica*	Bouhafs et al. (2009)
Captan	↑MN/CA ↑MN/CA	*Xenopus laevis* *Pleurodeles waltl*	Mouchet et al. (2006b)
PAHs:			
Phenanthrene	↑MN/CA	*Euphlyctis cyanophlyctis*	Bhuyan et al. (2020)
Benzo(*a*)pyrene	↑MN ↑MN	*Pleurodeles waltl* *Pleurodeles walt*	Djomo et al. (1995) Fernandez and Jaylet (1987)
Benz(*a*)anthracene	↑MN	*Pleurodeles waltl*	Fernandez and L`Haridon (1992)
7,12-Benz(*a*)anthraquinone	↑MN	*Pleurodeles waltl*	Fernandez and L`Haridon (1992)
9,10-Dimethylanthracene	↑MN	*Pleurodeles waltl*	Fernandez and L`Haridon (1992)
Metals:			
Fe and Mn	↑MN/CA	*Aquarana catesbeiana*	Veronez et al. (2016)
Sr	↑MN	*Bufo gargarizans*	Kai et al. (2020)
Cd	↑MN ↑MN/CA ↑NA ↑NA ↑MN/CA ↑MN	*Xenopus laevis* *Rana limnocharis* *Rana zhenhaiensis* *Microhyla fissipes* *Pleurodeles waltl* *Bufo raddei*	Mouchet et al. (2006a) Patar et al. (2016) Lu et al. (2021) Hu et al. (2019) Mouchet et al. (2007) Zhang et al. (2007)
Cr	↑MN ↑MN/CA	*Aquarana catesbeiana* *Duttaphrynus melanostictus*	Monteiro et al. (2018) Fernando et al. (2016)
As	↑CA	*Rana limnocharis*	Singha et al. (2014)
Cu	↑MN ↑MN/NA ↑MN	*Aquarana catesbeiana* *Aquarana catesbeiana* *Aquarana catesbeiana*	Ossana et al. (2010) Rocha et al. (2012) Ferreira et al. (2003)
Pb	↑MN	*Euphlyctis cyanophlyctis*	Kour et al. (2013)
Emerging contaminants:			
Triclosan	↑MN	*Aquarana catesbeiana*	Emery (2012)
Nonylphenol	↑NA ↑NA	*Aquarana catesbeiana*	Scaia et al. (2017) Gregorio et al. (2019)
Dyes:			
Sandopel Basic Black BHLN, Negrosine, Dermapel Black FNI, and Turquoise Blue	↑CA	*Rana hexadactyla*	Rajaguru et al. (2001)
Other compounds:			
Formaldehyde	↑MN	*Aquarana catesbeiana*	Santana et al. (2015)
Graphene Oxide	↑MN	*Xenopus laevis*	Evariste et al. (2019)
Polyoxyethylene amine	↑MN/NA	*Dendropsophus minutus*	Lopes et al. (2021)
Ni-Al nanoceramics	↑MN	*Rhinella arenarum*	Svartz et al. (2020)
Poultry litter Ti-O_2 nanoparticle	↑CA ↑CA	*Leptodactylus chaquensis* *Dendropsophus minutus*	Curi et al. (2017) Amaral et al. (2022)

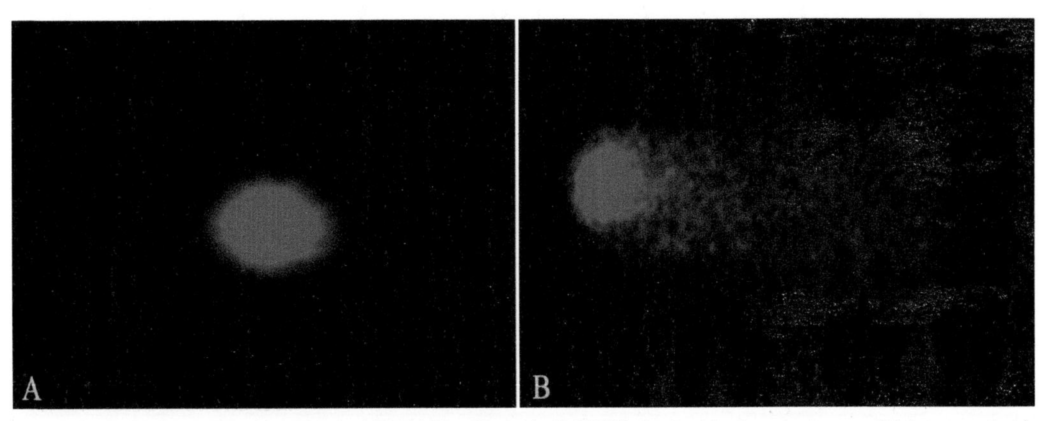

Figure 1. *D. minutus* nucleoids stained with Ethidium Bromide (0.002%) obtained in the comet assay. (A) intact nucleoid, without genomic lesions; (B) nucleoid with the "tail" demonstrating the migrated DNA fragments, representing the intensity of genomic lesions. Source: Gonçalves 2019. Doctoral thesis.

will have no tail (Fig. 1a), while nucleoids that have suffered more damage will have larger tails (Fig. 1b) (Olive et al. 1990, Collins et al. 1997, Azqueta et al. 2013a).

The comet assay is a simple, rapid, relatively inexpensive technique that can be applied to any eukaryotic cell. Due to its recognized sensitivity in detecting DNA damage in individual cells, it is possible to use the comet assay in small cell populations (Hartmann 2004). Since genotoxic compounds are often tissue-specific, a technique such as the comet assay detects genomic damage in individualized cells and is highly relevant in genotoxicity studies (Tice et al. 2000). Due to all the advantages and applications of this technique, the number of publications involving the comet assay has grown considerably in recent years (Collins 2015), making the comet assay a field of great interest (Del Bo et al. 2015, Neri et al. 2015). Over the last 10 years, PubMed recorded more than 6,000 publications about the comet assay, reinforcing the importance of this technique.

Some software makes it possible to quantify the comet assay results automated. The advantage of using this analysis is the significant reduction of the subjective character obtained in the visual quantification of the DNA damage (Kumaravel et al. 2006), also allowing the measurement of lesions occurring in the genome, even when the nucleoids present apparent visual integrity (Hartmann et al. 2003). Although software could be an excellent tool for analyzing comets, a limiting factor is that many are sold at a high cost, working only with the specific brand of the microscope purchased. To circumvent these inconveniences, public domain programs run on different platforms in different operating systems (Kumaravel and Jha et al. 2006). One can cite as an example of free software the Comet Score®, developed for analyzing images from the comet assay. This program provides 17 parameters, given in arbitrary units, representing the DNA damage in each nucleoid. Among these parameters, the most frequently used are Tail Length (CC), percentage of DNA in the tail (% DNA), and Olive's Tail Moment (MCO) (Lovell et al. 2008, Gonçalves 2019). However, the parameter that best represents a dose-response effect is % DNA(Collins 2004, Carvalho et al. 2018, Gonçalves 2019).

3. The Micronucleus Test

The micronucleus test is one of the most consolidated biomarkers used in ecotoxicology and mutagenesis to evaluate populations exposed to some genotoxic agent (Araldi et al. 2013). The term micronucleus was introduced in 1951, referring to acentric fragments dissociated from the central nucleus in the final phase of cell division (Kirsch-Volders et al. 2003). Initially, the micronucleus test was proposed for *in vivo* studies on polychromatic erythrocytes from the bone marrow of the femur of mice (Schmid 1975). Subsequently, some adaptations were made in the test, allowing its

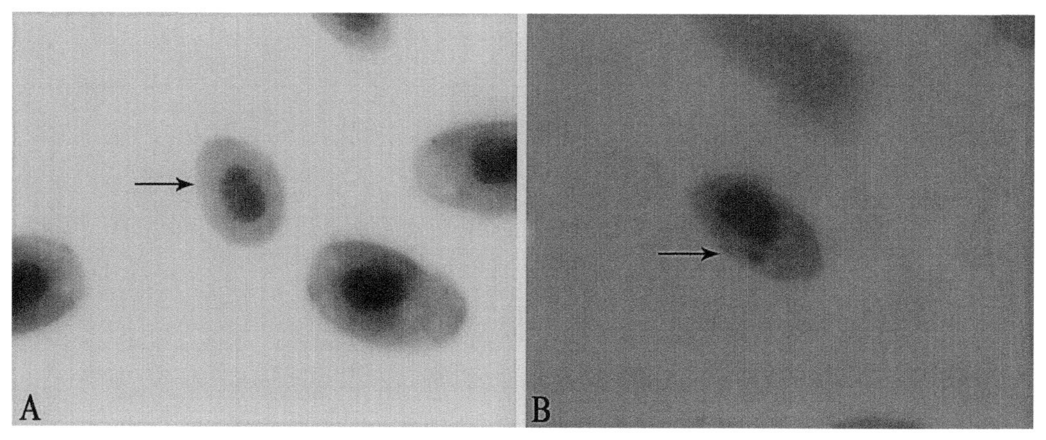

Figure 2. Red cells from *D. minutus* tadpoles stained with 10% Giemsa. (A) Normal cell; (b) Micronucleated cell.

use for other cell types and with other model organisms (Fenech 2000, Ferreira et al. 2004, Cabagna et al. 2006, Carvalho et al. 2018, Gonçalves 2019, Lopes et al. 2021).

Micronuclei are extranuclear bodies formed from fragments (Clastogenesis) or entire chromosomes (Aneugenesis) that were not incorporated into the daughter cell during cell division (Fenech 2000, Ribeiro et al. 2003, Bonassi et al. 2006, Samanta and Dey 2012). It is also important to note that not all genotoxic products are clastogenic or aneugenic. According to Heddle et al. (1991), many xenobiotics induce micronuclei formation due to their interference in spindle fibers, not directly causing damage to DNA.

However, micronuclei represent not only chromosomal losses but also the result of DNA amplification (Samanta and Dey 2012). DNA amplification is commonly observed in oncogenic processes, resulting in double-minute chromosomes removed from the central cell nucleus, giving rise to micronuclei (Shimizu et al. 2000, Samanta and Dey 2012). The micronuclei formation is also associated with allelic loss, contributing to carcinogenic processes (Bonassi et al. 2006, Terradas et al. 2010). Figure 2 shows two examples of red blood cells from tadpoles of the species *D. minutus*, with a cell without a micronucleus (Fig. 2a) and with a micronucleus (Fig. 2b).

As with all laboratory techniques, some essential criteria must be considered when using the micronucleus test. The diameter of the micronucleus must be less than one-third of the central nucleus; it should have the same refringence as the central core, distinguishable edges, its coloration should not be darker than the core, and it should not touch the core (Carrasco et al. 1990). It is known that the micronucleus test can only be performed on cell populations that can divide. Preferably, the analysis should be done after a single division cycle, as there is uncertainty about the permanence of these structures for more than one division cycle. In this way, it is evident that the frequency of micronuclei will depend on the time each cell takes to start dividing. This time may vary according to the tissue and species, making prior knowledge of the cell division kinetics of the model organism used essential (Fenech 2000).

4. Nuclear Abnormalities

Alterations in the normal shape of the nucleus of red blood cells (nuclear abnormalities, NA) have been also observed in erythrocytes of amphibian tadpoles as a consequence of exposure to environmental and chemical contaminants (Lajmanovich et al. 2014). Although the mechanism responsible for the formation of all NA types has not been totally explained, these abnormalities are considered to be indicators of genotoxic damage and therefore may complement the scoring of MN in routine assays for genotoxicity screening.

Carrasco et al. (1990) photographed and quantified the frequencies of the NA in fish erythrocytes and proposed the following classification, which is also used for other vertebrate groups, including amphibians:

(a) Blebbed: nuclei with a small evagination of the nuclear membrane, appearing to include euchromatin or heterochromatin (darker). The sizes of these evaginations are in the range of small protuberances to completely circumscribed structures, similar to micronuclei, but still connected to the main nucleus.

(b) Lobed: nuclei with wider evaginations than those described for Blebbed. Its structure is not as clearly outlined as above. Some nuclei have several of these structures.

(c) Vacuolated: nuclei that present a region resembling the vacuoles inside. These "vacuoles" are shown devoid of all visible material therein.

(d) Notched: nuclei that have a well-defined indentation on their surface, in general with an appreciable depth in the nucleus. These grooves do not seem to possess any nuclear material and seem to be delimited by the nuclear envelope.

Binucleated or even anucleated cells are also often reported when evaluating NA. Pictures of these NA taken from erythrocytes of bullfrog tadpoles are shown in Fig. 3.

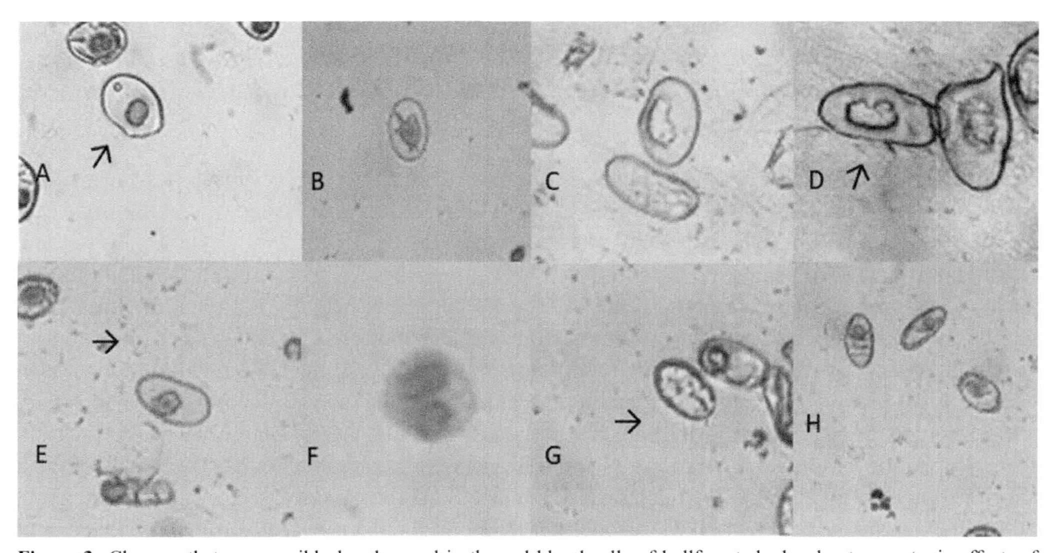

Figure 3. Changes that can possibly be observed in the red blood cells of bullfrog tadpoles due to genotoxic effects of xenobiotics. (A) Micronucleus; (B) Blebbed; (C) Lobed; (D) Notched; (E) Vacuolated nucleus; (F) Binucleated Cell; (G) Anucleated Cell; (H) Normal erythrocytes.

5. Genotoxic Biomarkers in Environmental Monitoring Studies Using Tadpoles

The analysis of the genotoxicity and mutagenicity of xenobiotics is essential to assess the genetic risk of tadpole populations from a given location. Moreover, the detection of DNA damage in cells of organisms from natural environments can be used in environmental biomonitoring to characterize impacted areas (Monserrat et al. 2007).

Numerous studies have used the evaluation of genotoxic markers in erythrocytes from tadpoles to characterize the negative effects of contaminants in animals from polluted areas (some of them are depicted in the beginning of Table 1). Santos et al. (2021) evaluated water samples from 9 sites along the Marrecas River (southern Paraná, Brazil) covering areas with rural and urban hydrological contribution. The results showed mutagenic effect in 4 sampling sites, possibly due to the use of different agrochemicals across the hydrographic basin region. Gonçalves et al. (2017) used MN frequencies and the comet assay to evaluate genotoxic effect in 3 species of tadpoles (*Dendropsophus minutus*, *Boana albopunctata*, and *Physalaemus cuvieri*) collected in areas dominated by soybean and corn. Compared to tadpoles from nonagricultural lands, those from agricultural areas showed higher DNA damage. Peluso et al. (2019) exposed *R. arenarum* tadpoles to surface waters samples from agricultural areas of Buenos Aires (Argentina), which resulted in increased incidence of micronucleus. *D. minutus*, *Physalaemus cuvieri*, and *Scinax fuscovarius* tadpoles collected in soybean/corn in the Brazilian Cerrado showed increased NA when compared to animals from conservation units (Borges et al. 2019). Babini et al. (2015, 2016) also showed *R. arenarum* tadpoles collected in agriculture-dominated areas presented increased MN and NA frequencies, compared to tadpoles from site without crops or livestock. Ossana et al. (2013) analyzed multiple biomarkers in *A. catesbeiana* tadpoles to evaluate the water quality of the Reconquista River (Argentina), a river that is affected by agricultural, urban, and industrial discharges. Among the different parameters, MN frequencies showed to be a key biomarker to indicate the impact of pollutants on tadpoles.

To better demonstrate the applicability of genotoxicity tests to evaluate DNA damage associated with environmental stressors, the following section will demonstrate a case study carried out in a Conservation Unit from Goias State, Central Brazil, with a native anuran species known as *Boana albopucntata* (Anura, Hylidae).

5.1 Case Study: The comet assay and Boana albopunctatus (Anura: Hylidae) Tadpoles as Tools for the Evaluation of Environmental Quality in a Federal Protected Area Situated at Cerrado of Central Brazil

5.1.1 Study Area

The Emas National Park (ENP) is situated in the central of the Cerrado Biome and the western portion of the Paraná Basin, bordered by the states of Goiás, Mato Grosso and Mato Grosso do Sul. Its altitude is between 650 and 1,000 m and its extension is 132,000 hectares. To protect representative areas of some Cerrado phytophysiognomies, the ENP was created in the 1960s and currently is surrounded by crops and pastures. The hydrographic network is formed by the Jacuba and Formoso rivers, which join later forming Correntes, a tributary of Paranaíba and a member of the Prata Basin. The region goes through a dry season from May to September and a rainy season from October to April, the average annual temperature varies between 22°C and 24°C and annual rainfall ranges from 1,500 to 1,700 mm, concentrated from October to March. In the dry season, the precipitation is always lower than 60 mm (Ramos-Neto and Pivello 2000). The amphibian species from ENP is relatively well known (Kopp et al. 2010, Alves-Ferreira et al. 2021) with about 30 species occurring in this area. To conduct this study, we considered *Boana albopunctata* as our model, because it is a species widely distributed throughout the Cerrado (Frost 2023) and commonly found in the ENP and surrounding areas (Kopp et al. 2010).

5.1.2 The Study Species

Boana albopunctata Spix, 1824 is a new species described by Pinheiro et al. (2018). Belonging to the Hylidae family, *B. albopunctata* is considered a medium to large tree frog (30 to 65 mm CRC), with a color that can vary from yellow to light brown (Fig. 4). It presents a larval stage (Fig. 4) evaluated in this case study.

An important feature that helps to distinguish the species *B. albopunctata* is the yellow spots on the back of the thighs. Although pre-pole is present in males and females, this structure is often more developed in males. Its diet is generalist, being more representative of invertebrates of the orders Hymenoptera, Araneae, and Coleoptera.

B. albopunctata behaves as a generalist species in the use of habitats and has a wide geographic distribution, being found in the Brazilian Southeast, South, and Central Brazil (Eterovich and Sazima 2004). It occurs in vegetation, the rainforest, and open areas and reproduces in permanent

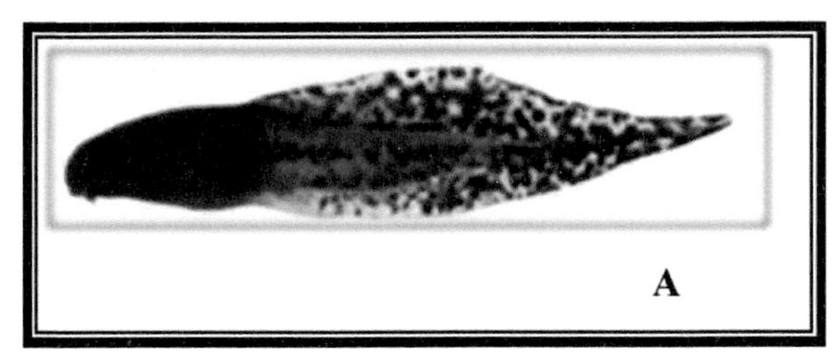

Figure 4. Representative of the species *Boana albopunctata* (Anura, Hylidae). (A) A tadpole. Source: Vieira (2017); (B). Male adult. Image courtesy by Rogério Pereira Bastos.

and temporary water bodies (Ribeiro et al. 2005). According to Aquino et al. (2010), this species adapts well to anthropogenic disturbance and is commonly seen in human settlements, with no reported threat to this species. However, habitat loss and degradation due to agricultural activity may have affected populations of this species in Paraguay (Toledo et al. 2006).

5.1.3 Field Data Collection

This study was authorized by the UFG Ethics Committee on the Experimental Use of Animals (license number 008/2019), and the Chico Mendes Institute for Biodiversity Conservation, under protocol number 13040. Before the field study, we obtained the geographic coordinates corresponding to the centroid (Latitude: –18.122, Longitude: –52.908) of the ENP. We assume that this represents the most central portion of this protected area and use it as a reference to select water bodies within (n = 3) of the ENP and around it (n = 3). After selecting the sample points, we calculated their distances concerning the UC centroid (Fig. 5). Thus, the points inside the ENP ranged from

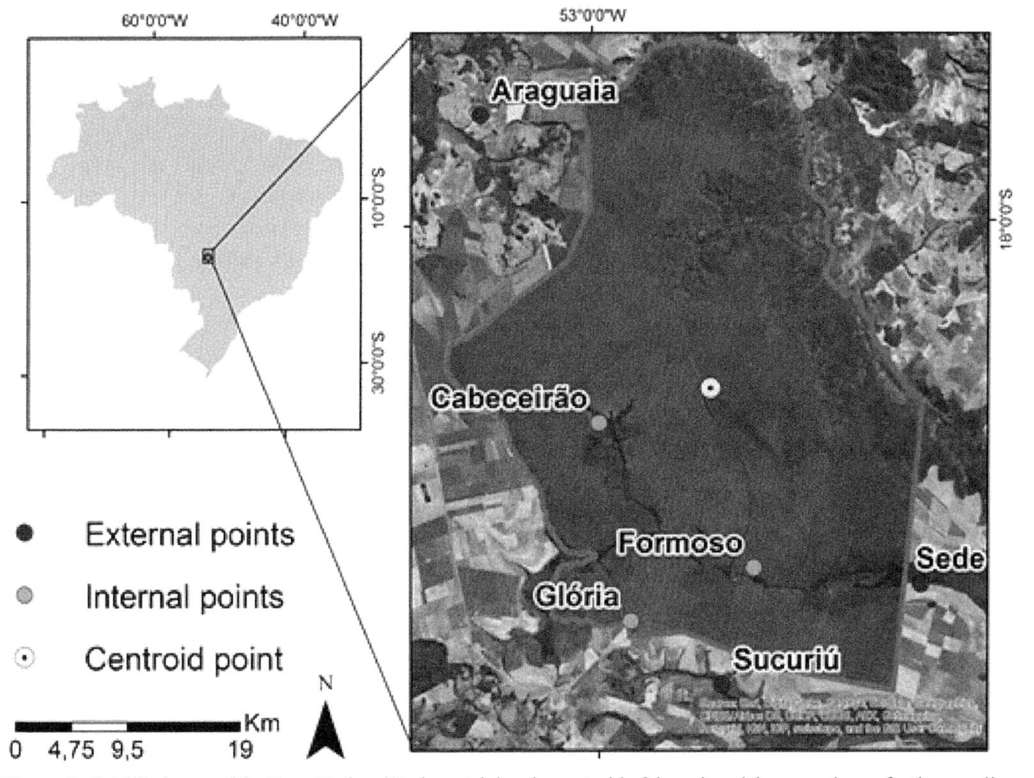

Figure 5. Satellite image of the Emas National Park containing the centroid of the unit and the areas chosen for the sampling. Three of them within the park (Cabeceirão, Formoso and Gloria) and three more exposed to contamination (Araguai, Sede and Sucuriú).

10.7 to 20.5 km to the innermost portion (centroid) of the UC, while the selected points around the ENP ranged from 18.6 to 33.2 km. The sampling points were generally characterized as open environments, consisting of ponds with predominant herbaceous and shrubby vegetation. As seen in Fig. 5, the points outside the ENP were inserted in an agriculture matrix, with a predominance of soybean monoculture.

In the field, we considered only *Boana albopunctata* in the larval stage. To collect the tadpoles, we used 3 mm^2 wire mesh nets, passed along the banks and inside the pools for 1 hour. The specimens were previously identified with a field identification manual (Rossa-Feres and Nomura 2006) and later confirmed the identification in the laboratory with a magnifying glass and specialists. The tadpoles in phase 30 (Gosner 1960) were kept alive until arrival at the laboratory, where they were euthanized for blood sample collection; such samples are analyzed for DNA damage. Some of the specimens collected were listed in the zoological collection of Goiás Federal University-UFG under voucher numbers 3031, 3032, 3034, 3036, 3038 and 3039. In each body of water, physicochemical data was obtained using a YSI 6600 multiparametric probe.

5.1.4 Comet Assay

The animals were euthanized with a 5% benzocaine anesthetic solution followed by cooling and soon after death, a transverse incision was made at the tail base of each specimen, so as to section the caudal artery to obtain samples of blood, a pool of 5 tadpoles per sample. Four slides were made per sampling point, totaling 20 tadpoles per sampling point, and 100 cells were analyzed for each of the 8points investigated. The comet assay was performed using the alkaline method with a few modifications as described by Singh et al. (1988). Slides previously coated with agarose with a normal melting point (1.5%) were prepared with 15 µl of blood and diluted in 1 ml of PBS buffer (pH 7.0) in a 2 ml Eppendorf tube. Ten microliters were removed, and 130 µl of agarose with a low melting point (0.5%) were embedded in this suspension in a water bath at 37°C. Then the slides were kept in cuvettes (protected from light) containing cold lysis solution (Triton X-100, DMSO, and Stock Lysis Solution) for 4 hours. Electrophoresis was carried out at 25V, and the current was adjusted to 300 mA. The slides were routinely exposed to this electrical current in the dark for 30 minutes. After electrophoresis, the slides were placed in a staining tray, covered with a neutralizing buffer (0.4 M, Tris-HCL, pH 7.5), and kept in the dark for 5 minutes. The slides were stained with 10 µl SYBR® Green I solution (0.02 mg/ml) and covered with a coverslip. 100 nucleotides per sample were analyzed. The analysis was performed by an Axioplan-Imaging® fluorescence microscopy system, using the Comet Imager software, with a 510–560 nm excitation filter and a 590 nm barrier filter, with a 200x magnification. DNA damage assessment uses the OpenComet plug-in for the popular image processing platform, ImageJ. ImageJ software can display, process and analyze images and is primarily intended for use with microscopy images (Gyori et al. 2014).

5.1.5 Statistical analysis

The variables collected in the field and those obtained in the laboratory are presented as mean and standard deviation. Furthermore, to test the hypothesis that Parque Nacional das Emas is efficient in containing genetic damage in the target species, we used Kruskal-Wallis nonparametric analysis of variance, followed by a post hoc Dunn test for pairwise comparisons. Thus, our response variable was the % of DNA in tail, while the location of the points about the ENP was our predictor variable. The analyses were performed according to Zar (2007), with a significance level of 5%.

5.1.6 Results

Physical-chemical Parameters

Regarding the physicochemical data, we detected a high concentration of chlorides at one of the sample points, external to the park (Table 2). This is the Sucuriú River, where the mean concentration of the above chemical element reached values well above the other collection points, of 115.65 mg/L,

Table 2. Mean and standard deviation obtained from water physicochemical data.

Areas	pH	Conductivity (uS/cm)	NH_4^+ (mg/L)	Cl (mg/L)	NH_3 (mg/L)
Araguaia	7.94 ± 0.2	6.1 ± 0.5	0.45 ± 0.2	0.837 ± 0.3	0.026 ± 0.04
Sede-Chapadão	5.43 ± 0.4	9.1 ± 0.5	0.13 ± 0.01	0.523 ± 0.2	0.387 ± 0.82
Sucuriú	5.08 ± 0.6	4.1 ± 0.5	0.04 ± 0.01	115.65 ± 0.3	0.395 ± 0.82
Cabeceirão	7.06 ± 0.2	4.2 ± 0.5	0.18 ± 0.03	0.252 ± 0.3	0.403 ± 0.82
Formoso	9.25 ± 0.3	9.4 ± 1.6	0.18 ± 0.01	0.724 ± 0.2	0.385 ± 0.82
Glória	7.65 ± 0.2	4.1 ± 0.5	0.04 ± 0.01	1.401 ± 0.3	0.030 ± 0.04

but this value is still below the maximum limit tolerated by the CONAMA (National Environment Council) resolution number 357/2005, which is 250 mg/L. The values found for ammonia (NH_3) were also above the limits used as reference by the Management Plan of the park at Formoso (0.384 mg/L, well above the Management Plan reference) and Araguaia (0.026 mg/L, at the limit of the reference used by the Management Plan). Regarding pH, the Management Plan uses values between 6.6 and 9.0 as reference, but the points Sucuriú e Sede-Chapadão presented values below the established minimum of 5.08 and 5.43, respectively. The Formoso river track presented a value of 9.25, above, therefore, the maximum established as reference.

DNA damage in Boana albopunctata

We recorded the amount of DNA damage in *Boana albopunctata* tadpoles collected inside the Emas National Park differed from those collected outside the park ($H = 26.4382$, $p < 0.0001$, Table 3). Specifically, we observed a higher frequency of DNA damage in tadpoles collected at sites outside the PNE (mean ± standard error of 46.37 ± 3.91; $P < 0.05$) when compared to sites inside the park. Concomitantly, areas outside the park did not differ in DNA damage nor between sites inside the PNE ($P > 0.05$).

Table 3. Mean ± standard deviation of DNA damage in tadpoles of *Boana albopunctatus*.

Mean ± SE			
ENP		Outside the park	
Cabeceirão (n = 11)	24.77 ± 2.22	Araguaia (n = 8)	35.21 ± 4.10
Formoso (n = 07)	24.96 ± 1.71	Sede-Chapadão (n = 9)	35.00 ± 2.45
Glória (n = 10)	25.47 ± 2.08	Sucuriú (n = 9)	46.37 ± 3.91*

SE = Standard error; ENP = Emas National Park; N = Sample number.
* Significant differences.

5.1.7 Discussion

Our results indicate that anuran reproductive sites close to the centroid of the PNE are more protected by the natural barrier of space to agricultural genotoxic agents. This was indicated when we observed the damage average slightly increasing (Cabeceirão and Formoso) concerning that site plus the edge of the protected area (Glória). On the other hand, specimens outside the PNE directly receive anthropic pressures, favoring a higher frequency of DNA damage. Given these scenarios, it seems that even if there is pollution outside the PNE, the distance to the interior of the Park is the best driving force to reduce contamination. In this case, we consider natural vegetation and a few rivers with direct entry to the center of the UC. Our study is also in agreement with other reports that related the genotoxic impact of pesticides on wild fauna in Goiás, such as studies with fish (Silva et al. 2016), anurans (Gonçalves et al. 2015, Borges et al. 2019) and mammals (Benvindo-Souza et al. 2019).

The most peripheral point within the protected area (Gloria) is more exposed to contaminants capable of being carried by the wind (pesticides). Once in the surroundings of the park, the rural workers spray insecticides and herbicides constantly, by means of aircraft, a practice originally prohibited by the management of the park. But it was used after the rural producers overturned such a ban in court. This activity should further potentiate the phenomenon of drift, because aircrafts perform spraying at a higher height than agricultural machines, further exposing pesticides to wind action. However, in this study, the DNA damage observed in tadpoles at Ponto do Gloria was low compared to the sampling sites within the Park and the cropland areas.

The greater intensity of DNA damage in samples from areas outside the park is probably caused by the presence of pesticides, such as glyphosate and atrazine, widely used in soybean and sugarcane crops in the region. Brazil is the Latin American country that consumes the most atrazine, approximately 30 thousand tons per year, according to the Brazilian Institute for the Environment and Renewable Natural Resources (IBAMA 2016), for weed control, for example (Perez- Iglesias et al. 2019). Recent reports have indicated high levels (5350 g/L) of the herbicide atrazine in Southwest Goiás, in soybean growing areas similar to those found near the PNE, whose authors report malformations in the tail and oral disc in tadpoles of *Boana albopunctata* (Borges et al. 2018).

According to Alvarez-Moya et al. (2014), comparisons of *in vivo* and *in vitro* cells of different organisms through the comet assay may be important in attempting to measure the genotoxicity of glyphosate. Glyphosate-based herbicides are the most widely distributed in the world, being the most used, including in Brazil (Cavalcante et al. 2008). The authors also affirmed that the genotoxic effect of these pesticides is not yet fully known, and there are conflicting studies of the effect of this product on the genetic material of several species. Alvarez-Moya et al. (2014) investigated the genotoxic effect of glyphosate in humans, tilapia (*Oreochromis niloticus*) and a plant of the genus Tradescantia, both *in vitro* and *in vivo* conditions. The authors found a positive relationship between comet tail size and the concentration of glyphosate to which the organisms were exposed.

In addition to the negative effects of agrochemicals, another relevant factor that should be taken into account in the attempt to explain a greater amount of DNA damage of the specimens originating near the park would be the suppression of native vegetation. It is possible that in deforested environments, with greater insolation, increase in the exposure to UV-B radiation in tadpoles is causing this increase in genetic damages. According to Assis (2018), climate change associated with other anthropogenic impacts, such as deforestation, may be causing microclimatic changes in amphibian habitats. There are an increasing number of cases of epidemics in amphibians spreading across the planet; probably such epidemics are associated with these environmental modifications, which may be weakening the repair system of these animals.

Tevini (1993) pointed out that UV-B radiation causes the greatest damage to organisms on the earth's surface. Such radiation is able to reduce growth rates, cause dysfunctions in the immune system and cause sublethal damage at the molecular level. UV-B radiation can cause mutations and cell death, but there is evidence suggesting that UV-A radiation can also be harmful, especially when combined with other stress factors (Blaustein et al. 1994). According to Cockell (2001), UV radiation is an important causative factor of stress in living organisms. Solar ultraviolet B (UVB) radiation is an essential environmental stressor for amphibian populations due to its genotoxicity (Londero et al. 2017). Such statements demonstrate that pesticides may not be an isolated cause of increased DNA damage in the tadpoles sampled in the periphery of the park. Therefore, it may be occurring as an association of pesticides with almost total suppression of vegetation native in the cultivated areas, exposing tadpoles to a higher incidence of solar radiation.

In this context, changes in the Brazilian forest code (Law no. 12.651, of May 25, 2012) weaken the protection of the environment by reducing permanent preservation areas (PPAs) on rural properties, exposing the wild species to UVB radiation. Toledo et al. (2010) criticized these changes in legislation, even before they were implemented, stating that such changes in the law encourage deforestation, generating the main causes of amphibian declines in the world: habitat reduction and fragmentation.

In general, agricultural practices around the PNE are currently the main threat to the biological integrity of the ecosystem. With the current Brazilian government wanting to reduce federal PAs, these problems will likely increase more in Brazil. In this study, the chlorine levels in Vereda do Sucuriú were higher than in the other sampling sites. In this case, we do not disregard the possibility of contamination of the water by activities of the sugarcane industry near this region. Vinasse has high concentrations of potassium, chloride, total nitrogen, calcium, sulphate, total phosphorus, significant concentrations of some metals and high biochemical oxygen demand, to the point of causing severe damage to surface watercourses and groundwater (Ribeiro et al. 2007). However, future studies are encouraged in this region to clarify this higher level of genotoxic damage in anuran tadpoles. In addition, the increase in sampling points inside and outside the Park and the increase in species may indicate a greater spectrum of the real problem facing anurans in aquatic ecosystems in the PNE.

In summary, the comet assay was applied to *Boana albopunctata* tadpoles. Tadpoles outside the Emas National Park present more significant DNA damage when compared to the same species inside the Park. The Emas National Park seems to be partially efficient in protecting biodiversity in terms of the impact of pesticides. Thus, these data highlight the importance of conservation units for the protection of biodiversity to the detriment of environmental health compared to unprotected areas. In addition, this study highlights the importance of buffer zone planning in protected areas to delay the entry of xenobiotic agents.

6. Conclusions and Perspectives

Classical genotoxicity tests such as the comet assay and the evaluation of the frequencies of MN and NA have shown to be efficient tools to assess the genotoxic effects on environmental contaminants in amphibian tadpoles, both in laboratory and field researches. These tests are cheap, rapid and efficient in evidencing injuries in the genetic material of nucleated cells, being suggested as important biomarkers in risk assessment of xenobiotics. There are increasing number of papers in scientific literature evaluating genotoxicity in tadpoles, however, given the great diversity of amphibian species in the world, it is reasonable to say that an immense lack of information exists yet to better characterize the genotoxic effects of pollutants on the development of these animals. More studies are needed regarding (i) the influences of abiotic factors on the genotoxic effect (water pH, temperature, food and oxygen availability, presence of predators, etc.), (ii) comparisons among different stages of development, (iii) influence of different habits (e.g., tadpoles with rapid × low metamorphosis time, tadpoles with direct development in the egg, tadpoles from temperate × tropical environments, benthonic × nektonic tadpoles, etc.), (iv) effects of mixtures of pollutants, (v) studies comparing different tadpole species exposed to same xenobiotics.

Acknowledgements

The authors thank the Coordination for the Improvement of Higher Education Personnel (CAPES). M.B.S. thanks the Brazilian Fund for Biodiversity (FUNBIO) and the Humaniza Institute. A.R.M. and D.M.S. thank the National Council for Scientific and Technological Development (CNPq) for their scholarships.

References Cited

Alvarez-Moya, C., M.R. Silva, C.V. Ramirez, D.G. Gallardo, R.L. Sanchez, A.C. Aguirre and A.F. Velasco. 2014. Comparison of the *in vivo* and *in vitro* genotoxicity of glyphosate isopropylamine salt in three different organisms. Genetics and Molecular Biology 37: 105–110.

Alves-Ferreira, G., I.B.F. da Paixão and F. Nomura. 2021. Morphological characterization and diversity of tadpoles (Amphibia: Anura) at Emas National Park and its surrounding, Goiás State, Brazil. Biota Neotropica 21: 1–11.

Amaral, D.F., M.F. Montalvão, B.O. Mendes, A.L.S. Castro and G. Malafaia. 2018. Behavioral and mutagenic biomarkers in tadpoles exposed to different abamectin concentrations. Environ. Sci. Pollut. Res. 25: 12932–12946.

Amaral, D.F., V. Guerra, K.L. Almeida, L. Signorelli, T.L. Rocha and D. de Melo e Silva. 2022. Titanium dioxide nanoparticles as a risk factor for the health of Neotropical tadpoles: A case study of *Dendropsophus minutus* (Anura: Hylidae). Environ. Sci. Pollut. Res Int. 29: 50515–50529.

Aquino, L., R. Bastos, A. Kwet, S. Reichle, D. Silvano, C. Azevedo-Ramos, N. Scott and D. Baldo. 2010. *Hypsiboas albopunctatus*. IUCN Red List of Threatened Species. Version 2012.2. www.iucnredlist.org.

Araldi, R.P., D.G. de Oliveira, D.F. da Silva, T.B. Mendes and E.B. Souza. 2013. Análise do potencial mutagênico dos esteroides anabólicos androgênicos (EAA) e da l-carnitina mediante o teste do micronúcleo em eritrócitos policromáticos. Rev. Bras. Med. do Esporte 19: 448–451.

Assis, A.B. de. 2018. Microbiota, secreções cutâneas e microclima: Consequências para os anfíbios. Revista Da Biologia 8(1): 45–48.

Azqueta, A. and A.R. Collins. 2013a. The essential comet assay: A comprehensive guide to measuring DNA damage and repair. Arch. Toxicol. 87: 949–968.

Azqueta, A., A.R. Collins, G.D.D. Jones, R.W.L. Kwok, D.H. Phillips, O. Sozeri et al. 2013b. DNA-repair measurements by use of the modified comet assay: An inter-laboratory comparison within the European Comet Assay Validation Group (ECVAG). Mutat. Res. Toxicol. Environ. Mutagen. 757: 60–67.

Babini, M.S., C.L. Bionda, N.E. Salas and A.L. Martino. 2015. Health status of tadpoles and metamorphs of *Rhinella arenarum* (Anura, Bufonidae) that inhabit agroecosystems and its implications for land use. Ecotoxicol Environ Saf. 118: 118–125.

Babini, M.S., C.L. Bionda, N.E. Salas and A.L. Martino. 2016. Adverse effect of agroecosystem pond water on biological endpoints of common toad (*Rhinella arenarum*) tadpoles. Environ Monit Assess. 188: 459.

Bastos, R.P., L.P. Lima and M.S. Pasquali. 2003. Sapos, rãs e pererecas: desvendando o segredo dos anfíbios, 1ª ed. R.P. Bastos, Goiânia.

Benvindo-Souza, M., R.E. Borges, S.M. Pacheco and L.R.S. Santos. 2019. Genotoxicological analyses of insectivorous bats (Mammalia: Chiroptera) in central Brazil: The oral epithelium as an indicator of environmental quality. Environ Pollut 245: 504–509.

Bhuyan, K., A. Patar, U. Singha, S. Giri and A. Giri. 2020. Phenanthrene alters oxidative stress parameters in tadpoles of *Euphlyctis cyanophlyctis* (Anura, Dicroglossidae) and induces genotoxicity assessed by micronucleus and comet assay. Environ. Sci. Pollut. Res. Int. 27: 20962–20971.

Blaustein, A.R., P.D. Hoffman, D.G. Hokit, J.M. Kiesecker, S.C. Walls and J.B. Hays. 1994. UV repair and resistance to solar UV-B in amphibian eggs: A link to population declines? Proc. Natl. Acad. Sci. 91: 1791–1795.

Borges, R.E., L.R.D. Santos, R.A. Assis, M. Benvindo-Souza, L. Franco-Belussi and C. de Oliveira. 2018. Monitoring the morphological integrity of neotropical anurans. Environ. Sci. Pollut. Res. 26: 2623–2634.

Borges, R.E., L.R.S. Santos, M. Benvindo-Souza, R.S. Modesto, R.A. Assis and C. de Oliveira. 2019. Genotoxic evaluation in tadpoles associated with agriculture in the central cerrado, Brazil. Arch. Environ. Contam. Toxicol. 77: 22–28.

Bosch, B., F. Mañas, N. Gorla and D. Aiassa. 2011. Micronucleus test in post metamorphic *Odontophrynus cordobae* and *Rhinella arenarum* (Amphibia: Anura) for environmental monitoring. J. Toxicol. Env. Health Sci. 3: 155–163.

Bonassi, S., A. Znaor, M. Ceppi, C. Lando, W.P. Chang, N. Holland et al. 2006. An increased micronucleus frequency in peripheral blood lymphocytes predicts the risk of cancer in humans. Carcinogenesis 28: 625–631.

Bouhafs, N., H. Berrebbah, A. Devaux, R. Rouabhi and M.R. Djebar. 2009. Micronucleus Induction in Erythrocytes of Tadpole *Rana saharica* (Green Frog of North Africa) Exposed to Artea 330EC. Am-Eur. J. Toxicol. Sci. 1: 07–12.

Cabagna, M.C., R.C. Lajmanovich, P.M. Peltzer, A.M. Attademo and E. Ale. 2006. Induction of micronuclei in tadpoles of *Odontophrynus americanus* (Amphibia: Leptodactylidae) by the pyrethroid insecticide cypermethrin. Toxicol. Environ. Chem. 88: 729–737.

Campana, M.A., A.M. Panzeri, V.J. Moreno and F.N. Dulout. 2003. Micronuclei induction in *Rana catesbeiana* tadpoles by the pyrethroid insecticide lambda-cyhalothrin. Genet. Mol. Biol. 26: 99–103.

Candioti, J.V., G.S. Natale, S. Soloneski, A.E. Ronco and M.L. Larramendy. 2010. Sublethal and lethal effects on *Rhinella arenarum* (Anura, Bufonidae) tadpoles exerted by the pirimicarb-containing technical formulation insecticide Aficida. Chemosphere 78: 249–255.

Carrasco, K.R., K.L. Tilbury and M.S. Myers. 1990. Assessment of the piscine micronucleus test as an *in situ* biological indicator of chemical contaminant effects. Can. J. Fish. Aquat. Sci. 47: 2123–2136.

Carvalho, W.F. 2018. Avaliação mutagênica e genotóxica de herbicidas em organismos aquáticos. Tese (Doutorado em Ciências Ambientais). Universidade Federal de Goiás. Goiânia, 130f.

Carvalho, W.F., F.C. Franco, F.R. Godoy, D. Folador, J.B. Avelar, F. Nomura et al. 2018. Evaluation of genotoxic and mutagenic effects of glyphosate roundup original® in *Dendropsophus minutus* Peters, 1872 tadpoles. South Am. J. Herpetol. 13: 220–229.

Cavalcante, D.G., C.B. Martinez and S.H. Sofia. 2008. Genotoxic effects of Roundup on the fish *Prochilodus lineatus*. Mutat Res. 655(1-2): 41–6.

Clements, C., S. Ralph and M. Petras. 1997. Genotoxicity of select herbicides in *Rana catesbeiana* tadpoles using the alkaline single-cell gel DNA electrophoresis (Comet) assay. Environ. Mol. Mutagen. 29: 277–288.

Cockell, C.S. 2001. A photobiological history of earth. Ecosystems evolution and ultraviolet radiation pp. 1–35. Springer, New York

Collins, A. 2015. The Problem with non-mechanical trading. pp. 1–4. *In*: Beating the Financial Futures Market. John Wiley & Sons, Inc., Hoboken, NJ, USA.

Collins, A.R., V.L. Dobson, M. Dušinská, G. Kennedy and R. Štětina. 1997. The comet assay: what can it really tell us? Mutat. Res. Mol. Mech. Mutagen. 375: 183–193.

Collins, A.R. 2004. The comet assay for DNA damage and repair: Principles, applications, and limitations. Mol. Biotechnol. 26: 249–61.

Curi, L.M., P.M. Peltzer, C. Martinuzzi, M.A. Attademo, S. Seib, M.F. Simoniello et al. 2017. Altered development, oxidative stress and DNA damage in *Leptodactylus chaquensis* (Anura: Leptodactylidae) larvae exposed to poultry litter. Ecotoxicol. Environ. Saf. 143: 62–71.

Del Bo', C., D. Fracassetti, C. Lanti, M. Porrini and P. Riso. 2015. Comparison of DNA damage by the comet assay in fresh versus cryopreserved peripheral blood mononuclear cells obtained following dietary intervention. Mutagenesis 30: 29–35.

Dhawan, A., M. Bajpayee and D. Parmar. 2009. Comet assay: A reliable tool for the assessment of DNA damage in different models. Cell Biol. Toxicol. 25: 5–32.

Djomo, J.E., V. Ferrier, L. Gauthier, C. Zoll-Moreux and J. Marty. 1995. Amphibian micronucleus test *in vivo*: evaluation of the genotoxicity of some major polycyclic aromatic hydrocarbons found in a crude oil. Mutagenesis 10: 223–226.

Emery, D. 2012. Assessing the genotoxicity of triclosan in tadpoles of the American bullfrog, Lithobates Catesbeianus. PhD Thesis, Virginia Commonwealth University, 50.

Eterovick, P.C. and I. Sazima. 2004. Anfíbios da Serra do Cipó – Minas Gerais – Brasil - Amphibians of the Serra do Cipó. Belo Horizonte: PUC Minas. p. 39.

Evariste, L., L. Lagier, P. Gonzalez, A. Mottier, F. Mouchet, S. Cadarsi et al. 2019. Thermal reduction of graphene oxide mitigates its *in vivo* genotoxicity toward *Xenopus laevis* tadpoles. Nanomaterials 9: 584.

Fenech, M. 2000. The *in vitro* micronucleus technique. Mutat. Res. Mol. Mech. Mutagen 455: 81–95.

Fernandez, M. and A. Jaylet. 1987. An antioxidant protects against the clastogenic effects of benzo[a]-pyrene in the newt *in vivo*. Mutagenesis. 2(4): 293–6.

Fernandez, M. and J. L'Haridon. 1992. Influence of lightning conditions on toxicity and genotoxicity of various PAH in the newt *in vivo*. Mutat. Res. Gen. Toxicol. 298: 31–41.

Fernando, V.A., J. Weerasena, G.P. Lakraj, I.C. Perera, C.D. Dangalle, S. Handunnetti et al. 2016. Lethal and sub-lethal effects on the Asian common toad *Duttaphrynus melanostictus* from exposure to hexavalent chromium. Aquat. Toxicol. 177: 98–105.

Ferreira, C.M., H.M. Bueno-Guimarães, M.J.T. Ranzani-Paiva, S.R.C. Soares, D.H.R. Rivero and P.H.N. Saldiva. 2003. Marcadores hematologicos do toxicidade de cobre em *Rana catesbeiana* girinos (Rã-touro). Rev. Bras. Toxicol. 16: 83–88.

Ferreira, C.M., J.V. Lombardi, J.G. Machado-Neto, H.M. Bueno-Guimarães, S.R.C. Soares and P.H.N. Saldiva. 2004. Effects of copper oxychloride in *Rana catesbeiana* Tadpoles: Toxicological and bioaccumulative aspects. Bull. Environ. Contam. Toxicol. 73: 465–470.

Frost, D.R. 2023. Amphibian Species of the World: An Online Reference. Version 6.2. Electronic Database accessible at https://amphibiansoftheworld.amnh.org/. American Museum of Natural History, New York, USA.

Giri, A., S.S. Yadav, S. Giri and G.D. Sharma. 2012. Effect of predator stress and malathion on tadpoles of Indian skittering frog. Aquat Toxicol. 106-107: 157–163.

Godschalk, R.W.L., C. Ersson, P. Riso, M. Porrini, S. Langie, F.J. van Schooten et al. 2013. DNA repair measurements using the modified comet assay: An inter-laboratory comparison within the European Comet Assay Validation Group (ECVAG). Mutagenesis 757: 60–67.

Gonçalves, M.W., T.B. Vieira, N.M. Maciel, W.F. Carvalho, L.S. Lima, P.G. Gambale, A.D. da Cruz, F. Nomura, R.P. Bastos and D.M. Silva. 2015. Detecting genomic damages in the frog *Dendropsophus minutus*: Preserved versus perturbed areas. Environ. Sci. Pollut. Res. Int. 22: 3947–54.

Gonçalves, M.W., P.G. Gambale, F.R. Godoy, A.A. Alves, P.H. Rezende and A.D. Cruz et al. 2017. The agricultural impact of pesticides on *Physalaemus cuvieri* tadpoles (Amphibia: Anura) ascertained by comet assay. Zoologia 34: 1–8.

Gonçalves, M.W. 2019. Análises de danos ao DNA em girinos de anfíbios anuros expostos a agrotóxicos: Abordagens *in vivo* e *in situ*. Tese (Doutorado em Genética e Biologia Molecular). Universidade Federal de Goiás, Goiânia, 74f.

Gosner, K. 1960. A simplified table for staging anuran embryos and larvae with notes on identification. Herpetologica 16: 183–190.

Gregorio, L.S., L. Franco-Belussi and C. De Oliveira. 2019. Genotoxic effects of 4-nonylphenol and Cyproterone Acetate on *Rana catesbeiana* (anura) tadpoles and juveniles. Environ. Pollut. 251: 879–884.

Gyori, B.M., G. Venkatachalam, P.S. Thigarajan, D. Hsu and M. Clement. 2014. Open comet: An automated tool for comet assay image analysis. Redox Biology 9(2): 457–465. doi: 10.1016/j.redox.2013.12.020.

Hartmann, A., E. Agurell, C. Beevers, S. Brendler-Schwaab, B. Burlinson, P. Clay, A. Collins, A. Smith, G. Speit, V. Thybaud and R.R. Tice. 2003. Recommendations for conducting the *in vivo* alkaline Comet assay. Mutagenesis 18: 45–51.

Hartmann, A. 2004. Use of the alkaline *in vivo* Comet assay for mechanistic genotoxicity investigations. Mutagenesis 19: 51–59.

Heddle, J.A., M.C. Cimino, M. Hayashi, F. Romagna, M.D. Shelby and J.D. Tucker et al. 1991. Micronuclei as an index of cytogenetic damage: Past, present, and future. Environ. Mol. Mutagen. 18: 277–291.

Herek, J.S., L. Vargas, S.A. Rinas Trindade, C.F. Rutkoski, N. Macagnan, P.A. Hartmann and M.T. Hartmann. 2021. Environ. Toxicol. Pharmacol. 81: 103516.

Hu, Y.C., Y. Tang, Z.Q. Chen, J.Y. Chen and G.H. Ding. 2019. Evaluation of the sensitivity of *Microhyla fissipes* tadpoles to aqueous cadmium. Ecotoxicology 28: 1150–1159.

IBAMA. Relatórios de comercialização de agrotóxicos (2016) Available from: http://www.ibama.gov.br/agrotoxicos/relatorios-de-comercializacao-de-agrotoxicos.

Jing, X., G. Yao, D. Liu, C. Liu, F. Wang and P. Wang et al. 2017. Exposure of frogs and tadpoles to chiral herbicide fenoxaprop-ethyl. Chemosphere 186: 832–838.

Kai, Y., N. Xiaojuan, W. Xu and L. Caiyun. 2020. Growth and genotoxicity effects induced by strontium exposure in tadpoles of *B. gargarizans*. Asian J. Ecotoxicol. 4: 233–239.

Kirsch-Volders, M., T. Sofuni, M. Aardema, S. Albertini, D. Eastmond, M. Fenech et al. 2003. Report from the *in vitro* micronucleus assay working group. Mutat. Res. Toxicol. Environ. Mutagen. 540: 153–163.

Kopp, K., L. Signorelli and R.P. Bastos. 2010. Distribuição temporal e diversidade de modos reprodutivos de anfíbios anuros no Parque Nacional das Emas e entorno, estado de Goiás, Brasil. Iheringia, Série Zoologia 100(3): 192–200.

Kour, P., N.K. Tripathi and Poonam. 2013. Induction of micronuclei due to cytotoxic and genotoxic effect of lead acetate metal salt in *Euphlyctis cyanophlyctis* (Amphibia: Anura). Int. J. Rec. Sci. Res. 4: 1528–1532.

Kumaravel, T.S. and A.N. Jha. 2006. Reliable Comet assay measurements for detecting DNA damage induced by ionizing radiation and chemicals. Mutat. Res. Toxicol. Environ. Mutagen. 605: 7–16.

Lajmanovich, R.C., M. Cabagna, P.M. Peltzer, G.A. Stringhini and A.M. Attademo. 2005. Micronucleus induction in erythrocytes of the *Hyla pulchella* tadpoles (Amphibia: Hylidae) exposed to insecticide endosulfan. Mutat Res. 587: 67–72.

Lajmanovich, R.C., M.C. Cabagna-Zenklusen, A.M. Attademo, C.M. Junges, P.M. Peltzer, A. Bassó et al. 2014. Induction of micronuclei and nuclear abnormalities in tadpoles of the common toad (*Rhinella arenarum*) treated with the herbicides Liberty® and glufosinate-ammonium. Mutat. Res. Genet. Toxicol. Environ. Mutagen. 769: 7–12.

Li, X., S. Li, S. Liu and G. Zhu. 2010. Lethal effect and *in vivo* genotoxicity of profenofos to Chinese native amphibian (*Rana spinosa*) tadpoles. Arch. Environ. Contam. Toxicol. 59: 478–483.

Liu, W.Y., C.Y. Wang, T.S. Wang, G.M. Fellers, B.C. Lai and Y.C. Kam. 2011. Impacts of the herbicide butachlor on the larvae of a paddy field breeding frog (*Fejervarya limnocharis)* in subtropical Taiwan. Ecotoxicology. 20: 377–384.

Londero, J.E.L., C.P. dos Santos, A.L.A. Segatto and A.P. Schuch. 2017. Impacts of UVB radiation on food consumption of forest specialist tadpoles. Ecotoxicol. Environ. Saf. 143: 12–18.

Lopes, A., M. Benvindo-Souza, W.F. Carvalho, H.F. Nunes, P.N. de Lima, M.S. Costa et al. 2021. Evaluation of the genotoxic, mutagenic, and histopathological hepatic effects of polyoxyethylene amine (POEA) and glyphosate on *Dendropsophus minutus* tadpoles. Environ. Pollut. 289: 117911.

Lovell, D.P. and T. Omori. 2008. Statistical issues in the use of the comet assay. Mutagenesis 23: 171–182.

Lu, H., Y. Hu, C. Kang, Q. Meng and Z. Lin. 2021. Cadmium-induced toxicity to amphibian tadpoles might be exacerbated by alkaline not acidic pH level. Ecotoxicol. Environ. Saf. 218: 112288.

Ma, Y., B. Li, Y. Ke and Y. Zhang. 2019. Effects of low doses Trichlorfon exposure on *Rana chensinensis* tadpoles. Environ. Toxicol. 34: 30–36.

Montalvão, M.F. and G. Malafaia. 2017. Effects of abamectin on bullfrog tadpoles: Insights on cytotoxicity. Environ. Sci. Pollut. Res. 24: 23411–23416.

Monteiro, J.A.D.N., L.A.D. Cunha, M.H.P.D. Costa, H.S.D. Reis, A.C.D.S. Aguiar, V.R.L. Oliveira-Bahia et al. 2018. Mutagenic and histopathological effects of hexavalent chromium in tadpoles of *Lithobates catesbeianus* (Shaw, 1802) (Anura, Ranidae). Ecotoxicol. Environ. Saf. 163: 400–407.

Monserrat, J.M., P.E. Martínez, L.A. Geracitano, L.L. Amado, C.M.G. Martins, G.L.L. Pinho, I.S. Chaves, M. Ferreira-Cravo, J. Ventura-Lima and A. Bianchini. 2007. Pollution biomarkers in estuarine animals: Critical review and new perspectives. Comp. Biochem. Physiol. Part - C: Toxicol. 146: 221–234.

Mouchet, F., M. Baudrimont, P. Gonzalez, Y. Cuenot, J.P. Bourdineaud, A. Boudou et al. 2006a. Genotoxic and stress inductive potential of cadmium in *Xenopus laevis* larvae. Aquat. Toxicol. 78: 157–166.

Mouchet, F., L. Gauthier, C. Mailhes, V. Ferrier and A. Devaux. 2006b. Comparative evaluation of genotoxicity of captan in amphibian larvae (*Xenopus laevis* and *Pleurodeles waltl*) using the comet assay and the micronucleus test. . Toxicol. 21: 264–277.

Mouchet, F., L. Gauthier, M. Baudrimont, P. Gonzalez, C. Mailhes, V. Ferrier and A. Devaux. 2007. Comparative evaluation of the toxicity and genotoxicity of cadmium in amphibian larvae (*Xenopus laevis* and *Pleurodeles waltl*) using the comet assay and the micronucleus test. Environ. Toxicol. 22: 422–435.

Natale, G.S., J. Vera-Candioti, C. Ruiz de Arcaute, S. Soloneski, M.L. Larramendy and A.E. Ronco. 2018. Lethal and sublethal effects of the pirimicarb-based formulation Aficida® on *Boana pulchella* (Duméril and Bibron, 1841) tadpoles (Anura, Hylidae) Ecotoxicol. Environ. Saf. 147: 471–479.

Neri, M., D. Milazzo, D. Ugolini, M. Milic, A. Campolongo, P. Pasqualetti et al. 2015. Worldwide interest in the comet assay: A bibliometric study. Mutagenesis 30: 155–163.

Nikoloff, N., G.S. Natale, D. Marino, S. Soloneski and M.L. Larramendy. 2014. Flurochloridone-based herbicides induced genotoxicity effects on *Rhinella arenarum* tadpoles (Anura: Bufonidae). Ecotoxicol. Environ. Saf. 100: 275–281.

Olive, P.L., J.P. Banáth, R.E. Durand and J.P. Banath. 1990. Heterogeneity in Radiation-Induced DNA Damage and Repair in Tumor and Normal Cells Measured Using the “Comet” Assay. Radiat. Res. 122: 86.

Oliveira, C.R., T.D. Garcia, L. Franco-Belussi, R.F. Salla, B.F.S. Souza, N.F.S. de Melo et al. 2019. Pyrethrum extract encapsulated in nanoparticles: Toxicity studies based on genotoxic and hematological effects in bullfrog tadpoles. Environ. Pollut. 253: 1009–1020.

Ossana, N.A., P.M. Castañé, G.L. Poletta, M.D. Mudry and A. Salibián. 2010. Toxicity of waterborne Copper in premetamorphic tadpoles of *Lithobates catesbeianus* (Shaw, 1802). Bull. Environ. Contam. Toxicol. 84: 712–715.

Ossana, N.A., P.M. Castañé and A. Salibián. 2013. Use of *Lithobates catesbeianus* tadpoles in a multiple biomarker approach for the assessment of water quality of the Reconquista River (Argentina). Arch. Environ. Contam Toxicol. 65: 486–97.

Ostling, O. and K.J. Johanson. 1984. Microelectrophoretic study of radiation-induced DNA damages in individual mammalian cells. Biochem. Biophys. Res. Commun. 123: 291–298.

Patar, A., A. Giri, F. Boro, K. Bhuyan, U. Singha and S. Giri. 2016. Cadmium pollution and amphibians—Studies in tadpoles of *Rana limnocharis*. Chemosphere 144: 1043–1049.

Peluso, J., C.M. Aronzon and C.S. Pérez Coll. 2019. Assessment of environmental quality of water bodies next to agricultural areas of Buenos Aires province (Argentina) by means of ecotoxicological studies with *Rhinella arenarum*. J. Environ. Sci. Health B 54: 655–664.

Pérez-Iglesias, J.M., C. Ruiz de Arcaute, N. Nikoloff, L. Dury, S. Soloneski, G.S. Natale et al. 2014. The genotoxic effects of the imidacloprid-based insecticide formulation Glacoxan Imida on Montevideo; tree frog *Hypsiboas pulchellus* tadpoles (Anura, Hylidae). Ecotoxicol. Env. Saf. 104: 120–126.

Pérez-Iglesias, J.M., G.S. Natale, S. Soloneski and M.L. Larramendy. 2018. Are the damaging effects induced by the imazethapyr formulation Pivot® H in *Boana pulchella* (Anura) reversible upon ceasing exposure? Ecotoxicol. Environ. Saf. 148: 1–10.

Pérez-Iglesias, J.M., J.C. Brodeur and M.L. Larramendy. 2020. An imazethapyr-based herbicide formulation induces genotoxic, biochemical, and individual organizational effects in *Leptodactylus latinasus* tadpoles (Anura: Leptodactylidae). Environ. Sci Pollut. Res. Int. 27: 2131–2143.

Pinheiro, P.D.P., C.E.D. Cintra, P.H. Valdujo, H.L.R. Silva, I.A. Martins, N.J. da Silva et al. 2018. A new species of the *Boana albopunctata* group (Anura: Hylidae) from the Cerrado of Brazil. South American Journal of Herpetology 13: 170–182.

Rabello-Gay, M.N., M. Rodrigues and R. Monteleone-Neto. 1991. Mutagênese, carcinogênese e teratogênese: métodos e critérios de avaliação. In Mutagênese, carcinogênese e teratogênese: métodos e critérios de avaliação (pp. 246–246).

Rajaguru, P., R. Kalpana, A. Hema, S. Suba, B. Baskarasethupathi, P.A. Kumar et al. 2001. Genotoxicity of some sulfur dyes on tadpoles (*Rana hexadactyla*) measured using the comet assay. Environ. Mol. Mutagen. 38: 316–322.

Ramos-Neto, M.B. and V.R. Pivello. 2000. Lightning fires in a Brazilian savanna National Park: Rethinking management strategies. Environmental Management 26(6): 675–684.

Ribeiro, L.R., D.M.F. Salvadori and E.K. Marques. 2003. Mutagênese ambiental. Canoas: Ed. ULBRA, 356p.

Ribeiro, M.L., C. Lourencetti, S.Y. Pereira and M.R.R. Marchi. 2007. Contaminação de águas subterrâneas por pesticidas: avaliação preliminar. Quimica Nova 30: 688–694.

Ribeiro, R.S., G.T.B.T. Egito and C.F.B. Haddad. 2005.Chave de identificação: anfíbios anuros da vertente de Jundiaí da Serra do Japi, Estado de São Paulo. Biota Neotrop. 5: 235–247.

Rocha, C.A.M., V.H.C. Almeida, A.S. Costa, J.C. Sagica-Júnior, J.A.N. Monteiro, Y.S.R. Souza et al. 2012. Induction of micronuclei and other nuclear abnormalities in bullfrog tadpoles (*Lithobates catesbeianus*) treated with copper sulphate. Int. J. Gen. 2: 06–11.

Rossa-Feres, D.C. and F. Nomura. 2006. Characterization and taxonomic key for tadpoles (Amphibia: Anura) from the northwestern region of São Paulo State, Brazil. Biota Neotropica. Campinas, SP, Brazil: Instituto Virtual da Biodiversidade | BIOTA – FAPESP 6: 1–26.

Said, R., A. Said, S. Saber and B. ElSalkh. 2022. Biomarker Responses in *Sclerophrys regularis* (Anura: Bufonidae) Exposed to Atrazine and Nitrate. Pollution 8: 1387–1397.

Samanta, S. and P. Dey. 2012. Micronucleus and its applications. Diagn. Cytopathol. 40: 84–90.

Samojeden, C.G., F.A. Pavan, C.F. Rutkoski, A. Folador, S.P. da Fré, C. Müller et al. 2022. Toxicity and genotoxicity of imidacloprid in the tadpoles of *Leptodactylus luctator* and *Physalaemus cuvieri* (Anura: Leptodactylidae). Sci. Rep. 12: 11926.

Santana, J.M., A. Dos Reis, P.C. Teixeira, F.C. Ferreira and C.M. Ferreira. 2015. Median lethal concentration of formaldehyde and its genotoxic potential in bullfrog tadpoles (*Lithobates catesbeianus*). J. Environ. Sci. Health B 50: 896–900.

Santos, A.T., B.S.L. Valverde, C. De Oliveira and L. Franco-Belussi. 2021. Genotoxic and melanic alterations in *Lithobates catesbeianus* (anura) tadpoles exposed to fipronil insecticide. Environ. Sci. Pollut. Res. Int. 28: 20072–20081.

Scaia, M.F., L.S. de Gregorio, L. Franco-Belussi, D.B. Provete, M. Succi Domingues and C. de Oliveira. 2017. Systemic effects of nonylphenol on tadpoles of Lithobates catesbeianus: Gonadal, genotoxic and color alterations. figshare. https://doi.org/10.6084/ m9.figshare.4689910.v3.

Schmid, W. 1975. The micronucleus test. Mutat. Res. Mutagen. Relat. Subj. 31: 9–15.

Shimizu, N., T. Shimura and T. Tanaka. 2000. Selective elimination of acentric double minutes from cancer cells through the extrusion of micronuclei. Mutat. Res. Mol. Mech. Mutagen. 448: 81–90.

Silva, M.B., R.E. Fraga, P.B. Nishiyama, N.L.B. Costa, I.S.S. Silva, D.A. Brandão et al. 2021. *In situ* assessment of genotoxicity in tadpoles (Amphibia: Anura) in impacted and protected areas of Chapada Diamantina, Brazil. Scientia Plena 17: 021701.

Silva, S.V.S., A.H.C. Dias, E.S. Dutra, A.L. Pavanin, S. Morelli and B.B. Pereira. 2016. The impact of water pollution on fish species in southeast region of Goiás, Brazil. J. Toxicol. Environ. Health Part A 79: 8–16.

Singha, U., N. Pandey, F. Boro, S. Giri, A. Giri and S. Biswas. 2014. Sodium arsenite induced changes in survival, growth, metamorphosis and genotoxicity in the Indian cricket frog (*Rana limnocharis*). Chemosphere 112: 333–339.

Singh, N.P., M.T. McCoy, R.R. Tice and E.L. Schneider. 1988. A simple technique for quantitation of low levels of DNA damage in individual cells. Exp. Cell Res. 175: 184–191.

Soloneski, S., C. Ruiz de Arcaute and M.L. Larramendy. 2016. Genotoxic effect of a binary mixture of dicamba- and glyphosate-based commercial herbicide formulations on *Rhinella arenarum* (Hensel, 1867) (Anura, Bufonidae) late-stage larvae. Environ. Sci. Pollut. Res. Int. 23: 17811–17821.

Speit, G. and A. Hartmann. 1999. The comet assay (single-cell gel test). In DNA repair protocols (pp. 203–212). Humana Press.

Storrs, S.I. and R.D. Semlitsch. 2008. Variation in somatic and ovarian development: Predicting susceptibility of amphibians to estrogenic contaminants. Gen. Comp. Endocrinol. 156: 524–530.

Svartz, G., C. Aronzon, S. Pérez Catán, S. Soloneski and C. Pérez Coll. 2020. Oxidative stress and genotoxicity in *Rhinella arenarum* (Anura: Bufonidae) tadpoles after acute exposure to Ni-Al nanoceramics. Environ. Toxicol. Pharmacol. 80: 103508.

Terradas, M., M. Martín, L. Tusell and A. Genescà. 2010. Genetic activities in micronuclei: Is the DNA entrapped in micronuclei lost for the cell? Mutat. Res. 705: 60–67.

Tevini, M. 1993. UV-B Radiation and ozone depletion: Effects on humans, animals, plants, microorganisms, and materials. Lewis Publishers. http://www.ciesin.columbia.edu/docs/001-540/001-540.html.

Tice, R.R., E. Agurell, D. Anderson, B. Burlinson, A. Hartmann, H. Kobayashi, Y. Miyamae et al. 2000. Single cell gel/comet assay: Guidelines for *in vitro* and *in vivo* genetic toxicology testing. Environ. Mol. Mutagen. 35: 206–221.

Toledo, L.F., F.B. Britto, O.G.S. Araujo, L.O. Giasson and C. Haddad. 2006. The occurrence of *Batrachochytrium dendrobatidis* in Brazil and the inclusion of 17 new cases of infection. South American of Herpetology.

Toledo, L.F., S.P. Carvalho-E-Silva, C. Sánchez, M.A. Almeida and C.F.B. Haddad. 2010. A revisão do Código Florestal Brasileiro: impactos negativos para a conservação dos anfíbios. Biota Neotrop. 10: 4.

Veronez, A.C.S., R.V. Salla, V.D. Baroni, I.F. Barcarolli, A. Bianchini, C.B.R. Martinez et al. 2016. Genetic and biochemical effects induced by iron ore, Fe and Mn exposure in tadpoles of the bullfrog *Lithobates catesbeianus*. Aquat. Toxicol. 174: 101–108.

Vieira, T.Q. 2017. O ensaio cometa e a espécie *Hypsiboas albopunctatus* (Spix, 1824) como ferramentas de avaliação de qualidade ambiental em uma unidade de conservação federal inserida no Cerrado goiano. 2017. 63 p. Dissertação (Mestrado em Biodiversidade Animal) - Universidade Federal de Goiás, Goiânia.

Yadav, S.S., S. Giri, U. Singha, F. Boro and A. Giri. 2013. Toxic and genotoxic effects of Roundup on tadpoles of the Indian skittering frog (*Euflictis cyanophlyctis*) in the presence and absence of predator stress. Aquat Toxicol. 132-133: 1–8.

Yin, X.H., S.N. Li, L. Zhang, G.N. Zhu and H.S. Zhuang. 2008. Evaluation of DNA damage in Chinese toad (*Bufo bufo gargarizans*) after *in vivo* exposure to sublethal concentrations of four herbicides using the comet assay. Ecotoxicology 17: 280–286.

Zar, J.H. 2007. Biostatistical Analysis. Pearson. Upper Saddle River.

Zhang, Y.M., D.J Huang, D.Q. Zhao, J. Long, G. Song and A. Li. 2007. Long-term toxicity effects of cadmium and lead on *Bufo raddei tadpoles*. Bull. Environ. Contam. Toxicol. 79: 178–183.

8

Pollutant Effects on Immune Defense Cells of Anuran Tadpoles

Camila Fatima Rutkoski[1] and *Eduardo Alves de Almeida*[2,*]

1. Introduction

Immune system in vertebrates is composed by innate and adaptive immune systems, which are interconnected (Robert and Ohta 2009) and can induce local and systemic responses (Flajnik and Kasahara 2010, Hooper et al. 2012). Innate immunity is the first line of immunological defense, composed mainly of effector cells that can eliminate pathogens by phagocytosis or by direct cytotoxicity without the need for previous exposure. This innate immunity also plays a role in initiating adaptive immune responses specific to a particular foreign antigen (Robert and Ohta 2009). In contrast, adaptive immune system is mainly composed by B and T cells that are able to express specific antigen (Ag) receptors on the cell surface in response to different exogenous molecules and pathogens. The adaptative response is also referred to as the "anticipatory system", since it can be stimulated by any non-self-molecule and often requires a pre-exposure of the cells to an Ag (Robert and Ohta 2009).

Morphology of immune cells of the innate system in amphibians is similar to those observed in mammals and comprises neutrophils, eosinophils and basophils (polymorphonuclear cells), monocytes, macrophages, and natural killer cells (NK) (Robert and Ohta 2009). In vertebrates, innate immune response is activated by the interaction between pattern recognition receptors (PRRs) and pathogen-specific effector cells, a mechanism known as pathogen-associated molecular patterns (PAMPs) (Robert and Ohta 2009). This interaction induces biochemical cascades that stimulate effector cells against various pathogens, in addition to secreting antimicrobial peptides in the skin, as well as serum proteins and complement components by the liver (Robert and Ohta 2009).

[1] Environmental Engineering Post-Graduation Program, University of Blumenau, Blumenau, SC, Brazil. Email: cfrutkoski@furb.br

[2] Department of Natural Science, University of Blumenau, Blumenau, SC, Brazil.

* Corresponding author: eduardoalves@furb.br

The adaptive biology of B and T cells is conserved in vertebrates (Grogan et al. 2018). Thus, adaptative immunity in *Xenopus* spp. is composed by B and T cells, and by most of the molecules associated with adaptive immunity such as immunoglobulins (Igs), T cell receptor (TCR), major histocompatibility complex (MHC), recombinant-activating genes (RAG), and activation-induced cytidine deaminase (AID) (Robert and Ohta 2009). More specific information on the innate and adaptive immune system of *Xenopus* spp. can be assessed in the study of Robert and Ohta (2009).

Leukocytes are involved in both innate and adaptive immune responses (Rollins-Smith and Woodhams 2011), with different functions (Arikan and Çiçek 2014). In amphibians, agranulocytes consist of lymphocytes and monocytes, and granulocytes of neutrophils, basophils and eosinophils (Arikan and Çiçek 2014). Lymphocytes are composed by B and T cells, with type B containing B cell receptors and synthesizing the immunoglobulin antibody in response to an antigen, while type T lymphocytes have T cell receptors and participate in cell-mediated immunity (Grogan et al. 2018). Lymphocytes and neutrophils are the leukocytes most often found in the blood of amphibians (Davis et al. 2008). Most leukocytes are related to the innate immune system, except for lymphocytes, which are involved in the adaptive response and antibody production (Davis et al. 2008, Shutler et al. 2009). Monocytes have a dual function as antigen-presenting cells (Robert and Ohta 2009). Neutrophils respond to inflammation, disease and other stressful situations (Davis et al. 2008, Shutler et al. 2009). Eosinophils play a phagocytic function and are associated with hypersensitivity events (Klion and Nutman 2004). Basophils have the chemotactic factor histamine for eosinophils and neutrophils (Junqueira and Carneiro 2004), despite not having a well-defined function in amphibians (Davis 2008).

Lymphocytes have a large nucleus that occupies most of the cytoplasm (Oliveira et al. 2019) (Fig. 1, A). On the other hand, monocytes have a smaller nucleus with different shapes, presenting a visible cytoplasm area (Oliveira et al. 2019) (Fig. 1, B). Neutrophils do not have noticeable granules in the cytoplasm and have a remarkably lobed nucleus (Oliveira et al. 2019) (Fig. 1, C). Eosinophils, in turn, have more visible cytoplasmic granules in addition to a large nucleus (Oliveira et al. 2019) (Fig. 1, D). Finally, in basophils, granules are more evident in the cytoplasm than the two previous leukocytes, but the nucleus is smaller and barely visible (Oliveira et al. 2019, Gregorio et al. 2021) (Fig. 1, D).

In several vertebrate, transient signaling molecules known as cytokines play a key role in innate immunity (Rayl and Allender 2020). Cytokines include macrophage activating factor (MAF), interleukins (ILs), chemokines, interferons (IFNs), tumor necrosis factors (TNFs), transforming growth factors (TGFs), colony stimulating factors (CSFs), growth proteins (GFs), and stress proteins such as heat shock (HSPs) (Murphy 2012, Grogan et al. 2018); they perform their function by binding to specific receptors present on the surface of target cells (Scapigliati et al. 2006). Cytokines participate in a multitude of processes such as local and systemic inflammation, cell proliferation, metabolism, chemotaxis and tissue repair (Arango Duque and Descoteaux 2014).

In addition, amphibians rely heavily on macrophage lineage cells for immune defense (Grayfer and Robert 2016), whose functions range from body formation to ingestion and elimination of pathogens and fragmented cells (apoptosis) (Arango Duque and Descoteaux 2014). Melanomacrophages (Fig. 2) are important cells to assess adaptive immune responses, consisting

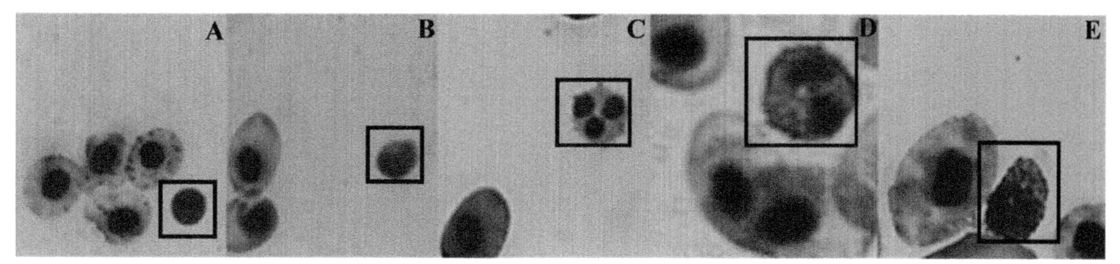

Figure 1. Leukocytes present in the blood of *Lithobates catesbeianus* tadpoles. Leukocytes are highlighted with a rectangle. The other cells are blood erythrocytes. (A) lymphocyte, (B) monocyte, (C) neutrophil, (D) eosinophil (E) basophil.

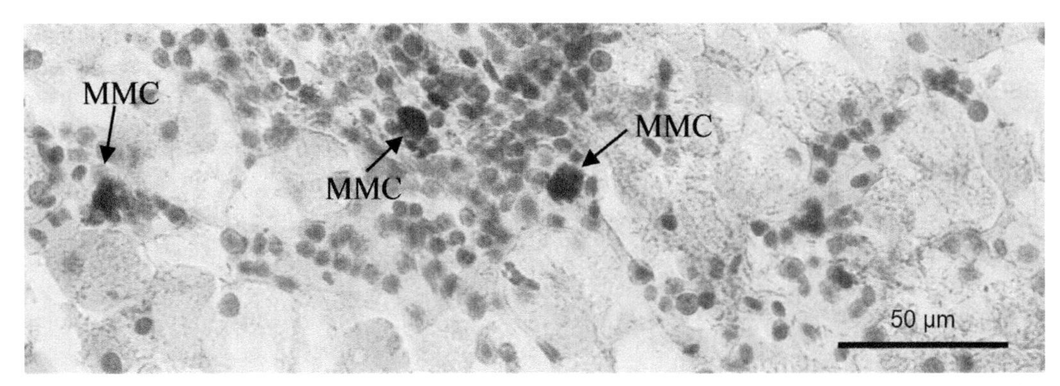

Figure 2. Melanomacrophage (MMC) present in an inflammatory infiltrate in the liver of *Lithobates catesbeianus* tadpoles.

of phagocytes with pigment that are usually present in lymphoid tissues of poikilothermic animals such as liver, kidney and spleen (Steinel and Bolnick 2017), including the liver of amphibians. (Fig. 2) (De Oliveira et al. 2017). These cells have three types of pigment: melanin, hemosiderin and lipofuscin (Agius and Roberts 2003), and perform non-immunological and immunological functions such as phagocytosis (Agius and Roberts 2003), pathogen deposition (Roberts 1975) and immune responses to antigens (Agius 1985).

2. Development of the Immune System in Anurans

Development of the immune system in Anura has been mainly studied in *Xenopus laevis* (Robert and Ohta 2009, Grogan et al. 2018). In this species, thymus starts developing 3 days after fertilization (stage NF 40, Nieuwkoop and Faber 1967) and after a few days, colonization by embryonic stem cells occurs (Tochinai 1980, Kau and Turpen 1983, Flajnik et al. 1985, Robert and Ohta 2009). Around stage NF 48 (6 to 8 days of development), the cortex-medulla architecture becomes perceptible (Du Pasquier and Flajnik 1990) but the thymus reaches its maximum size at larval stages NF 58–63 (Du Pasquier and Weiss 1973, Rollins-Smith et al. 1984, Robert and Ohta 2009). The spleen develops 4 to 5 days after fertilization (stage NF 47), but with blood cells mainly associated with hematopoiesis (Hsu 1998, Du Pasquier et al. 2000). Up to 12 days after fertilization, the embryo's defense occurs through an innate type of myeloid cells or non-lymphocyte leukocytes (Hansen and Zapata 1998, Robert and Ohta 2009). Lymphocytes (leukocytes) begin to accumulate in the spleen around 12 to 14 days after fertilization (stage NF49) (Hsu 1998, Du Pasquier et al. 2000, Robert and Ohta 2009). During the larval period, the lymphopoiesis occurs primarily in the liver (Du Pasquier et al. 2000). The spleen size and the number of lymphocytes increase throughout the tadpole´ development, reaching maximum size at stage NF 58, that corresponds to the beginning of the metamorphosis (Robert and Ohta 2009).

In the larvae, liver colonization by stem cells occurs after the establishment of blood circulation, so the erythrocytes, leukocytes and B cells are formed (Chen and Turpen 1995, Robert and Ota 2009). Tadpole immune system undergoes changes during metamorphosis, ranging from thymus involution with subsequent redevelopment to renewal of lymphocyte populations, with 50–90% reduction in cells (Flajnik et al. 1987, Rollins-Smith 1998, Robert and Ohta 2009). Davis (2008) studied the leukocyte profile throughout the development of *Lithobates catesbeianus* and verified these modifications. In bullfrog tadpoles (*L. catesbeianus*), the number of lymphocytes increased in the blood during the early stages of development, but its abundance decreased during metamorphic climax (Davis 2008). A similar pattern was found for neutrophils, except for a peak in these cells at stages 38–39 of Gosner (1960). Lymphocytes are known to be involved in tissue growth, but not neutrophils (Jordan and Speidel 1923, 1924, Davis 2008). Further studies are needed to clarify the role of neutrophils in tadpole development (Davis 2008).

Eosinophils are less abundant in the early stages of development, with peaks at metamorphic climax, and declines (to basal levels) in metamorphic and adult frogs (Davis 2008). This pattern occurs because at metamorphic climax, tissues are heavily remodeled, suggesting that these cells play a role in the tissue lysis during metamorphosis (Davis 2008), in addition to producing several factors that initiate and modulate immune and inflammatory responses (Adamko et al. 2005, Rothenberg and Hogan 2006, Davis 2008). Monocytes also have a significant increase during the final stages of metamorphosis, acting to eliminate cellular debris from the amphibian's body (Davis 2008). The role of basophils in the innate immune system is still unclear. However, Davis (2008) observed that these cells gradually increase throughout development, although they seem to not have specific role in metamorphosis process, being more abundant in adults.

Studies monitoring changes of leukocyte number and function throughout amphibian development are important for understanding the role of these cells during tissue growth and degeneration during metamorphosis (Davis 2008). Juvenile frogs are particularly susceptible to diseases because their adaptive immune defenses are not yet mature (Rollins-Smith 2017). Indeed, compared to adult frogs, lymphocyte-mediated responses in tadpoles have limitations in recognizing Ag, implying poorer allorecognition and fewer antibody responses (Rollins-Smith 1998, Robert and Ohta 2009, Rollins-Smith 2017). The adaptative immune system becomes more efficient in adults as the number of lymphocytes increases rapidly after metamorphosis (Rollins-Smith 2017).

The immune system of adult amphibians is similar to that of other jawed vertebrates (Du Pasquier et al. 1989), being also dependent on major lymphoid organs, such as thymus and bone marrow (Colombo et al. 2015, Grogan et al. 2018), in addition to the spleen. T lymphocytes source in larval and adult amphibians is the thymus. However, B-cell differentiation arises into the liver, while bone marrow, which is more rudimentary than in mammals, essentially supports neutrophil differentiation and contains macrophages precursors (Grayfer and Roberts 2013, Colombo et al. 2015). In adult *Xenopus*, the spleen is the main site of B-cell differentiation, while in tadpoles, lymphopoiesis takes place in the liver and spleen (Hadji-Azimi et al. 1982, 1990). Moreover, spleen represents the main peripheral lymphoid organ in amphibians, where B and T cells accumulate inside the white pulp. Most of the splenic B cells produce IgM, with few producing IgY or IgX (Pasquier et al. 1989).

Epidermis of amphibians also constitutes an immediate innate physical defense barrier against pathogen invasion. In addition to internal organs, amphibians' skin has several layers of leukocytes and secretory factors that function as an immune barrier, especially in adults (Yaparla et al. 2017), where epidermis and dermis are constituted by T and B lymphocytes, as well as macrophages and dendritic cells (Rollins-Smith et al. 2011). Adult amphibian skin also has granular glands that secrete compounds such as antimicrobial peptides that act in defense against several microorganisms (Siano et al. 2014). In contrast, the skin of pre- and prometamorphic tadpoles is much less developed, often composed of two or three layers of cells, with no substantial presence of glands or keratin, which possibly turn tadpoles much more susceptible to pathogen infection than adults.

The mucus layer that covers the body surface is also an important component to protect epithelia from infections. The production of mucus helps to trap pathogens, preventing their entry into the body. In *X. tropicalis* tadpoles, Dubaissi et al. (2018) characterized the major structural component of the mucus epithelial barrier and showed that it is composed by a mucin glycoprotein with structural properties similar to human gel-forming mucins, which protects tadpoles against infection by retaining pathogens and avoiding contact with epithelial cell membranes. In the small intestine, the mucus of adult amphibians contains high concentrations of antibacterial peptides, such as defensins and lysozymes (Johansson et al. 2013, Kim and Ho 2010). Similar to mammals, amphibians produce high levels of lysozymes with antimicrobial activity. Several lysozyme genes are expressed in several tissues including the skin and the egg (Colombo et al. 2015).

It is also important to mention that symbiotic microorganisms associated with skin and gastrointestinal tract of tadpoles can also exert immune defense roles in tadpoles. More details on the importance of tadpoles' microbiota and its beneficial functions for host animals are discussed in Chapter 9.

3. Immune System Alterations in Tadpoles due to Pollutant Exposure

One of the most important innate protective mechanisms is the potential for leukocytes to kill pathogenic organisms (Rymuszka and Adaszek 2013). Thus, alterations in these cells can affect the defense of animals against pathogens and infectious diseases (Rymuszka and Adaszek 2013). Reduced number of leukocytes may make amphibians more susceptible to certain types of diseases and parasites, which would result in important negative impacts at population level (Silva et al. 2020). In this way, analysis of amphibian leukocyte cells can be used as interesting biomarkers of exposure to environmental stressors (Davis et al. 2008), indicating immunosuppression (Mann et al. 2009). However, studies evaluating the effects of environmental contaminants on amphibian leukocytes are still recent and, therefore, scarce to date.

Silva et al. (2020) found that the insecticide chlorpyrifos caused immunosuppression in tadpoles of *Odontophrynus carvalhoi,* evidenced by lymphopenia at concentrations of 100, 200 e 400 mg.L^{-1}. Neutrophilia and eosinophilia were also observed, a common response in tadpoles after pesticide poisoning (Silva et al. 2020). Similar results, i.e., decreased lymphocytes, were observed in *Rana sylvatica* tadpoles after exposure to clothianidin neonicotinoid insecticide (1 and 10 mg.L^{-1}) and diquat herbicide (532 mg.L^{-1}). Increased ratio of neutrophils to lymphocytes in tadpoles exposed to 1 mg.L^{-1} of clothianidin also indicated a stress response in the animals (Robinson et al. 2021). The herbicide glyphosate reduced the number of lymphocytes and increased the neutrophils in *Pelobates cultripes* tadpoles (Burraco et al. 2013). *L. catesbeianus* tadpoles exposed to 1 mgL^{-1} of 4-nonylphenol had reduced lymphocytes and increased neutrophils (Gregorio et al. 2021). The same tadpoles exposed to 0.25 and 2.5 ngL^{-1} cyproterone acetate showed increased eosinophils rates (Gregorio et al. 2021). *L. catesbeianus* tadpoles under acute exposure (48 h) to pyrethrum (PYR) extract and solid lipid nanoparticles loaded with PYR presented increased number of leukocytes, specifically eosinophils in the nanoparticle-exposed groups. Augmented basophil number were also reported in PYR-exposed group, indicating that the compound may have triggered an inflammatory and allergic response, activating the recruitment of eosinophils and basophils, respectively (Oliveira et al. 2019).

Furthermore, Rutkoski et al. (2022) observed lymphocytosis and neutropenia in *L. catesbeianus* tadpoles exposed to the antibiotics sulfamethoxazole (460 ngL^{-1}) and oxytetracycline (20, 90 and 460 ng.L^{-1}), suggesting that pharmaceuticals are able to affect the tadpoles' immune system. Other environmental stressors, such as changes in abiotic factors, can also change amphibian immune system. Franco-Belussi et al. (2018) found that *L. catesbeianus* tadpoles exposed to UV radiation presented decreased number of leukocytes. Also, non-pigmented tadpoles (albinos) had lymphopenia earlier than pigmented tadpoles, in addition to presenting DNA damage (Franco-Belussi et al. 2018). A study simulating the effects of climate changes showed that increases in temperature (32°C) and CO_2 (dropping water pH in the range of 5.5–5.6) decreased the number of leukocytes in the blood of *Polypedates cruciger* tadpoles, resulting in greater susceptibility to infections (Manasee et al. 2020).

As mentioned, cytokines play important roles in inflammatory events (Scapigliati et al. 2006). Pro-inflammatory cytokines, for example, can attract leukocytes to the site of infection, participate in tissue repair and activate pathways associated with blood clotting (Murphy 2012, Grogan et al. 2018). Exposure of *X. laevis* tadpoles to residues of oils and gas extraction significantly altered homeostatic expression of myeloid lineage genes and compromised tadpole responses to virus infection through alteration in mRNA transcript levels of pro-inflammatory cytokine genes (Robert et al. 2018). Tadpoles of the same species exposed to a mixture of 23 chemical compounds associated with unconventional oil and gas (UOG) operations (concentration of 1 μg.L^{-1}) resulted in increased levels of macrophage colony-stimulating factor transcription (CSF-1) and interleukin 34 (IL-34) in liver and kidney (Robert et al. 2018). For the same tadpoles, an increase in the expression of the CSF-1 gene and of the granulocyte colony stimulating receptor (GCSF-R1) was observed in the kidneys (Robert et al. 2018). There was an increase in the expression of GCSF-R and IL-34 in the kidneys (Robert et al. 2018).

Moreover, exposure during larval phase also resulted in perturbation of immune homeostasis in adult frogs, as indicated by significantly decreased number of leukocytes, B and T lymphocytes, which increased the susceptibility of the adults to viral infection (Robert et al. 2019). Similarly, Cary et al. (2014) exposed *R. pipiens* pre-metamorphic tadpoles to 0–634 $ng.g^{-1}$ of a polybrominated diphenyl ether (PBDE) mixture until they reached the metamorphic climax. PBDE exposure during larval life significantly lowered the ability of postmetamorphic frogs of establish an adaptive humoral response, presenting up to 92.4% lower levels of antibodies. Similar effects on the immune system of postmetamorfic *R. pipiens* were observed after exposing prometamorphic tadpoles to PCB-126 (Cary and Karasov 2021). Moreover, exposure of pre-metamorphic *X. laevis* tadpoles to the insecticide carbaryl (0.1 and 1.0 mgL^{-1}) impaired tadpole innate antiviral immune responses, as evidenced by significantly decreased transcript levels of pro-inflammatory cytokines (Andino et al. 2017). Based on these findings, the authors concluded that the agriculture-associated carbaryl at ecologically relevant concentrations had the potential to induce long term alterations in host-pathogen interactions and antiviral immunity. Furthermore, Langerveld et al. (2009) found 7 genes associated with immune system function significantly downregulated in female *X. laevis* tadpoles chronically exposed to 400 $mg.L^{-1}$ of atrazine, supporting the idea that this herbicide compromises immunity during frog development. Atrazine at 0.1 $mg.L^{-1}$ also altered cytokine responses in *X. laevis* tadpoles, disrupting the adult immune response to viral infection (Sifkarovski et al. 2014). Exposure to 3 and 30 $mg.L^{-1}$ of atrazine also significantly increased trematode parasitism in *R. sylvatica* tadpoles (Koprivnikar et al. 2009).

When pre-exposed for 96 h to the organophosphate insecticide malathion (60 and 600 $mg.L^{-1}$), *Rana palustris* tadpoles presented increased infections with the trematode parasite *Echinostoma trivolvis*, a consequence that was attributed to an altered immune response of tadpoles after pesticide exposure (Budischak et al. 2009). Similarly, higher infestation of trematode parasites was observed in *R. pipiens* tadpoles that were previously exposed for 10 d to 10 $mg.L^{-1}$ of perfluorohexanesulfonic acid (Brown et al. 2020). Moreover, pre-exposure of *Lithobates sylvaticus* tadpoles for 7 to 250 $mg.L^{-1}$ of the neonicotinoid insecticide thiametoxan produced juvenile with elevated neutrophil-to-lymphocyte and neutrophil-to-white blood cell ratios, but with decreased lymphocyte-to-white blood cell ratios (Gavel et al. 2019).

With respect to pollutant effects on the tadpoles' immune system associated to skin barrier alteration, only one study has been found so far. Shu et al. (2022) exposed *L. catesbeianus* to a cyanobacteria toxin, microcystin-leucine arginine (MC-LR), for 30 days. Exposure to 0.5 mgL^{-1} MC-LR significantly altered epidermis structures, damaging the physical barrier of the skin. Increased skin eosinophils and upregulated transcriptions of inflammation-related genes were also found in the exposed tadpoles, suggesting that skin inflammation resulted from MC-LR exposure. Moreover, antimicrobial peptides (brevinin-1PLc, brevinin-2GHc, and ranatuerin-2PLa) and lysozyme were down-regulated in the exposed groups, while the complement system, pattern recognition receptor, and specific immune processes were up-regulated.

Regarding melanomacrophages (MMs), they are considered excellent and simple biomarkers to assess effects on adaptive immunity (Steinel and Bolnick 2017) and may undergo changes in tadpoles exposed to contaminants (Grott et al. 2021, Rutkoski et al. 2022). Exposure of *Osteopilus septentrionalis* tadpoles to the fungicide chlorothalonil at 17.6 $mg.L^{-1}$ reduced the number of liver granulocytes and MMs in tadpoles, whereas further increases in chlorothalonil concentrations caused increased numbers of these cells in the liver (McMahon et al. 2011). Finally, *R. pipiens* tadpoles exposed for 21 d to a mixture of 6 pesticides (atrazine, metribuzin, aldicarb, endosulfane, lindane, and dieldrin) at environmentally realistic concentrations had significantly reduced lymphocyte proliferation (Christin et al. 2009). On the other hand, increases in MMs may be related to the occurrence of oxidative stress and activation/recruitment of immune cells (Da Silva et al. 2012). *L. catesbeianus* tadpoles exposed to linear alkylbenzene sulfonate surfactant (0.5 mgL^{-1}) showed an increase in MMs area of the liver (Franco-Belussi et al. 2021). Tadpoles of the same species exposed to the antibiotics sulfamethoxazole (90 and 460 $ng.L^{-1}$) and oxytetracycline (20, 90 and 460 $ng.L^{-1}$) (Rutkoski et al. 2022), as well as to the pesticide tebuthiuron (50 and 200 ngL^{-1}) (Grott et al. 2021) also had increased MMs. A similar effect was observed in tadpoles of *Rana esculenta* exposed to hexavalent chromium (5 mgL^{-1}) (Boncompagni et al. 2004) and in *Bufo gargarizans* treated with the antibacterial agent triclosam (60 μgL^{-1}) (Chai et al. 2017).

4. Conclusions and Perspectives

The immune system of amphibians undergoes significant changes along the animal's development and is not yet mature in the larval stage. This makes tadpoles more susceptible to parasites, diseases and the effect of contaminants. Pollutants can disrupt tadpoles' immunity by decreasing circulating immune defense cells and/or by altering the capacity of the organism to produce transient signaling molecules of the immune response, such as cytokines and immunoglobulins. Studies evaluating changes in the immune system of tadpoles as a result of environmental pollutants are still limited. This limitation is even more evident if considering studies regarding the effects of contaminant mixtures, associations with abiotic stressors and comparisons between species with different life history. Diseases and pollutant exposure are among the main causes of amphibian declines in the world, and both factors are possibly acting in combination in most impacted natural environments. Therefore, studies pointing the relationships between immune depression due to pollutants exposure and susceptibility to infectious diseases are still necessary for better understanding the health consequences of such disturbed environments for tadpoles. These studies can be also helpful in refining action plans for the conservation of these organisms.

Acknowledgements

We thank "Conselho Nacional de Desenvolvimento Científico e Tecnológico, CNPq", "Fundação de Amparo à Pesquisa e Inovação do Estado de Santa Catarina, FAPESC" and "Coordenação de Aperfeiçoamento de Pessoal de Nível Superior, CAPES" (001) for the financial support.

References Cited

Adamko, D.J., S.O. Odemuyiwa, D. Vethanayagam and R. Moqbel. 2005. The rise of the phoenix: The expanding role of the eosinophil in health and disease. Allergy 60: 13–22.

Agius, C. 1985. The melano-macrophage centres in fish: A review. pp. 85–105. *In*: Manning, M.J. and M.F. Tatner [eds.]. Fish Immunology. Academic Press, London.

Agius, C. and R.J. Roberts. 2003. Melano-macrophage centres and their role in fish pathology. J. Fish Dis. 26: 499–509.

Andino, F. de J., B.P. Lawrence and J. Robert. 2017. Long term effects of carbaryl exposure on antiviral immune responses in *Xenopus laevis*. Chemosphere 170: 169–175.

Arango Duque, G. and A. Descoteaux. 2014. Macrophage cytokines: Involvement in immunity and infectious diseases. Front. Immunol. 5: 1–12.

Arikan, H. and K. Çiçek. 2014. Haematology of amphibians and reptiles: A review. North-West J. Zool, 10: 190–209.

Boncompagni, E., C. Fenoglio, R. Vaccarone, P. Chiari, G. Milanesi, M. Fasola and S. Barni. 2004. Toxicity of chromium and heptachlor epoxide on liver of Rana kl. esculenta: A morphological and histochemical study. Ital. J. Zool. 71: 63–167.

Brown, S.R., R.W. Flynn and J.T. Hoverman. 2020. Perfluoroalkyl substances increase susceptibility of northern leopard frog tadpoles to trematode infection. Environ. Toxicol. Chem. 40: 689–694.

Budischak, S.A., L.K Belden and W.A. Hopkins. 2009. Effects of malathion on embryonic development and latent susceptibility to trematode parasites in ranid tadpoles. Environ. Toxicol. Chem. 27: 2496–2500.

Burraco, P., L.J. Duarte and I. Gomez-Mestre. 2013. Predator-induced physiological responses in tadpoles challenged with herbicide pollution. Curr. Zool. 59: 475–484.

Cary, T.L. and W.H. Karasov. 2021. Larval exposure to polychlorinated biphenyl-126 led to a long-lasting decrease in immune function in postmetamorphic juvenile northern leopard frogs, lithobates pipiens. Environ. Toxicol. Chem. 41: 81–94.

Cary, T.L., M.E. Ortiz-Santaliestra and W.H. Karasov. 2014. Immunomodulation in post-metamorphic northern leopard frogs, lithobates pipiens, following larval exposure to polybrominated diphenyl ether. Environ. Sci. Technol. 48: 5910–5919.

Chai, L., A. Chen, P. Luo, H. Zhao and H. Wang. 2017. Histopathological changes and lipid metabolism in the liver of Bufo gargarizans tadpoles exposed to Triclosan. Chemosphere 182: 255–266.

Chen, X.D. and J.B. Turpen. 1995. Intraembryonic origin of hepatic hematopoiesis in *Xenopus laevis*. J. Immunol. 154: 2557–2567.

Christin, M.-S., A.D. Gendron, P. Brousseau, L. Ménard, D.J. Marcogliese, D. Cyr, S. Ruby and M. Fournier. 2009. Effects of agricultural pesticides on the immune system of Rana pipiens and on its resistance to parasitic infection. Environ. Toxicol. Chem. 22: 1127–1133.

Colombo, B.M., T. Scalvenzi, S. Benlamara and N. Pollet. 2015. Microbiota and mucosal immunity in amphibians. Front. Immunol. 6: 1–15.

Da Silva, G.S., F.F. Neto, H.C.S. De Assis, W.R. Bastos and C.A. De Oliveira Ribeiro. 2012. Potential risks of natural mercury levels to wild predator fish in an Amazon reservoir. Environ. Monit. Assess. 184: 4815–4827.

Davis, A.K. 2008. Metamorphosis-related changes in leukocyte profiles of larval bullfrogs (Rana catesbeiana). Comp. Clin. Pathol. 18: 181–186.

Davis, A.K., D.L. Maney and J.C. Maerz. 2008. The use of leukocyte profiles to measure stress in vertebrates: A review for ecologists. Funct. Ecol. 22: 760–772.

De Oliveira, C., L. Franco-Belussi, L.Z. Fanali and L.R.S. Santos. 2017. Use of melanin-pigmented cells as a new tool to evaluate effects of agrochemicals and other emerging contaminants in Brazilian anurans. Marcelo L. Larramendy [ed.]. Ecotoxicology and Genotoxicology: Non-traditional Terrestrial Models. 1ed, vol. 1, The Royal Society of Chemistry, Londres, pp. 125–142.

Du Pasquier, L. and N. Weiss. 1973. The thymus during the ontogeny of the toad *Xenopus laevis*: Growth, membrane-bound immunoglobulins and mixed lymphocyte reaction. Eur. J. Immunol. 3: 773–777.

Du Pasquier, L., J. Schwager and M.F. Flajnik. 1989. The immune system of xenopus. Annu. Rev. Immunol. 7: 251–275.

Du Pasquier, L. and M.F. Flajnik. 1990. Expression of mhc class ii antigens during xenopus development. Dev. Immunol. 1: 85–95.

Du Pasquier, L., J. Robert, M. Courtet and R. Mußmann. 2000. B-cell development in the amphibian Xenopus. Immunol. Rev. 175: 201–213.

Dubaissi, E., K. Rousseau, G.W. Hughes, C. Ridley, R.K. Grencis, I.S. Roberts and D.J. Thornton. 2018. Functional characterization of the mucus barrier on the *Xenopus tropicalis* skin surface. PNAS 115: 726–731.

Flajnik, M.F., L. Du Pasquier and N. Cohen. 1985. Immune responses of thymusf/ymphocyte embryonic chimeras: Studies on tolerance and major histocompatibility complex restriction in Xenopus. Eur. J. Immunol. 15: 540–547.

Flajnik, M.F., E. Hsu, J.F. Kaufman and L. Du Pasquier. 1987. Changes in the immune system during metamorphosis of Xenopus. Immunology Today 8: 58–64.

Flajnik, M.F. and M. Kasahara. 2010. Origin and evolution of the adaptive immune system: Genetic events and selective pressures. Nat. Rev. Genet. 11: 47–59.

Franco-Belussi, L., L.Z. Fanali and C. De Oliveira. 2018. UV-B affects the immune system and promotes nuclear abnormalities in pigmented and non-pigmented bullfrog tadpoles. J. Photochem. Photobiol. B. 180: 109–117.

Franco-Belussi, L., M. Jones-Costa, R.F. Salla, B.F.S. Souza, F.A. Pinto-Vidal, C.R. Oliveira, E.C.M. Silva-Zacarin, F.C. Abdalla, L.C.S. Duarte and C. De Oliveira. 2021. Hepatotoxicity of the anionic surfactant linear alkylbenzene sulphonate (LAS) in bullfrog tadpoles. Chemosphere 266: 1–10.

Gavel, M.J., S.D. Richardson, R.L. Dalton, C. Soos, B. Ashby, L. McPhee, M.R. Forbes and S.A. Robinson. 2019. Effects of 2 neonicotinoid insecticides on blood cell profiles and corticosterone concentrations of wood frogs (lithobates sylvaticus). Environ. Toxicol. Chem. 38: 1273–1284.

Gosner, K.L. 1960. A simplified table for staging anuran embryos and larvae with notes on identification. Herpetologica. 16: 183–190.

Grayfer, L. and J. Robert. 2013. Colony-stimulating factor-1-responsive macrophage precursors reside in the amphibian (*Xenopus laevis*) bone marrow rather than the hematopoietic subcapsular liver. J. Innate Immun. 5: 531–542.

Grayfer, L. and J. Robert. 2016. Amphibian macrophage development and antiviral defenses. Dev. Comp. Immunol. 58: 60–67.

Gregorio, L.S., L. Franco-Belussi and C. Oliveira. 2021. Leukocyte profile of tadpoles and juveniles of lithobates catesbeianus shaw, 1802 (Anura) and the effects of nonylphenol and cyproterone acetate. S. Am. J. Herpetol. 20: 1–10.

Grogan, L.F., J. Robert, L. Berger, L.F. Skerratt, B.C. Scheele, J.G. Castley, D.A. Newell and H.I. Mccallum. 2018. Review of the amphibian immune response to chytridiomycosis, and future directions. Front. Immunol. 9: 1 20.

Grott, S.C., D. Bitschinski, N.G. Israel, G. Abel, S.P. Silva, T.C. Alves, D. Lima, A.C.D. Bainy, J.J. Mattos, E.B. Silva, C.A.C. Albuquerque and E.A. Almeida. 2021. Influence of temperature on biomarker responses and histology of the liver of American bullfrog tadpoles (Lithobates catesbeianus, Shaw, 1802) exposed to the herbicide Tebuthiuron. Sci. Total Environ. 771: 1–14.

Hadji-Azimi, I., J. Schwager and C. Thiebaud. 1982. B-lymphocyte differentiation in *Xenopus laevis* larvae. Dev. Biol. 90: 253–258.

Hadji-Azimi, I., V. Coosemans and C. Canicatti. 1990. B-lymphocyte populations in *Xenopus laevis*. Dev. Comp. Immunol. 14:69–84.

Hansen, J.D. and A.G. Zapata. 1998. Lymphocyte development in fish and amphibians. Immunol Rev. 166: 199–220.

Hooper, L.V., D.R. Littman and A.J. Macpherson. 2012. Interactions between the microbiota and the immune system. Science 336: 1268–1273.

Hsu, E. 1998. Mutation, selection, and memory in B lymphocytes of exothermic vertebrates. Immunol Rev. 162: 25–36.

Johansson, M.E.V., H. Sjövall and G.C. Hansson. 2013. The gastrointestinal mucus system in health and disease. Nat. Rev. Gastroenterol. Hepatol. 10: 352–361.

Jordan, H.E. and C.C. Speidel. 1923. Leucocytes in relation to the mechanism of thyroid-accelerated metamorphosis in the larval frog. Exp. Biol. Med. 20: 380–383.

Jordan H.E. and C.C. Speidel. 1923. Blood cell formation and distribution in relation to the mechanism of thyroid-accelerated metamorphosis in the larval frog. J. Exp. Med. 38: 529–543.

Jordan, H.E. and C.C. Speidel. 1924. The behavior of the leucocytes during coincident regeneration and thyroid-induced metamorphosis in the frog larva, with a consideration of growth factors. J. Exp. Med. 40: 1–11.

Junqueira, L.C. and J. Carneiro. 2004. Célula do sangue. pp. 224–237. *In*: Junqueira, L.C. and J. Carneiro [eds.]. Histologia básica. 10 ed. Rio de Janeiro: Guanabara Koogan.

Kau, C.L. and J.B. Turpen. 1983. Dual contribution of embryonic ventral blood island and dorsal lateral plate mesoderm during ontogeny of hematopoietic cells in *Xenopus laevis*. J. Immunol. 131: 2262–2269.

Kim, Y.S. and S.B. Ho. 2010. Intestinal goblet cells and mucins in health and disease: Recent insights and progress. Curr. Gastroenterol. Rep. 12: 319–330.

Klion, A.D. and T.B. Nutman. 2004. The role of eosinophils in host defense against helminth parasites. J. Allergy Clin. Immunol. 113: 30–37.

Koprivnikar, J., M.R. Forbes and R.L. Baker. 2009. Contaminant effects on host–parasite interactions: Atrazine, frogs, and trematodes. Environ. Toxicol. Chem. 26: 2166–2170.

Langerveld, A.J., R. Celestine, R. Zaya, D. Mihalko and C.F. Ide. 2009. Chronic exposure to high levels of atrazine alters expression of genes that regulate immune and growth-related functions in developing *Xenopus laevis* tadpoles. Environ. Res. 109: 379–389.

Manasee, W.A., T. Weerathunga and G. Rajapaksa. 2020. The impact of elevated temperature and CO2 on growth, physiological and immune responses of *Polypedates cruciger* (common hourglass tree frog). Front. Zool. 17: 1–25.

Marr, S., H. Morales, A. Bottaro, M. Cooper, M. Flajnik and J. Robert. 2007. Localization and differential expression of activation-induced cytidine deaminase in the amphibian xenopus upon antigen stimulation and during early development. J. Immunol. 179: 6783–6789.

McMahon, T.A., N.T. Halstead, S. Johnson, T.R. Raffel, J.M. Romansic, P.W. Crumrine, R.K. Boughton, L.B. Martin and J.R. Rohr. 2011. The fungicide chlorothalonil is nonlinearly associated with corticosterone levels, immunity, and mortality in amphibians. Environ. Health Perspect. 119: 1098–1103.

Murphy, K.P. 2012. Janeway's Immunobiology, 8th, New York, NY: Garland Science.

Nieuwkoop, P.D. and J. Faber. 1967. Normal tables of *Xenopus laevis* (Daudin) Amsterdam: North-Holland.

Oliveira, C.R., T.D. Garcia, L. Franco-Belussi, R.F. Salla, B.F.S. Souza, N.F.S. De Melo, S.P. Irazusta, M. Jones-Costa, E.C.M. Silva-Zacarin and L.F. Fraceto. 2019. Pyrethrum extract encapsulated in nanoparticles: Toxicity studies based on genotoxic and hematological effects in bullfrog tadpoles. Environ. Pollut. 253: 1009–1020.

Pasquier, L.D., J. Schwager and M.F. Flajnik. 1989. The immune system of xenopus. Annu. Rev. Immunol. 7: 251–275.

Perf Weerathunga, W.A.M.T. and G. Rajapaksa. 2020. The impact of elevated temperature and CO2 on growth, physiological and immune responses of *Polypedates cruciger* (Common hourglass tree frog). Front. Zool. 17: 1–25.

Rayl, J.M. and M.C. Allender. 2020. Temperature affects the host hematological and cytokine response following experimental ranavirus infection in red-eared sliders (Trachemys scripta elegans). PLOS ONE 15: 1–15.

Robert, J. and Y. Ohta. 2009. Comparative and developmental study of the immune system in Xenopus. Dev. Dyn. 238: 1249–1270.

Robert, J., C.C. McGuire, F. Kim, S.C. Nagel, S.J. Price, B.P. Lawrence and F. De Jesús Andino. 2018. Water contaminants associated with unconventional oil and gas extraction cause immunotoxicity to amphibian tadpoles. Toxicol. Sc. 166: 39–50.

Robert, J., C.C. McGuire, S. Nagel, B.P. Lawrence and F.D.J. Andino. 2019. Developmental exposure to chemicals associated with unconventional oil and gas extraction alters immune homeostasis and viral immunity of the amphibian Xenopus. Sci. Total Environ. 671: 644–654.

Roberts, R.J. 1975. Melanin-containing cells of the teleost fish and their relation to disease. pp. 399–428. *In*: Ribelin, W.E. and G. Migaki [eds.]. The Pathology of Fishes. University of Wisconsin Press, Madison, WI.

Robinson, S.A., R.J. Chlebak, S.D. Young, R.L. Dalton, M.J. Gavel, R.S. Prosser, A.J. Bartlett and S.R. de Solla. 2021. Clothianidin alters leukocyte profiles and elevates measures of oxidative stress in tadpoles of the amphibian, Rana pipiens. Environ. Pollut. 284: 1–9.

Rollins-Smith, L.A., S.C. Parsons and N. Cohen. 1984. During frog ontogeny, PHA and Con A responsiveness of splenocytes precedes that of thymocytes. Immunology 52: 491–500.

Rollins-Smith, L.A., K.S. Barker and A.T. Davis. 1997. Involvement of glucocorticoids in the reorganization of the amphibian immune system at metamorphosis. Dev. Immunol. 5: 145–152.

Rollins-Smith, L.A. 1998. Metamorphosis and the amphibian immune system. Immunological Reviews 166: 221–230.

Rollins-Smith, L.A. and D.C. Woodhams. 2011. Amphibian immunity—Staying in tune with the environment. pp. 92–119. *In*: Demas, G.E. and R.J. Nelson [eds.]. Ecoimmunology. Oxford University Press, New York.

Rollins-Smith, L.A., J.P. Ramsey, J.D. Pask, L.K. Reinert and D.C. Woodhams. 2011. Amphibian immune defenses against chytridiomycosis: Impacts of changing environments. Integr. Comp. Biol. 51: 552–562.

Rollins-Smith, L.A. 2017. Amphibian immunity–stress, disease, and climate change. Dev. Comp. Immunol. 66: 111–119.

Rothenberg, M.E. and S.P. Hogan. 2006. The eosinophil. Annu. Rev. Immunol. 24: 147–174.

Rutkoski, C.F., S.C. Grott, N.C. Israel, F.E. Carneiro, F. de Campos Guerreiro, S. Santos, P.A. Horn, A.A. Trentini, E. Barbosa da Silva, C.A. Coelho de Albuquerque, T.C. Alves and E.A. Almeida. 2022. Hepatic and blood alterations in *Lithobates catesbeianus* tadpoles exposed to sulfamethoxazole and oxytetracycline. Chemosphere 307: 1–11.

Rymuszka, A. and L. Adaszek. 2013. Cytotoxic effects and changes in cytokine gene expression induced by microcystin-containing extract in fish immune cells—An *in vitro* and *in vivo* study. Fish Shellfish Immunol. 34: 1524–1532.

Scapigliati, G., F. Buonocore and M. Mazzini. 2006. Biological activity of cytokines: An evolutionary perspective. Curr. Pharm. Des. 12: 3071–3081.

Shu, Y., H. Jiang, C.N.T. Yuen, W. Wang, J. He, H. Zhang, G. Liu, L. Wei, L. Chen, H. Wu. 2022. Microcystin-leucine arginine induces skin barrier damage and reduces resistance to pathogenic bacteria in *Lithobates catesbeianus* tadpoles. Ecotoxicol. Environ. Saf. 238: 1–10.

Shutler, D., T.G. Smith and S.R. Robinson. 2009. Relationships between leukocytes and *Hepatozoon* spp. In green frogs, rana clamitans. J. Wildl. Dis. 45: 67–72.

Siano, A., M.V. Húmpola, E. de Oliveira, F. Albericio, A.C. Simonetta, R. Lajmanovich and G.G. Tonarelli. 2014. Antimicrobial peptides from skin secretions of *Hypsiboas pulchellus* (Anura: Hylidae). J. Nat. Prod. 77: 831–841.

Sifkarovski, J., L. Grayfer, F. De Jesús Andino, B.P. Lawrence and J. Robert. 2014. Negative effects of low dose atrazine exposure on the development of effective immunity to FV3 in *Xenopus laevis*. Dev. Comp. Immunol. 47: 52–58.

Silva, M.B. da, R.E. Fraga, P.B. Nishiyama, I.S.S. Silva, N.L.B. Costa, L.A.A. de Oliveira, M.A. Rocha and F.A. Juncá. 2020. Leukocyte profiles in odontophrynus carvalhoi (Amphibia: Odontophrynidae) tadpoles exposed to organophosphate chlorpyrifos pesticides. Water Air Soil Pollut. 231: 1–11.

Steinel, N.C. and D.I. Bolnick. 2017. Melanomacrophage centers as a histological indicator of immune function in fish and other poikilotherms. Front. Immunol. 8: 1–8.

Tochinai, S. 1980. Direct observation of cell migration into *Xenopus thymus* rudiments through mesenchyme. Dev. Comp. Immunol. 4: 273–282.

Yaparla, A., D.V. Koubourli, E.S. Wendel and L. Grayfer. 2017. Immune system organs of amphibians. Reference Module in Life Sciences. 1–7.

9

Pollutant Effects on Tadpole's Microbiota

Gustavo Henrique Pereira Gonçalves and *Eduardo Alves de Almeida**

1. Introduction

Symbiosis between bacteria and multicellular organisms play a pivotal role in the evolution, health, and physiology of animals. Symbiotic bacterial communities of animals range from autochthonous (or resident) groups adapted to the host, and allochthonous (or transient) groups that are generalists and depend on extrinsic factors for their colonization (Derrien and van Hylckama Vlieg 2015). The "microbiota" comprises of a set of microorganisms in a given environment, while the "microbiome" is the set of genes in these organisms. Notably, the term "holobiont" was assigned to the unit formed by the multicellular eukaryote and its persistent symbiotic colonies (Gilbert et al. 2012, Alberdi et al. 2016). Therefore, it is important to understand the evolutionary aspects related to the development, physiology, metabolism, and immunology of the holobiont complex.

The application of high-throughput DNA sequencing methodology (next-generation sequencing—NGS) combined with the known "omics" techniques such as genomics, metagenomics, transcriptomics, proteomics, metabolomics and epigenomics, has potentiated studies on the microbiome constitution and its functional roles in animals (Derrien and van Hylckama Vlieg 2015, Rebollar et al. 2016a, Thambirajah et al. 2019).

Currently, studies regarding taxonomic microbiota profiles obtained by culture-dependent analysis use classical biochemical tests combined with robust tools for the identification of specific macromolecules, such as matrix-assisted laser desorption/ionization-time-of-flight (MALDI-TOF) (Boccioni et al. 2021). Culture-independent techniques have higher screening power, with most research focusing on16S rRNA gene, an excellent marker of bacterial taxa that allows phylogenetic and functional analysis of the host microbiome (Bahrndorff et al. 2016, Edgar 2018). Microbiome characterization studies present data as operational taxonomic units (OTUs) or amplicon sequence

Centro de Estudos em Toxicologia Aquática, Universidade Regional de Blumenau, Rua São Paulo, 3366, Bloco Q, setor Q101, Blumenau, SC, 9030-000, Brazil. Email ghpgoncalves@furb.br

* Corresponding author: eduardoalves@furb.br

variants (ASVs), which are representations of bacterial taxa from different taxonomic levels (Callahan et al. 2017, Schloss 2021).

The increased interest in characterizing the microbiome in wild animals emerge from evidence of the intimate relationship between host health and the evolutionary processes resulting from this bacterial symbiosis, as well as the ability of the microbiota to adapt to adverse environmental conditions (Amato 2013). Amphibians are animals that have complex bacterial communities inhabiting their bodies, especially in the gut and skin. This group have marked stages throughout their lifecycles, from embryos to larval phase, until reaching the adult stage (see Chapter 2). In particular, amphibians experience natural morphophysiological changes during metamorphosis and drastic changes in their living conditions and behavior (Kohl et al. 2013, Prest et al. 2018). Thus, they present themselves as excellent candidate models for investigations on microbiota plasticity.

Amphibians represent the most threatened group of vertebrates in the world, mainly due to anthropogenic pressures and emerging diseases, which may be directly associated with an imbalance of their mucosal surfaces' microbiota (dysbiosis) (Jiménez and Sommer 2017, Rebollar et al. 2016a, Scheele et al. 2019). Currently, chemical pollution threatens global health, evidencing the importance of ecotoxicological research (Evariste et al. 2019). In this context, the effects of contaminants on the symbiotic microbiota of organisms, e.g., the characterization of alterations in bacterial taxa through metagenomics methods, helps to clarify the causes of dysbiosis and its negative impacts on amphibians' health at early stages (Fackelmann and Sommer 2019, Venâncio et al. 2022).

In this chapter, we will focus on the main aspects involving the complex microbiota of amphibians, with emphasis on the larval stage (tadpoles). Emerging issues on factors that affect the natural microbiome of these animals, such as the dependence of the developmental stage, pollutant exposure, influence of experimental conditions during investigation (captivity × laboratory × wildlife), as well as the best strategies for identifying microbiome stressors, are also discussed. The importance of the microbial community in the evolutionary processes and health of amphibians also permeate the issues discussed throughout the chapter. Initially, we will examine the natural modifiers of the tadpole microbiome, focusing on the particularities of bacterial colonization, especially in the intestinal and cutaneous mucosa (Fig. 1). The influence of biotic and abiotic factors on the establishment of symbiotic microbiomes, including genetic characteristics of the host animal, habitat, seasonality, temperature, and diet is then addressed. Finally, information available on anthropogenic microbiota disrupting chemicals (MDCs) are explored, as well as the causes of dysbiosis in the gut and skin of these animals. The application of omics tools for identifying the effects of MDCs and for conservation strategies for highly endangered amphibians has also been included in the text.

2. Importance of the Symbiotic Microbiota for Tadpoles

During evolution, interaction between animals and their resident microbiota were reflected in their genomes. Several animal genes are homologs to bacterial genes, mainly derived by descent, but occasionally by horizontal transfer from bacterial genes (Keeling and Palmer 2008, McFall-Ngai et al. 2013). The sharing of these genes ranges from encoding central pathways of metabolism to cell signaling mechanisms, including hormonal and short-chain fatty acid receptors (Hughes and Sperandio 2008, Ang et al. 2018). Since bacterial repertoire is very diverse, this intertwining genome potentiate the metabolic versatility of the animals (Lapierre and Gogarten 2009).

Symbiosis between animals and bacteria has shaped the evolutionary success of animals in adapting to new environments (Alberdi et al. 2016, Shapira 2016). In combination with the theory of endosymbiosis that gave rise to mitochondria, with better energy yield for multicellular organisms, the establishment of symbiosis with external bacteria was also fundamental for speciation processes, enhancing the ability of animals to withstand biotic and abiotic pressures. Interestingly, the microbiota is also involved in cell signaling during animals' development (McFall-Ngai et al. 2013).

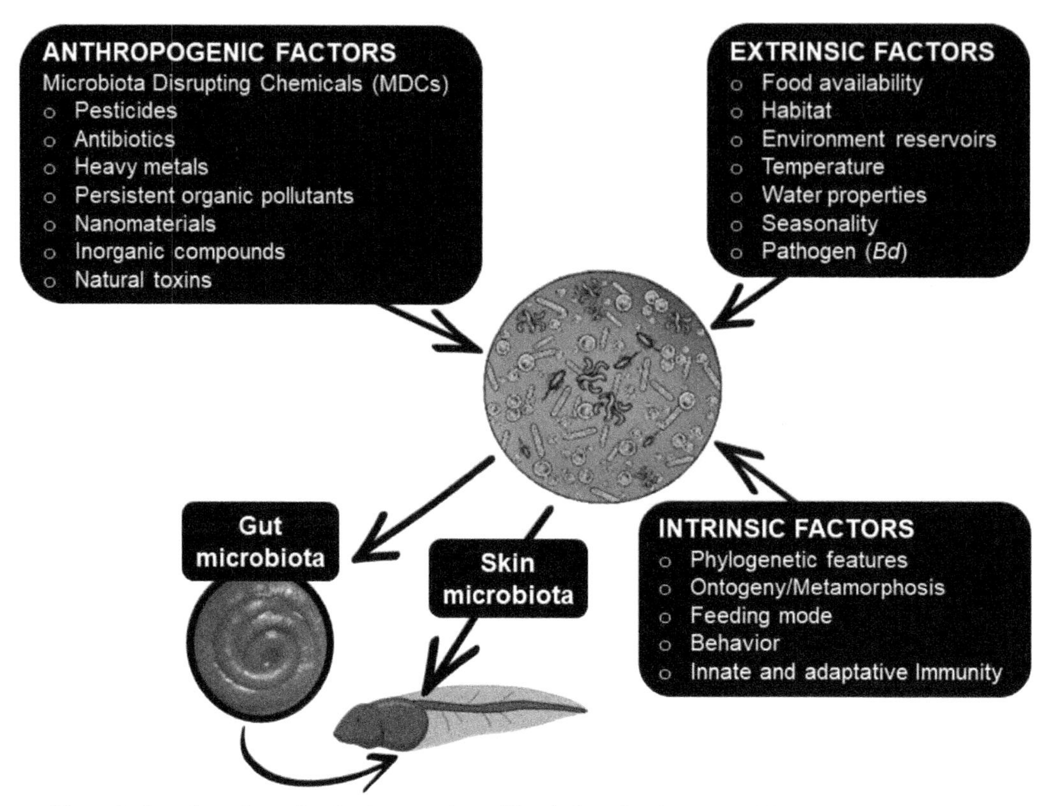

Figure 1. Overview of natural and anthropogenic modifiers influencing the gut and skin microbiome of tadpoles.

Biphasic life of amphibians stands out among vertebrates, especially regarding animal development (Lyra et al. 2018). Metamorphosis in amphibians is induced by thyroid hormones in synergy with corticosteroids, growth hormone and prolactin (Denver et al. 2002, Kulkarni and Buchholz 2014, Kikuyama et al. 2021), which cause profound changes in morphology, physiology, metabolism, and behavior as tadpoles progress through their development (Tata 2006, Sterner and Buchholz 2022). These metamorphic processes involve not only the emergence of hind and fore limbs, tail and gills resorption, but also changes in energy and structural metabolism to maintain the life of juveniles and adults outside the water (Paris and Laudet 2008).

Ontogenetic transformations of the gastrointestinal system fostered by metamorphosis induce morphological changes on stomach and gut, as well as in the intestinal bacterial diversity (Vences et al. 2016, Chai et al. 2018, Fabrezi and Cruz 2020). Tadpoles are usually herbivorous or omnivorous, although feeding strategies are extremely varied, including microcarnivory (detritus, predation and cannibalism), oophagy, coprophagy, and filter-feeding habits (Steinwascher 1978, Pryor and Bjorndal 2005a, Altig et al. 2007, Pryor 2014, Fabrezi and Cruz 2020). During larval phase, they breathe through the gills and excrete their nitrogen compounds mainly as ammonia (ammonotelic) (McIndoe and Smith 1984, Chen and Atkinson 1997, Ultsch et al. 1999). While adults, amphibians become exclusively carnivorous, are ureotelic (excrete nitrogen as urea) and have a pulmonary respiratory system (Atkinson et al. 1998, Sterner et al. 2020, Méndez-Narváez and Warkentin 2022). All these physiological changes can imply significant alterations in their symbiotic microbiota.

Environment can also influence the microbiota composition in both tadpoles and adults. Tadpoles have symbiotic communities similar to fish, while adults with more terrestrial habits have the microbiota similar to amniotes (Kohl et al. 2013). Although dietary changes between tadpoles and adults alter macronutrient contents and consequently the symbiotic microbiota (Knutie et al. 2017a), two disturbance events during amphibian development drive the main composition of bacterial communities: egg hatching and metamorphosis (Prest et al. 2018). Amphibian metamorphosis is closely related to the composition of symbiotic bacterial community, which in turn is involved in signaling mechanisms during development (McFall-Ngai et al. 2013). Thus, an interruption in the microbiota restructuring process during metamorphosis can be fatal for these animals (Knutie et al. 2017b, Jiménez and Sommer 2017, Long et al. 2020).

Drastic physiological and morphological changes during metamorphosis regulate and impact the bacterial community of the skin and gut, the main targets of amphibian microbiota studies. Metagenomic approaches can be employed to elucidate the functional transition of the microbiome during metamorphosis (Vences et al. 2016). While the intestinal microbiota plays a pivotal role in the health and energy metabolism of the host, bacterial skin community in amphibians is a key factor for their survival, as it is associated with immunity and metabolism (Bletz et al. 2016, Bernardo-Cravo et al. 2020, Song et al. 2021, Jiménez et al. 2022). Some bacterial taxa are found both in the gut and skin in amphibians (Vences et al. 2016, Tong et al. 2020).

2.1 Gut Microbiome

Bacterial community from the intestinal mucosa in amphibians have phenotypic plasticity, with taxonomic fluctuations that follow modifications of the gastrointestinal tract during metamorphosis (Fedewa et al. 2006, Chai et al. 2018, Yang et al. 2022). For example, there is a significant reduction in the intestinal length during the metamorphosis climax. In tadpoles, the gut microbiota assists the nutrient digestion and can influence the immune system, consequently affecting their ability to adapt to environmental changes (Colombo et al. 2015, Alberdi et al. 2016, Knutie et al. 2017b, Bernardo-Cravo et al. 2020, Fontaine and Kohl 2020). However, its composition is dynamic and may vary due to intrinsic and extrinsic factors such as diet, food availability, habitat and temperature (Benson et al. 2010, Wu et al. 2011, Kohl et al. 2014, Bletz et al. 2016, Kohl and Yahn 2016, Tong et al. 2019, Wang et al. 2021a, Fontaine et al. 2022).

Although the tadpole gut is a simple tubular structure, composed of a single layer of epithelium, its microbiota has greater diversity compared to adults, where a more complex gut is observed (Shi and Ishizuya-Oka 1996, Heimeier et al. 2010, Kohl et al. 2013, Shi et al. 2021). As omnivorous/herbivorous tadpoles develop into carnivorous adults, changes in intestinal bacterial taxa are expected (Vences et al. 2016). This natural restructuring of the gut microbiota during metamorphosis is essential for amphibian nutrition and health (Kohl et al. 2013, Knutie et al. 2017a).

Among the environmental conditions that can alter gut microbiota, food availability may shape and determine the composition of the central gut microbiome (Chang et al. 2016, Knutie et al. 2017a), once syntopic animals inhabiting similar areas and with same feeding mode were shown to present similar intestinal bacterial communities (Lyra et al. 2018). In addition, several gut bacterial taxa are shared among tadpoles from different continents, but absent in adults (Vences et al. 2016), which suggest a key role of these microorganisms only during the larval phase of amphibians.

The understanding of the functional pattern of the central bacteriome of tadpoles and their enzymatic specializations according to the dietary preferences of the hosts are fundamental to clarify the hypotheses that the composition of gut microbiome is dependent on diet and tadpole capacity on using available food. Products of symbiotic fermentation are formed from metabolic reactions of the gut bacterial community, which is influenced by food substrates provided by the host (Bletz et al. 2016, Chen et al. 2022). In this sense, we can suppose that (i) algal eaters can be highly colonized by agarases-encoding bacteria and (ii) insect-eating microcarnivorous tadpoles have high prevalence of chitinolytic bacteria (Hehemann et al. 2010, Delsuc et al. 2014, Vences et al. 2016). Tadpoles with coprophagic habits can also increase colonization by bacteria specializing in nutrients of complex degradation (Pryor and Bjorndal 2005b, Pryor 2014).

During the larval stage, intestine in amphibians is a large structure and contains a robust symbiotic bacterial community that drive fermentation of complex carbohydrates into short-chain fatty acids (SCFAs) (Pryor and Bjorndal 2005a, Flint et al. 2012). Symbiotic fermentation provides approximately 20% of daily energy requirements in tadpoles through SCFAs, mainly acetate, butyrate and propionate, the latter precursor of glucose from the gluconeogenic pathway (Pryor and Bjorndal 2005b, Nicholson et al. 2012, Bhagavan and Ha 2015). Therefore, alterations on the intestinal microbiota can cause dysbiosis, inducing an imbalance in SCFAs concentrations, and consequently a disorder in energy metabolism of tadpoles (Huang et al. 2021, Shen et al. 2022).

Despite the evident influence of environmental factors on gut microbiota composition, tadpoles may present specific bacterial community patterns according to the host species (Vences et al. 2016). Phylosymbiosis, i.e., host-specific microbial signatures, appears to flirt with the structural features of the tadpoles' intestinal bacterial communities (Brooks et al. 2016). The increased interest in characterizing intestinal bacterial profiles of amphibian larvae is increasingly contributing to the construction of a central bacteriome of different species, also allowing to understand the relationships between bacterial colonization and physiological characteristics of distinct species.

The main intestinal bacterial phyla of tadpoles are Proteobacteria, Firmicutes, Bacteroidetes and Fusobacteria (Kohl et al. 2015, Vences et al. 2016, Chai et al. 2018, Liu et al. 2020, Xie et al. 2020, Evariste et al. 2021, Gutierrez-Villagomez et al. 2021, Yang et al. 2022). Similar characteristics are also found in the intestinal microbiota of teleost fish (Sullam et al. 2012). In addition, the phylum Acidobacteria seems to be related only to larval stages of amphibians, since it was undetectable in adults (Kohl et al. 2013). Establishing the profile of the central bacteriome that are common among tadpole species is essential for health, conservation, and for understanding evolutionary processes in amphibians. A profile of the gut microbiota in tadpoles is then necessary to elucidate the factors that cause dysbiosis (Jiménez and Sommer 2017, Rosenfeld 2017).

2.2 Skin Microbiome

Skin microbiota of tadpoles can effectively protect animals against pathogens or environmental disturbances to which they are exposed (Grice and Segre 2011, Bataille et al. 2018, Chapman et al. 2022). Mucosome represents the symbiotic microbiome of the mucus present on amphibians' skin that are essential for both host health and vulnerability to infections (Woodhams et al. 2014, Rebollar et al. 2016b, Sanchez et al. 2017, Bates et al. 2018, Jiménez et al. 2022).

As an important component of the innate immune system of amphibians, the cutaneous bacterial community is dynamic (Davis et al. 2017, Jiménez et al. 2019) and strongly influenced by intrinsic factors of each species, which include ontogenetic characteristics, body surface areas and developmental stages (McKenzie et al. 2012, Walke et al. 2014, Kueneman et al. 2016a, Prado-Irwin et al. 2017, Sanchez et al. 2017, Griffiths et al. 2018, Belasen et al. 2021). Abiotic factors such as seasonality, temperature and habitat also directly influence the skin microbiome (Belden et al. 2015, Costa et al. 2016, Kueneman et al. 2016b, Sabino-Pinto et al. 2017, Longo and Zamudio 2017, Edwards et al. 2017, Varela et al. 2018, Kueneman et al. 2019, Ruthsatz et al. 2020, Jervis et al. 2021, Piccinni et al. 2021).

Metamorphosis can cause profound changes in bacterial skin community of amphibians (Kueneman et al. 2014, Sanchez et al. 2017), which are closely related to variations in their susceptibility to skin infections (Bataille et al. 2018). Tadpoles apparently have less diversity in their skin microbiota than adults (Bresciano et al. 2015, Sabino-Pinto et al. 2017). In contrast, they exhibit a rich bacterial community that inhibits pathogens, including the fungus *Batrachochytrium dendrobatidis* (*Bd*) that commonly infects amphibians' skin and is involved in global population declines of these animals (Jani and Briggs 2014, Kueneman et al. 2014, Garner et al. 2016, Kueneman et al. 2017, Sabino-Pinto et al. 2017, Scheele et al. 2019).

Life stage may be a major factor to *Bd* susceptibility in some amphibian species. Post-metamorphic tadpoles of *Anaxyrus boreas* have a microbiome poor in *Bd*-inhibitory bacteria and, consequently, higher vulnerability to chytridiomycosis infection in relation to pre-metamorphic individuals (Kueneman et al. 2016a). In contrary, a study with the species *Lithobates vibicarius* observed that changes in bacterial diversity naturally changed by metamorphosis did not influence taxa with putative anti-*Bd* capacity in pre- and post-metamorphic tadpoles (Jiménez et al. 2019). Therefore, these data suggest that anti-*Bd*-linked microbial skin signatures do not only depend on pre-metamorphic stages, but also on intrinsic characteristics of the species. In addition, as the metamorphosis progresses, the amphibian epithelium becomes more complex and has higher levels of keratin, which may allow greater susceptibility to fungal infection. This can be evidenced by the higher frequency of amphibian mortality by chytridiomycosis infections after metamorphosis (Bosch et al. 2001, Jani and Briggs 2014, Piovia-Scott et al. 2015).

It is unclear whether the natural dynamics of chytridiomycosis directly affect tadpole skin microbiome. Although post-metamorphic tadpoles present changes in cutaneous bacterial community due to *Bd* infection (Jani and Briggs 2014, Bataille et al. 2018, Jani and Briggs 2018), in several species there are no such disorders in the skin microbiome (Kruger 2020). Thus, the host genotype plays a major role in the maintenance and colonization of the bacterial community, which in turn can influence the susceptibility to chytridiomycosis infection (Griffiths et al. 2018).

The ability to inhibit *Bd* infections is closely related to antifungal secondary metabolites produced by a group of approximately 10 bacterial genera present in amphibians' skin (Harris et al. 2006, Becker and Harris 2010, Becker et al. 2015). In the mucosa, the set of potentially anti-*Bd* bacterial taxa can potentiate the production of secondary metabolites and provide greater synergistic protection against chytridiomycosis (Loudon et al. 2014a, Davis et al. 2017). Combinations of different bacterial taxa are often found in the skin microbiome of pre-metamorphic tadpoles (Kueneman et al. 2016a), which are strongly influenced by environmental factors and dependent on the bacteria present in the aquatic environment (Colombo et al. 2015). Bacterial community of the skin is not only affected by the intrinsic characteristics of the host, but also by the reservoirs of environmental bacteria in the habitats of these animals (Loudon et al. 2014b, Kueneman et al. 2016b, Loudon et al. 2016, Longo and Zamudio 2017, Sanchez et al. 2017, Bates et al. 2018, Jervis et al. 2021, Piccinni et al. 2021).

Determining the microbiome composition primarily in the larval stage can assist in amphibian conservation efforts against infectious diseases (Griffiths et al. 2018). In tadpoles of several species, the most abundant phyla of skin bacteria are Proteobacteria, Actinobacteria and Bacteroidetes (McKenzie et al. 2012, Kueneman et al. 2014, Kueneman et al. 2016a, Vences et al. 2016, Sanchez et al. 2017, Bataille et al. 2018, Jiménez et al. 2019). The antifungal properties are phylogenetically widespread among the bacterial taxa of the amphibian mucous membranes (Lauer et al. 2007, Becker et al. 2015, Woodhams et al. 2015, Bernardo-Cravo et al. 2020). The bacterial species *Janthinobacterium lividum*, found naturally in amphibians, is likely the most widespread taxon associated to production of antifungal metabolites, including the capacity of acting as a probiotic in chytridiomycosis infected animals (Brucker et al. 2008, Becker et al. 2009, Harris et al. 2009, Kueneman et al. 2016b, Rebollar et al. 2016c, Rubio et al. 2018, Knapp et al. 2022). Other genera found in tadpoles' skin, such as *Lysobacter*, *Stenotrophomonas*, *Aeromonas*, *Pseudomonas*, *Chryseobacterium* and *Bacillus*, also have high capacity to inhibit *Bd* infection (Kueneman et al. 2014, Becker et al. 2015, Bresciano et al. 2015, Woodhams et al. 2015, Kueneman et al. 2016a, Jiménez et al. 2019).

About 53% of the skin bacteria in amphibians are also found in the gut (Vences et al. 2016). This occurs because adults of several species of amphibians ingest their skin after episodes of peeling, conferring a complex cycle between the skin and gut microbiome (the importance of skin microbiota establishment during tadpole phase to this transference in adult phase remains yet to be studied). Although these mucous membranes provide different metabolic contexts for the microbiota (e.g., the skin is exposed to O_2 and the intestine is poor in O_2), some facultative bacterial taxa manage to survive in both tissues (Wiggins et al. 2011, Tong et al. 2020). In this sense, intestine can act as a reservoir of skin-protective bacteria, as this tissue shares some bacterial groups with the skin (Vences et al. 2016, Tong et al. 2020). The comprehension of relationships between the skin microbiota and tadpoles' survival is important to understand indirect negative impacts of anthropogenic stressors in these organisms, since any disturbances in the microbiota can lead to imbalances on tadpole physiology (Kueneman et al. 2016a, Kueneman et al. 2017, Hernández-Gómez et al. 2020).

3. Influence of Environmental Factors on Amphibians' Microbiota

Structure of symbiotic bacterial community is shaped and influenced by the environment of the host (Krynak et al. 2016, Kueneman et al. 2016b, Jiménez and Sommer 2017, Greenspan et al. 2019, Kueneman et al. 2019). Symbiont microbiota is primordial for ecological acclimatization of amphibians to new environments, mainly due to immune functions and action of its enzymes in nutrient uptake (Krynak et al. 2015, Bletz et al. 2016, Bernardo-Cravo et al. 2020, Ujszegi et al. 2020).

Composition of amphibian gut microbiome is influenced by diet content, which is closely related to food availability, nutrient diversity in food webs, and species distribution in the ecosystems (Chang et al. 2016). The environmental bacterial community helps to model the skin microbiome with greater inhibitory capacity against pathogens, providing a primary protection for these animals, mainly in tadpoles (Jani and Briggs 2018, Jiménez et al. 2022).

Some bacterial taxa are able to metabolize natural toxins produced by amphibians into non-toxic compounds (Kamalakkannan et al. 2017, Ujszegi et al. 2020). In this case, amphibians colonized by these bacteria would "lose" their natural toxin protection against predators (Kamalakkannan et al. 2017). This seems to be especially relevant for non-native species inhabiting new habitats, hypothesizing that environmental bacterial communities may increase vulnerability of invasive animal species (Santos et al. 2021). In contrast, environmental influence on the microbiome may allow essential physiological adaptation in these animals, as found in hibernating adult amphibians that have a wealth of intestinal bacteria with ureolytic capacity. Bacterial hydrolysis of urea in a urea-nitrogen recycling system for the hosts is a valuable adaptation mechanism for amphibians in their environments (Wiebler et al. 2018).

Tadpole microbiome is sensitive to changes in water properties, and as it is involved in immunity and metabolism of the animals, the environmental quality is essential to maintain microbial homeostasis. While microbiome composition in adult amphibians seems to be more dependent on physiological characteristics of the animal, symbiotic bacterial community in tadpoles are more general and variable, with higher influence of environmental conditions (Piccinni et al. 2021). Therefore, building a protective symbiotic microbiota in tadpoles is strongly dependent on the health of the ecosystem (Greenspan et al. 2019, Ujszegi et al. 2020).

Amphibians are experiencing population declines worldwide, and the main associated causes include climate change and exposure to pathogens and chemicals from anthropic origins, as well as the synergistic effects of these factors (Davidson et al. 2007, Hayes et al. 2010, Koprivnikar 2010, Costa et al. 2016, Jiménez and Sommer 2017, Boccioni et al. 2021). In addition to the direct effects on microbiome, chemical compounds, which are well known to alter ecosystems quality and consequently the availability of nutrients, indirectly affect diet and intestinal microbiome of amphibians (Relyea et al. 2005, Chang et al. 2016, Huang et al. 2018, Venâncio et al. 2022). Therefore, mitigation of impacts due to environmental stressors on the microbiome of amphibians, especially in the early stages of life, is essential for conservation measures for these animals (Rohr et al. 2013, Krynak et al. 2015, Costa et al. 2016, Rohr et al. 2017, Hernández-Gómez et al. 2020).

4. Microbiota Disrupting Chemicals (MDCs) in Tadpoles

Contaminants of emerging concern such as pharmaceuticals, personal care products, veterinary drugs (antibiotics, anti-fungals, and hormones), endocrine-disrupting chemicals, persistent organic pollutants, pesticides, and nanomaterials are constantly released into the environment (Ankley et al. 2008, Kuzmanović et al. 2015, Alves et al. 2017, Dulio et al. 2018, Fackelmann and Sommer 2019, Patterson et al. 2022). Metals and other emerging natural toxins also cause environmental issues of global concern (Linzey et al. 2003, Paerl 2008, Todd et al. 2011, Aiman et al. 2016, Zhang et al. 2016, Girotto et al. 2020, Pierozan et al. 2020, Shu et al. 2022). Amphibians inhabit both aquatic and terrestrial habitats during their life, so they are doubly exposed to contaminants from these environments during their development (Hayes et al. 2010, Todd et al. 2011, Wagner et al. 2013, Dornelles and Oliveira 2016, Hughey et al. 2016, Loutfy and Kamel 2018, Boccioni et al. 2021). Chemical pollution can affect the microbiota in many ways, thus making species more vulnerable to emerging diseases (Rosenfeld 2017, McCoy and Peralta 2018, Evariste et al. 2019). So far, few studies have investigated the interaction between pollutants, diseases, and microbiome in amphibians (McCoy and Peralta 2018). Accordingly, the evaluation of the effects of chemical pollutants on microbiota have been considered as an essential tool for emerging ecotoxicological studies for amphibians (Hughey et al. 2016, Jiménez and Sommer 2017, Evariste et al. 2019).

Xenobiotics that modify the composition or metabolic function of the symbiotic bacterial community are categorized as microbiota disrupting chemicals (MDCs). MDCs usually change bacterial communities increasing animal susceptibility to diseases (Aguilera et al. 2020). Unlike other natural modifiers of amphibian microbiota, exposure to chemical pollutants can decrease the abundance of potentially chytridiomycosis-inhibiting skin bacteria, even at low concentrations (Jiménez et al. 2021). The effects of these contaminants, eventually and depending on the tadpole species, can affect both the skin and gut microbiome, decreasing immunity and increasing susceptibility to infectious diseases (Gutierrez-Villagomez et al. 2021, Jiménez et al. 2021, Knapp et al. 2022). In this scenario, MDCs are important to be identified in environmental monitoring studies on amphibians. Furthermore, approaches on the combination of different environmental stressors are essential to elucidate the synergistic effects of contaminants with amphibian microbiome, as well as studies regarding effects of environmentally realistic concentrations of MDCs (Kuzmanović et al. 2015, Boccioni et al. 2021, Alves et al. 2022, Brack et al. 2022, Finckh et al. 2022, Fontes et al. 2022).

Some recent studies have included the assessment of the potential impacts that environmental and anthropogenic factors can have on the intestinal and skin microbiome of amphibians (Krynak et al. 2016, Zhang et al. 2016, McCoy and Peralta 2018, Jiménez et al. 2020). However, information on this topic is still scarce, especially for larval stage (Knutie et al. 2018, Gust et al. 2021, Shen et al. 2022). Tadpoles have exclusively aquatic habits, so they are directly affected by a range of chemicals that can occur at different aquatic environments (Relyea et al. 2005, Gao et al. 2012, Girotto et al. 2020, da Costa Araujo et al. 2021, Patterson et al. 2022). Existing studies on the toxic effects of MDCs on amphibian larvae to date will be discussed in the next topics and are summarized in Table 1.

4.1 Pesticides

Agricultural activities affect health of wildlife, particularly organisms from aquatic ecosystems (Rohr et al. 2013, Chang et al. 2016, Shuman-Goodier et al. 2017, Boccioni et al. 2021, Finckh et al. 2022). Pesticides can trigger a series of direct or indirect negative effects in amphibians, resulting in increased infections by trematodes or chytridiomycosis (Davidson et al. 2007, Rohr et al. 2008, Rohr and McCoy 2010, Koprivnikar 2010, Rohr et al. 2017), metabolic disorder (Boone et al. 2013, Dornelles and Oliveira 2016, Loutfy and Kamel 2018, Viriato et al. 2021) and immunosuppression (Forson and Storfer 2006, Langerveld et al. 2009), potentiated by disruption of the protective microbiota. Pesticides also induce significant changes in the composition, diversity, and richness of the cutaneous and intestinal bacterial community of tadpoles, including the appearance or suppression of bacterial taxa indicative of dysbiosis (Krynak et al. 2017, Boccioni et al. 2021, Gutierrez-Villagomez et al. 2021, Huang et al. 2021, Jiménez et al. 2021).

It is unclear how pesticides can affect the microbiome of amphibians (Huang et al. 2018), particularly in the early stages. To date, studies on exposure to pesticides in tadpoles include the herbicides atrazine (gut microbiota) and glyphosate (gut and skin microbiota) (Krynak et al. 2017,

Table 1. Main gut and skin bacterial taxa of tadpoles affected in their relative abundance by microbiota disrupting chemicals (MDCs).

Species	Microbiota disrupting chemicals	Sample type	Affected taxa	References
Pelophylax nigromaculatus	Atrazine	G	(P) - Bacteroidetes (+), Fusobacteria (+), Verrucomicrobia (–) and Firmicutes (–). (G) - presence of *Cetobacterium*.	Huang et al. 2021
Osteopilus septentrionalis	Atrazine	G	It did not significantly alter the gut microbiota taxa.	Knutie et al. 2018
Acris blanchardi	Glyphosate	S	Not specified.	Krynak et al. 2017
Lithobates vibicarius	Chlorothalonil	S	(G) - *Sulfuricurvum* (–), *Janthinobacterium* (–), *Acinetobacter* (–), *Novosphingobium* (-), *Nevskia* (+), *Flavobacterium* (+), and *Runella* (+). 13 suppressed ASVs after exposure.	Jiménez et al. 2021
Lithobates sylvaticus	Biopesticide *Bacillus thuringiensis* var. *israelensis*	G	It did not significantly alter the gut microbiota taxa.	Gutierrez-Villagomez et al. 2021
Anaxyrus americanus			(P) - Verrucomicrobia (+), Firmicutes (+), Bacteroidetes (+), and Actinobacteria (+).	
Rhinella Arenarum	Glyphosate	G	(G) - *Aeromonas* spp. (-), and presence of *Yersinia* spp. and *Proteus* spp.	Boccioni et al.2021
	Ciprofloxacin		(G) - *Aeromonas* spp. (+), and presence of *Leclercia* spp.	
	Glyphosate and Ciprofloxacin (mix)		(G) - *Aeromonas* spp. (+), *Escherichia coli* (–), *Klebsiella* spp. (–), *Shewanella* spp. (-), and *Pseudomonas* spp. (–).	
Lithobates pipiens	Sulfadimethoxine	S	(G) - *Flavobacterium* (+), *Acinetobacter* (+), and presence of *Leadbeterella*.	Hernández-Gómez et al. 2020
Rana omeimontis	Tetracycline	G	(G) - *Cyanobacteria* (+), *Rickettsiales* (+), *Mycobacterium* (+), *Clostridiales* (–), *Bilophila* (–), and *Actinobacteria PeM15* (–).	Zhu et al. 2022
	Erythromycin		(G) - *Cyanobacteria* (+), *Rickettsiales* (+), *Clostridiales* (–), *Bilophila* (–), and *Actinobacteria PeM15* (–).	
	Sulfamethoxazole		(G) - *Cyanobacteria* (–) and *Rickettsiales* (–).	
Rana chensinensis	Cadmium	G	(P) - Fusobacteria (–), Spirochaetae (–), Firmicutes (–). (G) - vanished *Succinispira*, *Desulfovibrio* and *Fusobacterium*.	Mu et al. 2018
Bufo gargarizans	Cadmium	G	(P) - Proteobacteria (+), Bacteroidetes (–), Firmicutes (–).	Ya et al. 2019
Bufo gargarizans	Cadmium	G	(P) - Firmicutes (+), Bacteroidetes (+), Proteobacteria (–), Fusobacteria (–). (G) - presence of 226 bacterial genera found only in exposure group.	Ya et al.2020

Rana chensinensis	Copper	G	(P) - Fusobacteria (–), Proteobacteria (–), Cyanobacteria (+), and Actinobacteria (+). (G) - *Flavobacterium* (+), and *Rahnella* (+).	Yang et al. 2020
Bufo gargarizans	Hexavalent chromium	G	(P) - Proteobacteria (+), Fusobacteria (+), Bacteroidetes (–). (G) - presence of *Saccharibacteria* e *TM6_Dependentiae*.	Yao et al. 2019
Bufo gargarizans	Copper	G	(P) - Proteobacteria (+), Bacteroidetes (+), Firmicutes (+), Fusobacteria (–). (G) - *Kluyvera* (+) and *Aeromonas* (–).	Zheng et al. 2021a
Bufo gargarizans	Cadmium	G	(P) - Proteobacteria (+), Firmicutes (+), Fusobacteria (–), Verrucomicrobia (–), and Firmicutes/Bacteroidetes ratio (+).	Zheng et al. 2021b
	Lead		(P) - Proteobacteria (+), Firmicutes (+), Fusobacteria (–), Verrucomicrobia (–), and Firmicutes/Bacteroidetes ratio (+).	
	Cadmium and Lead (mix)		(P) - Proteobacteria (+), Firmicutes (+), Fusobacteria (–), Verrucomicrobia (–), and Firmicutes/Bacteroidetes ratio (+).	
Bufo gargarizans	Copper	G	(P) - Proteobacteria (+), Bacteroidetes (+), Fusobacteria (–). (G) - *Flavobacterium* (+).	Zheng et al. 2020
	Chromium		(P) - Proteobacteria (–), Bacteroidetes (+). (G) - *Flavobacterium* (+).	
	Cadmium		(P) - Proteobacteria (+), Bacteroidetes (+), Fusobacteria (–), Firmicutes/Bacteroidetes ratio (+). (G) - *Flavobacterium* (+).	
	Nitrate		(P) - Proteobacteria (+), Bacteroidetes (+), Firmicutes/Bacteroidetes ratio (+). (G) - *Flavobacterium* (+), *Shewanella* (+) and *Azospira* (+).	
Rana chensinensis	Cadmium	G	(P) - Firmicutes (–), Verrucomicrobia (–), Actinobacteria (–), Fusobacteria (–).	Shen et al. 2022
	Diethylhexyl phthalate		(P) - Firmicutes (–).	
	Cadmium and Diethylhexyl phthalate (mix)		(P) - Firmicutes (–), Verrucomicrobia (–), Actinobacteria (–), Fusobacteria (–), and Proteobacteria (+).	
Rana chensinensis	Octylphenol	G	(P) - Bacteroidetes (+), Cyanobacteria (+), Proteobacteria (–), Firmicutes (–), and Firmicutes/Bacteroidetes ratio (–).	Liu et al. 2020
Lithobates pipiens	Polychlorinated biphenyl 126 (PCB-126)	G	(G) - *Aminobacter* (–), and *Pseudomonas* (+).	Kohl et al. 2015

Table 1 contd. ...

...Table 1 contd.

Species	Microbiota disrupting chemicals	Sample type	Affected taxa	References
Rana pipiens	2,4,6-trinitrotoluene (TNT)	S	(P) - Proteobacteria (+).	Gust et al. 2021
	2,4-dinitroanisol (DNAN) + 3-nitro-1,2,4-triazol-5-ona (NTO)		(P) - Proteobacteria (+).	
	Nitroguanidina (NQ)		(P) - Proteobacteria (+), Acidobacteria (–), Fiacidobacteria (–), Firmicutes (–), and Yerrucomicrobia. (F) - Aeromonadaceae (+) and Pseudomonadaceae (+).	
	1-metil-3-nitroguanidina (MeNQ)		(P) - Proteobacteria (+).	
Xenopus laevis	Boron nitride nanotubes	G	(P) - Bacteroidetes (+), and Proteobacteria (–).	Evariste et al.2021
Lithobates catesbeianus	Microcystin-leucine arginine	S	(P) - Bacteroidetes (–), and Actinobacteria (+). (G) - *unclassified_f__Lachnospiraceae* (+), *Gordonibacter* (+), *Akkermansia* (+), *Pygmaiobacter* (+).	Shu et al. 2022
Bufo gargarizans	Nitrate	G	(P) - Proteobacteria (+), Firmicutes (+), Bacteroidetes (–), Fusobacteria (–), Bacteroidetes/Firmicutes ratio (–).	Xie et al. 2020

Publications until September 28, 2022. Sample type (G: gut, S: skin), Affected taxa (P: phylum, F: family, G: genus and +/– indicates increase or decrease of the relative abundance of taxa affected by MDCs).

Knutie et al. 2018, Boccioni et al. 2021, Huang et al. 2021), the fungicide chlorothalonil (skin microbiota) (Jiménez et al. 2021) and the bioinsecticide *Bacillus thuringiensis* var. *israelensis* (gut microbiota) (Gutierrez-Villagomez et al. 2021). In general, these pollutants can be considered MDCs, although more studies are needed to strengthen this claim.

However, some aspects about pesticides are decisive to deregulate the microbiota, such as concentration, reactivity, route of exposure and biodegradability. For example, the gut microbiota profile was not altered in the species *Osteopilus septentrionalis* exposed to atrazine, at concentration up to 200 µg/L (Knutie et al. 2018). Another determining factor is the intrinsic characteristic of the amphibian species. Tadpoles of *Lithobates sylvaticus* had no changes in the intestinal bacterial community exposed to the biopesticide VectoBac®, while larvae of *Anaxyrus americanus* increased their relative abundance of some bacterial taxa (Gutierrez-Villagomez et al. 2021).

4.2 Antibiotics

Antibiotics can affect the structure and function of the bacterial community associated to tadpoles, either by causing its mortality or by promoting adaptive responses that lead to tolerance and resistance of some bacterial taxa (Gao et al. 2012, Bernier and Surette 2013). Amphibian skin microbiota are modified when animals are exposed to antibiotics, but in some cases, they have the plasticity necessary to tolerate these contaminants over time (Costa et al. 2016, Huang et al. 2018). Interestingly, gut microbiota of adult amphibians treated with antibiotics showed a greater capacity to degrade xenobiotics (Ampatzoglou et al. 2022, Zhu et al. 2022). However, in tadpoles, this detoxification plasticity of the microbiome has not yet been characterized.

Studies regarding the evaluation of antibiotic effects on the symbiotic microbiome of tadpoles used ciprofloxacin, tetracycline, erythromycin, and sulfamethoxazole in gut microbiota (Boccioni et al. 2021, Zhu et al. 2022), and sulfadimethoxine in the skin microbiota (Hernández-Gómez et al. 2020). In tadpoles of the species *Rhinella arenarum* and *Rana omeimontis*, antibiotics decreased the richness and diversity of the intestinal bacterial community (Boccioni et al. 2021, Zhu et al. 2022). The synergistic effects of ciprofloxacin and the herbicide glyphosate were also tested, showing a decrease in several bacterial genera (Boccioni et al. 2021). Intestinal dysbiosis was also associated with weight loss in tadpoles exposed to these contaminants, indicating indirect deleterious effects in the animals.

In an opposite way, the effects of sulfadimethoxine on the skin microbiota of *Lithobates pipiens* larvae resulted in increased bacterial diversity (Hernández-Gómez et al. 2020). However, there was a change in the composition of the skin's natural microbiota, which allowed the colonization of opportunistic bacterial taxa. These changes may indicate dysbiosis and possible disruption of cutaneous immunity, which would compromise tadpole development in early life stages.

4.3 Metals

Amphibian habitats are continuously contaminated by heavy metals from anthropic actions or even by natural origins (Andreu and Gimeno-García 1999, Blaustein et al. 2003, Aiman et al. 2016, Yang et al. 2020, Huang et al. 2022). A variety of metals can cause a range of lethal and sub-lethal effects on tadpoles (Todd et al. 2011, Chai et al. 2014, Wang et al. 2016, Girotto et al. 2020, Ye et al. 2022) (see Chapter 13). These contaminants can also interact and change the structure of the microbiome (Arun et al. 2021).

Metals are the pollutants with the largest number of investigations on the effects on tadpoles' microbiota. All studies on tadpoles evaluating microbiota-disrupting metals were carried out in China, using two species of anurans (*Rana chensinensis* and *Bufo gargarizans*) and targeting the intestinal microbiota. Metals evaluated by them were cadmium (Mu et al. 2018, Ya et al. 2019, Ya et al. 2020, Zheng et al. 2020, Zheng et al. 2021a, Shen et al. 2022), copper (Yang et al. 2020, Zheng et al. 2020, Zheng et al. 2021b), chromium (Yao et al. 2019, Zheng et al. 2020), and lead (Zheng et al. 2021a). In general, all of them caused intestinal dysbiosis in tadpoles and exhibited patterns of dose-response effects depending on the metal used (Ya et al. 2019, Zheng et al. 2020, Zheng et al. 2021a). The main results showed a significant decrease in diversity of the intestinal bacterial community, the emergence of potentially opportunistic bacterial taxa, and the extinction of bacterial genera considered beneficial for tadpoles (Mu et al. 2018, Ya et al. 2019, Yao et al. 2019, Ya et al. 2020, Zheng et al. 2020, Zheng et al. 2021a, Shen et al. 2022). Additional details about changes on microbiota composition observed in these studies can be found in Table 1.

4.4 Persistent Organic Pollutants (POPs)

Persistent organic pollutants are considered contaminants of emerging concern, precisely due to the diversity of substances categorized in this group (Kuzmanović et al. 2015, Su et al. 2020). These contaminants are potentially MDCs and comprise industrial chemicals such as polybrominated diphenyl ethers (PBDEs; used in flame retardants, furniture foam and plastics), perfluorinated organic acids, plasticizers, surfactants, and even munitions waste (Ankley et al. 2008, Lotufo et al. 2021).

Currently, studies involving the effects of POPs on tadpole's gut microbiota were performed using diethylhexyl phthalate (plasticizers) (Shen et al. 2022), octylphenol (surfactant) (Liu et al. 2020) and polychlorinated biphenyl 126 (PCB-126) (Kohl et al. 2015). POPs had long-lasting effects on gut microbiota of tadpoles, which spanned generations (i.e., natural metamorphic changes) and persisted in adults after larval stage exposure (Kohl et al. 2015). Alterations in intestinal bacterial composition induced by POPs significantly modify the metabolic function of the microbiome (Liu et al. 2020, Shen et al. 2022).

Effects on skin microbiome composition of *Rana pipiens* tadpoles were also evaluated after exposure to the munition constituents 2,4,6-trinitrotoluene, 2,4-dinitroanisole, 3-nitro-1,2,4-triazol-5-one, Nitroguanidine, and 1-Methyl-3-nitroguanidine (Gust et al. 2021). These POPs disrupted the microbiome and decreased cutaneous bacterial diversity. Skin dysbiosis promoted indirect negative effects on transcriptional responses of tadpoles, such as altered mucus properties, decreased production of antimicrobial compounds and immunological factors, and altered the skin keratinization process (Gust et al. 2021).

4.5 Nanomaterials

Nanomaterials are emerging threats to ecological balance of aquatic ecosystems (Ali et al. 2021, Biagi et al. 2021, da Costa Araujo et al. 2021). It is highlighted that the pollution by micro(nano)plastics can disrupt microbiota, promoting dysbiosis and compromising the health of several vertebrates (Jin et al. 2018, Fackelmann and Sommer 2019, Qiao et al. 2019, Biagi et al. 2021, Rai 2022, Tamargo et al. 2022). However, to date, this evidence has not been reported for amphibians.

Due to their strictly aquatic, mostly herbivorous, and severely endangered habit, tadpoles may be particularly vulnerable to exposure to nanomaterials (Venâncio et al. 2022). Currently, a single study has evaluated the toxic effect of nanomaterials on the gut microbiome of tadpoles so far, indicating its absence of toxicity in the microbiota with slight alterations on bacterial composition (Evariste et al. 2021). Nanomaterial used in the study, boron nitride nanotubes (0, 0.1, 1, and 10 mg L^{-1}), is considered biocompatible due to the absence of genotoxic effects, which can be also associated with the lack of negative effects on the microbiota. This demonstrates how we still need to expand knowledge of the impacts of nanomaterials (including microplastics) on symbiotic microbiota and amphibian health (Fackelmann and Sommer 2019, da Costa Araujo et al. 2021, Venâncio et al. 2022).

4.6 Inorganic Compounds and Natural Toxins

Eutrophication of aquatic environments has been exacerbated by anthropogenic activities on agriculture, industry, and urbanization (Diaz and Rosenberg 2008, Costa et al. 2018, Pierozan et al. 2020). Changes in natural environments can promote increase in nutrients such as nitrogen (N) and alterations in flowering dynamics of cyanobacteria (Paerl 2008, Paerl 2009). This web of factors can trigger emerging toxicity problems for amphibians, especially during their larval stages.

In aquatic ecosystems, inorganic nitrogen is available for absorption by organisms mainly in the stable form of nitrate (NO_3^-) (Rabalais 2002). However, contamination by this inorganic compound occurs by increasing concentrations mainly in areas adjacent to industrial and agricultural effluents (Paerl 2009, Costa et al. 2018). Toxic effects of nitrate were verified in the gut microbiota of tadpoles of the species *Bufo gargarizans* (Xie et al. 2020, Zheng et al. 2020). The study indicated that nitrate induced intestinal dysbiosis in amphibians. Changes in bacterial composition at high nitrate concentrations increased the risks of metabolic disorders, causing dysregulation of fatty acid metabolism and amino acid metabolism in tadpoles (Xie et al. 2020, Zheng et al. 2020).

Considering that nutrients that enhance eutrophication cause toxicity to amphibians, the consequent disturbance of the bloom dynamics of cyanobacteria also deserves to be highlighted for aquatic ecosystems (Trung et al. 2018, Costa et al. 2018). High levels of toxins produced by cyanobacteria has become an emerging environmental problem (Pierozan et al. 2020, Wang et al. 2021b). Effects of the cyanobacterial toxin microcystin-leucine arginine was tested on the skin microbiome of *Lithobates catesbeianus* tadpoles (Shu et al. 2022). Cyanobacterial toxin caused dysbiosis in the skin microbiota, decreasing the diversity of the bacterial community. In addition, it was observed that a reduction in bacterial taxa with antipathogenic ability, then favoring pathogenic opportunistic bacteria (Shu et al. 2022). These results indicate that emerging ecotoxicological studies should include, or at least consider, these contaminants as potential disruptors of the microbiota in amphibians.

5. Conclusions and Perspectives

This chapter presents several studies on amphibian microbiome highlighting the role of the bacterial community in tadpoles. Microbiota-disrupting chemicals can affect the health of tadpoles and induce harmful imbalances in the microbiome throughout their development. Sublethal effects of pollutants on metabolism, physiology and reproduction have been the most common focus in ecotoxicological studies, but in recent years the effects on amphibian microbiome have also been receiving increasing attention. Implications caused by MDCs in tadpoles involve direct effects such as changes in richness, diversity, and composition of intestinal and cutaneous bacterial taxa. However, dysbiosis induces indirect effects such as energy metabolism disorder, increases on intestinal permeability, inflammation, decreased immunity, resistance to pathogens and skin protective toxins.

Based on the data presented here, we propose important points that deserve attention in future studies involving chemical pollutants and tadpole microbiome. (i) Investigations of the effects of MDCs in amphibians should use pre-metamorphic tadpoles as a model, as they include the interpretation of natural modifiers of the microbiota (metamorphosis and/or ontogeny), in addition to changes induced by xenobiotics. (ii) Evaluation of changes in the microbiota of amphibians exposed to contaminants in natural environments is essential to understand symbiotic phenomena, since captive laboratorial conditions can change the natural microbiota in tadpoles, mainly due to the alterations in water and diet characteristics. (iii) Synergistic effects of MDCs in tadpoles should be studied using environmentally realistic concentrations, promoting the expansion of knowledge with the natural context in which these animals are exposed. (iv) Knowledge on microbiome composition data should be expanded, since studies with tadpoles often present bacterial community composition data in OTUs, although some recent investigations have brought the ASVs approach, which is sufficiently more robust. The agglutination of data is essential for the knowledge about the amphibian microbiome to be increasingly solid and comparable. (v) Consolidate information on the bacterial community profile of tadpoles is essential for a better prediction of changes in genes involved in the metabolic function of the microbiome, recently proposed as endobolome for human gut microbiota (see Aguilera et al. 2020). (vi) A better identification of both pathogenic and symbiotic bacterial groups that potentially acts as biotransformers (degraders) of xenobiotics are necessary, to better understand their roles in adaptive response of tadpoles to MDCs.

Acknowledgements

We thank "Conselho Nacional de Desenvolvimento Científico e Tecnológico, CNPq", "Fundação de Amparo à Pesquisa e Inovação do Estado de Santa Catarina, FAPESC" and "Coordenação de Aperfeiçoamento de Pessoal de Nível Superior, CAPES" (001) for the financial support.

References Cited

Aguilera, M., Y. Gálvez-Ontiveros and A. Rivas. 2020. Endobolome, a new concept for determining the influence of microbiota disrupting chemicals (MDC) in relation to specific endocrine pathogenesis. Front. Microbiol. 11: 578007.

Aiman, U., A. Mahmood, S. Waheed and R.N. Malik. 2016. Enrichment, geo-accumulation and risk surveillance of toxic metals for different environmental compartments from Mehmood Booti dumping site, Lahore city, Pakistan. Chemosphere 144: 2229–2237.

Alberdi, A., O. Aizpurua, K. Bohmann, M.L. Zepeda-Mendoza and M.T.P. Gilbert. 2016. Do vertebrate gut metagenomes confer rapid ecological adaptation? Trends Ecol. Evol. 31(9): 689–699.

Ali, I., Q. Cheng, T. Ding, Q. Yiguang, Z. Yuechao, H. Sun et al. 2021. Micro-and nanoplastics in the environment: Occurrence, detection, characterization and toxicity–A critical review. J. Clean. Prod. 313: 127863.

Altig, R., M.R. Whiles and C.L. Taylor. 2007. What do tadpoles really eat? Assessing the trophic status of an understudied and imperiled group of consumers in freshwater habitats. Freshw. Biol. 52(2): 386–395.

Alves, T.C., R. Girardi and A. Pinheiro. 2017. Micropoluentes orgânicos: Ocorrência, remoção e regulamentação. REGA 14: e1.

Alves, T.C., G. Rozza and A. Pinheiro. 2022. Evaluation of concerning emergent compounds characteristics and simultaneous biosorption through multivariate technique. Eng. Sanit. e Ambient. 27: 403–412.

Amato, K.R. 2013. Co-evolution in context: The importance of studying gut microbiomes in wild animals. MICSM 1(1): 10–29.

Ampatzoglou, A., A. Gruszecka-Kosowska, A. Torres-Sánchez, A. López-Moreno, K. Cerk, P. Ortiz et al. 2022. Incorporating the gut microbiome in the risk assessment of xenobiotics and identifying beneficial components for One Health. Front. Microbiol. 13: 872583.

Andreu, V. and E. Gimeno-García. 1999. Evolution of heavy metals in marsh areas under rice farming. Environ. Pollut. 104(2): 271–282.

Ang, Z., D. Xiong, M. Wu and J.L. Ding. 2018. FFAR2-FFAR3 receptor heteromerization modulates short-chain fatty acid sensing. FASEB J. 32(1): 289–303.

Ankley, G.T., D.J. Hoff, D.R. Mount, J. Lazorchak, J. Beaman, T.K. Linton et al. 2008. Aquatic life criteria for contaminants of emerging concern. OW/ORD Emerging Contaminants Workgroup, 1–46.

Arun, K.B., A. Madhavan, R. Sindhu, S. Emmanual, P. Binod, A. Pugazhendhi et al. 2021. Probiotics and gut microbiome—Prospects and challenges in remediating heavy metal toxicity. J. Hazard. Mater. 420: 126676.

Atkinson, B.G., A.S. Warkman and Y. Chen. 1998. Thyroid hormone induces a reprogramming of gene expression in the liver of premetamorphic *Rana catesbeiana* tadpoles. Wound Repair Regen 6(4): S-323.

Bahrndorff, S., T. Alemu, T. Alemneh and J.L. Nielsen. 2016. The microbiome of animals: Implications for conservation biology. Int. J. Genomics. 2016: 5304028.

Bataille, A., L. Lee-Cruz, B. Tripathi and B. Waldman. 2018. Skin bacterial community reorganization following metamorphosis of the fire-bellied toad (*Bombina orientalis*). Microb. Ecol. 75(2): 505–514.

Bates, K.A., F.C. Clare, S. O'Hanlon, J. Bosch, L. Brookes, K. Hopkins et al. 2018. Amphibian chytridiomycosis outbreak dynamics are linked with host skin bacterial community structure. Nat. Commun. 9(1): 1–11.

Becker, M.H., R.M. Brucker, C.R. Schwantes, R.N. Harris and K.P. Minbiole. 2009. The bacterially produced metabolite violacein is associated with survival of amphibians infected with a lethal fungus. Appl. Environ. Microbiol. 75(21): 6635–6638.

Becker, M.H. and R.N. Harris. 2010. Cutaneous bacteria of the redback salamander prevent morbidity associated with a lethal disease. PloS one 5(6): e10957.

Becker, M.H., J.B. Walke, L. Murrill, D.C. Woodhams, L.K. Reinert, L.A. Rollins-Smith et al. 2015. Phylogenetic distribution of symbiotic bacteria from Panamanian amphibians that inhibit growth of the lethal fungal pathogen *Batrachochytrium dendrobatidis*. Mol. Ecol. 24(7): 1628–1641.

Belasen, A.M., M.A. Riolo, M.C. Bletz, M.L. Lyra, L.F. Toledo and T.Y. James. 2021. Geography, host genetics, and cross-domain microbial networks structure the skin microbiota of fragmented brazilian Atlantic Forest frog populations. Ecol. Evol. 11(14): 9293–9307.

Belden, L.K., M.C. Hughey, E.A. Rebollar, T.P. Umile, S.C. Loftus, E.A. Burzynski et al. 2015. Panamanian frog species host unique skin bacterial communities. Front. Microbiol. 6: 1171.

Benson, A.K., S.A. Kelly, R. Legge, F. Ma, S.J. Low, J. Kim et al. 2010. Individuality in gut microbiota composition is a complex polygenic trait shaped by multiple environmental and host genetic factors. PNAS 107(44): 18933–18938.

Bernardo-Cravo, A.P., D.S. Schmeller, A. Chatzinotas, V.T. Vredenburg and A. Loyau. 2020. Environmental factors and host microbiomes shape host-pathogen dynamics. Trends Parasitol. 36(7): 616–633.

Bernier, S.P. and M.G. Surette. 2013. Concentration-dependent activity of antibiotics in natural environments. Front. Microbiol. 4: 20.

Bhagavan, N.V. and C.E. Ha. 2015. Chapter 14—Carbohydrate Metabolism II: Gluconeogenesis, glycogen synthesis and breakdown, and alternative pathways. pp. 205–225. *In*: Bhagavan, N.V. and C.E. Ha [eds.]. Essentials of Medical Biochemistry. Academic Press, Cambridge, MA, USA.

Biagi, E., M. Musella, G. Palladino, V. Angelini, S. Pari, C. Roncari et al. 2021. Impact of plastic debris on the gut microbiota of *Caretta caretta* from Northwestern Adriatic Sea. Front. Mar. Sci. 8: 637030.

Blaustein, A.R., J.M. Romansic, J.M. Kiesecker and A.C. Hatch. 2003. Ultraviolet radiation, toxic chemicals and amphibian population declines. Divers. Distrib. 9(2): 123–140.

Boccioni, A.P.P.C., G. García-Effron, P.M. Peltzer and R.C. Lajmanovich. 2021. The effect of glyphosate and ciprofloxacin exposure on the gut bacterial microbiota diversity of *Rhinella arenarum* (Anura: Bufonidae) tadpoles. Res. Sq. (in press).

Boone, M.D., S.A. Hammond, N. Veldhoen, M. Youngquist and C.C. Helbing. 2013. Specific time of exposure during tadpole development influences biological effects of the insecticide carbaryl in green frogs (*Lithobates clamitans*). Aquat. Toxicol. 130: 139–148.

Bosch, J., I. Martínez-Solano and M. García-París. 2001. Evidence of a chytrid fungus infection involved in the decline of the common midwife toad (*Alytes obstetricans*) in protected areas of central Spain. Biol. Conserv. 97(3): 331–337.

Brack, W., D. Barcelo Culleres, A. Boxall, H. Budzinski, S. Castiglioni, A. Covaci et al. 2022. One Planet: One Health. A call to support the initiative on a global science-policy body on chemicals and waste. Environ. Sci. Eur. 34(1): 1–10.

Bresciano, J.C., C.A. Salvador, C. Paz-y-Mino, A.M. Parody-Merino, J. Bosch and D.C. Woodhams. 2015. Variation in the presence of anti-*Batrachochytrium dendrobatidis* bacteria of amphibians across life stages and elevations in Ecuador. Ecohealth 12(2): 310–319.

Bletz, M.C., D.J. Goedbloed, E. Sanchez, T. Reinhardt, C.C. Tebbe, S. Bhuju et al. 2016. Amphibian gut microbiota shifts differentially in community structure but converges on habitat-specific predicted functions. Nat. Commun. 7(1): 1–12.

Bletz, M.C., D.J. Goedbloed, E. Sanchez, T. Reinhardt, C.C. Tebbe, S. Bhuju, R. Geffers, M. Jarek, M. Vences and S. Steinfartz. 2016. Amphibian gut microbiota shifts differentially in community structure but converges on habitat-specific predicted functions. Nat. Commun. 7(1): 13699.

Brooks, A.W., K.D. Kohl, R.M. Brucker, E.J. van Opstal and S.R. Bordenstein. 2016. Phylosymbiosis: Relationships and functional effects of microbial communities across host evolutionary history. PLOS Biol. 14(11): e2000225.

Brucker, R.M., R.N. Harris, C.R. Schwantes, T.N. Gallaher, D.C. Flaherty, B.A. Lam et al. 2008. Amphibian chemical defense: Antifungal metabolites of the microsymbiont *Janthinobacterium lividum* on the salamander *Plethodon cinereus*. J. Chem. Ecol. 34(11): 1422–1429.

Callahan, B.J., P.J. McMurdie and S.P. Holmes. 2017. Exact sequence variants should replace operational taxonomic units in marker-gene data analysis. ISME J. 11(12): 2639–2643.

Chai, L., H. Wang, H. Deng, H. Zhao and W. Wang. 2014. Chronic exposure effects of copper on growth, metamorphosis and thyroid gland, liver health in Chinese toad, *Bufo gargarizans* tadpoles. Chem Ecol. 30(7): 589–601.

Chai, L., Z. Dong, A. Chen and H. Wang. 2018. Changes in intestinal microbiota of *Bufo gargarizans* and its association with body weight during metamorphosis. Arch. Microbiol. 200(7): 1087–1099.

Chang, C.W., B.H. Huang, S.M. Lin, C.L. Huang and P.C. Liao. 2016. Changes of diet and dominant intestinal microbes in farmland frogs. BMC Microbiol. 16(1): 1–13.

Chen, Y. and B.G. Atkinson. 1997. Role for the *Rana catesbeiana* homologue of C/EBP α in the reprogramming of gene expression in the liver of metamorphosing tadpoles. Dev. Genet. 20(2): 152–162.

Chen, Z., J.Q. Chen, Y. Liu, J. Zhang, X.H. Chen and Y.F. Qu. 2022. Comparative study on gut microbiota in three Anura frogs from a mountain stream. Ecol. Evol. 12(4): e8854.

Chapman, P.A., C.B. Gilbert, T.J. Devine, D.T. Hudson, J. Ward, X.C. Morgan et al. 2022. Manipulating the microbiome alters regenerative outcomes in *Xenopus laevis* tadpoles via lipopolysaccharide signalling. Wound Repair Regen.

Colombo, B.M., T. Scalvenzi, S. Benlamara and N. Pollet. 2015. Microbiota and mucosal immunity in amphibians. Front. Immunol. 6: 111.

Costa, S., I. Lopes, D.N. Proença, R. Ribeiro and P.V. Morais. 2016. Diversity of cutaneous microbiome of *Pelophylax perezi* populations inhabiting different environments. Sci. Total Environ. 572: 995–1004.

Costa, J.A.D., J.P.D. Souza, A.P. Teixeira, J.C. Nabout and F.M. Carneiro. 2018. Eutrophication in aquatic ecosystems: a scientometric study. Acta Limnol. Bras. 30.

da Costa Araújo, A.P., T.L. Rocha, D.D.M. Silva and G. Malafaia. 2021. Micro (nano) plastics as an emerging risk factor to the health of amphibian: A scientometric and systematic review. Chemosphere 283: 131090.

Davis, L.R., L. Bigler and D.C. Woodhams. 2017. Developmental trajectories of amphibian microbiota: Response to bacterial therapy depends on initial community structure. Environ. Microbiol. 19(4): 1502–1517.

Davidson, C., M.F. Benard, H.B. Shaffer, J.M. Parker, C. O'Leary, J.M. Conlon et al. 2007. Effects of chytrid and carbaryl exposure on survival, growth and skin peptide defenses in foothill yellow-legged frogs. Environ. Sci. Technol. 41(5): 1771–1776.

Delsuc, F., J.L. Metcalf, L. Wegener Parfrey, S.J. Song, A. González and R. Knight. 2014. Convergence of gut microbiomes in myrmecophagous mammals. Mol. Ecol. 23(6): 1301–1317.

Denver, R.J., K.A. Glennemeier and G.C. Boorse. 2002. Endocrinology of complex life cycles: Amphibians. pp. 469–513. *In*: Pfaff, D.W. [ed.]. Hormones, brain and behavior. Academic press, Cambridge, MA, USA.

Derrien, M. and J.E. van Hylckama Vlieg. 2015. Fate, activity, and impact of ingested bacteria within the human gut microbiota. Trends Microbiol. 23(6): 354–366.

Diaz, R.J. and R. Rosenberg. 2008. Spreading dead zones and consequences for marine ecosystems. Science 321(5891): 926–929.

Dornelles, M.F. and G.T. Oliveira. 2016. Toxicity of atrazine, glyphosate, and quinclorac in bullfrog tadpoles exposed to concentrations below legal limits. Environ. Sci. Pollut. Res. 23(2): 1610–1620.

Dulio, V., B. van Bavel, E. Brorström-Lundén, J. Harmsen, J. Hollender, M. Schlabach et al. 2018. Emerging pollutants in the EU: 10 years of NORMAN in support of environmental policies and regulations. Environ. Sci. Eur. 30(1): 1–13.

Edgar, R.C. 2018. Updating the 97% identity threshold for 16S ribosomal RNA OTUs. Bioinformatics 34(14): 2371–2375.

Edwards, C.L., P.G. Byrne, P. Harlow and A.J. Silla. 2017. Dietary carotenoid supplementation enhances the cutaneous bacterial communities of the critically endangered southern corroboree frog (*Pseudophryne corroboree*). Microb. Ecol. 73(2): 435–444.

Evariste, L., M. Barret, A. Mottier, F. Mouchet, L. Gauthier and E. Pinelli. 2019. Gut microbiota of aquatic organisms: A key endpoint for ecotoxicological studies. Environ. Pollut. 248: 989–999.

Evariste, L., E. Flahaut, C. Baratange, M. Barret, F. Mouchet, E. Pinelli et al. 2021. Ecotoxicological assessment of commercial boron nitride nanotubes toward *Xenopus laevis* tadpoles and host-associated gut microbiota. Nanotoxicology 15(1): 35–51.

Fabrezi, M. and J.C. Cruz. 2020. Evolutionary and developmental considerations of the diet and gut morphology in ceratophryid tadpoles (Anura). BMC Dev. Biol. 20(1): 1–17.

Fackelmann, G. and S. Sommer. 2019. Microplastics and the gut microbiome: How chronically exposed species may suffer from gut dysbiosis. Mar. Pollut. Bull. 143: 193–203.

Fedewa, L.A. 2006. Fluctuating gram-negative microflora in developing anurans. J. Herpetol. 40(1): 131–135.

Finckh, S., L.M. Beckers, W. Busch, E. Carmona, V. Dulio, L. Kramer et al. 2022. A risk based assessment approach for chemical mixtures from wastewater treatment plant effluents. Environ. Int. 164: 107234.

Flint, H.J., K.P. Scott, S.H. Duncan, P. Louis and E. Forano. 2012. Microbial degradation of complex carbohydrates in the gut. Gut Microbes 3(4): 289–306.

Fontaine, S.S. and K.D. Kohl. 2020. Gut microbiota of invasive bullfrog tadpoles responds more rapidly to temperature than a noninvasive congener. Mol. Ecol. 29(13): 2449–2462.

Fontaine, S.S., P.M. Mineo and K.D. Kohl. 2022. Experimental manipulation of microbiota reduces host thermal tolerance and fitness under heat stress in a vertebrate ectotherm. Nat. Ecol. Evol. 6(4): 405–417.

Fontes, M.K., P.L.R. Dourado, B.G. de Campos, L.A. Maranho, E.A. de Almeida, D.M. de Souza Abessa et al. 2022. Environmentally realistic concentrations of cocaine in seawater disturbed neuroendrocrine parameters and energy status in the marine mussel *Perna perna*. Comp. Biochem. Physiol. C Toxicol. Pharmacol. 251: 109198.

Forson, D.D. and A. Storfer. 2006. Atrazine increases ranavirus susceptibility in the tiger salamander, *Ambystoma tigrinum*. Ecol. Appl. 16(6): 2325–2332.

Gao, P., D. Mao, Y. Luo, L. Wang, B. Xu and L. Xu. 2012. Occurrence of sulfonamide and tetracycline-resistant bacteria and resistance genes in aquaculture environment. Water Res. 46(7): 2355–2364.

Garner, T.W., B.R. Schmidt, A. Martel, F. Pasmans, E. Muths, A.A. Cunningham et al. 2016. Mitigating amphibian chytridiomycoses in nature. Philos. Trans. R. Soc. Lond., B, Biol. Sci. 371(1709): 20160207.

Gilbert, S.F., J. Sapp and A.I. Tauber. 2012. A symbiotic view of life: We have never been individuals. Q. Rev. Biol. 87(4): 325–341.
Girotto, L., E.L.G. Espíndola, R.C. Gebara and J.S. Freitas. 2020. Acute and chronic effects on tadpoles (*Lithobates catesbeianus*) exposed to mining tailings from the dam rupture in Mariana, MG (Brazil). Water Air Soil Pollut. 231(7): 1–15.
Greenspan, S.E., M.L. Lyra, G.H. Migliorini, M.F. Kersch-Becker, M.C. Bletz, C.S. Lisboa et al. 2019. Arthropod–bacteria interactions influence assembly of aquatic host microbiome and pathogen defense. Proc. Royal Soc. B. 286(1905): 20190924.
Grice, E.A. and J.A. Segre. 2011. The skin microbiome. Nat. Rev. Microbiol. 9(4): 244–253.
Griffiths, S.M., X.A. Harrison, C. Weldon, M.D. Wood, A. Pretorius, K. Hopkins et al. 2018. Genetic variability and ontogeny predict microbiome structure in a disease-challenged montane amphibian. ISME J. 12(10): 2506–2517.
Gust, K.A., K.J. Indest, G. Lotufo, S.J. Everman, C.M. Jung, M.L. Ballentine et al. 2021. Genomic investigations of acute munitions exposures on the health and skin microbiome composition of leopard frog (*Rana pipiens*) tadpoles. Environ. Res. 192: 110245.
Gutierrez-Villagomez, J.M., G. Patey, T.A. To, M. Lefebvre-Raine, L.R. Lara-Jacobo, J. Comte et al. 2021. Frogs respond to commercial formulations of the biopesticide *Bacillus thuringiensis* var. *israelensis*, especially their intestine microbiota. Environ. Sci. Technol. 55(18): 12504–12516.
Harris, R.N., T.Y. James, A. Lauer, M.A. Simon and A. Patel. 2006. Amphibian pathogen *Batrachochytrium dendrobatidis* is inhibited by the cutaneous bacteria of amphibian species. EcoHealth 3(1): 53–56.
Harris, R.N., R.M. Brucker, J.B. Walke, M.H. Becker, C.R. Schwantes, D.C. Flaherty et al. 2009. Skin microbes on frogs prevent morbidity and mortality caused by a lethal skin fungus. ISME J. 3(7): 818–824.
Hayes, T.B., P. Falso, S. Gallipeau and M. Stice. 2010. The cause of global amphibian declines: a developmental endocrinologist's perspective. J. Exp. Biol. 213(6): 921–933.
Hehemann, J.H., G. Correc, T. Barbeyron, W. Helbert, M. Czjzek and G. Michel. 2010. Transfer of carbohydrate-active enzymes from marine bacteria to Japanese gut microbiota. Nature 464(7290): 908–912.
Heimeier, R.A., B. Das, D.R. Buchholz, M. Fiorentino and Y.B. Shi. 2010. Studies on *Xenopus laevis* intestine reveal biological pathways underlying vertebrate gut adaptation from embryo to adult. Genome Biol. 11(5): 1–20.
Hernández-Gómez, O., V. Wuerthner and J. Hua. 2020. Amphibian host and skin microbiota response to a common agricultural antimicrobial and internal parasite. Microb. Ecol. 79(1): 175–191.
Huang, B.H., C.W. Chang, C.W. Huang, J. Gao and P.C. Liao. 2018. Composition and functional specialists of the gut microbiota of frogs reflect habitat differences and agricultural activity. Front. Microbiol. 8: 2670.
Huang, M.Y., Q. Zhao, R.Y. Duan, Y. Liu and Y.Y. Wan. 2021. The effect of atrazine on intestinal histology, microbial community and short chain fatty acids in *Pelophylax nigromaculatus* tadpoles. Environ. Pollut. 288: 117702.
Huang, M., Y. Liu, W. Dong, Q. Zhao, R. Duan, X. Cao et al. 2022. Toxicity of Pb continuous and pulse exposure on intestinal anatomy, bacterial diversity, and metabolites of *Pelophylax nigromaculatus* in pre-hibernation. Chemosphere 290: 133304.
Hughes, D.T. and V. Sperandio. 2008. Inter-kingdom signalling: communication between bacteria and their hosts. Nat. Rev. Microbiol. 6(2): 111–120.
Hughey, M.C., J.B. Walke, M.H. Becker, T.P. Umile, E.A. Burzynski, K.P. Minbiole et al. 2016. Short-term exposure to coal combustion waste has little impact on the skin microbiome of adult spring peepers (*Pseudacris crucifer*). Appl. Environ. Microbiol. 82(12): 3493–3502.
Jani, A.J. and C.J. Briggs. 2014. The pathogen *Batrachochytrium dendrobatidis* disturbs the frog skin microbiome during a natural epidemic and experimental infection. PNAS 111(47): E5049–E5058.
Jani, A.J. and C.J. Briggs. 2018. Host and aquatic environment shape the amphibian skin microbiome but effects on downstream resistance to the pathogen *Batrachochytrium dendrobatidis* are variable. Front. Microbiol. 9: 487.
Jervis, P., P. Pintanel, K. Hopkins, C. Wierzbicki, J.M. Shelton, E. Skelly et al. 2021. Post-epizootic microbiome associations across communities of neotropical amphibians. Mol. Ecol. 30(5): 1322–1335.
Jiménez, R.R. and S. Sommer. 2017. The amphibian microbiome: Natural range of variation, pathogenic dysbiosis, and role in conservation. Biodivers. Conserv. 26(4): 763–786.
Jiménez, R.R., G. Alvarado, J. Estrella and S. Sommer. 2019. Moving beyond the host: Unraveling the skin microbiome of endangered Costa Rican amphibians. Front. Microbiol. 10: 2060.
Jiménez, R.R., G. Alvarado, J. Sandoval and S. Sommer. 2020. Habitat disturbance influences the skin microbiome of a rediscovered neotropical-montane frog. BMC Microbiol. 20(1): 1–14.
Jiménez, R.R., G. Alvarado, C. Ruepert, E. Ballestero and S. Sommer. 2021. The fungicide chlorothalonil changes the amphibian skin microbiome: A potential factor disrupting a host disease-protective trait. Appl. Microbiol. 1(1): 26–37.
Jiménez, R.R., A. Carfagno, L. Linhoff, B. Gratwicke, D.C. Woodhams, L.S. Chafran et al. 2022. Inhibitory bacterial diversity and mucosome function differentiate susceptibility of Appalachian salamanders to Chytrid fungal infection. Appl. Environ. Microbiol. 88(8): e01818–21.

Jin, Y., J. Xia, Z. Pan, J. Yang, W. Wang and Z. Fu. 2018. Polystyrene microplastics induce microbiota dysbiosis and inflammation in the gut of adult zebrafish. Environ. Pollut. 235: 322–329.

Kamalakkannan, V., A.A. Salim and R.J. Capon. 2017. Microbiome-mediated biotransformation of cane toad bufagenins. J. Nat. Prod. 80(7): 2012–2017.

Keeling, P.J. and Palmer, J.D. 2008. Horizontal gene transfer in eukaryotic evolution. Nat. Rev. Genet. 9(8): 605–618.

Kikuyama, S., I. Hasunuma and R. Okada. 2021. Development of the hypothalamo–hypophyseal system in amphibians with special reference to metamorphosis. Mol. Cell. Endocrinol. 524: 111143.

Knapp, R.A., M.B. Joseph, T.C. Smith, E.E. Hegeman, V.T. Vredenburg, J.E. Erdman Jr. et al. 2022. Effectiveness of antifungal treatments during chytridiomycosis epizootics in populations of an endangered frog. PeerJ. 10: e12712.

Knutie, S.A., L.A. Shea, M. Kupselaitis, C.L. Wilkinson, K.D. Kohl and J.R. Rohr. 2017a. Early-life diet affects host microbiota and later-life defenses against parasites in frogs. Integr. Comp. Biol. 57(4): 732–742.

Knutie, S.A., C.L. Wilkinson, K.D. Kohl and J.R. Rohr. 2017b. Early-life disruption of amphibian microbiota decreases later-life resistance to parasites. Nat. Commun. 8(1): 1–8.

Knutie, S.A., C.R. Gabor, K.D. Kohl and J.R. Rohr. 2018. Do host-associated gut microbiota mediate the effect of an herbicide on disease risk in frogs? J. Anim. Ecol. 87(2): 489–499.

Kohl, K.D., T.L. Cary, W.H. Karasov and M.D. Dearing. 2013. Restructuring of the amphibian gut microbiota through metamorphosis. Environ. Microbiol. Rep. 5(6): 899–903.

Kohl, K.D., J. Amaya, C.A. Passement, M.D. Dearing and M.D. McCue. 2014. Unique and shared responses of the gut microbiota to prolonged fasting: A comparative study across five classes of vertebrate hosts. FEMS Microbiol. Ecol. 90(3): 883–894.

Kohl, K.D., T.L. Cary, W.H. Karasov and M.D. Dearing. 2015. Larval exposure to polychlorinated biphenyl 126 (PCB-126) causes persistent alteration of the amphibian gut microbiota. Environ. Toxicol. Chem. 34(5): 1113–1118.

Kohl, K.D. and J. Yahn. 2016. Effects of environmental temperature on the gut microbial communities of tadpoles. Environ. Microbiol. 18(5): 1561–1565.

Koprivnikar, J. 2010. Interactions of environmental stressors impact survival and development of parasitized larval amphibians. Ecol. Appl. 20(8): 2263–2272.

Krynak, K.L., D.J. Burke and M.F. Benard. 2015. Larval environment alters amphibian immune defenses differentially across life stages and populations. PLoS One 10(6): e0130383.

Krynak, K.L., D.J. Burke and M.F. Benard. 2016. Landscape and water characteristics correlate with immune defense traits across Blanchard's cricket frog (*Acris blanchardi*) populations. Biol. Conserv. 193: 153–167.

Krynak, K.L., D.J. Burke and M.F. Benard. 2017. Rodeo™ herbicide negatively affects Blanchard's cricket frogs (*Acris blanchardi*) survival and alters the skin-associated bacterial community. J. Herpetol. 51(3): 402–410.

Kruger, A. 2020. Frog skin microbiota vary with host species and environment but not chytrid infection. Front. Microbiol. 1330.

Kueneman, J.G., L.W. Parfrey, D.C. Woodhams, H.M. Archer, R. Knight and V.J. McKenzie. 2014. The amphibian skin-associated microbiome across species, space and life history stages. Mol. Ecol. 23(6): 1238–1250.

Kueneman, J.G., D.C. Woodhams, W. Van Treuren, H.M. Archer, R. Knight and V.J. McKenzie. 2016a. Inhibitory bacteria reduce fungi on early life stages of endangered Colorado boreal toads (*Anaxyrus boreas*). ISME J. 10(4): 934–944.

Kueneman, J.G., D.C. Woodhams, R. Harris, H.M. Archer, R. Knight and V.J. McKenzie. 2016b. Probiotic treatment restores protection against lethal fungal infection lost during amphibian captivity. Proc. Royal Soc. B 283(1839): 20161553.

Kueneman, J.G., S. Weiss and V.J. McKenzie. 2017. Composition of micro-eukaryotes on the skin of the cascades frog (*Rana cascadae*) and patterns of correlation between skin microbes and *Batrachochytrium dendrobatidis*. Front. Microbiol. 8: 2350.

Kueneman, J.G., M.C. Bletz, V.J. McKenzie, C.G. Becker, M.B. Joseph, J.G. Abarca et al. 2019. Community richness of amphibian skin bacteria correlates with bioclimate at the global scale. Nat. Ecol. Evol. 3(3): 381–389.

Kulkarni, S.S. and D.R. Buchholz. 2014. Corticosteroid signaling in frog metamorphosis. Gen. Comp. Endocrinol. 203: 225–231.

Kuzmanović, M., A. Ginebreda, M. Petrović and D. Barceló. 2015. Risk assessment based prioritization of 200 organic micropollutants in 4 Iberian rivers. Sci. Total Environ. 503: 289–299.

Langerveld, A.J., R. Celestine, R. Zaya, D. Mihalko and C.F. Ide. 2009. Chronic exposure to high levels of atrazine alters expression of genes that regulate immune and growth-related functions in developing *Xenopus laevis* tadpoles. Environ. Res. 109(4): 379–389.

Lapierre, P. and J.P. Gogarten. 2009. Estimating the size of the bacterial pan-genome. Trends Genet. 25(3): 107–110.

Lauer, A., M.A. Simon, J.L. Banning, E. André, K. Duncan and R.N. Harris. 2007. Common cutaneous bacteria from the eastern red-backed salamander can inhibit pathogenic fungi. Copeia 2007(3): 630–640.

Linzey, D., J. Burroughs, L. Hudson, M. Marini, J. Robertson, J. Bacon et al. 2003. Role of environmental pollutants on immune functions, parasitic infections and limb malformations in marine toads and whistling frogs from Bermuda. Int. J. Environ. Health Res. 13(2): 125–148.

Liu, R., Y. Zhang, J. Gao and X. Li. 2020. Effects of octylphenol exposure on the lipid metabolism and microbiome of the intestinal tract of *Rana chensinensis* tadpole by RNAseq and 16s amplicon sequencing. Ecotoxicol. Environ. Saf. 197: 110650.

Long, J., J. Xiang, T. He, N. Zhang and W. Pan. 2020. Gut microbiota differences during metamorphosis in sick and healthy giant spiny frogs (*Paa spinosa*) tadpoles. Lett. Appl. Microbiol. 70(2): 109–117.

Longo, A.V. and K.R. Zamudio. 2017. Temperature variation, bacterial diversity and fungal infection dynamics in the amphibian skin. Mol. Ecol. 26(18): 4787–4797.

Lotufo, G.R., M.L. Ballentine, L.R. May, L.C. Moores, K.A. Gust and P. Chappell. 2021. Multi-species aquatic toxicity assessment of 1-Methyl-3-Nitroguanidine (MeNQ). Arch. Environ. Contam. Toxicol. 80(2): 426–436.

Loudon, A.H., J.A. Holland, T.P. Umile, E.A. Burzynski, K.P. Minbiole and R.N. Harris. 2014a. Interactions between amphibians' symbiotic bacteria cause the production of emergent anti-fungal metabolites. Front. Microbiol. 5: 441.

Loudon, A.H., D.C. Woodhams, L.W. Parfrey, H. Archer, R. Knight, V. McKenzie et al. 2014b. Microbial community dynamics and effect of environmental microbial reservoirs on red-backed salamanders (*Plethodon cinereus*). ISME J. 8(4): 830–840.

Loudon, A.H., A. Venkataraman, W. Van Treuren, D.C. Woodhams, L.W. Parfrey, V.J. McKenzie et al. 2016. Vertebrate hosts as islands: Dynamics of selection, immigration, loss, persistence, and potential function of bacteria on salamander skin. Front. Microbiol. 7: 333.

Loutfy, N.M. and M.S. Kamel. 2018. Effects of two neonicotinoids insecticides on some anti-oxidant enzymes and hematological parameters in egyptian frogs, *Bufo regularis*. Egypt. Acad. J. Biolog. Sci. (F-Toxicology & Pest Control) 10(1): 25–36.

Lyra, M.L., M.C. Bletz, C.F. Haddad and M. Vences. 2018. The intestinal microbiota of tadpoles differs from those of syntopic aquatic invertebrates. Microb. Ecol. 76(1): 121–124.

McCoy, K.A. and A.L. Peralta. 2018. Pesticides could alter amphibian skin microbiomes and the effects of *Batrachochytrium dendrobatidis*. Front. Microbiol. 9: 748.

McFall-Ngai, M., M.G. Hadfield, T.C. Bosch, H.V. Carey, T. Domazet-Lošo, A.E. Douglas et al. 2013. Animals in a bacterial world, a new imperative for the life sciences. PNAS 110(9): 3229–3236.

McIndoe, R. and D.G. Smith. 1984. Functional anatomy of the internal gills of the tadpole of *Litoria ewingii* (Anura, Hylidae). Zoomorphology 104(5): 280–291.

McKenzie, V.J., R.M. Bowers, N. Fierer, R. Knight and C.L. Lauber. 2012. Co-habiting amphibian species harbor unique skin bacterial communities in wild populations. ISME J. 6(3): 588–596.

Méndez-Narváez, J. and K.M. Warkentin. 2022. Reproductive colonization of land by frogs: Embryos and larvae excrete urea to avoid ammonia toxicity. Ecol. Evol. 12(2): e8570.

Mu, D., J. Meng, X. Bo, M. Wu, H. Xiao and H. Wang. 2018. The effect of cadmium exposure on diversity of intestinal microbial community of *Rana chensinensis* tadpoles. Ecotoxicol. Environ. Saf. 154: 6–12.

Nicholson, J.K., E. Holmes, J. Kinross, R. Burcelin, G. Gibson, W. Jia et al. 2012. Host-gut microbiota metabolic interactions. Science 336(6086): 1262–1267.

Paerl, H. 2008. Nutrient and other environmental controls of harmful cyanobacterial blooms along the freshwater–marine continuum. pp. 217–237. *In*: Hudnell, H.K. [ed.]. Cyanobacterial Harmful Algal Blooms: State of the Science and Research Needs. Springer, New York, NY, USA.

Paerl, H.W. 2009. Controlling eutrophication along the freshwater–marine continuum: Dual nutrient (N and P) reductions are essential. Estuaries Coast 32(4): 593–601.

Paris, M. and V. Laudet. 2008. The history of a developmental stage: metamorphosis in chordates. Genesis 46(11): 657–672.

Patterson, S.A., D.T. Denton, C.T. Hasler, J.M. Blais, M.L. Hanson, B.P. Hollebone et al. 2022. Resilience of larval wood frogs (*Rana sylvatica*) to hydrocarbons and other compounds released from naturally weathered diluted bitumen in a boreal lake. Aquat. Toxicol. 245: 106128.

Piccinni, M.Z., J.E. Watts, M. Fourny, M. Guille and S.C. Robson. 2021. The skin microbiome of *Xenopus laevis* and the effects of husbandry conditions. Animal Microbiome 3(1): 1–13.

Pierozan, P., D. Cattani and O. Karlsson. 2020. Hippocampal neural stem cells are more susceptible to the neurotoxin BMAA than primary neurons: Effects on apoptosis, cellular differentiation, neurite outgrowth, and DNA methylation. Cell Death Dis. 11(10): 1–14.

Piovia-Scott, J., K. Pope, S. Joy Worth, E.B. Rosenblum, T. Poorten, J. Refsnider et al. 2015. Correlates of virulence in a frog-killing fungal pathogen: evidence from a California amphibian decline. ISME J. 9(7): 1570–1578.

Prado-Irwin, S.R., A.K. Bird, A.G. Zink and V.T. Vredenburg. 2017. Intraspecific variation in the skin-associated microbiome of a terrestrial salamander. Microb. Ecol. 74(3): 745–756.

Prest, T.L., A.K. Kimball, J.G. Kueneman and V.J. McKenzie. 2018. Host-associated bacterial community succession during amphibian development. Mol. Ecol. 27(8): 1992–2006.

Pryor, G.S. and K.A. Bjorndal. 2005a. Effects of the nematode *Gyrinicola batrachiensis* on development, gut morphology, and fermentation in bullfrog tadpoles (*Rana catesbeiana*): A novel mutualism. J. Exp. Zool. A. Comp. Exp. Biol. 303(8): 704–712.

Pryor, G.S. and K.A. Bjorndal. 2005b. Symbiotic fermentation, digesta passage, and gastrointestinal morphology in bullfrog tadpoles (*Rana catesbeiana*). Physiol. Biochem. Zool. 78(2): 201–215.

Pryor, G.S. 2014. Tadpole nutritional ecology and digestive physiology: Implications for captive rearing of larval anurans. Zoo Biol. 33(6): 502–507.

Qiao, R., Y. Deng, S. Zhang, M.B. Wolosker, Q. Zhu, H. Ren et al. 2019. Accumulation of different shapes of microplastics initiates intestinal injury and gut microbiota dysbiosis in the gut of zebrafish. Chemosphere 236: 124334.

Rabalais, N.N. 2002. Nitrogen in aquatic ecosystems. AMBIO 31(2): 102–112.

Rai, A. 2022. Nanoplastics, Gut Microbiota, and Neurodegeneration. pp. 211–234. *In*: Tripathi, A.K. and M. Kotak [eds.]. Gut Microbiome in Neurological Health and Disorders. Springer Nature Singapore, Singapore, Malaysia.

Rebollar, E.A., R.E. Antwis, M.H. Becker, L.K. Belden, M.C. Bletz, R.M. Brucker et al. 2016a. Using "omics" and integrated multi-omics approaches to guide probiotic selection to mitigate chytridiomycosis and other emerging infectious diseases. Front. Microbiol. 7: 68.

Rebollar, E.A., M.C. Hughey, D. Medina, R.N. Harris, R. Ibáñez and L.K. Belden. 2016b. Skin bacterial diversity of Panamanian frogs is associated with host susceptibility and presence of *Batrachochytrium dendrobatidis*. ISME J. 10(7): 1682–1695.

Rebollar, E.A., S.J. Simonetti, W.R. Shoemaker and R.N. Harris. 2016c. Direct and indirect horizontal transmission of the antifungal probiotic bacterium *Janthinobacterium lividum* on green frog (*Lithobates clamitans*) tadpoles. Appl. Environ. Microbiol. 82(8): 2457–2466.

Relyea, R.A., N.M. Schoeppner and J.T. Hoverman. 2005. Pesticides and amphibians: the importance of community context. Ecol. Appl. 15(4): 1125–1134.

Rohr, J.R., A.M. Schotthoefer, T.R. Raffel, H.J. Carrick, N. Halstead, J.T. Hoverman et al. 2008. Agrochemicals increase trematode infections in a declining amphibian species. Nature 455(7217): 1235–1239.

Rohr, J.R. and K.A. McCoy. 2010. A qualitative meta-analysis reveals consistent effects of atrazine on freshwater fish and amphibians. Environ. Health Perspect. 118(1): 20–32.

Rohr, J.R., T.R. Raffel, N.T. Halstead, T.A. McMahon, S.A. Johnson, R.K. Boughton et al. 2013. Early-life exposure to a herbicide has enduring effects on pathogen-induced mortality. Proc. Royal Soc. B 280(1772): 20131502.

Rohr, J.R., J. Brown, W.A. Battaglin, T.A. McMahon and R.A. Relyea. 2017. A pesticide paradox: Fungicides indirectly increase fungal infections. Ecol. Appl. 27(8): 2290–2302.

Rosenfeld, C.S. 2017. Gut dysbiosis in animals due to environmental chemical exposures. Front. Cell. Infect. Microbiol. 7: 396.

Rubio, A.O., S.J. Kupferberg, V.V. García, A. Ttito, A. Shepack and A. Catenazzi. 2018. Widespread occurrence of the antifungal cutaneous bacterium *Janthinobacterium lividum* on Andean water frogs threatened by fungal disease. Dis. Aquat. Org. 131(3): 233–238.

Ruthsatz, K., M.L. Lyra, C. Lambertini, A.M. Belasen, T.S. Jenkinson, D. da Silva Leite et al. 2020. Skin microbiome correlates with bioclimate and *Batrachochytrium dendrobatidis* infection intensity in Brazil's Atlantic Forest treefrogs. Sci. Rep. 10(1): 1–15.

Sabino-Pinto, J., P. Galán, S. Rodríguez, M.C. Bletz, S. Bhuju, R. Geffers et al. 2017. Temporal changes in cutaneous bacterial communities of terrestrial-and aquatic-phase newts (Amphibia). Environ. Microbiol. 19(8): 3025–3038.

Sanchez, E., M.C. Bletz, L. Duntsch, S. Bhuju, R. Geffers, M. Jarek et al. 2017. Cutaneous bacterial communities of a poisonous salamander: A perspective from life stages, body parts and environmental conditions. Microb. Ecol. 73(2): 455–465.

Santos, B., M.C. Bletz, J. Sabino-Pinto, W. Cocca, J.F.S. Fidy, K.L. Freeman et al. 2021. Characterization of the microbiome of the invasive Asian toad in Madagascar across the expansion range and comparison with a native co-occurring species. PeerJ. 9: e11532.

Scheele, B.C., F. Pasmans, L.F. Skerratt, L. Berger, A.N. Martel, W. Beukema et al. 2019. Amphibian fungal panzootic causes catastrophic and ongoing loss of biodiversity. Science 363(6434): 1459–1463.

Schloss, P.D. 2021. Amplicon sequence variants artificially split bacterial genomes into separate clusters. mSphere 6(4): e00191–21.

Shapira, M. 2016. Gut microbiotas and host evolution: scaling up symbiosis. Trends Ecol. Evol. 31(7): 539–549.

Shen, Y., Z. Jiang, X. Zhong, H. Wang, Y. Liu and X. Li. 2022. Manipulation of cadmium and diethylhexyl phthalate on *Rana chensinensis* tadpoles affects the intestinal microbiota and fatty acid metabolism. Sci. Total Environ. 821: 153455.

Shi, Y.B. and A. Ishizuya-Oka. 1996. 7 Biphasic intestinal development in amphibians: eEmbryogenesis and remodeling during metamorphosis. Curr. Top. Dev. Biol. 32: 205–235.

Shi, Y.B., Y. Shibata, Y. Tanizaki and L. Fu. 2021. The development of adult intestinal stem cells: Insights from studies on thyroid hormone-dependent anuran metamorphosis. pp. 269–293. *In*: Litwack, G. [ed.]. Hormones and Stem Cells. Academic Press, Cambridge, MA, USA.

Shu, Y., H. Jiang, C.N. Yuen, W. Wang, J. He, H. Zhang et al. 2022. Microcystin-leucine arginine induces skin barrier damage and reduces resistance to pathogenic bacteria in *Lithobates catesbeianus* tadpoles. Ecotoxicol. Environ. Saf. 238: 113584.

Shuman-Goodier, M.E., G.R. Singleton and C.R. Propper. 2017. Competition and pesticide exposure affect development of invasive (*Rhinella marina*) and native (*Fejervarya vittigera*) rice paddy amphibian larvae. Ecotoxicology 26(10): 1293–1304.

Song, X., J. Zhang, J. Song and Y. Zhai. 2021. Decisive effects of life stage on the gut microbiota discrepancy between two wild populations of hibernating asiatic toads (*Bufo gargarizans*). Front. Microbiol. 12: 665849.

Steinwascher, K. 1978. The effect of coprophagy on the growth of *Rana catesbeiana* tadpoles. Copeia 130–134.

Sterner, Z.R., L.H. Shewade, K.M. Mertz, S.M. Sturgeon and D.R. Buchholz. 2020. Glucocorticoid receptor is required for survival through metamorphosis in the frog *Xenopus tropicalis*. Gen. Comp. Endocrinol. 291: 113419.

Sterner, Z.R. and D.R. Buchholz. 2022. Glucocorticoid receptor mediates corticosterone-thyroid hormone synergy essential for metamorphosis in *Xenopus tropicalis* tadpoles. Gen. Comp. Endocrinol. 315: 113942.

Su, C., Y. Cui, D. Liu, H. Zhang and Y. Baninla. 2020. Endocrine disrupting compounds, pharmaceuticals and personal care products in the aquatic environment of China: Which chemicals are the prioritized ones? Sci. Total Environ. 720: 137652.

Sullam, K.E., S.D. Essinger, C.A. Lozupone, M.P. O'Connor, G.L. Rosen, R.O.B. Knight et al. 2012. Environmental and ecological factors that shape the gut bacterial communities of fish: A meta-analysis. Mol. Ecol. 21(13): 3363–3378.

Tamargo, A., N. Molinero, J.J. Reinosa, V. Alcolea-Rodriguez, R. Portela, M.A. Bañares et al. 2022. PET microplastics affect human gut microbiota communities during simulated gastrointestinal digestion, first evidence of plausible polymer biodegradation during human digestion. Sci. Rep. 12(1): 1–15.

Tata, J.R. 2006. Amphibian metamorphosis as a model for the developmental actions of thyroid hormone. Mol. Cell. Endocrinol. 246(1-2): 10–20.

Thambirajah, A.A., E.M. Koide, J.J. Imbery and C.C. Helbing. 2019. Contaminant and environmental influences on thyroid hormone action in amphibian metamorphosis. Front. Endocrinol. 10: 276.

Todd, B.D., C.M. Bergeron, M.J. Hepner and W.A. Hopkins. 2011. Aquatic and terrestrial stressors in amphibians: A test of the double jeopardy hypothesis based on maternally and trophically derived contaminants. Environ. Toxicol. Chem. 30(10): 2277–2284.

Tong, Q., X.P. Du, Z.F. Hu, L.Y. Cui, J. Bie, Q.Z. Zhang et al. 2019. Comparison of the gut microbiota of *Rana amurensis* and *Rana dybowskii* under natural winter fasting conditions. FEMS Microbiol. Lett. 366(21): fnz241.

Tong, Q., Z.F. Hu, X.P. Du, J. Bie and H.B. Wang. 2020. Effects of seasonal hibernation on the similarities between the skin microbiota and gut microbiota of an amphibian (*Rana dybowskii*). Microb. Ecol. 79(4): 898–909.

Trung, B., T.S. Dao, E. Faassen and M. Lürling. 2018. Cyanobacterial blooms and microcystins in Southern Vietnam. Toxins 10(11): 471.

Ujszegi, J., B. Vajna, Á.M. Móricz, D. Krüzselyi, K. Korponai, G. Krett et al. 2020. Relationships between chemical defenses of common Toad (*Bufo bufo*) tadpoles and bacterial community structure of their natural aquatic habitat. J. Chem. Ecol. 46(5): 534–543.

Ultsch, G.R., D.F. Bradford and J. Freda. 1999. Physiology: coping with the environment. pp. 189–214. *In*: McDiarmid, R.W. and R. Altig [eds.]. Tadpoles: The biology of anuran larvae. University of Chicago Press, Chicago, IL, USA.

Varela, B.J., D. Lesbarrères, R. Ibáñez and D.M. Green. 2018. Environmental and host effects on skin bacterial community composition in Panamanian frogs. Front. Microbiol. 9: 298.

Venâncio, C., A. Gabriel, M. Oliveira and I. Lopes. 2022. Feeding exposure and feeding behaviour as relevant approaches in the assessment of the effects of micro (nano) plastics to early life stages of amphibians. Environ. Res. 113476.

Vences, M., M.L. Lyra, J.G. Kueneman, M.C. Bletz, H.M. Archer, J. Canitz et al. 2016. Gut bacterial communities across tadpole ecomorphs in two diverse tropical anuran faunas. Sci. Nat. 103(3): 1–14.

Viriato, C., F.M. Franca, D.S. Santos, A.S. Marcantonio, C. Badaro-Pedroso and C.M. Ferreira. 2021. Evaluation of the potential teratogenic and toxic effect of the herbicide 2, 4-D (DMA® 806) in bullfrog embryos and tadpoles (*Lithobates catesbeianus*). Chemosphere 266: 129018.

Wagner, N., W. Reichenbecher, H. Teichmann, B. Tappeser and S. Lötters. 2013. Questions concerning the potential impact of glyphosate-based herbicides on amphibians. Environ. Toxicol. Chem. 32(8): 1688–1700.

Walke, J.B., M.H. Becker, S.C. Loftus, L.L. House, G. Cormier, R.V. Jensen et al. 2014. Amphibian skin may select for rare environmental microbes. ISME J. 8(11): 2207–2217.

Wang, C., G. Liang, L. Chai and H. Wang. 2016. Effects of copper on growth, metamorphosis and endocrine disruption of *Bufo gargarizans* larvae. Aquat. Toxicol. 170: 24–30.

Wang, Y., H.K. Smith, E. Goossens, L. Hertzog, M.C. Bletz, D. Bonte et al. 2021a. Diet diversity and environment determine the intestinal microbiome and bacterial pathogen load of fire salamanders. Sci. Rep. 11(1): 1–11.

Wang, H., C. Xu, Y. Liu, E. Jeppesen, J.C. Svenning, J. Wu et al. 2021b. From unusual suspect to serial killer: Cyanotoxins boosted by climate change may jeopardize megafauna. The Innovation 2(2): 100092.

Wiebler, J.M., K.D. Kohl, R.E. Lee Jr. and J.P. Costanzo. 2018. Urea hydrolysis by gut bacteria in a hibernating frog: Evidence for urea-nitrogen recycling in Amphibia. Proc. Royal Soc. B. 285(1878): 20180241.

Wiggins, P.J., J.M. Smith, R.N. Harris and K.P. Minbiole. 2011. Gut of red-backed salamanders (*Plethodon cinereus*) may serve as a reservoir for an antifungal cutaneous bacterium. J. Herpetol. 45(3): 329–332.

Woodhams, D.C., H. Brandt, S. Baumgartner, J. Kielgast, E. Küpfer, U. Tobler et al. 2014. Interacting symbionts and immunity in the amphibian skin mucosome predict disease risk and probiotic effectiveness. PloS one 9(4): e96375.

Woodhams, D.C., R.A. Alford, R.E. Antwis, H. Archer, M.H. Becker, L.K. Belden et al. 2015. Antifungal isolates database of amphibian skin-associated bacteria and function against emerging fungal pathogens: Ecological Archives E096-059. Ecology 96(2): 595–595.

Wu, G.D., J. Chen, C. Hoffmann, K. Bittinger, Y.Y. Chen, S.A. Keilbaugh et al. 2011. Linking long-term dietary patterns with gut microbial enterotypes. Science 334(6052): 105–108.

Xie, L., Y. Zhang, J. Gao, X. Li and H. Wang. 2020. Nitrate exposure induces intestinal microbiota dysbiosis and metabolism disorder in *Bufo gargarizans* tadpoles. Environ. Pollut. 264: 114712.

Ya, J., Z. Ju, H. Wang and H. Zhao. 2019. Exposure to cadmium induced gut histopathological damages and microbiota alterations of Chinese toad (*Bufo gargarizans*) larvae. Ecotoxicol. Environ. Saf. 180: 449–456.

Ya, J., X. Li, L. Wang, H. Kou, H. Wang and H. Zhao. 2020. The effects of chronic cadmium exposure on the gut of *Bufo gargarizans* larvae at metamorphic climax: Histopathological impairments, microbiota changes and intestinal remodeling disruption. Ecotoxicol. Environ. Saf. 195: 110523.

Yang, Y., X. Song, A. Chen, H. Wang and L. Chai. 2020. Exposure to copper altered the intestinal microbiota in Chinese brown frog (*Rana chensinensis*). Environ. Sci. Pollut. Res. 27(12): 13855–13865.

Yang, B., Z. Cui, M. Ning, Y. Chen, Z. Wu and H. Huang. 2022. Variation in the intestinal microbiota at different developmental stages of *Hynobius maoershanensis*. Ecol. Evol. 12(3): e8712.

Yao, Q., H. Yang, X. Wang and H. Wang. 2019. Effects of hexavalent chromium on intestinal histology and microbiota in *Bufo gargarizans* tadpoles. Chemosphere 216: 313–323.

Ye, H., Y. Zhang, L. Wei, H. Feng, Q. Fu and Z. Guo. 2022. Waterborne Cr3+ and Cr 6+ exposure disturbed the intestinal microbiota homeostasis in juvenile leopard coral grouper *Plectropomus leopardus*. Ecotoxicol. Environ. Saf. 239: 113653.

Zhang, W., R. Guo, Y. Yang, J. Ding and Y. Zhang. 2016. Long-term effect of heavy-metal pollution on diversity of gastrointestinal microbial community of *Bufo raddei*. Toxicol. Lett. 258: 192–197.

Zheng, R., X. Chen, C. Ren, Y. Teng, Y. Shen, M. Wu et al. 2020. Comparison of the characteristics of intestinal microbiota response in *Bufo gargarizans* tadpoles: Exposure to the different environmental chemicals (Cu, Cr, Cd and NO3–N). Chemosphere 247: 125925.

Zheng, R., P. Wang, B. Cao, M. Wu, X. Li, H. Wang et al. 2021a. Intestinal response characteristic and potential microbial dysbiosis in digestive tract of *Bufo gargarizans* after exposure to cadmium and lead, alone or combined. Chemosphere 271: 129511.

Zheng, R., M. Wu, H. Wang, L. Chai and J. Peng. 2021b. Copper-induced sublethal effects in *Bufo gargarizans* tadpoles: Growth, intestinal histology and microbial alternations. Ecotoxicology 30(3): 502–513.

Zhu, W., D. Yang, L. Chang, M. Zhang, L. Zhu and J. Jiang. 2022. Animal gut microbiome mediates the effects of antibiotic pollution on an artificial freshwater system. J. Hazard. Mater. 425: 127968.

10

Effects of Pharmaceutical Compounds in Tadpoles

Claudia Bueno dos Reis Martinez[1,2,*] and *Aline Aguiar*[2,3]

1. Introduction

Pharmaceuticals are considered one of the most potent emerging micropollutants of current times and their possible adverse effects on non-target organisms are gaining increased attention (Aliko et al. 2021). The term micropollutants describes organic contaminants detected in the environment at concentrations ranging from ng L^{-1} to μg L^{-1} and includes emerging organic contaminants (Figuière et al. 2022). The "emergent" status is attributable to the recent rise in their recognition as important contaminants whose environmental risks remain largely unknown, and, consequently, their release into receiving environments remains unregulated (Batucan et al. 2022). Pharmaceuticals enter freshwater bodies via discharge from wastewater treatment facilities, agricultural and aquaculture activities, landfill leachate, and groundwater contaminated by leaking sewage pipes (Fedaku et al. 2019) (Fig. 1).

The concepts of pharmaceutical compounds and drugs are similar since both are chemical compounds used to prevent, diagnose, treat, or cure a disease or other type of disorder (US-FDA 2022). Commonly, some of these products are denominated as medicines and used for diseases, while others are used for hygiene and beauty care. In general, drugs are chemical substances with a formula recognized by an official pharmacopoeia or formulary, which can promote biological effects in living organisms; while medicines or pharmaceutical drugs contain controlled doses of one or more drugs in their composition for therapeutic ends (Rang et al. 2011, US-FDA 2022). Historically, before the knowledge of pharmaceutical compounds and chemical molecules, people from Mesopotamia in ancient times, 2600 BC, used plants and herbs to treat patients (Cragg and Newman 2013, Atanasov et al. 2015). One of the main milestones in pharmacology comprises the

[1] Department of Physiological Sciences, PostGraduate Program in Biological Sciences, Londrina State University, Rod. Celso Garcia Cid, Km 380, 86051-970, Londrina, Paraná, Brazil. Email: cbueno@uel.br

[2] PostGraduate Program in Biological Sciences, Londrina State University, Rod. Celso Garcia Cid, Km 380, 86051-970, Londrina, Paraná, Brazil.

[3] Department of Biodiversity, Institute of Biosciences, São Paulo State University, Avenida 24A, 1515, 13506-900, Rio Claro, São Paulo, Brazil.

* Corresponding author: aline.aguiarr@gmail.com

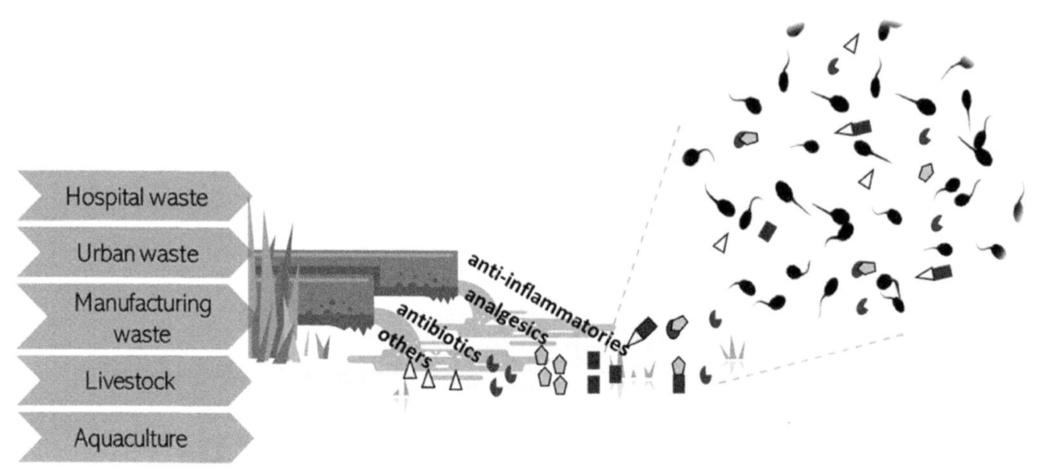

Figure 1. Main sources of pharmaceuticals into freshwater bodies.

beginning of the 19th century, with the discovery of opium as an analgesic and sleep-inducing agent, named morphine by Friedrich Sertürner (Sertürner 1817, Atanasov et al. 2015). In 1826, in Germany, pharmaceutical companies such as H.E. Merck started extracting morphine and other alkaloids (Atanasov et al. 2015). Another important discovery was the antibiotic Penicillin from the fungi *Penicillium* in 1928 by the British biologist Alexander Fleming, followed by the production of this antibiotic by pharmaceutical industries.

Therefore, with advances in chemistry and biology, numerous substances from microbial, animal, and plant sources have been discovered and synthetized by the pharmaceutical industry for use as medicines and other health care products. On the other side, their disposal has become an environmental issue in recent decades (Kümmerer 2009, Rivera-Utrilla et al. 2013). Since the 1970s evidence has been collected, and several studies have confirmed the presence of numerous pharmaceutical compounds in aquatic ecosystems as well as in wastewater treatment plants, effluent, and sludge around the world, including the Polar regions (Patel et al. 2019). Although recent data shows that 71 countries present some type of pharmaceutical compound in the natural environment, there is no specific regulation about the disposal of this contaminant of emerging concern. According to Patel et al. (2019), only Australia has implemented guidelines for these compounds in drinking water (Pomati 2007). Therefore, public laws are urgently needed for the disposal of several types of contaminants, including pharmaceutical compounds, which have been used intensively by the world population for many years, pushing the pharmaceutical industries to increase their production. The pharmaceutical industry and animal husbandry (use of veterinary pharmaceutical substances) have been recorded as the main contributors for contaminant dumping in the environment due to the lack of suitable waste management (der Beek et al. 2016). In addition, drug residues are often not fully eliminated during wastewater treatment, besides not being biodegraded (Amaral et al. 2019). Consequently, as a result of constant discharges, many components are able to accumulate in the environment and biota. These residuals can affect human health through the consumption of contaminated water as well as having consequences for environmental imbalance and loss of biodiversity (Rivera-Utrilla et al. 2013). Moreover, pharmaceuticals reaching water bodies and soil could be incorporated into plants, as already reported in cabbages, cucumbers, corn, and others (Kumar et al. 2005, Patel et al. 2019).

More than 600 pharmaceutical compounds are recorded in natural environments around the world (der Beek et al. 2016). Santos et al. (2010) inventoried the most frequent pharmaceutical compounds worldwide between 1997 and 2009, and found: anti-inflammatory drugs (16%), antibiotics (15%), lipid regulators (12%), hormones (9%), antiepileptic drugs (8%), and β-blockers (8%). Although hormones represent only 9%, they are the compound with the highest concentrations found in the environment (Quadra et al. 2017). According to Patel et al. (2019), the most used 100 pharmaceuticals include analgesics (i.e., Acetaminophen – Paracetamol) and antibiotics (Amoxicillin) which are presented by the authors in an extensive list with their classes, therapeutic applications, and important physicochemical properties.

There are different ways to classify a drug, for example, by mode of action, location on the body or type of tissue, or type of disease. According to the United States Food and Drugs Administration (US-FDA 2022) there are general drug categories such as *analgesics*, *antacids*, *antibiotics*, *corticosteroids*, *diuretics*, *laxatives*, and others. However, for scientific perspectives or medical orientations there are 51 drug classes and thousands of sub-classes and sub-categories (US-FDA 2022). This classification is important to drive treatment decisions and avoid pharmacological interactions. In this sense, a drug is classified using one of the following three attributes: (i) mechanism of action (pharmacological action at the receptor, membrane, or tissue level); (ii) physiological effect (pharmacological effect at the organ, system, or whole-body level), and (iii) chemical structure. For example, ibuprofen is a drug classified as (i) a cyclooxygenase inhibitor; (ii) for decreased prostaglandin production; and (iii) nonsteroidal anti-inflammatory drug (the most common designation for ibuprofen).

Assuming pharmaceutical compounds or drugs as chemical compounds, they have a great ability to interact and react, resulting in other compounds and making a "complex pharmaceutical pool" in the ecosystem (der Beek et al. 2016, Patel et al. 2019). In the same way, just as some medicines are activated in the body, by reaching acid or basic organs (i.e., stomach or intestines), and some vitamins are more absorbent when taken with lipids (i.e., vitamin D), these reactions may occur in the environment since different types of drugs can react in several ways according to the abiotic and biotic features of each ecosystem. From these interactions, several consequences in the environment can occur on different scales, from individual to the whole ecosystem. Although some interaction mechanisms are still not completely elucidated, we know that, depending on the dose, a drug can be curative or toxic to the individual. However, many drugs commonly used for therapeutic ends are indiscriminately reaching aquatic ecosystems, where they are accumulating and mixing with each other, directly and indirectly affecting all biodiversity.

In this context, many studies have been conducted with different organism models (i.e., microcrustaceans, fishes, crayfishes, algae, etc.) (see Patel et al. 2019). However, there are still many questions on this topic to be understood and evaluated, considering the long-term effects and interactions of the contaminant with relevant environmental variables, as well as the use of sentinel organisms in the experiments. In this sense, amphibians are interesting organisms due to their lifestyle, as they use both aquatic and terrestrial habitats, and because of their high sensitivity to many types of environmental impacts, such as forest fragmentation, habitat loss, and contaminants (Green et al. 2020, Langlois 2021). The sensitivity of amphibians is related to their complex and specific reproductive modes, which depend on specific sites for reproduction (Haddad and Prado 2005). Additionally, their permeable skin and physiological demands require a limited range of temperature and humidity which are commonly found in preserved forests.

Besides the habitat loss, deforestation can affect the transition between terrestrial and aquatic ecosystems where riparian forests have an essential role (Turunen et al. 2021). The vegetation of these forests works as a biological filter that can help to mitigate the impacts of contaminants on water bodies. Thus, the lack of riparian forests promotes a greater flow to aquatic ecosystems, which also increases the input of contaminants such as pharmaceutical compounds and, therefore, their effects on amphibians and other aquatic organisms. As a consequence, the populational decline in amphibians has increased and local extinctions have been confirmed owing to the direct and indirect effects of contaminants. Several studies have revealed contaminant effects on behavioral, physiological, immunological, and reproductive aspects, as well as the potential to increase the risk of infectious diseases by parasites in adults and tadpoles (Rohr et al. 2008, Ehrsam et al. 2016). Inherently, to reach the adult stage, tadpoles and froglets are subject to critical hormonal, morphological, and immunological alterations which, when combined with contaminants, can be lethal, so that they may not reach the reproductive stage and cannot sustain the population ratio. In this way, many populations could become extinct before they are discovered.

Studies on the effects of pharmaceutical compounds on tadpoles began less than two decades ago (Swiacka et al. 2022). Currently there are about 30 studies using tadpoles as models to investigate the effects of numerous pharmaceuticals in different species (Tables 1 and 2). From these studies, 31 pharmaceuticals in different concentrations were tested in experiments using 16 tadpole species, belonging to the families: Bufonidae—*Anaxyrus terrestris* (Bonnaterre 1789), *Bufo bufo* (Linnaeus 1758), *Rhinella arenarum* (Hensel 1867), and *Sclerophrys arabica* (Heyden 1827) (= *Bufo arabicus*); Hylidae—*Trachycephalus typhonius* (Linnaeus 1758); Leptodactylidae—*Physalaemus albonotatus* (Steindachner 1864), *Physalaemus cuvieri* Fitzinger 1826; Limnodynastidae—*Limnodynastes peronii* (Duméril and Bibron 1841); Pipidae—*Xenopus laevis* (Daudin1802), *Xenopus tropicalis* (Gray 1864), and *Xenopus* sp.; and Ranidae—*Lithobates catesbeianus* (Shaw 1802) (= *Rana catesbeiana*), *Lithobates pipiens* (Schreber 1782), *Rana dalmatina* Fitzinger 1838, *Rana omeimontis* Ye and Fei 1993, and *Pelophylax shqipericus* (Hotz et al. 1987) (Table 1). Among these species, the most widely tested, with numerous pharmaceuticals are *L. catesbeianus, L. peronii*, and *R. arenarum*, while the most studied compounds are analgesic and non-steroidal anti-inflammatory drugs (i.e., acetylsalicylic acid, diclofenac, ibuprofen, metamizole, and naproxen) and psychiatric drugs (fluoxetine and sertraline). Fluoxetine (from 0.03 μg L^{-1} to 10 μg L^{-1}) and diclofenac (from 0.1 μg L^{-1} to 64 mg L^{-1}) are the most tested drugs in five and four tadpole species, respectively, inducing different effects (Tables 1 and 2); whereas exposure to low concentrations of the antibiotic amoxicillin (0.0045 μg L^{-1}) was enough to promote behavior and morphological alterations in *L. catesbeianus* (Table 2) (Amaral et al. 2019).

In general, findings such as changes in growth and development, behavior alterations, and loss of tactile response are the main effects that have been found in different amphibian species under different pharmaceutical expositions (see Table 2). However, experiments conducted with embryos of *Xenopus* sp. exposed to diclofenac (1 to 64 mg L^{-1}) for 96 h showed mortality and teratogenic effects on tadpoles (Chae et al. 2015). Cory et al. (2019) also reported mortality and toxicity in tadpoles of *A. terrestris* when they studied the effects of naproxen and the products of its photo-transformation, showing that they are more toxic than the original pharmaceutical. This study highlights the importance of considering the products of a chemical reaction, as they are commonly neglected. Another important factor was studied by Melvin et al. (2018) considering the effects of methanol, which is commonly used for the compound dilution in the experiments. Their findings revealed considerable interactions between the pharmaceutical and methanol, as well as alterations in some metabolites of *L. peronii* tadpoles.

2. Effects of Pharmaceuticals on Amphibian Tadpoles

Human and veterinary pharmaceuticals target a range of physiological and metabolic processes, many of which are highly conserved across vertebrate species, including amphibians (Melvin 2016). Herein, we report some of these processes according to the pharmaceutical classes.

2.1 Analgesics and Anti-Inflammatories

Because of their volume of consumption and incomplete removal during wastewater treatment processes, steroidal and nonsteroidal anti-inflammatory drugs are among the most frequently detected pharmaceuticals in treatment plants and surface waters worldwide (Ribas et al. 2016). The occurrence of nonsteroidal anti-inflammatory drugs (NSAIDs) in waterbodies and the significant risk they pose to non-target organisms have received considerable attention in recent years (Sehonova et al. 2017).

Among them, **diclofenac** ([2-(2,6- dichloroaniline) phenyl acetate]), used as an analgesic and anti-inflammatory in human and veterinary medicine, has been identified as one of the

Table 1. Some pharmaceutical compounds (A = alone; M = in mixture) used in ecotoxicological studies on anurans tadpoles of 16 species, under laboratorial conditions. Reference codes: 1 (Amaral et al. 2019); 2 (Chae et al. 2015); 3 (Melvin 2016); 4 (Melvin et al. 2018); 5 (Peltzer et al. 2019); 6 (Cuzziol Boccioni et al. 2021); 7 (Aliko et al. 2021); 8 (Turani et al. 2019); 9 (Veldhoen et al. 2014); 10 (Melvin et al. 2014); 11 (Cory et al. 2019); 12 (Cuzziol Boccioni et al. 2020); 13 (Luz et al. 2021); 14 (Zhu et al. 2022); 15 (Carlsson et al. 2010); 16 (Rutkoski et al. 2022); 17 (Melvin et al. 2017); 18 (Bókony et al. 2020); 19 (Pablos et al. 2020); 20 (Barry 2018); 21 (Sehonova et al. 2019); 22 (Barry 2014); 23 (Conners et al 2009); 24 (Foster et al. 2010); 25 (Fernández et al. 2022); 26 (Amaral et al. 2018); 27 (Montalvão and Malafaia 2017). ([#]) Experiments started with embryos.

Pharmaceuticals and their classes	***Anaxyrus terrestris***	***Bufo bufo***	***Rhinella arenarum***	***Sclerophrys arabica***	***Trachycephalus typhonius***	***Physalaemus albonotatus***	***Physalaemus cuvieri***	***Limnodynastes peronii***	***Xenopus laevis***	***Xenopus tropicalis***	***Xenopus* sp.**	***Lithobates catesbeianus***	***Lithobates pipiens***	***Pelophylax shqipericus***	***Rana dalmatina***	***Rana omeimontis***	**References**
Analgesics and anti-inflammatories																	
Acetylsalicylic acid												M					1
Diclofenac					A	A		M			A	M					1; 2[#]–5
Dexamethasone			A														6
Ibuprofen		M/A										M/A		A			1; 7–9
Metamizole												M					1
Naproxen	A							M/A									3;10;11
Antacid																	
Ranitidine												M					1
Antibiotics																	
Amoxicillin			M									M					1; 12
Azithromycin							M/A										13
Ciprofloxacin			M/A														12
Erythromycin																A	14
Fluoroquinolone effluent										A							15
Oxytetracycline												A					16
Sulfamethoxazole								M/A				A				A	10; 14; 16
Tetracycline																A	14
Psychiatric drugs																	
Amitriptyline										A							21[#]
Carbamazepine		A						M/A	A						A		10; 18; 19
Clonazepam												M					1
Fluoxetine		M/A		M/A					A			M	A				1; 7; 20; 22–24
Sertraline				M/A					A	A							20; 21[#]; 23
Valproic Acid								M									4
Venlafaxine				M/A						A							20; 21[#]
Antidiabetic																	
Metformin								M									4; 17
Antihypertensive																	

Table 1 contd. ...

...Table 1 contd.

Pharmaceuticals and their classes	*Anaxyrus terrestris*	*Bufo bufo*	*Rhinella arenarum*	*Sclerophrys arabica*	*Trachycephalus typhonius*	*Physalaemus albonotatus*	*Physalaemus cuvieri*	*Limnodynastes peronii*	*Xenopus laevis*	*Xenopus tropicalis*	*Xenopus* sp.	*Lithobates catesbeianus*	*Lithobates pipiens*	*Pelophylax shqipericus*	*Rana dalmatina*	*Rana omeimontis*	References
Atenolol								M									3
Anthelmintic																	
Abamectin												A					26; 27
Anti-rheumatic																	
Hydroxychloroquine							M/A										13
Antiviral																	
Abacavir			A														25
Efavirenz			A														25
Lipid regulator																	
Atorvastatin								M									17
Bezafibrate								M									17
Gemfibrozil								M									3
Total of pharmaceutical for species	**1**	**3**	**5**	**3**	**1**	**1**	**2**	**10**	**3**	**4**	**1**	**11**	**1**	**1**	**1**	**3**	

most important NSAIDs present in water bodies. After consumption, diclofenac is generally not completely metabolized and is non-biodegradable and resistant to wastewater treatments, contributing to its presence in surface waters, at concentrations ranging from 2 ng L^{-1} to 100 μg L^{-1} (Table 3). In a study carried out by Peltzer et al. (2019), larvae of two Neotropical anuran species, *T. typhonius* and *P. albonotatus* were chronically exposed to sublethal concentrations of diclofenac (125–2000 μg L^{-1}). After exposure (20–22 days), the authors described abnormal development and growth rates, decreased body condition, morphological, visceral, and organ abnormalities, altered swimming behavior, variations in the activities of acetylcholinesterase and glutathione-S transferase, and neurotoxic and cardiotoxic alterations. The authors highlight that the widespread presence of diclofenac in aquatic environments and its deleterious effects at concentrations below 2000 μg L^{-1}, show the urgent need for research, pharmacovigilance, and exposure approaches using different biological models.

Ibuprofen (α-methyl-4-(isobutyl)phenylacetic acid) is another broadly used NSAID. However, little is known about its effects on aquatic organisms during sensitive developmental periods, which is concerning given the multiplicity of molecular pathways ibuprofen targets and its abundance in global freshwater environments (Thambirajah et al. 2019). Thus, using a combination of DNA microarray, quantitative real-time polymerase chain reaction, and quantitative nuclease protection assays, Veldhoen et al. (2014) assessed the ability of environmentally relevant concentrations of ibuprofen (15 μg L^{-1}) to function as a disruptor of endocrine-mediated post-embryonic development of *L. catesbeiana* tadpoles. The authors observed that animals exposed to ibuprofen demonstrated disruption of thyroid hormone-mediated reprogramming in the liver transcriptome, affecting constituents of several metabolic, developmental, and signaling pathways. Additionally, ibuprofen (at 5 μg L^{-1}) caused several hematological damages, especially erythrocyte-related, in tadpoles of *P. shqipericus* after 48 h of exposure (Turani et al. 2019). These authors suggested that ibuprofen

Table 2. Concentrations (min-max), time exposure, and main findings of pharmaceutical compounds used in ecotoxicological studies on 16 species of tadpoles under laboratorial conditions.

References	Pharmaceutical	Concentrations (μg L^{-1})	Alone or in mixture	Time of exposure	Main findings
Aliko et al. (2021)	Fluoxetine	5	alone	7 days	changes in growth and development
	Ibuprofen	5	alone	7 days	
	Fluoxetine	5	in mixture	7 days	
	Ibuprofen	5			
Amaral et al. (2018)	Abamectin	18–72	alone	12 days	behavioral and morphological changes; cytotoxic effects
Amaral et al. (2019)	Acetylsalicylic acid	0.34	in mixture	15 days	changes in mandibular sheath pigmentation, dentition, and behavior; nuclear abnormalities in erythrocytes
	Amoxicillin	0.0045			
	Clonazepam	0.053			
	Diclofenac	0.15			
	Fluoxetine	0.030			
	Ibuprofen	0.07			
	Metamizole	5			
	Ranitidine	0.01			
Barry (2014)	Fluoxetine	0.03 – 3	alone	14 days	vulnerability to predation
Barry (2018)	Fluoxetine	0.5 – 2.0	alone	14 days	reduced swimming speed
	Sertraline	0.5 – 2.0	alone	14 days	
	Venlafaxine	0.5 – 2.0	alone	14 days	
	Fluoxetine	0.5 – 2.0	in mixture	14 days	
	Sertraline	0.5 – 2.0			
	Venlafaxine	0.5 – 2.0			
Cuzziol Boccioni et al. (2021)	Ciprofloxacin	10	alone	14 days	morphological abnormalities
	Ciprofloxacin	10 – 100	in mixture	14 days	morphological abnormalities; decrease in T4 levels and lower development
	Glyphosate	1,250 – 2,500			
	Amoxicillin	10 – 100	in mixture	28 days	increased T4 levels, inhibited AChE activities, and lower development
	Chlorpyrifos	1.25 – 2.5			
	Ciprofloxacin	10 – 100			
	Glyphosate	1,250 – 2,500			
Cuzziol Boccioni et al. (2020)	Dexamethasone	1 – 1,000	alone	22 days	changes in development; increased number and size of melanomacrophages; increased levels of glutathione S-transferase

Table 2 contd. ...

...*Table 2 contd.*

References	Pharmaceutical	Concentrations (μg L^{-1})	Alone or in mixture	Time of exposure	Main findings
Bókony et al. (2020)	Carbamazepine	0.5 – 50	alone	21 days	decreased spleen size and increased pigmentation
Chae et al. (2015)	Diclofenac	1,000 – 64,000	alone	96 hours	mortality and teratogenic effects
Conners et al. (2009)	Fluoxetine	0.1 – 10	alone	70 days	reduced growth at metamorphosis; acceleration of development
	Sertraline	0.1 – 10	alone	70 days	
Cory et al. (2019)	Naproxen and its phototransformation products	3,000 – 5,000	alone	96 hours	toxicity and mortality
Fernández et al. (2022)	Abacavir	0.5 – 10	alone	96 hours	high bioaccumulation levels
	Efavirenz	0.5 – 10	alone	96 hours	
Foster et al. (2010)	Fluoxetine	0.1 – 1	alone	45 days or more (until metamorphosis)	delayed development
Luz et al. (2021)	Azithromycin	12.5	alone	72 hours	increased activity of the enzymes SOD and catalase; increase of BChE
	Hydroxychloroquine	12.5	alone	72 hours	
	Azithromycin	12.5	in mixture	72 hours	
	Hydroxychloroquine	12.5			
Melvin (2016)	Atenolol	0.1 – 1,000	in mixture	30 days	reduced developmental rates and liver-somatic index; altered levels of hepatic triglycerides and peroxidase activity
	Diclofenac	0.1 – 1,000			
	Gemfibrozil	0.1 – 1,000			
	Naproxen	0.1 – 1,000			
Melvin et al. (2014)	Carbamazepine	10 – 100	alone	Gosner 26 until limb emergence	loss of tactile response
	Naproxen	10 – 100	alone	Gosner 26 until limb emergence	
	Sulfamethoxazole	10 – 100	alone	Gosner 26 until limb emergence	
	Carbamazepine	10 – 100	in mixture	Gosner 26 until limb emergence	
	Naproxen	10 – 100			
	Sulfamethoxazole	10 – 100			
Melvin et al. (2017)	Atorvastatin	0.5 – 500	in mixture	30 days	eliciting physiological and developmental effects
	Bezafibrate	0.5 – 500			
	Metformin	0.5 – 500			
Melvin et al. (2018)	Diclofenac	5	in mixture	30 days	differences for leucine, acetate, glutamine, citrate, glycogen, tyrosine, arginine, purine nucleotides
	Metformin	50			
	Valproic Acid	500			

Table 2 contd. ...

...Table 2 contd.

References	Pharmaceutical	Concentrations ($\mu g\ L^{-1}$)	Alone or in mixture	Time of exposure	Main findings
Montalvão and Malafaia (2017)	Abamectin	36 – 72	alone	24, 48, and 72 hours	cytotoxic effects
Pablos et al. (2020)	Carbamazepine	9 – 90	alone	21 days	accelerated development and histological alterations in the thyroid gland
Peltzer et al. (2019)	Diclofenac	125 – 2,000	alone	20 days	changes in morphology, development, growth rates, swimming, and activity of acetylcholinesterase
	Diclofenac	125 – 2,000	alone	22 days	
Rutkoski et al. (2022)	Sulfamethoxazole	0.02 – 0.46	alone	16 days	increased number of lymphocytes; decreased number of neutrophils and activity of glutathione S-transferase; liver damage
	Oxytetracycline	0.02 – 0.46	alone	16 days	
Sehonova et al. (2019)	Amitriptyline	0.3 – 3,000	alone	48 hours	lethal and sublethal effects; swimming alterations; effects on mRNA expression of genes
	Sertraline	0.1 – 1,000	alone	48 hours	
	Venlafaxine	0.3 – 3,000	alone	48 hours	
Turani et al. (2019)	Ibuprofen	5	alone	48 hours	high frequency of micronucleated erythrocytes; cellular and nuclear vacuolization, collapse and rupture of the cell membrane
Veldhoen et al. (2014)	Ibuprofen	15	alone	6 days	changes in postembryonic development
Zhu et al. (2022)	Tetracycline	0.1 – 1	alone	60 days	reduced fitness with increased mortality and physiological abnormality; reshaped gut bacterial and fungi diversity and composition
	Erythromycin	0.1 – 1	alone	60 days	
	Sulfamethoxazole	0.05 – 0.2	alone	60 days	

causes oxidative stress followed by eryptosis and health impairment. More recently, the work by Aliko et al. (2021) with tadpoles of *B. bufo* exposed to ibuprofen (5 $\mu g\ L^{-1}$) reinforced that this NSAID negatively affected the tadpoles' growth and development, increased the frequency of erythrocytic cellular and nuclear abnormalities, and also impaired behavioral responses, with a significant increase in unresponsiveness to different stimuli.

Naproxen (*S-2-(6-Methoxynaphthalen-2-yl) propanoic acid)* is an NSAID frequently detected in environmental samples (Table 3), which, under sunlight, is quickly transformed into two primary phototransformation products. Bearing this in mind, Cory et al. (2019) used tadpoles of the southern toad (*A. terrestris*) to evaluate the acute toxicity (EC50) of naproxen and two of its photo-

Table 3. Environmental occurrence of the pharmaceuticals cited in the present work whose effects were determined in amphibian tadpoles.

Pharmaceutical	Concentration range (µg L^{-1})	Water type	Continent or country	References	Reviewed by
Abacavir	0.0026	Surface water	Europe	Aminot et al. (2015)	Nannou et al. (2020)
Acetylsalicylic acid	0.13 – 20.96	Surface water	Worldwilde	-	der Beek et al (2016)
Amoxicillin	0.017	Surface water	Brazil	Locatelli et al. (2011)	Quadra et al. (2017)
Atenolol	0.0142 – 0.18	Surface water	Worldwilde	-	Patel et al. (2019)
	1.44	Surface water	Brazil	Gonçalves (2012)	Quadra et al. (2017)
	0.195	Tap water	USA	Subedi et al. (2015)	Patel et al. (2019)
	0.0036	Ground water	Taiwan	Lin et al. (2015)	Patel et al. (2019)
Atorvastatin	0.0006 – 0.0015	Surface water	Asia and Europe	-	Patel et al. (2019)
	1.02	Drinking water sources	Brazil	Reis et al. (2019)	Reis et al. (2021)
Azithromycin	1.1	Surface water	Croatia	Ivešić et al. (2017)	Fekadu et al. (2019)
	0.03 – 0.167	Surface water	Asia and Europe	-	Patel et al. (2019)
Bezafibrate	0.007 – 0.314	Surface water	Asia and Europe	-	Patel et al. (2019)
	0.606	Surface water	Brazil	Gonçalves (2012)	Quadra et al. (2017)
Carbamazepine	4.5 – 8.1	Surface water	Asia	-	der Beek et al. (2016)
	0.000144 – 0.214	Surface water	Asia and Europe	-	Patel et al. (2019)
	0.310	Surface water	Brazil	Gonçalves (2012)	Quadra et al. (2017)
	8	Surface water	Nigeria	Ogunbanwo et al. (2022)	-
Ciprofloxacin	0.031 – 6,500	Surface water	Worldwilde	-	der Beek et al. (2016)
	0.0015 – 0.0887	Surface water	Europe	-	Patel et al. (2019)
	0.119	Surface water	Brazil	Locatelli et al. (2011)	Quadra et al. (2017)
	14.33	Surface water	South African	Gezahegn et al. (2019)	Omotola et al. (2022)
Dexamethasone	0.031 – 0.352	Hospital effuents	Portugal	Santos et al. (2013)	Patel et al. (2019)

Table 3 contd. ...

...Table 3 contd.

Pharmaceutical	Concentration range (μg L^{-1})	Water type	Continent or country	References	Reviewed by
Diclofenac	0.0002 – 10.2	surface water	Worldwide	-	Patel et al. (2019)
	1.52 – 18.74	Surface water	Worldwide	-	der Beek et al. (2016)
Efavirenz	0.020 – 0.56	Surface water	Africa	-	Fedaku et al. (2019)
Erythromycin	0.0004 – 0.292	Surface water	Asia and Europe	-	Patel et al. (2019)
	5.30	Surface water	Croatia	Ivešić et al. (2017)	Fekadu et al. (2019)
Fluoxetine	0.0004 – 0.0195	Surface water	Asia and Europe	Wu et al. (2015)	Patel et al. (2019)
	0.0003 – 0.41	Surface water	Worldwide	Sumpter and Margiotta-Casaluci (2022)	-
Gemfibrozil	0.00005 – 0.0023	Surface water	Asia	-	Patel et al. (2019)
Ibuprofen	0.0001 – 1.32	Surface water	Asia and USA	-	Patel et al. (2019)
	0.524 – 17.60	Surface water	South Africa	Gumbi et al. (2017)	Patel et al. (2019)
	0.333	Drinking water sources	Brazil	Reis et al. (2019)	Reis et al. (2021)
	20.5 – 303	Surface water	Worldwide	-	der Beek et al. (2016)
	0.00023 – 303.0	Surface water	Worldwide	Sumpter and Margiotta-Casaluci (2022)	-
Metformin	0.203	Drinking water sources	Brazil	Reis et al. (2019)	Reis et al. (2021)
	3.1	Surface water	Germany	Scheurer et al. (2012)	Fekadu et al. (2019)
	0.0084	Surface water	Sweden	Lindim et al. (2016)	Patel et al. (2019)
	10	Surface water	Nigeria	Ogunbanwo et al. (2022)	-
Naproxen	0.07 – 32	Surface water	Worldwide	-	der Beek et al. (2016)
	0.00022 – 59.3	Surface water	Worldwide	-	Patel et al. (2019)
Ranitidine	0.0544	Surface water	Serbia	Petrović et al. (2014)	Patel et al. (2019)
Sertraline	0.0001 – 0.088	Surface water	Worldwide	Sumpter and Margiotta-Casaluci (2022)	-

Table 3 contd. ...

...Table 3 contd.

Pharmaceutical	Concentration range (μg L^{-1})	Water type	Continent or country	References	Reviewed by
Sulfamethoxazole	0.11 – 29	Surface water	Worldwide		der Beek et al. (2016)
	0.000075 – 0.115	Surface water	Worldwide	-	Patel et al. (2019)
	0.467	Surface water	Brazil	Monteiro et al. (2016)	Quadra et al. (2017)
	129	Surface water	Nigeria	Ogunbanwo et al. (2022)	-
Valproic Acid	0.019	Surface water	Worldwide	Sumpter and Margiotta-Casaluci (2022)	-
Venlafaxine	0.0053 – 0.159	surface water	Europe	-	Patel et al. (2019)
	0.00002 – 2.61	surface water	Worldwide	Sumpter and Margiotta-Casaluci (2022)	-
	0.0401	Surface water	Brazil	Gonçalves (2012)	Quadra et al. (2017)

transformation products. The authors showed that both products, with mean values of EC_{50} (96 h) of 20.7 mg L^{-1} and 8.4 mg L^{-1}, were more toxic than the naproxen itself (EC_{50} of 129 mg L^{-1}). These results indicate that the photo-transformation products may be of greater environmental concern compared to naproxen itself and point out that the release of a relatively safe pharmaceutical into the aquatic environment, when followed by exposure to solar radiation in surface waters, may be transformed into compounds that are many times more toxic (Cory et al. 2019).

The synthetic glucocorticoid **dexamethasone** is a steroidal anti-inflammatory drug commonly prescribed in human and veterinary medicine. It has been detected at relatively high levels in sewage effluents and rivers (Table 3) due to its relative stability in the environment (Shen et al. 2020). In a study conducted by Cuzziol Boccioni et al. (2020), tadpoles of *R. arenarum* were exposed to dexamethasone (1–1000 μg L^{-1}) for 22 days. The authors showed that dexamethasone exposure affected normal larval development and growth, being teratogenic to general body configuration. Preneoplastic processes, such as dysplastic cells of digestive and central nervous systems, and misplaced organs, such as the heart, kidney, and eyes were also determined. These results, based on a variety of biomarkers, provided clear evidence of toad larvae sensitivity to dexamethasone, and the ecotoxicological risk of this pharmaceutical, commonly found in different water bodies worldwide to aquatic animals (Cuzziol Boccioni et al. 2020).

2.2 Antibiotics

Data on the occurrence of antibiotics in fresh waters at several locations in the world (Table 3) indicate concentrations of up to 15 μg L^{-1} in the Americas, higher concentrations in Europe and Africa (over 10 μg L^{-1} and 50 μg L^{-1} respectively), and concentrations of over 450 μg L^{-1} in China (Danner et al. 2019). Antibiotics in the environment, besides causing concern because of the introduction and spread of resistant bacteria, can also adversely affect the physiology of exposed vertebrates and invertebrates (Rutkoski et al. 2022). The fitness of amphibian larvae to evaluate the influences of antibiotic pollution on aquatic ecosystems should be highly valued (Zhu et al. 2022).

Among the antibiotics frequently detected in surface waters at relative high concentrations are **sulfamethoxazole**, **(oxy)tetracycline**, and **erythromycin** (Zhu et al. 2022). Tadpoles of *L. catesbeianus* exposed to sulfamethoxazole (90 and 460 ng L^{-1}) and oxytetracycline (90 and 460 ng L^{-1}) for 16 days showed decreased activity of glutathione S-transferase and histopathological alterations in the liver, indicating hepatic damage (Rutkoski et al. 2022). The longer exposure (60 days) of *R. omeimontis* tadpoles to current environmental concentrations of sulfamethoxazole (50 and 200 ng L^{-1}), tetracycline (100 and 1000 ng L^{-1}), and erythromycin (100 and 1000 ng L^{-1}) led to malformation, or death, and affected the progression of metamorphosis (Zhu et al. 2022). The authors depicted that sulfamethoxazole was the most detrimental among the 3 antibiotics tested, causing a mortality rate of over 10% and a malformation rate of over 20% at the higher concentration (200 ng L^{-1}). In addition, it was shown that reshaping of the bacterial and fungi diversity and composition of the tadpole gut partly accounted for the tadpole's health condition (Zhu et al. 2022).

Enrofloxacin and **ciprofloxacin** belong to another class of antibiotics (fluoroquinolones), widely used in human and veterinary medicine, and which persist for a long time in the aquatic environment. They can be found in numerous water matrices worldwide due to their massive doses and frequent applications in humans and livestock (Table 3). Ciprofloxacin has been reported at concentrations up to 16 μg L^{-1} in several wastewaters and effluents (Danner et al. 2019). Peltzer et al. (2017) exposed tadpoles of the South American common toad *R. arenarum* to environmentally relevant concentrations of both enrofloxacin and ciprofloxacin (1–1000 μg L^{-1}) for 96 h. The authors showed that both antibiotics induced impairments in development, growth, and antioxidant enzyme activities. These effects might exert potential damage to long-term maintenance of larvae populations continuously exposed to the input of antibiotics in their natural habitat (Peltzer et al. 2017). More recently, in a study conducted by Cuzziol Boccioni et al. (2021), eggs of *R. arenarum* were exposed to ciprofloxacin (10 μg L^{-1} and 100 μg L^{-1}) for two weeks, until the tadpoles reached Gosner Stage 26 (indicating the beginning of premetamorphosis). The authors observed morphological abnormalities in the tadpoles, suggesting a teratogenic effect. The most frequent abnormalities included body alterations (bilateral asymmetry and swollen body or edemas), as well as abnormalities in specific organs or regions, such as in the eye shape, mouth and gill chamber structure, visceral organization, and tail orientation and shape.

The COVID-19 pandemic led to a dramatic increase in the production of hospital waste, amplifying the arrival and dispersion of some pharmaceuticals into the aquatic environment, among them **azithromycin**, a macrolide antibiotic which inhibits bacterial protein synthesis (Table 3). Tadpoles of the Neotropical species *P. cuvieri*, exposed to azithromycin (12.5 μg L^{-1}) for 72 h, showed increased antioxidant defenses (Luz et al. 2021). In addition, the authors suggested that a strong interaction of azithromycin with acetylcholinesterase (AChE) was associated with the anticholinesterase effect observed in the groups exposed to the antibiotic.

2.3 Anthelmintics

Anthelmintics comprise the largest sector of the animal pharmaceutical industry by volume and value. They are anti-parasitic and mainly used for the prevention and control of stomach worms that are found primarily in cattle. **Abamectin** is one of the most commonly used compounds in the avermectin group (macrocyclic lactones), both in crop protection and animal husbandry (Omotola et al. 2022).

In a study carried out by Montalvão and Matafaia (2017), tadpoles of *L. catesbeianus* were exposed to low abamectin concentrations (36 and 72 μg L^{-1}) and the presence of micronucleus and other nuclear abnormalities in blood erythrocytes were checked during 72 h of exposure. The results evidenced that exposure of the tadpoles to abamectin, even for a short period of time and at low concentrations, caused a cytotoxic effect, indicated by the significant increase in the number of nuclear abnormalities in the erythrocytes of the animals exposed to this anthelmintic.

2.4 Antivirals

The presence of antiviral drugs in the environment is an emerging issue that raises concerns, as some of them are highly bioactive and may negatively affect non-target organisms and persist in aquatic environments. Despite their high consumption rates, and the fact that various antiviral drugs have been detected, they are not often systematically monitored in the aquatic environment to the same extent as other pharmaceuticals (Nannou et al. 2020). Among the antiretrovirals, two, **abacavir** and **efavirenz**, that are usually prescribed in combination in the treatment of human immunodeficiency virus (HIV) have been detected in water bodies (Table 3). However, data on the bioaccumulation and possible environmental risks posed by these drugs to aquatic organisms are still very scarce (Fernández et al. 2022).

The bioaccumulation of abacavir and efavirenz was determined in tadpoles of *R. arenarum* exposed to both drugs separately, at environmentally relevant concentrations (0.5, 1.0, and 10.0 µg L^{-1}) for 96 h (Fernández et al. 2022). The authors demonstrated the high bioaccumulation levels of these antiretrovirals, particularly efavirenz, in amphibian tadpoles after only a few days of exposure at concentrations similar to those found in the environment, indicating an ecological risk for populations of *R. arenarum* and probably to other aquatic organisms exposed to these drugs in water bodies.

2.5 Psychiatric Drugs

Anti-depressants are one of the most widely prescribed classes of medication and often found in different types of water bodies around the world (Table 3). Among them, selective serotonin reuptake inhibitors (SSRIs), such as **fluoxetine** (*N*-methyl-3-phenyl-3-[4-(trifluoromethyl)phenoxy]propan-1-amine) and **sertraline** ((1*S*,4*S*)-4-(3,4-Dichlorophenyl)-*N*-methyl-1,2,3,4-tetrahydronaphthalen-1-amine) are widely prescribed for the treatment of depression, obsessive compulsive behaviors, and anxiety (Conners et al. 2009, Barry 2014). Although the effects of SSRIs on select organisms have been reported, relatively few studies have investigated the effects of these drugs on amphibians (Foster et al. 2010, Barry 2018). **Venlafaxine** (1-[2-(dimethylamino)-1-(4 methoxyphenyl)ethyl] cyclohexanol) is a selective serotonin and norepinephrine reuptake inhibitor (SSNRI), but its affinity for the receptor norepinephrine is 10 times lower than for serotonin. Venlafaxine is also used to treat anxiety and depression, like SSRIs, however it may have different physiological effects on non-target species (Barry 2018). As the serotonergic system is highly conserved throughout the animal kingdom, it is very likely that trace quantities of these drugs could affect aquatic biota.

Tadpoles of the African clawed frog (*X. laevis*) exposed to fluoxetine (10 µg L^{-1}) and sertraline (0.1, 1, and 10 µg L^{-1}) for 70 d throughout metamorphosis, exhibited reduced growth and accelerated metamorphosis (Conners et al. 2009). These effects of SSRIs were likely driven by reduced food intake, as nutritional status can influence growth and development in amphibians via effects on the neuroendocrine system. In a study performed by Foster et al. (2010), tadpoles of *L. pipiens*, from stages 21 to 22 until the completion of metamorphosis, were exposed to low concentrations of fluoxetine both in the laboratory (0.029 and 0.29 µg L^{-1}) and in a mesocosm (ranging from 0.1 to 0.3 µg L^{-1}). The authors also showed that tadpoles exposed to fluoxetine in laboratory experiments exhibited delays in development and gained weight more slowly, which may be explained by reduced food intake. The developmental delays caused by fluoxetine may put tadpoles at greater risk of desiccation and overcrowding as temporal ponds dry, as well as size-specific predation.

In addition to the effects on development, fluoxetine can also interfere in the behavioral responses of tadpoles, as shown by Barry (2014). This author measured the effects of fluoxetine (0.03, 0.3, and 3 µg L^{-1}) on the swimming and behavioral responses of *S. arabica* tadpoles to alarm chemicals from predatory dragonfly larvae (*Anax imperator*). The results showed that fluoxetine reduced the swimming speed of these tadpoles at 0.3 µg L^{-1} and completely eliminated predator-avoidance behavior at 3 µg L^{-1}, indicating that it has the potential to make the tadpoles more vulnerable to predation at concentrations close to those that have been reported in natural ecosystems (Table 3). More recently, Barry (2018) investigated the effects of two SSRIs (fluoxetine and sertraline) and one SSNRI (venlafaxine), at 0.5 µg L^{-1} and 2.0 µg L^{-1}, alone or as a mixture, for 14 days, on swimming behavior of *S. arabica* tadpoles exposed to predator alarm cues. The lower concentration of fluoxetine and venlafaxine inhibited the normal response to alarm chemicals, while 2 µg L^{-1} of any of the 3 pharmaceuticals also reduced swimming speed, reinforcing the effect of these anti-depressants on the predatory avoidance response of anuran tadpoles (Barry 2014, 2018).

A greater number of endpoints were analyzed by Aliko et al. (2021) in tadpoles of *B. bufo* exposed to fluoxetine (5 µg L^{-1}) for 7 days. Their results showed that besides impaired behaviors, delayed metamorphosis, and reduced body weight, tadpoles showed an increased frequency of cellular and nuclear abnormalities in blood cells. These erythrocyte abnormalities negatively influence the blood oxygen-carrying capacities and erythrocyte deformability, compromising the blood flow, cardiac function, and hemodynamics of these tadpoles (Aliko et al. 2021).

Amitriptyline (3-(10,11-dihydro-5*H*-dibenzo[*a*,*d*]cycloheptene-5-ylidene)-*N*,*N*-dimethylpropan-1-amine) is a representative of the tricyclic antidepressants (TCAs) that has been used for decades, despite its connection to a large number of side effects. Its mode of action is the inhibition of serotonin and noradrenaline uptake in presynaptic nerve endings. In order to assess the effect of TCA on amphibian early life stages, Sehonova et al. (2019) exposed eggs of *X. tropicalis*, collected immediately after spawning, to 0.3, 3, 30, 300, and 3000 µg L^{-1} of amitriptyline until 48 h post fertilization. The authors recorded mortality, eye or head deformation, reduced pigmentation, the presence of edemas, and lack of circulation as well as heart rate. The results showed that amitriptyline caused decreased heart rate in embryos of *X. tropicalis* only at the highest tested concentration.

Carbamazepine (5H-dibenzo[b,f]azepine-5-carboxamide) is a pharmaceutical drug with anticonvulsant, analgesic, and mood-stabilizing properties, considered a marker of anthropogenic pollution, as it is frequently detected in surface waters (Table 3), is difficult to be removed by waste water treatments (WWTPs), and persistent to photolysis and biodegradation (Bókony et al. 2018, Fekadu et al. 2019). In a study conducted by Pablos et al. (2020), tadpoles of *X. laevis* were exposed to reclaimed water spiked with 9 µg L^{-1} and 90 µg L^{-1} of carbamazepine for 21 days. The authors reported that, at these exposure concentrations, carbamazepine seems to have a minor impact on the thyroid axis. On the other hand, the effects of environmentally relevant concentrations of carbamazepine (0.5 and 50 µg L^{-1}) were assessed throughout the larval development in 2 anuran species abundant in Europe, the agile frog (*R. dalmatina*) and the common toad (*B. bufo*) (Bókony et al. 2020). The study revealed that carbamazepine (50 µg L^{-1}) had effects on the spleen of both species, suggesting that this chemical may modulate immune system activity in amphibians, as it does in mammals. Moreover, carbamazepine reduced the feeding activity of toad tadpoles, decreased their production of anti-predatory toxins, and increased their body mass at metamorphosis. Taken together, these results show that carbamazepine can have several sub-lethal effects on anurans, which may be detrimental to individual fitness and population persistence in natural conditions (Bókony et al. 2020).

2.6 Mixtures of Pharmaceuticals

A multitude of pharmaceuticals from different therapeutic classes are used in human and veterinary medicine, thus pharmaceuticals do not occur as isolated contaminants, and organisms in the environment are exposed to multi-component pharmaceutical mixtures (Backhaus 2014, Amaral et al. 2019). The need for research characterizing toxicological responses and mechanisms of toxicity of pharmaceutical mixtures in non-target organisms was identified as the number one priority in a global survey of leading environmental scientists (Rudd et al. 2014). The ecotoxicity of pharmaceutical mixtures is usually higher than the effects of each single component and considerable adverse effects can occur even if all the components of the mixture are present below their individual non-observed effect concentration (NOEC) (Backhaus 2016). However, amphibians are still much less commonly addressed in pharmaceutical mixture toxicity studies compared to primary producers, invertebrates, fungi, and bacteria (Godoy and Kummrow 2017).

2.6.1 Binary Mixtures

Studies with binary mixtures offer the advantage of allowing the elucidation of the effect of one specific chemical on the biological action of another, while little can be concluded from mixtures with several components (Godoy and Kummrow 2017). The toxic effects of a mixture of an antidepressant (fluoxetine) with a non-steroidal anti-inflammatory drug (ibuprofen) using the Bullfrog tadpole as a model organism were investigated by Aliko et al. (2021). The authors showed that the two drugs interacted, impairing development and fitness in tadpoles and affecting long-term species perpetuation and population dynamics. Effects of fluoxetine and ibuprofen on the development of common toad tadpoles suggest that in natural conditions, reduced growth and delayed metamorphosis may produce smaller tadpoles, which can have negative effects on the fitness of the population as well as on survival in the wintertime.

The use of azithromycin together with hydroxychloroquine has increased considerably due to the COVID-19 pandemic, amplifying the presumed concentrations of these pharmaceuticals in the aquatic environment. Giving this, Luz et al. (2021) exposed tadpoles of *P. cuvieri* to water containing azithromycin and hydroxychloroquine (both at 12.5 μg L^{-1}) in order to assess the effects of both drugs, alone and in a mixture. The authors demonstrated that the short exposure (72 h) of these tadpoles to azithromycin and hydroxychloroquine induced an adaptive physiological response marked by increased activity of antioxidant enzymes and increased butyrylcholinesterase, possibly to counteract the anticholinesterase effect induced by azithromycin.

2.6.2 Complex Mixtures

The effects of a mixture of two SSRIs (fluoxetine and sertraline) with one SSNRI (venlafaxine) were investigated on the swimming behavior of tadpoles of *S. arabica* when exposed to dragonfly larvae alarm chemicals (Barry 2018). Tadpoles typically reduce their activity and show intense lateralization of responses when exposed to alarm chemicals. However, these tadpoles when exposed to fluoxetine or venlafaxine did not respond to the alarm cues. On the other hand, the tadpoles exposed to the mixture of antidepressants showed the expected responses (similar to that of the control), but stronger lateralization, indicating that these drugs have an additive mode of action (Barry 2018).

The effects of a mixture combining 4 pharmaceuticals, 2 NSAIDs (ibuprofen and diclofenac), 1 beta-blocker (atenolol), and 1 lipid regulator (gemfibrozil), at 0.1, 1, 10, 100, and 1000 μg L^{-1}, were evaluated on tadpoles of the Australian striped marsh frog (*L. peronii*) throughout the larval developmental period (Melvin 2016). The author showed that the mixture caused a reduction in developmental rates and the liver-somatic index, altered levels of hepatic triglycerides, increased peroxidase activity, and a trend to decreased swimming velocity. These results demonstrate that mixtures of common non-steroidal pharmaceuticals, at environmental concentrations (Table 3), can elicit a range of physiological, metabolic, and morphological responses in larval amphibians (Melvin 2016).

In another work, Melvin et al. (2017) exposed the same tadpole species (*L. peronii*) to a mixture of 3 drugs widely prescribed to treat metabolic syndrome, the anti-diabetic metformin and the lipid regulators atorvastatin and bezafibrate, and analyzed the effects on growth and development, and energy reserves (triglycerides and cholesterol). Despite the lack of effects on the expected bioenergetic endpoints, morphometric data indicated that the mixture used to treat human metabolic syndrome may elicit developmental effects in amphibian larvae at concentrations approaching environmental relevance (5 μg L^{-1}). More recently, Melvin et al. (2018) investigated whether methanol, a common carrier solvent used in standard aquatic toxicity bioassays, influences sub-lethal physiological and metabolic endpoints in larval amphibians. The results demonstrated the potential for interactive effects between a low dose of methanol (0.003%) and a sublethal pharmaceutical mixture. Based on this finding, the authors (Melvin et al. 2018) highlighted the necessity to consider the influence of this interaction when interpreting biochemical data, since solvents may influence sublethal responses.

Using a more complex mixture, composed of 8 pharmaceuticals of different classes (antibiotic, anti-inflammatory, antidepressant, anxiolytic, analgesic, and antacid), at environmentally relevant concentrations (Tables 2 and 3), Amaral et al. (2019) evaluated whether this mixture affects the oral morphology, triggers behavioral disorders, and has mutagenic effects on erythrocyte cells of tadpoles of *L. castesbeianus*. The authors showed that the mixture of drugs had harmful effects on the tadpoles that may have resulted from synergistic, antagonistic, and/or additive interactions between the mixture components. They identified changes in mandibular sheath pigmentation, dentition, and swimming activity, atypical behavior in social aggregation, and an anti-predatory defensive response deficit, as well as mutagenic effects on erythrocytes. These morphological, behavioral, and mutagenic abnormalities may impair fitness-related traits in tadpoles, besides having a potential impact on the dynamic of their populations and species perpetuation (Amaral et al. 2019).

The effects of a mixture containing 2 pharmaceuticals (ciprofloxacin and amoxicillin) plus an herbicide (glyphosate) and an insecticide (chlorpyrifos) were evaluated by Cuzziol Boccioni et al. (2021) on the development of tadpoles of *R. arenarum*. The results showed that combinations of these 4 potential environmental toxic compounds promoted several sublethal effects on amphibian development and highlight the importance of the analysis of biomarkers from different levels of organization, as well as the long-term evaluation of organisms. Thus, knowledge on the interaction of pollutants and their combined effect is essential to approximate the real situation that tadpoles experience in contaminated environments.

3. Conclusions and Perspectives

The study of pharmaceuticals and their effects on tadpoles has increased in recent years, especially after dissemination of the alarming rates of amphibian population decline. Moreover, recent studies with extensive sampling (Wilkinson et al. 2022) and important reviews (e.g., Patel et al. 2019) show the presence of high levels of these contaminants in aquatic ecosystems around the world.

Although the use of tadpoles as biological models is still emerging in ecotoxicology, compared with other aquatic organisms (e.g., microcrustaceans, bivalves, fishes), there are already several indicators showing that these organisms could be considered a sentinel organism in ecological issues, since they can respond to the contaminant effects before other organisms from the ecosystem (Falfushynska et al. 2017). In order to improve this important field, researchers are also encouraged to explore other approaches, such as the pharmaceutical potential to bioaccumulate and to remain longer in tadpole tissues, as well as to assess the influence of intrinsic abiotic variables from a given environment type.

Studies on the bioaccumulation potential of pharmaceuticals must be encouraged, especially in food web approaches. Although pharmaceutical analysis of tadpole tissues under both laboratorial and natural conditions are scarce, they could provide important findings and predictions with respect to human and ecosystem health. The presence of pharmaceuticals was already reported in algae and microcrustaceans (Rodríguez-Mozaz et al. 2015) which are potential prey for tadpoles, as most tadpoles scrape substrate and feed on periphyton, while others, such as those of the genus *Xenopus*, are filter-feeders (Altig and Johnston 1989). Thus, pharmaceutical effects are not only related to exposure but also to their potential for bioaccumulation.

Like other contaminants, some pharmaceuticals are able to remain longer in the environment, as can their effects, which could decrease local aquatic populations. Pharmaceuticals such as naproxen, sulfamethoxazole, and erythromycin can persist for almost a year, while others can remain stable for several years (Patel et al. 2019). These findings reinforce the urgent need for more ecotoxicological long-term studies, mainly in aquatic environments with high contaminant risks.

As a chemical compound, pharmaceuticals can also be affected by physical-chemical conditions of the aquatic environment. Depending on these conditions, some pharmaceuticals may be more soluble, while others are more lipophilic, which could imply high affinity to lipid membranes and bioaccumulation in organisms (Rivera-Utrilla et al. 2013, Swicka et al. 2022). Moreover, products of photo-transformation may be produced from the UV-incidence on certain pharmaceuticals and their toxicity could be higher than the original compound, as shown by Cory et al. (2019). Thus, it is essential to assess changes in temperature, pH, UV-incidence, oxygen, and other parameters which could increase the pharmaceutical effects on organisms.

Some authors highlighted the importance of the exposure time in some ecotoxicological studies, as well as the number of replicates and the monitoring of contaminant levels during the experiments (Batucan et al. 2022). Long-term exposures are interesting because several pharmaceuticals last for a long time in the environment and their main effects could be observed after a considerable time. Moreover, for statistical power and consistency, an adequate number of replicates is essential, and the exposed animals must be acquired from egg mass in a laboratory, ensuring the absence of previous contaminant effects. In addition, we encourage researchers to provide precise exposure time (e.g., hours or days) and consider the tadpole development (Gosner 1960), allowing study replication by other researchers.

Clearly, several researchers agree that mitigation of the pharmaceutical discharge to the environment is urgently needed, as undoubtable evidence of negative effects has been reported. At least one given pharmaceutical was found at high concentrations in more than 25% of the sampling sites, in a broad study including more than 100 countries (Wilkinson et al. 2022). This and other important findings must be considered both for further research and to shape the regulatory polices of governments for aquatic ecosystem conservation, since water is essential for life.

Acknowledgments

The authors acknowledge the editors for the invitation to participate in this book. A. Aguiar thanks to São Paulo Research Foundation (FAPESP - grants 2018/25554-2 and 2021/09892-8) for the post-doctoral fellowship. CBR Martinez thanks to National Council for Scientific and Technological Development (CNPq - grant 307146/2019-7) for the research fellowship.

References Cited

Aliko, V., R.S. Korriku, M. Pagano and C. Faggio. 2021. Double-edged sword: Fluoxetine and ibuprofen as development jeopardizers and apoptosis' inducers in common toad, *Bufo bufo*, tadpoles. Sci. Total Environ. 776: 145945.

Altig, R. and F. Johnston. 1989. Guilds of anuran larvae: Relationships among developmental modes, morphologies, and habitat. Herpetol. Monogr. 3: 81–109.

Amaral, D.F., M.F. Montalvão, B.O. Mendes, A.L.S. Castro and G. Malafaia. 2018. Behavioral and mutagenic biomarkers in tadpoles exposed to different abamectin concentrations. Environ. Sci. Pollut. Res. 25: 12932–12946.

Amaral, D.F., M.F. Montalvão, B.O. Mendes, A.P.C. Araújo, A.S.L. Rodrigues and G. Malafaia. 2019. Sub-lethal effects induced by a mixture of different pharmaceutical drugs in predicted environmentally relevant concentrations on *Lithobates catesbeianus* (Shaw, 1802) (Anura, ranidae) tadpoles. Environ. Sci. Pollut. Res. 26: 600–616.

Aminot, Y., X. Litrico, M. Chambolle, C. Arnaud, P. Pardon and H. Budzindki. 2015. Development and application of a multi-residue method for the determination of 53 pharmaceuticals in water, sediment, and suspended solids using liquid chromatography-tandem mass spectrometry. Anal. Bioanal. Chem. 407: 8585–8604.

Atanasov, A.G., B. Waltenberger, E.M. Pferschy-Wenzig, T. Linder, C. Wawrosch, P. Uhrin, V. Temml, L. Wang, S. Schwaiger, E.H. Heiss, J.M. Rollinger, D. Schuster, J.M. Breuss, V. Bochkov, M.D. Mihovilovic, B. Kopp, R. Bauer, V.M. Dirsch and H. Stuppner. 2015. Discovery and resupply of pharmacologically active plant-derived natural products: A review. Biotechnol. Adv. 33(8): 1582–614.

Backhaus T. 2014. Medicines, shaken and stirred: A critical review on the ecotoxicology of pharmaceutical mixtures. Phil. Trans. R. Soc. B 369: 20130585.

Backhaus, T. 2016. Environmental risk assessment of pharmaceutical mixtures: Demands, gaps, and possible bridges. AAPS J. 18: 804–813.

Barry, M. 2014. Fluoxetine inhibits predator avoidance behavior in tadpoles. Toxicol. Environ. Chem. 96(4): 641–649.

Barry, M. 2018. Effects of three pharmaceuticals on the responses of tadpoles to predator alarm cues. Toxicol. Environ. Chem. 100(2): 205–213.

Batucan, N.S.P., L.A. Tremblay, G.L. Northcott and C.D. Matthaei. 2022. Medicating the environment? A critical review on the risks of carbamazepine, diclofenac and ibuprofen to aquatic organisms. Environmental Advances 7: 1–14.

Bókony, V., N.B. Üveges, N. Ujhegyi, V. Verebélyi, E. Nemesházi, O. Csíkvári and A. Hettyey. 2018. Endocrine disruptors in breeding ponds and reproductive health of toads in agricultural, urban and natural landscapes. Sci. Total Enviro. 634: 1335–1345.

Bókony, V., V. Verebélyi, N. Ujhegyi, Z. Mikó, E. Nemesházi, M. Szederkényi, S. Orf, E. Vitányi and A.M. Móricz. 2020. Effects of two little-studied environmental pollutants on early development in anurans. Environ. Pollut. 260: 114078.

Carlsson, G., S. Örn and D.G.J. Larsson. 2010. Effluent from bulk drug production is toxic to Aquatic vertebrates. Environ. Toxicol. Chem. 28(12): 2656–2662.

Chae J.P., M.S. Park, Y.S. Hwang, B.H. Min, S.H. Kim, H.S. Lee and M.J. Park. 2015. Evaluation of developmental toxicity and teratogenicity of diclofenac using *Xenopus embryos*. Chemosphere 120: 52–58.

Conners, D.E., E.D. Rogers, K.L. Armbrust, J.W. Kwon and M.C. Black. 2009. Growth and development of tadpoles (*Xenopus laevis*) exposed to selective serotonin reuptake inhibitors, Fluoxetine and Sertraline, throughout metamorphosis. Environ. Toxicol. Chem. 28: 2671–2676.

Cory, W.C., A.M. Welch, J.N. Ramirez and L.C. Rein. 2019. Naproxen and its phototransformation products: Persistence and ecotoxicity to toad tadpoles (*Anaxyrus terrestris*), individually and in mixtures. Environ. Toxicol. Chem. 38(9): 2008–2019.

Cragg, G.M. and D.J. Newman. 2013. Natural products: A continuing source of novel drug leads. Biochim. Biophys. Acta 1830: 3670–3695.

Cuzziol Boccioni, A.P., P.M. Peltzer, C.S. Martinuzzi, A.M. Attademo, E.J. León and R.C. Lajmanovich. 2020. Morphological and histological abnormalities of the neotropical toad, *Rhinella arenarum* (Anura: Bufonidae) larvae exposed to dexamethasone. J. Environ. Sci. Heal. B 56(1): 41–53.

Cuzziol Boccioni, A.P., R.C. Lajmanovich, P.M. Peltzer, A.M. Attademo and C.S. Martinuzzi. 2021. Toxicity assessment at different experimental scenarios with glyphosate, chlorpyrifos and antibiotics in *Rhinella arenarum* (Anura: Bufonidae) tadpoles. Chemosphere 273: 128475.

Danner, M.C., A. Robertson, V. Behrends and J. Reiss. 2019. Antibiotic pollution in surface fresh waters: Occurrence and effects. Sci. Total Environ. 664: 793–804.

der Beek, T.A., F.A. Weber, A. Bergmann, S. Hickmann, I. Ebert, A. Hein and A. Küster. 2016. Pharmaceuticals in the Environment-Global Occurrences and Perspectives. Environ. Toxicol. Chem. 35: 823–835.

Ehrsam, M., S.A. Knutie and J.R. Rohr. 2016. The herbicide atrazine induces hyperactivity and compromises tadpole detection of predator chemical cues. Environ. Toxicol. Chem. 35: 2239–2244.

Falfushynska, H.I., L.L. Gnatyshyna, O. Horyn and O.B. Stoliar. 2017. Vulnerability of marsh frog *Pelophylax ridibundus* to the typical wastewater effluents ibuprofen, triclosan and estrone, detected by multi-biomarker approach. Comp. Biochem. Phys. 202: 26–38.

Fekadu, S., E. Alemayehu, R. Dewil and B. Van der Bruggen. 2019. Pharmaceuticals in freshwater aquatic environments: A comparison of the African and European challenge. Sci. Total Environ. 654: 324–337.

Fernández, L.P., R. Brasca, M.R. Repetti, A.M. Attademo, P.M. Peltzer, R.C. Lajmanovich and M.J. Culzoni. 2022. Bioaccumulation of abacavir and efavirenz in *Rhinella arenarum* tadpoles after exposure to environmentally relevant concentrations. Chemosphere 301: 134631.

Figuière, R., S. Waara, L. Ahrens and O. Golovko. 2022. Risk-based screening for prioritisation of organic micropollutants in Swedish freshwater. Journal of Hazardous Materials 429: 1–11.

Foster, H.R., G.A. Burton, N. Basu and E.E. Werner. 2010. Chronic exposure to fluoxetine (Prozac) causes developmental delays in *Rana pipiens* larvae. Environ. Toxicol. Chem. 29: 2845–2850.

Gezahegn, T., B. Tegegne, F. Zewge and B.S. Chandravanshi. 2019. Salting-out assisted liquid-liquid extraction for the determination of ciprofloxacin residues in water samples by high performance liquid chromatography-diode array detector. BMC Chem. 13: 28–28.

Gonçalves, E.S. 2012. Occurrence and distribution of drugs, caffeine and bisphenol in some water bodies in the state of Rio de Janeiro (In Portuguese). Ph.D. Thesis, Federal University Fluminense, Rio de Janeiro, Brazil.

GOSNER, K.L. 1960. A simplified table for staging anuran embryos and larvae with notes on identification. Herpetologica 16: 183–190.

Green, D.M., M.J. Lannoo, D. Lesbarrères and E. Muths. 2020. Amphibian population declines: 30 years of progress in confronting a complex problem. Herpetologica 76(2): 97–100.

Godoy, A.A. and F. Kummrow. 2017. What do we know about the ecotoxicology of pharmaceutical and personal care product mixtures? A critical review. Crit. Rev. Env. Sci. Tec. 47(16): 1453–1496.

Gumbi, B.P., B. Moodley, G. Birungi and P.G. Ndungu. 2017. Detection and quantification of acidic drug residues in South African surface water using gas chromatography-mass spectrometry. Chemosphere 168: 1042–1050.

Haddad, C.F.B. and C.P.A. Prado. 2005. Reproductive modes in frogs and their unexpected diversity in the atlantic forest of brazil. BioScience 55: 207–217.

Ivešić, M., A. Krivohlavek, I. Žuntar, S. Tolić, S. Šikić, V. Musić, I. Pavlić, A. Bursik and N. Galić. 2017. Monitoring of selected pharmaceuticals in surface waters of Croatia. Environ. Sci. Pollut. Res. 24: 23389–23400.

Kumar, K., S. Gupta, S. Baidoo, Y. Chander and C. Rosen. 2005. Antibiotic uptake by plants from soil fertilized with animal manure. J. Environ. Qual. 34: 2082–2085.

Kümmerer, K. 2009. The Presence of pharmaceuticals in the environment due to human use–present knowledge and future challenges. J. Environ. Manage. 90: 2354–2366.

Langlois, V.S. 2021. Amphibian toxicology: A rich but underappreciated model for ecotoxicology research. Arch. Environ. Con. Tox. 80: 661–662.

Lin, Y.C., W.W.P. Lai, H.H. Tung and A.Y.C. Lin. 2015. Occurrence of pharmaceuticals, hormones, and perfluorinated compounds in groundwater in Taiwan. Environ. Monit. Assess. 187: 256.

Lindim, C., J. van Gils, D. Georgieva, O. Mekenyan and I.T. Cousins. 2016. Evaluation of human pharmaceutical emissions and concentrations in Swedish river basins. Sci. Total Environ. 572: 508–519.

Locatelli, M.A.F, F.F. Sodré and W.F. Jardim. 2011. Determination of antibiotics in Brazilian surface waters using liquid chromatography-electrospray tandem mass spectrometry. Arch. Environ. Contam. Toxicol. 60: 385–393.

Luz, T.M., A.P.C. Araújo, F.N. Estrela, H.L.B. Braz, R.J.B. Jorge, I. Charlie-Silva and G. Malafaia. 2021. Can use of hydroxychloroquine and azithromycin as a treatment of COVID-19 affect aquatic wildlife? A study conducted with neotropical tadpole. Sci. Total Environ. 780: 146553.

Melvin, S.D., M.C. Cameron and C.M. Lanctôt. 2014. Individual and mixture toxicity of pharmaceuticals naproxen, carbamazepine, and sulfamethoxazole to Australian Striped Marsh Frog Tadpoles (*Limnodynastes peronii*). J. Toxicol. Env. Heal. A 77(6): 337–345.

Melvin, S.D. 2016. Oxidative stress, energy storage, and swimming performance of *Limnodynastes peronii* tadpoles exposed to a sub-lethal pharmaceutical mixture throughout development. Chemosphere 150: 790–797.

Melvin, S.D., L.J. Habener, F.D.L. Leusch and A.R. Carroll. 2017. ^{1}H NMR-based metabolomics reveals sub-lethal toxicity of a mixture of diabetic and lipid-regulating pharmaceuticals on amphibian larvae. Aquat. Toxicol. 184: 123–132.

Melvin, S.D., O.A.H. Jones, A.R. Carroll and F.D.L. Leusch. 2018. ^{1}H NMR-based metabolomics reveals interactive effects between the carrier solvent methanol and a pharmaceutical mixture in an amphibian developmental bioassay with *Limnodynastes peronii*. Chemosphere 199: 372–381.

Montalvão, M.F. and G. Malafaia. 2017. Effects of abamectin on bullfrog tadpoles: Insights on cytotoxicity. Environ. Sci. Pollut. Res. Int. 24: 23411–23416.

Monteiro, M.A., B.F. Spisso, J.R.M.P. dos Santos, R.P. da Costa, R.G. Ferreira, M.U. Pereira, T.D.S. Miranda, B.R.G. de Andrade and L.A. d'Avila. 2016. Occurrence of antimicrobials in river water samples from rural region of the state of Rio de Janeiro. Brazil. J. Environ. Prot. 7: 230–241.

Nannou, C., A. Ofrydopoulou, E. Evgenidou, D. Heath, E. Heath and D. Lambropoulou. 2020. Antiviral drugs in aquatic environment and wastewater treatment plants: A review on occurrence, fate, removal and ecotoxicity. Sci. Total Environ. 699: 134322.

Ogunbanwo, O.M., P. Kay, A.B. Boxall, J. Wilkinson, C.J. Sinclair, R.A. Shabi, A.E. Fasasi, G.A. Lewis, O.A. Amoda and L.E. Brown. 2022. High concentrations of pharmaceuticals in a Nigerian river catchment. Environ. Toxicol. Chem. 41: 551–558.

Omotola, E.O., A.O. Oluwole, P.O. Oladoye, and O.S. Olatunji. 2022. Occurrence, detection and ecotoxicity studies of selected pharmaceuticals in aqueous ecosystems—a systematic appraisal. Environ. Toxicol. Phar. 91: 103831.

Pablos, M.V., E.M. Beltrán, M.A. Jiménez, P. García-Hortigüela, A. Fernández, M. González-Doncel and C. Fernández. 2020. Effect assessment of reclaimed water and carbamazepine exposure on the thyroid axis of *X. laevis*: apical and histological effects. Sci. Total Environ. 723: 138023.

Patel, M., R. Kumar, K. Kishor, T. Mlsna, C.U. Pittman and D. Mohan. 2019. Pharmaceuticals of emerging concern in aquatic systems: Chemistry, occurrence, effects, and removal methods. Chem. Rev. 119: 3510–3673.

Peltzer, P.M., R.C. Lajmanovich, A.M. Attademo, C.M. Junges, C.M. Teglia, C. Martinuzzi, L. Curi, M.J. Culzoni and H.C. Goicoechea. 2017. Ecotoxicity of veterinary enrofloxacin and ciprofloxacin antibiotics on anuran amphibian larvae. Environ. Toxicol. Phar. 51: 114–123.

Peltzer, P.M., R.C. Lajmanovich, C. Martinuzzi, A.M. Attademo, L.M. Curi and M.T. Sandoval. 2019. Biotoxicity of diclofenac on two larval amphibians: Assessment of development, growth, cardiac function and rhythm, behavior and antioxidant system. Sci. Total Environ. 683: 624–637.

Petrović, M., B. Škrbić, J. Živančev, L. Ferrando-Climent and D. Barcelo. 2014. Determination of 81 pharmaceutical drugs by high performance liquid chromatography coupled to mass spectrometry with hybrid triple quadrupole–linear ion trap in different types of water in Serbia. Sci. Total Environ. 468: 415–428.

Pomati, F. 2007. Pharmaceuticals in drinking water: Is the cure worse than the disease. Environ. Sci. Technol. 41: 8204–8204.

Quadra, G.R., H.O. Souza, R.S. Costa and M.A.S. Fernandez. 2017. Do pharmaceuticals reach and affect the aquatic ecosystems in Brazil? A critical review of current studies in a developing country. Environ Sci Pollut Res. 24: 1200–1218.

Rang, H.P., M.M. Dale, J.M. Ritter, R.J. Flower and G. Henderson. 2011. What is pharmacology. *In*: Rang & Dale's Pharmacology (7 ed.). Edinburgh: Churchill Livingstone.

Reis, E.O., A.F.S. Foureaux, J.S. Rodrigues, V.R. Moreira, Y.A. Lebron, L.V. Santos, M.C.S. Amaral and L.C. Lange. 2019. Occurrence, removal and seasonal variation of pharmaceuticals in Brazilian drinking water treatment plants. Environ. Pollut. 250: 773–781.

Reis, E.O., L.V.S. Santos and L.C. Lange. 2021. Prioritization and environmental risk assessment of pharmaceuticals mixtures from Brazilian surface waters. Environ. Pollut. 288: 117803.

Ribas, J.L.C., A.R. Zampronio and H.C.S. Assis. 2016. Effects of trophic exposure to diclofenac and dexamethasone on hematological parameters and immune response in freshwater fish. Environ. Toxicol. Chem. 35: 975–982.

Rivera-Utrilla, J., M. Sánchez-Polo, M.A. Ferro-García, G. Prados-Joya and R. Ocampo-Pérez. 2013. Pharmaceuticals as emerging contaminants and their removal from water. A review. Chemosphere 93: 1268–1287.

Rodríguez-Mozaz, S., B. Huerta and D. Barceló. 2015. Bioaccumulation of emerging contaminants in aquatic biota: Patterns of pharmaceuticals in Mediterranean river networks. Emerging Contaminants in River Ecosystems. Springer, Cham, pp. 121–141.

Rohr, J.R., A.M. Schotthoefer, T.R. Raffel, H.J. Carrick, N. Halstead, J.T. Hoverman, C.M. Johnson, L.B. Johnson, C. Lieske, M.D. Piwoni, P.K. Schoff and V.R. Beasley. 2008. Agrochemicals increase trematode infections in a declining amphibian species. Nature 455: 1235–1240.

Rudd, M.A., G.T. Ankley, A.B.A. Boxall and B.W. Brooks. 2014. International scientists' priorities for research on pharmaceutical and personal care products in the environment. Integr. Environ. Assess. Manag. 10: 576–587.

Rutkoski, C.F., S.C. Grott, N.G. Israel, F.E. Carneiro, F.C. Guerreiro, S. Santos, P.A. Horn, A.A. Trentini, E.B. Silva, C.A.C. Albuquerque, T.C. Alves and E.A. Almeida. 2022. Hepatic and blood alterations in *Lithobates catesbeianus* tadpoles exposed to sulfamethoxazole and oxytetracycline. Chemosphere 307: 136215.

Santos, L.H., A.N. Araújo, A. Fachini, A. Pena, C. Delerue-Matos and M.C.B.S.M. Montenegro. 2010. Ecotoxicological aspects related to the presence of pharmaceuticals in the aquatic environment. J. Hazard Mater 175: 45–95.

Santos, L.H., M. Gros, S. Rodriguez-Mozaz, C. Delerue- Matos, A. Pena, D. Barceló and M.C.B. Montenegro. 2013. Contribution of hospital effluents to the load of pharmaceuticals in urban wastewaters: identification of ecologically relevant pharmaceuticals. Sci. Total Environ. 461: 302–316.

Scheurer, M., A. Michel, H.J. Brauch, W. Ruck and F. Sacher. 2012. Occurrence and fate of the antidiabetic drug metformin and its metabolite guanylurea in the environment and during drinking water treatment. Water Res. 46: 4790–4802.

Sehonova, P., L. Plhalova, J. Blahova, V. Doubkova, M. Prokes, F. Tichy, E. Fiorino, C. Faggio and Z. Svobodova. 2017. Toxicity of naproxen sodium and its mixture with tramadol hydrochloride on fish early life stages. Chemosphere 188: 414e423.

Sehonova, P., N. Hodkovicova, M. Urbanova, S. Orn, S.J. Blahova, Z. Svobodova, M. Faldyna, P. Chloupek, K. Briedikova and G. Carlsson. 2019. Effects of antidepressants with different modes of action on early life stages of fish and amphibians. Environ. Pollut. 254: 112999.

Sertürner, F.W. 1817. Über das Morphium, eine neue salzfähige Grundlage, und die Mekonsäure, als Hauptbestandteile des Opiums. Ann. Phys. 25: 56–90.

Shen, X., H. Chang, Y. Sun and Y. Wan. 2020. Determination and occurrence of natural and synthetic glucocorticoids in surface Waters. Environ. Int. 134: 105278.

Subedi, B., N. Codru, D.M. Dziewulski, L.R. Wilson, J. Xue, S. Yun, E. Braun-Howland, C. Minihane and K. Kannan. 2015. A pilot study on the assessment of trace organic contaminants including pharmaceuticals and personal care products from on-site wastewater treatment systems along Skaneateles Lake in New York State, USA. Water Res. 72: 28–39.

Sumpter, J.P. and L. Margiotta-Casaluci. 2022. Environmental occurrence and predicted pharmacological risk to freshwater fish of over 200 neuroactive pharmaceuticals in widespread use. Toxics 10: 233.

Swiacka, K., J. Maculewicz, D. Kowalska, M. Caban, K. Smolarz and J. Swiezak. 2022. Presence of pharmaceuticals and their metabolites in wild-living aquatic organisms—Current state of knowledge. J. Hazard. Mater. 424: 127350.

Thambirajah, A.A., E.M. Koide, J.J. Imbery and C.C. Helbing. 2019. Contaminant and environmental influences on thyroid hormone action in amphibian metamorphosis. Front. Endocrinol. 10: 276.

Turani, B., V. Aliko and C. Faggio. 2019. Amphibian embryos as an alternative model to study the pharmaceutical toxicity of cyclophosphamide and ibuprofen. J. Biol. Res. 92: 8370.

Turunen, J., V. Elbrecht, D. Steinke and J. Aroviita. 2021. Riparian forests can mitigate warming and ecological degradation of agricultural headwater streams. Freshwater Biol. 66: 785–798.

US-FDA – US Food and Drug Administration. 2022. Accessed in 27th July 2022 on https://www.fda.gov/industry/structured-product-labeling-resources/pharmacologic-class.

Veldhoen, N., R.C. Skirrow, L.L.Y. Brown, G.V. Aggelen and C.C. Helbing. 2014. Effects of acute exposure to the non-steroidal anti-inflammatory drug Ibuprofen on the developing North American Bullfrog (*Rana catesbeiana*) tadpole. Environ. Sci. Technol. 48: 10439–10447.

Wilkinson, J.L., A.B.A. Boxalla, D.W. Kolpinb, K.M.Y. Leung, R.W.S. Lai, C. Galbán-Malagón et al. 2022. Pharmaceutical pollution of the world's rivers. PNAS 119(8): 1–10.

Wu, M., J. Xiang, C. Que, F. Chen and G. Xu. 2015. Occurrence and fate of psychiatric pharmaceuticals in the urban water system of Shanghai, China. Chemosphere 138: 486–493.

Zhu, W., D. Yang, L. Chang, M. Zhang, L. Zhu and J. Jiang. 2022. Animal gut microbiome mediates the effects of antibiotic pollution on an artificial freshwater system. J. Hazard. Mat. 425: 127968.

11

Pesticide Effects on Tadpole's Survival

Marilia Hartmann, Paulo Afonso Hartmann* and *Caroline Müller*

1. Introduction

Survival. This race for life in which individuals, populations, and species continually participate, but fewer and fewer amphibians have crossed the finish line. According to the International Union for Conservation of Nature (IUCN 2022), populations of 3,063 amphibian species are experiencing population decline, and the conservation status is unknown for another 2,333, considering a total of 7,296 known species. These values indicate that approximately 42% of the species would be in population decline, in addition to not knowing the conservation status of more than 30% of them. Of the total number of species described so far, 5,046 are endemic, which makes the decline in populations even more worrying, as they are restricted to one or a few areas. Only for 28 species, there is a tendency for population increase. Also, according to the IUCN and considering the identified conservation threats, at least one of the causes of population reduction is the interaction with agricultural effluents for 1,015 species, notably pesticides.

Brazil has the greatest richness of amphibians in the world, with 1,188 species described (Segalla et al. 2021), and at the same time is one of the world leaders in the use of pesticides (Porto and Kiill 2020). From 2009 to 2019, Brazil increased the commercialization of pesticides four times (Nunes et al. 2021) and authorized the use of 562 new agrochemicals in 2021 (Brasil 2022). These substances are primarily chemical rather than biological, and about a third contain active substances that have been listed by the European Chemicals Association as banned or severely restricted (Sarkar et al. 2021). The country is also one of the main destinations for pesticides that have been banned in the European Union precisely because they are considered dangerous to human or environmental health (Sarkar et al. 2021).

Laboratório de Ecologia e Conservação, Universidade Federal da Fronteira Sul, Campus Erechim, RS, Brazil.
Emails: paulo.hartmann@uffs.edu.br; carolinemulleram@gmail.com
* Corresponding author: marilia.hartmann@uffs.edu.br

2. Why Are Amphibians Good Indicators of Pesticide Contamination?

Most amphibian species are dependent on aquatic environments, mainly for reproduction and development during early life stages. The species require humid or aquatic environments to reproduce, and spawning may often coincide with the periods when pesticides are applied to agricultural areas, in the spring and summer (Adams et al. 2021), increasing the susceptibility of these vertebrates to pollutants. Because of their highly permeable skin, anurans are vulnerable to chemical contamination, mainly from water pollution by pesticides (Varga et al. 2019). A recent study showed that *Engystomops pustulosus* frogs from agricultural areas had lower hatching success, fewer eggs, and also morphological changes in males compared to individuals of the same species from non-agricultural areas (Orton et al. 2022). These data, as well as other studies that will be addressed throughout this chapter, report the need for ecotoxicological studies on the effects of different types of pesticides on the health of this group of vertebrates.

International protocols for ecotoxicological studies, such as ASTM E1192 (2014), ASTM E729 (2014), OECD 231 (2009), and OECD 241 (2015), indicate the use of tadpoles as test organisms. These protocols were developed using species of the genera *Rana* or *Bufo* (ASTM) and *Xenopus* (OECD). Despite being very well described and standardized, these protocols are not always adequate to the reality of research laboratories, both financially and technically. In addition, they use few species of amphibians, which may not reflect the reality of the great diversity of species in nature. Thus, to achieve the objective of understanding the toxicity of pesticides in amphibians, it is essential to explore several species and identify bioindicator species from a given region. Based on key species, it is possible to compare and discover how and how much pesticides are affecting the survival of amphibian populations.

It is important to note that the comparison of a test organism for exposure to pesticides must include (and distinguish) the type of product used. Pesticides are products of high technical complexity, involving a large number of active ingredients and formulated products, applied in different agricultural production systems (Moraes 2019). In general, pesticides are structured in active ingredient (AI), that is, the molecule that has the power of effectiveness, and technical product (TP), from which the company makes the commercial formulation (definitions in Decree 4074; Brasil 2002). Commercial formulation is composed of different proportions of the active ingredient, in addition to other inert components such as surfactants, dispersants, mineral oil, and others, which are added to provide chemical stability and increased agronomic effectiveness (Vargas and Roman 2006). A technical product can give rise to different commercial products; however, they must change the toxicological characteristics of the formulation (Rocha 2018).

The inert components of commercial formulations are considered "trade secret", and are not described in the product labels (Zaller 2020). For this reason, different commercial formulations of pesticides with the same AI may have different LC_{50} values (which indicates the mortality of 50% of the studied population) for the tested species. From the few studies that have been carried out to evaluate the effects of surfactants in pesticide formulations to date, it has been found that these constituents may be more toxic than the active ingredient itself or its final formulation for aquatic animals (Moore et al. 2012, Rodrigues et al. 2019). However, most research focuses on studying the technical product, which is toxicologically relevant but may not reflect what is chemically released into the environment that effectively affects the species. Farmers use formulated products of different brands and percentages of AI, which reach soil and water and come into contact with non-target species. In this scenario, the sensitivity of amphibians to pesticides may depend on the formulation, species, and specific life stage (Wagner et al. 2013).

2.1 Laboratory Assays

In the search for bioindicator species of wild fauna in Brazil, the team from the Laboratory of Ecology and Conservation (LABECO) at the Federal University of Fronteira Sul (UFFS), Campus

Erechim (Brazil) studied the acute and chronic toxicity of different pesticides in two native species of amphibians: *Physalaemus cuvieri* known as Cuvier's Foam Froglet and *Physalaemus gracilis*, called Graceful Dwarf Frog. Both species are abundant and their reproductive cycles and behaviors have been well studied, in addition to having stable and non-threatened populations (IUCN 2022). They reproduce in environments commonly associated with agricultural areas (Achaval and Olmos 2003, Lenhardt et al. 2015) and therefore, with a high possibility of contact with pesticides.

Physalaemus is one of the largest genera in the Leptodactylidae family, which contains 50 species (Frost 2021). *P. cuvieri* and *P. gracilis* are classified on the Red List as Least-Concern (LC) due to their wide range of habitats (IUCN 2022). *P. cuvieri* occurs in northeastern, central, and southern Brazil, barely into adjacent Uruguay; arguably from Misiones, Argentina; eastern and northern Paraguay; Departments of Beni and Santa Cruz in Bolivia; lowlands of southern Venezuela (Bolívar and Delta Amacuro states). *P. gracilis* is found in southern Brazil and Uruguay to adjacent Argentina, likely in Paraguay (Lavilla et al. 2010, Mijares et al. 2010). Both species have similar reproductive habits. *P. cuvieri* and *P. gracilis* prefer lentic water bodies, such as wetlands and temporary natural pools for their reproduction and larval development. These species lay their eggs near or among vegetation, in masses of foam that are produced when they beat their legs in the water during amplexus (Barreto and Andrade 1995).

The advantage of comparing these two species by the same research group is the standardization of the methodology used. For the experiments, spawns were collected in the environment with less than 24 hours of oviposition, that is, in the early embryonic stage. Immediately after collection, the spawns were taken to the laboratory and placed individually in aquariums with 10 liters of water. They were kept in aquariums until the tadpoles reached stage 25, according to Gosner (1960) classification, when ecotoxicological tests with pesticides were started. From entering the laboratory until the end of the tests, tadpoles remained on artificial aeration. The water used had the following parameters: dissolved oxygen (DO) between 4.0 and 6.0 mg/L, temperature 23°C ± 2, pH 6.8–7.2, conductivity 151–178 μS/cm, alkalinity 9.74 mg $CaCO_3$/L, turbidity < 5.0, Ca 6.76 mg/L, Na 44.1 mg/L, Mg 1.35 mg/L, Fe 0.08 mg/L, and Ni < 0.001 mg/L.

Tadpoles were fed daily *ad libitum* with commercial food for fish and aquatic animals (MEP 200 complex, Alcon, Brazil), composed of chelated organic minerals, digestive enzymes, and probiotics that favor the development of the intestinal flora, in addition to 45% of crude and vegetable protein in flakes.

2.1.1 Acute Toxicity

Toxicity has two main components: the effect caused and the exposure level (dose) at which the effect is observed (Nuffield Council on Bioethics 2005). To measure survival, the expected effect is the death of part of the exposed organisms, and the dose will be the concentration of the substance used. An efficient way to compare the effects of pesticides on amphibian survival is acute toxicity, as this test employs a dose-response curve model to estimate contaminant concentrations, which provides a good measure for comparisons between contaminants and species (Kerby et al. 2010). Acute toxicity tests are commonly used to determine the concentration of a test substance that produces a specific adverse effect (primarily lethality) on a percentage of test organisms during a short exposure (ASTM E729 2014). These studies aim to verify the mortality or immobility that is usually expressed for 50% of the organisms by effective concentration (EC_{50}), lethal dose (LD_{50}), and lethal concentration (LC_{50}) (GHS 2021). Through ecotoxicological tests using high concentrations or doses of chemical substances, such as pesticides, it is possible to have a value normally measured after 48 or 96 hours of exposure. In amphibians, these bioassays use mainly (but not exclusively) the tadpole phase 25 (Gosner 1960), measure the LC_{50}, and last for 96 hours. Acute assays are usually carried out under controlled (laboratory) conditions.

At stage 25 (Gosner 1960), tadpoles have a complete mouth (Mcdiarmid and Altig 1999), spiracle that allows breathing, internal gills, and specialized structures for swimming (Duellman and Trueb 1986). In our assays, 10 tadpoles were placed in a sterile, cylindrical glass container with 450 mL of solution (water + pesticide), in triplicate, totaling 30 tadpoles exposed to each pesticide concentration. The acute assay lasted 96 hours, and mortality, as well physicochemical parameters of water as temperature and water oxygenation, were monitored every 24 h. A control was also carried out without adding pesticides, for each species, equally with 30 tadpoles. The concentrations used for each assay were based on the literature and according to LC_{50} values found for other amphibians. The results were analyzed by the Trimmed Spearman-Karber method using the GBasic program (Hamilton et al. 1977).

Six commercial pesticide formulations were tested in *P. cuvieri* and 10 in *P. gracilis* in experiments carried out over 6 years. $LC_{50,96\ h}$ data presented in Table 1 are the results of these studies, where pesticides are ordered from the most toxic to the least toxic, according to the recorded $LC_{50,96\ h}$. The classification of acute toxicity was determined according to the Globally Harmonized System of Classification and Labeling of Chemicals (GHS 2021) for aquatic environment in fish, as it is the closest standardized organism to amphibians: acute 1, very toxic ($LC_{50} \leq 1$ mg/L), acute 2, toxic (> 1 but ≤ 10 mg/L) and acute 3, harmful (> 10 but ≤ 100 mg/L).

Only the commercial formulation Roundup original® DI (active ingredient: glyphosate) was tested for both species. Although the LC_{50} was different, glyphosate was toxic to *P. cuvieri* and *P. gracilis* tadpoles. Therefore, at least, in this case, the two species showed similar survival to the same commercial product.

When the species were exposed to different formulations of the same pesticide, the LC_{50} was discrepant. *P. cuvieri* tadpoles were more sensitive to atrazine AclamadoBR® (500 g/L a.i., 620 g/L inert ingredients) (toxic) than atrazine Nortox 500 SC (500 g/L a.i., 572 g/L inert ingredients) (harmful). *P. gracilis* tadpoles were more sensitive to cypermethrin Cyptrin 250 EC (250 g/L a.i., 723 g/L inert ingredients) (very toxic) than Cipermetrin 250 EC (250 g/L a.i., 120 g/L inert ingredients) (toxic). This variation in LC_{50} clearly shows that commercial formulations are different from each other, which makes it more difficult to understand how amphibians survive these compounds. Considering that the two *Physalaemus* species are expected to present similar sensitivity, only 4 of the 15 formulations tested showed low toxicity, so the other 10 formulations can impact the survival of individuals in a high (6 pesticides) or moderate (4 pesticides) way.

Currently in Brazil, according to the Ministry of Agriculture, Livestock and Food Supply, there are 75 authorized products formulated with atrazine and 22 with cypermethrin (AGROFIT 2022). These represent only two AI among the 309 chemical AI commercialized in the country in 2020, according to the reports from the Brazilian Institute of Environment and Renewable Natural Resources (IBAMA 2020). Therefore, the number of formulated products used in the environment can be much more impactful to amphibians than imagined.

Although it has high comparative potential and the result is effective and clear, the LC_{50} is a value obtained through mortality from exposure to high concentrations of the compounds. That is, a test with several high concentrations is necessary to get a single value, which can indicate the toxicity of the pesticide. The information obtained is primarily used to assign bands of acute toxic effect to a chemical, which restricts how materials can be used (Nuffield Council on Bioethics 2005). Even if the toxicity is high, these studies with tadpoles are still not considered to update the legislation to make it more protective for aquatic vertebrates, at least in Brazil.

2.1.2 Chronic Toxicity

Chronic toxicity data are generally less available than acute toxicity data, and the range of procedures for this test is less standardized. In chronic assays, low concentrations of the substances are used and progressive accumulation of damage is expected in one or more critical target organs in the test organism (Combes 2013). The value used for verification and comparison is the No Observed Effect Concentration (NOEC), i.e., the test concentration immediately below the lowest tested concentration with a statistically significant adverse effect (GHS 2021). The use of NOEC has been

Table 1. Acute lethal concentration of pesticides in tadpoles of *Physalaemus cuvieri* and *P. gracilis* from studies carried out at the Laboratory of Ecology and Conservation (LABECO), at the Federal University of Fronteira Sul.

Species	Pesticide	Commercial formulation	$LC_{50,96h}$ (mg/L)	Toxicity	References
Physalaemus cuvieri	Cypermethrin (I)	Cipermetrina Nortox 250 EC	0.24	very toxic	Wrubleswski et al. (2018)
	Tebuconazole (F)	Tebuconazole Nortox	0.98	very toxic	Wrubleswski et al. (2018)
	Glyphosate (H)	Roundup original® DI	1.006	toxic	Herek et al. (2020)
	Atrazine (H)	AclamadoBR®	9.86	toxic	Folador et al. (2022)
	2,4–D (H)	2,4–D Nortox	12.66	harmful	Folador et al. (2022)
	Atrazine (H)	Atrazine Nortox 500 SC	19.69	harmful	Wrubleswski et al. (2018)
Physalaemus gracilis	Clorotalonil (F)	Previnil®	0.036	very toxic	Da Fré (2021)
	Cipermetrine (I)	Cyptrin 250 EC	0. 273	very toxic	Macagnan (2018)
	Deltamethrin (I)	Decis 25 EC	0.5	very toxic	Macagnan et al. (2017)
	Chlorpyrifos (I)	Klorpan 480 EC	0.839	very toxic	Rutkoski et al. (2020)
	Glyphosate (H)	Roundup original® DI	1.131	toxic	Herek et al. (2020)
	Tebuconazole (F)	Rival 200 EC	1.64	toxic	Sturza (2017)
	Cypermetrhine (I)	Cipermetrin 250 EC	5.01	toxic	Macagnan et al. (2017)
	Picloram (H)	Padron®	15.81	harmful	Da Fré (2021)
	Atrazine (H)	Siptran 500 SC	47.9	harmful	Sturza (2017)
	Imidacloprid (I)	Imidacloprid Nortox	69.20	harmful	Da Fré (2021)

(I) Insecticide, (F) fungicide, (H) herbicide

criticized, but it is still a parameter calculated and used for effects comparisons, together with the Lowest Observed Effect Concentration (LOEC) and Maximum Acceptable Threshold Concentration (MATC—geometric mean of a NOEC and a LOEC) values (Warne and Van Dam 2008). Chronic assays to obtain NOEC, LOEC, and MATC values can only be performed under controlled conditions. When pesticide concentrations are similar to those found in the environment and native species are used, survival data can help to understand the effects on amphibian populations. The problem is that tadpoles undergoing chronic assays are expected to survive until the end of the experiment, so that potential damage can be analyzed. Therefore, it is very difficult to have NOEC, LOEC and MATC values referring to mortality. Even if mortality does occur, the difference in mortality between the concentrations tested is usually not statistically significant, making it impossible to obtain any index. This represents a major challenge in the analysis of amphibian survival.

To demonstrate the effects of sublethal concentrations of pesticides on tadpole survival, chronic assays were carried out also with *P. cuvieri* and *P. gracilis*, using the same methodology previously described in this chapter (i.e., type of water, number of tadpoles per treatment, and daily control). Tadpoles were exposed to various sublethal concentrations of commercial pesticide formulations, chosen based on what has already been recorded in aquatic environments or allowed by legislation. Duration of the assays was between 7 to 21 days, and the water solution was renewed at least once for each assay.

The chronic toxicity of tadpole survival in the trials was classified according to the GHS (2021) for fish: chronic 1, very toxic (NOEC ≤ 0.1 mg/L), and chronic 2, toxic ($0.1 <$ NOEC ≤ 1 mg/L) to aquatic life. When it was not possible to calculate NOEC, LOEC was considered, but only for species that had less than 50% survival. The data obtained are presented in Table 2. Pesticides were ranked in descending order of toxicity, according to NOEC, LOEC, or mean percentage of survival.

Table 2. No observed effect concentration (NOEC), lowest observed effect concentration (LOEC), survival and toxicity of pesticides in tadpoles of *P. cuvieri* and *P. gracilis* from studies carried out at the Laboratory of Ecology and Conservation (LABECO), at the Federal University of Fronteira Sul.

Species	Pesticide	Commercial formulation	Min-Max concentration (mg/L)	NOEC	LOEC	% survival average	Toxicity	References
Physalaemus cuvieri	Cypermethrin (I)	Nortox 250 EC	0.014 – 0.14	-	0.014	34.5	very toxic	Wrubleswski et al. (2018)
	2,4–D (H)	Nortox	0.004 – 0.1	0.03	0.05	80	very toxic	Santos (2020)
	Glyphosate (H)	Roundup original® DI	0.065 – 1	0.065	0.144	89	very toxic	Herek et al. (2020)
	Tebuconazole (F)	Nortox	0.01 – 0.11	-	-	94.5	-	Wrubleswski et al. (2018)
	Atrazine (H)	Nortox 500 SC	0.24 – 2.4	-	-	70.0	-	Wrubleswski et al. (2018)
	Imidacloprid (I)	Nortox	0.003 – 0.3	-	-	100.0	-	Samojeden et al. (2022)
Physalaemus gracilis	Cypermetrhin (I)	Cipermetrin 250 EC	0.001 – 0.1	-	0.001	17.8	very toxic	Vanzetto et al. (2019)
	Deltamethrin (I)	Decis 25 EC	0.001 – 0.009	-	0.001	23.9	very toxic	Vanzetto et al. (2019)
	Glyphosate (H)	Roundup original® DI	0.065 – 1	0.144	0.28	78.5	toxic	Herek et al. (2020)
	Chlorpyrifos (I)	Klorpan 480 EC	0.11 – 0.5	0.25	0.5	70	toxic	Rutkoski et al. (2020)
	Atrazine (H)	Siptran 500 SC	0.5 – 4.8	-	-	78.89	-	Sturza (2017)
	Tebuconazole (F)	Rival 200 EC	0.01 – 0.15		-	82.23	-	Sturza (2017)

(I) Insecticide, (F) fungicide, (H) herbicide.

Analyzing the amount of each commercial formulation that the tadpoles were able to withstand, combined with the percentage of survival, the order of toxicity of pesticides obtained in the studies was: cypermethrin > deltamethrin > 2,4–D > glyphosate > chlorpyrifos > atrazine ≥ tebuconazole ≥ imidacloprid. Cypermethrin was toxic to *Physalaemus* tadpoles in all formulations tested. This was demonstrated in both acute and chronic toxicity assays, with low survival rates in both cases. A similar result was observed for deltamethrin, which was considered very toxic for *P. gracilis*. These two pesticides are pyrethroid insecticides, and after chronic studies, they were found to cause over 70% mortality in tadpoles exposed for one week at environmentally relevant concentrations (Vanzetto et al. 2019). Formulations based on cypermethrin and deltamethrin are indicated on packaging label as very dangerous to the environment, but are considered low toxic in their toxicological classification (AGROFIT 2022).

The other two insecticides evaluated, despite being less toxic than cypermethrin and deltamethrin, also impaired at least one of the amphibian species. Chlorpyrifos, an organophosphate pesticide, showed high acute toxicity and was toxic in the chronic assay for *P. gracilis* tadpoles. The study by Rutkoski et al. (2020) showed that after chronic exposure, only 25% of the acute toxicity concentration of chlorpyrifos (0.25 mg/L) was able to kill almost half (41.7%) of the tadpoles exposed for 7 days, indicating that these concentrations could contribute to the decline of amphibian populations in their natural environments. Whereas, imidacloprid was harmful to *P. gracilis* in an acute assay (Da Fré 2021) and caused no mortality in *P. cuvieri* or little mortality in *Leptodactylus luctator* (Samojeden et al. 2022). However, as demonstrated by Samojeden et al. (2022), the issue of toxicity goes beyond survival, as this insecticide caused smaller body size, malformations, and genotoxic effects in these 2 species at concentrations from 0.003 mg/L. In addition, imidacloprid is a pesticide with a high risk of contamination of groundwater and surface water (Grandi et al. 2021), therefore it can be dangerous to all aquatic life in the long term.

Low doses of 2,4–D and glyphosate were also very toxic to *P. cuvieri* and glyphosate was toxic to *P. gracilis* (Table 2). Regarding 2,4–D, *P.cuvieri* was more sensitive than other amphibian species such as *Leptodactylus fuscus* (Leptodactylidae), *Lithobates catesbeianus* (Ranidae), *P. nattereri* and *P. albonotatus* (Leiuperidae) (Curi et al. 2019, Freitas et al. 2019), which may be related to the type of commercial formulation used in the different studies. Acute exposure to glyphosate has been reported to cause moderate toxicity ($LC_{50,96\,h}$ between 1.80 and 4.22 mg/L) for *Bufo fowleri*, *Rana catesbeiana*, *Hyla chrysoscelis*, and *R. clamitans* (Moore et al. 2012), *Scinax squalirostris* when associated with polyethylene microplastics ($LC_{50,24\,h}$ 6.25) (Lajmanovich et al. 2022), and genotoxic damage to *Dendropsophus minutus* tadpoles in the acute assay (Lopes et al. 2021), and in *Boana faber* and *Leptodactylus latrans* in chronic assays (Pavan et al. 2021).

The problem with glyphosate is extensive, as this is the most used herbicide in the world (Soares et al. 2021). At concentrations above 1 μg/L, glyphosate poses a high risk to aquatic organisms, but many countries do not have restrictive legislation on its concentration in water (Brovini et al. 2021). According to Herek et al. (2020), the MATC for *P. cuvieri* and *P. gracilis* is 104 μg/L glyphosate, below the concentration allowed for Brazilian waters, which is 280 μg/L for class III waters (Brasil 2005) and 500 μg/L for drinking water (Brasil 2011). This demonstrates that the continuous use of this herbicide is a challenge for the health of amphibians and the environment.

Atrazine was not very toxic to both *Physalaemus* species. Tadpoles exposed to this herbicide showed survival above 70% in the chronic assay and $LC_{50,96\,h} > 10$ mg/L in the acute assay (Wrubleswki et al. 2018). In a study performed in the early stages of development (starting at stage 19 and ending at 25 of Gosner) of *P. gracilis,* mortality mainly occurred between 135 and 165 mg/L atrazine (Rutkoski et al. 2018), still in the harmful category of toxicity.

On the other hand, the fungicide tebuconazole had different results in the acute and chronic tests for *P. cuvieri* and *P. gracilis*. While exposure to sublethal doses for 7 days resulted in an average survival of 88.4% of the tadpoles, acute doses demonstrated high toxicity ($LC_{50,96\,h} = 0.98$ mg/L) and moderate toxicity ($LC_{50,96\ h}$ = 1.64 mg/L) for *P. cuvieri* and *P. gracilis*, respectively (Sturza 2017, Wrubleswki et al. 2018). In addition to the direct toxicity caused by fungicides, previous exposure of tadpoles to these chemicals has led to an increased susceptibility to infectious skin diseases. The skin microbiome of tadpoles is important for their immune function (Kueneman et al. 2016) (see more details in Chapter 8), but the wet skin of these animals is subject to several fungal infections, such as chytridiomycosis, caused by the fungus *Batrachochytrium dendrobatidis* (Bd). Rohr et al. (2017) observed that previous exposure of *Osteopilus septentrionalis* tadpoles to 3 toxic fungicides to Bd, during the larval stage, allowed greater toxicity in later contamination by Bd, in the metamorphosis phase (71 days after the first exposure to the fungicide), causing greater individual mortality. This indicates that alterations in the healthy skin microbiome (McCoy and Peralta 2018, Jiménez et al. 2021), as well as developmental and behavioral alterations (Yu et al. 2013, Acquaroni et al. 2021), or even genotoxic alterations (Herek et al. 2021, Pavan et al. 2021) may compromise the future survival of the individuals.

Despite the recent and important increase in the number of published works regarding the toxicity of pesticides to tadpoles, there is still doubt as to whether laboratory studies reflect or not what happens in the environment, and a premise that tadpoles under controlled conditions are more sensitive than those in natural habitats. Thus, the study by Folador et al. (2022) compared the LC_{50} of 2,4–D (Nortox) and atrazine (AclamadoBR®) for *P. cuvieri* tadpoles conducted in both laboratory and field conditions and demonstrated that the values can vary. The tadpoles studied in the field were more resistant to 2,4–D, but less resistant to atrazine. When exposed to 2,4–D, the authors found $LC_{50;96\ h}$ of 12.66 mg/L (± 95% = 9.9 – 16.17) in the laboratory and 31.44 mg/L (28.4–34.8) in the field; when exposed to atrazine, the $LC_{50,96\ h}$ was 9.86 mg/L (5.8 – 16.5) in the laboratory and 6.8 mg/L (4.85 – 9.53) in the field. Atrazine LC_{50} values were within or very close to the confidence interval (± 95%) between laboratory and field tadpoles, so, in this case, they can be considered similar. Certainly, more comparative studies should be carried out, but this variability found is a good sign. If tadpoles under controlled conditions are more sensitive, the results found by laboratory assays can generate more restrictive regulatory legislation and, in fact, protect wildlife. If at least they are comparable, it indicates that we are on the right way. Also, in future studies, it will be important to relate other environmental characteristics with chemicals, such as high temperatures, since this condition can alter the tadpoles' metabolism and increase the toxicity caused by pesticides (Freitas et al. 2016, 2017, Grott et al. 2021).

3. Conclusions and Perspectives

The studies discussed in this chapter show that there is a diversity of effects caused by pesticides on tadpoles, and this depends on the active ingredients, commercial formulations and species used in each study. The observed toxic damages also demonstrate that there is a crucial need to restrict the global use of pesticides on living organisms, and their impacts on water and soil should be equally recognized (Ansari et al. 2019). In tadpoles, changes in growth and development caused by exposure to pesticides can contribute to the reduction of populations and communities. Thus, implementation of more sustainable and alternative methodologies to the use of pesticides would be of great importance for the protection and conservation of amphibians, in addition to expanding incentives for research related to the survival of amphibian populations. Finally, closer relations between science and technology, together with the strengthening of environmental agencies, are important to achieve a less contaminated environment.

Acknowledgments

The authors thank the Federal University of Fronteira Sul for financial support. C.M. is grateful to CAPES for fellowship. This study is financed in part by the Coordenação de Aperfeiçoamento de Pessoal de Nivel Superior—Brazil (CAPES), finance code 001.

References Cited

Achaval, F. and A. Olmos. 2003. Anfibios y reptiles del Uruguay. 2nd. ed. Graphis, Impresora, Montevideo, Uruguay.

Acquaroni, M., G. Svartz and C. Pérez Coll. 2021. Developmental toxicity assessment of a chlorothalonil-based fungicide in a native amphibian species. Arch. Environ. Contam. Toxicol. 80: 680–690.

Adams, E., C. Leeb and C.A. Brühl. 2021. Pesticide exposure affects reproductive capacity of common toads (*Bufo bufo*) in a viticultural landscape. Ecotoxicology 30: 213–223.

AGROFIT. 2022. Sistema de Agrotóxicos Fitosanitários. Ministério da Agricultura, Pecuária e Abastecimento. https://www.gov.br/agricultura/pt-br/assuntos/insumos-agropecuarios/insumos-agricolas/agrotoxicos/agrofit. Accessed 25 July 2022.

Ansari, M., B. Hatami and S.S. Khavida. 2019. Toxicity, biodegradability and detection methods of glyphosate; the most used herbicide: A systematic review. J. Environ. Health. Sustain. Dev. 4: 731–743.

ASTM E1192-97. 2014. Standard guide for conducting acute toxicity tests on aqueous ambient samples and effluents with fishes, macro invertebrates, and amphibians. United States: American Society for Testing and Materials, Washington DC.

ASTM E729. 2014. Standard Guide for Conducting Acute Toxicity Tests on Test Materials with Fishes, Macroinvertebrates, and Amphibians. United States: American Society for Testing and Materials, Washington DC.

Barreto, L. and G.V. Andrade. 1995. Aspects of the reproductive biology of *Physalaemus cuvieri* (Anura: Leptodactylidae) in northeastern Brazil. Amphib-Reptil 16: 67–76.

Brasil. 2005. Resolução CONAMA n°357 de 17 de março de 2005. http://www.mma.gov.br/port/conama/res/res05/res35705.pdf. Accessed 20 July 2022.

Brasil. 2011. Portaria do Ministério da Saúde, 2914 de 12 de dezembro de 2011. Health Ministry. http://bvsms.saude.gov.br/bvs/saudelegis/gm/ 2011/prt2914_12_12_2011.html. Accessed 20 July 2022.

Brasil. 2002. Decreto n° 4.074, de 4 de janeiro de 2002. Diário Oficial da República Federativa do Brasil, Poder Executivo, Brasília, DF, 4 Jan 2002. http://www.planalto.gov.br/ccivil_03/decreto/2002/d4074.htm. Accessed 25 June 2022.

Brasil. 2022. Registros concedidos 2000–2022. Ministério da Agricultura. http://www.gov.br/agricultura/pt-br/assuntos/insumos-agropecuarios/insumos-agricolas/agrotoxicos/RegistrosConcedidos20002022.xlsx. Accessed 25 June 2022.

Brovini, E.M., S.J. Cardoso, G.R. Quadra, J.A. Vilas-Boas, J.R Paranaíba, R.O. Pereira et al. 2021. Glyphosate concentrations in global freshwaters: Are aquatic organisms at risk? Environ. Sci. Pollut. Res. 28: 60635–60648.

Combes, R.D. 2013. Progress in the development, validation and regulatory acceptance of *in vitro* methods for toxicity testing. Reference Module in Chemistry, Molecular Sciences and Chemical Engineering, Elsevier, 2013.

Curi, L.M., P.M. Peltzer, M.T. Sandoval, M.T. and R.C. Lajmanovich. 2019. Acute toxicity and sublethal effects caused by a commercial herbicide formulated with 2,4–D on *Physalaemus albonotatus* tadpoles. Water Air Soil Pollut. 230: 22.

Da Fré, S. 2021. Avaliação do potencial genotóxico da exposição aguda de três tipos de agrotóxicos em girinos de *Physalaemus gracilis* (Anura: Leptodactylidae). Bachelor's Dissertation. Universidade Federal da Fronteira Sul, Erechim, BR.

Duellman, W.E. and L. Trueb. 1986. Biology of Amphibia. Mc Graw - Hill Book Company, New York, USA.

Folador, A., C.F. Rutkoski, N. Macagnan, V. Skovronski, P. Hartmann and M.T. Hartmann. 2022. Toxicidade aguda dos herbicidas 2,4-D e atrazina em girinos de *Physalaemus cuvieri*. pp. 118–128. *In*: Oliveira-Júnior, J.M.B. and Calvão, L.B. [eds.]. Ecologia e conservação da biodiversidade 2. Atena, Ponta Grossa, PR, Brazil.

Freitas, J.S., A. Kupsco, G. Diamante, A.A. Felicio, E.A. Almeida and D. Schlenk, 2016. Influence of temperature on the thyroidogenic effects of diuron and its metabolite 3,4-DCA in tadpoles of the american bullfrog (*Lithobates catesbeianus*). Environ. Sci. Technol. 50: 13095–13104.

Freitas, J.S., A.A. Felicio, F.T. Teresa and E.A. Almeida. 2017. Combined effects of temperature and clomazone (Gamit®) on oxidative stress responses and B-esterase activity of *Physalaemus nattereri* (Leiuperidae) and *Rhinella schneideri* (Bufonidae) tadpoles. Chemosphere. 185: 548–562.

Freitas, J.S., L. Girotto, B.V. Goulart, L.O.G. Alho, R.C. Gebara, C.C. Montagner et al. 2019. Effects of 2,4–D-based herbicide (DMA® 806) on sensitivity, respiration rates, energy reserves and behavior of tadpoles. Ecotoxicol. Environ. Saf. 182: 109446.

Frost, D.R. 2021. Amphibian Species of the World: an Online Reference. Version 6.1. Electronic Database accessible at https://amphibiansoftheworld.amnh.org/index.php. American Museum of Natural History, New York, USA. Accessed 25 July 2022.

GHS. 2021. Globally Harmonized System of Classification and Labeling of Chemicals. New York and Geneva, United Nations.

Gosner. K.L. 1960. A simplified table for staging anuran embryos and larvae with notes on identification. Herpetologica 16: 183–190.

Grandi, A.L., C. Müller, P. Hartmann and M. Hartmann. 2021. Avaliação do risco de contaminação de águas superficiais e subterrâneas por agrotóxicos no Brasil. pp. 199–211. *In:* Silva, C.E. [ed.]. Meio ambiente: Preservação, saúde e sobrevivência. Atena, Ponta Grossa, PR, Brazil.

Grott, S.C., D. Bitschinski, N.G. Israel, G. Abel, S.P. Silva, T.C. Alves et al. 2021. Influence of temperature on biomarker responses and histology of the liver of American bullfrog tadpoles (*Lithobates catesbeianus*, Shaw, 1802) exposed to the herbicide Tebuthiuron. Sci. Total Environ. 771: 144971.

Hamilton, M.A., R.C. Russo and R.V. Thurston. 1977. Trimmed Spearman-Karber method for estimating median lethal concentrations in toxicity bioassays. Environ. Sci. Technol. 11: 714–719.

Herek, J.S., L. Vargas, S.A.R. Trindade, C.F. Rutkoski, N. Macagnan, P.A. Hartmann et al. 2020. Can environmental concentrations of glyphosate affect survival and cause malformation in amphibians? Effects from a glyphosate-based herbicide on *Physalaemus cuvieri* and *P. gracilis* (Anura: Leptodactylidae). Environ. Sci. Pollut. Res. Int. 27: 22619–22630.

Herek, J.S., L. Vargas, S.A.R. Trindade, C.F. Rutkoski, N. Macagnan, P.A. Hartmann and M.T. Hartmann. 2021. Genotoxic effects of glyphosate on *Physalaemus tadpoles*. Environ. Toxicol. Pharmacol. 81: 103516.

IBAMA. 2020. Instituto Brasileiro do Meio Ambiente e dos Recursos Naturais Renováveis. Relatórios de comercialização de agrotóxicos. Boletim 2020. http://ibama.gov.br/agrotoxicos/relatorios-de-comercializacao-de-agrotoxicos. Accessed 25 July 2022.

IUCN 2022. International Union for Conservation of Nature and Natural Resources. The IUCN Red List of Threatened Species. Version 2022-1. https://www.iucnredlist.org. Accessed 25 July 2022.

Jiménez, R.R., G. Alvarado, C. Ruepert, E. Ballestero and S. Sommer. 2021. The fungicide chlorothalonil changes the amphibian skin microbiome: A potential factor disrupting a host disease-protective trait. Appl. Microbiol. 1: 26–37.

Kerby, J.L., K.L. Richards-Hrdlicka, A. Storfer and D.K. Skelly. 2010. An examination of amphibian sensitivity to environmental contaminants: Are amphibians poor canaries? Ecol. Lett. 13: 60–67.

Kueneman, J.G., D.C. Woodhams, W. Van Treuren, H.M. Archer, R. Knight and V.J. Mckenzie. 2016. Inhibitory bacteria reduce fungi on early life stages of endangered Colorado boreal toads (*Anaxyrus boreas*). ISME J. 10: 934–944.

Lajmanovich, R.C., A.M. Attademo, G. Lener, A.P.C. Boccioni, P.M. Peltzer, C.S. Martinuzzi et al. 2022. Glyphosate and glufosinate ammonium, herbicides commonly used on genetically modified crops, and their interaction with microplastics: Ecotoxicity in anuran tadpoles. Sci. Total Environ. 804: 150177.

Lavilla E., A. Kwet, M.V. Segalla, J. Langone and D. Baldo. 2010. *Physalaemus gracilis.* The IUCN Red List of Threatened Species 2010: e.T57258A11610839.

Lenhardt, P.P., C.A. Bruhla and G.E.R.T. Berger. 2015. Temporal coincidence of amphibian migration and pesticide applications on arable fields in spring. Basic Appl. Ecol. 16: 54–63.

Lopes, A., M. Benvindo-Souza, W.F. Carvalho, H.F. Nunes, P.N. Lima, M.S. Costa et al. 2021. Evaluation of the genotoxic, mutagenic, and histopathological hepatic effects of polyoxyethylene amine (POEA) and glyphosate on *Dendropsophus minutus* tadpoles. Environ. Pollut. 289: 117911.

Macagnan, N. 2018. Avaliação dos efeitos letais e subletais dos inseticidas Cipermetrina e Fipronil em girinos de *Physalaemus gracilis* (Anura: Leptodactylidae). M.Sc. Dissertation. Universidade Federal da Fronteira Sul, Erechim, BR.

Macagnan N., C.F. Rutkoski, C. Kolcenti, G.V. Vanzetto, L.P. Macagnan, P.F. Sturza et al. 2017. Toxicity of cypermethrin and deltamethrin insecticides on embryos and larvae of *Physalaemus gracilis* (Anura: Leptodactylidae). Environ. Sci. Pollut. Res. Int. 24: 20699–20704.

McCoy, K.A. and A.L. Peralta. 2018. Pesticides could alter amphibian skin microbiomes and the effects of *Batrachochytrium dendrobatidis*. Front. Microbiol. 9: 748.

Mcdiarmid, R.W. and R. Altig. 1999. The Biology of Anuran Larvae. The University of Chicago Press, Chicago, USA. 458 p.

Mijares, A., M.T. Rodrigues and D. Baldo. 2010. *Physalaemus cuvieri.* The IUCN Red List of Threatened Species 2010: e.T57250A11609155.

Moore, L.J., L. Fuentes, J.H. Rodgers Jr., W.W. Bowerman, G.K. Yarrow, W.Y. Chao et al. 2012. Relative toxicity of the components of the original formulation of Roundup® to five North American anurans. Ecotoxicol. Environ. Saf. 78: 128–133.

Moraes, R.F. 2019. Agrotóxicos no Brasil: Padrões de uso, política da regulação e prevenção da captura regulatória. No. 2506. Texto para Discussão. Instituto de Pesquisa Econômica Aplicada (IPEA), Brasília, BR.

Nuffield Council on Bioethics 2005. The ethics of research involving animals full report. London, Nuffield Council on Bioethics.

Nunes, A., C. Schmitz, S. Moura and M. Maraschin. 2021. The use of pesticides in Brazil and the risks linked to human health. Braz. J. Dev. 7: 37885–37904.

OECD. 2009. *Test No. 231*: Amphibian Metamorphosis Assay, OECD Guidelines for the Testing of Chemicals, Section 2, OECD Publishing, Paris.

OECD. 2015. *Test No. 241*: The Larval Amphibian Growth and Development Assay (LAGDA), OECD Guidelines for the Testing of Chemicals, Section 2, OECD Publishing, Paris.

Orton, F., S. Mangan, L. Newton and A. Marianes. 2022. Non-destructive methods to assess health of wild tropical frogs (túngara frogs: *Engystomops pustulosus*) in Trinidad reveal negative impacts of agricultural land. Environ. Sci. Pollut. Res. 29: 40262–40272.

Pavan, F.A., C.G. Samojeden, C.F. Rutkoski, A. Folador, S.P. Da Fré, C. Müller et al. 2021. Morphological, behavioral and genotoxic effects of glyphosate and 2,4–D mixture in tadpoles of two native species of South American amphibians. Environ. Toxicol. Pharmacol. 85: 103637.

Porto, D.D. and L.H.P. Kiill. 2020. Challenges to ensuring good health and well-being. pp. 19–27. *In:* Killl, L.H.P., H.C.A. Kato and F.F. Calegario [eds.]. Good health and well-being: Contributions of Embrapa. Embrapa, Brasília, DF, BR.

Rocha, G.M. 2018. Análise bioética das informações toxicológicas para fins de registros de agrotóxicos no Brasil: A ciência regulatória e o conflito de interesses. Ph.D. Thesis, Universidade de Brasília, Brasília, BR.

Rodrigues, L.B., G.G. Costa, E.L. Thá, L.R. Silva, R. Oliveira, D.M. Leme et al. 2019. Impact of the glyphosate-based commercial herbicide, its components and its metabolite AMPA on non-target aquatic organisms. Mutat. Res. Genet. Toxicol. Environ. Mutagen. 842: 94–101.

Rohr, J.R., J. Brown, W.A. Battaglin, T.A. McMahon and R.A. Relyea. 2017. A pesticide paradox: Fungicides indirectly increase fungal infections. Ecol. Appl. 27: 2290–2302.

Rutkoski, C., N. Macagnan, A. Folador, V.J. Skovronski, A.M.B. Amaral, J.W. Leitemperger et al. 2020. Morphological and biochemical traits and mortality in *Physalaemus gracilis* (Anura: Leptodactylidae) tadpoles exposed to the insecticide chlorpyrifos. Chemosphere 250: 126162.

Rutkoski, C.F., N. Macagnan, C. Kolcenti, G.V. Vanzetto, P.F. Sturza, P.A. Hartmann et al. 2018. Lethal and sublethal effects of the herbicide atrazine in the early stages of development of *Physalaemus gracilis* (Anura: Leptodactylidae). Arch. Environ. Contam. Toxicol. 74: 587–593.

Samojeden, C.G., F.A. Pavan, C.F. Rutkoski, A. Folador, S.P. Da Fré, C. Müller et al. 2022. Toxicity and genotoxicity of imidacloprid in the tadpoles of *Leptodactylus luctator* and *Physalaemus cuvieri* (Anura: Leptodactylidae). Sci. Rep. 12: 11926.

Santos, G. 2020. Efeitos subletais da formulação comercial do herbicida 2,4–D em *Physalaemus cuvieri* (Anura: Leptodactylidae). M.Sc. Dissertation. Universidade Federal da Fronteira Sul, Erechim, BR.

Sarkar, S., J.D.B. Gil, J. Keeley and K. Jansen. 2021. The use of pesticides in developing countries and their impact on health and the right to food. European Union. Policy Department for External Relations. Directorate General for External Policies of the Union. January 2021.

Segalla, M.V., B. Berneck, C. Canedo, U. Caramaschi, C.A.G. Cruz, P.C. Garcia et al. 2021. List of Brazilian amphibians. Herpetol. Bras. 10: 121–216.

Soares, D., L. Silva, S. Duarte, A. Pena and A. Pereira. 2021. Glyphosate use, toxicity and occurrence in food. Foods. 10: 2785.

Sturza, P. 2017. Toxidade aguda e crônica em girinos de *Physalaemus gracilis* (Anura: leptodactylidae) expostos à Atrazina e Tebuconazole. M.Sc. Dissertation. Universidade Federal da Fronteira Sul, Erechim, BR.

Vanzetto, G.V., J.G. Slaviero, P.F. Sturza, C.F. Rutkoski, N. Macagnan, C. Kolcenti et al. 2019. Toxic effects of pyrethroids in tadpoles of *Physalaemus gracilis* (Anura: Leptodactylidae). Ecotoxicology 28: 1105–1114.

Varga, J.F., M.P. Bui-Marinos and B.A. Katzenback. 2019. Frog skin innate immune defences: Sensing and surviving pathogens. Frontiers. Immunol. 9: 3128.

Vargas, L. and E.S. Roman. 2006. Conceitos e aplicações dos adjuvantes. Embrapa Trigo-Documentos (INFOTECA-E), Passo Fundo, RS, BR.

Wagner, N., W. Reichenbecher, H. Teichmann, B. Tappeser and S. Lötters. 2013. Questions concerning the potential impact of glyphosate-based herbicides on amphibians. Environ. Toxicol. Chem. 32: 1688–1700.

Warne, M.S.J. and R. Van Dam. 2008. NOEC and LOEC data should no longer be generated or used. Australas. J. Ecotoxicol. 14: 1–5.

Wrubleswski, J., F.W. Reichert, L. Galon, P.A. Hartmann and M.T. Hartmann. 2018. Acute and chronic toxicity of pesticides on tadpoles of *Physalaemus cuvieri* (Anura, Leptodactylidae). Ecotoxicology 27: 360–368.

Yu, S., M.R. Wages, G.P. Cobb and J.D. Maul. 2013. Effects of chlorothalonil on development and growth of amphibian embryos and larvae. Environ. Pollut. 181: 329–334.

Zaller, J.G. 2020. Daily Poison: Pesticides–an Underestimated Danger. Springer Nature, Switzerland.

12

Pesticide Effects on Growth and External Morphology of Larvae and Metamorphs (Amphibia, Anura)

Evidence From Experimental Studies

Renan Nunes Costa,[1,*] *Amanda Gomes dos Anjos,*[2] *Fausto Nomura*[3] and *Mirco Solé*[4]

1. Introduction

Pesticides can be classified according to their designation (e.g., herbicide, insecticide, fungicide), toxicity (ranging from highly toxic to non-toxic) and nature (e.g., organic, inorganic), and are highly effective to control agricultural pests, reducc damage to crops and increase food production (Devine and Furlong 2007). However, there is a consensus that many environmental and social problems are associated with pesticide contamination (Wilson and Tisdell 2001, Pimentel 2009, Pedlowski et al. 2012). Based on evidence of how damaging pesticides can be for human and wildlife, diverse active ingredients and commercial formulations have been banned in many countries (e.g., United States, European Union, Sri Lanka, Switzerland). Conversely, countries like Brazil, one of the world leaders in pesticide consumption, have followed a different path, allowing the use of a long list of active ingredients and pesticide formulations (Schiesari and Grillitsch 2011, Carneiro et al. 2015, Bombardi 2017, MAPA 2022).

[1] Departamento de Ciências Biológicas, Universidade do Estado de Minas Gerais-UEMG, Praça dos Estudantes, 23, Santa Emília, 36800-000, Carangola, Minas Gerais, Brasil.

[2] Programa de Pós-Graduação em Ecologia e Conservação da Biodiversidade, Universidade Estadual de Santa Cruz, Rodovia Jorge Amado, km 16, Ilhéus, Bahia, Brasil. Email: amandaanjos09@gmail.com

[3] Departamento de Ecologia, Universidade Federal de Goiás-UFG, Av. Esperança, s/n. 74690-900, Goiânia, Goiás, Brasil. Email: faustonomura@ufg.br

[4] Departamento de Ciências Biológicas, Universidade Estadual de Santa Cruz-UESC, Rodovia Jorge Amado, km 16, 45662-900, Ilhéus, Bahia, Brasil. Email: msole@uesc.br.

* Corresponding author: renan.costa@uemg.br

In extensive croplands, high levels of pesticide are indiscriminately applied, leading to habitat perturbation, water systems eutrophication and contamination of ground and surface water, air, soil, and biota, directly affecting different groups of non-target organisms and contributing to species loss (Foley et al. 2005, Devine and Furlong 2007, Schiesari and Grillitsch 2011, Schiesari et al. 2013, Pietrzak et al. 2019). Amphibians are among the most threatened by contamination and habitat loss resulting from agricultural expansion, and are the vertebrate group with the most population declines in the world (Sparling et al. 2001, Blaustein and Kiesecker 2002, Stuart et al. 2004, Gallant et al. 2007, Mann et al. 2009, Hayes et al. 2010, Alroy 2015). Amphibians are higly susceptible to environmental contamination mainly because they have permeable skin and a byphasic lifecycle. Thus, aquatic contamination is a special concern for species with indirect development with aquatic eggs and larvae (Bishop et al. 1999, Gallant et al. 2007, Schiesari et al. 2007, Mann et al. 2009, Allentoft and O'Brien 2010). Pesticides can act as endocrine disruptors affecting the production, metabolism, and action of natural hormones, which are responsible for growth, behavior and/or developmental regulation in amphibians (Hayes et al. 2006, Hayes et al. 2010). In addition to lethal effects (e.g., mortality), pesticide contamination often induces sublethal effects on amphibian attributes, particularly during the aquatic developmental stages (i.e., larvae and metamorphs) and metamorphosis process, when hormonal regulation is more critical (Hayes et al. 2006, Mann et al. 2009).

The high phenotypic plasticity of larvae favors ecotoxicological studies in laboratory and outdoor mesocosms (Van Buskirk and Relyea 1998, Steiner and Van Buskirk 2008, Van Buskirk 2009, Fusco and Minelli 2010), allowing to evaluate the indirect effects caused by pesticide contamination in a short time. Sublethal effects directly and indirectly affect survival and/or individual fitness, and can be observed by changes in different larval attributes, such as growth (mass and length) and morphology (deformities and variation in body traits) (e.g., Boone and Semlitsch 2002, Relyea 2012, Devi and Gupta 2013, Katzenberger et al. 2014, Freitas et al. 2019, Herek et al. 2020, Pavan et al. 2021).

Effects on growth and morphology of larvae and metamorphs are easily observed in experimental or field surveys because these traits are relatively easier to measure compared to behavioral or developmental responses. Also, their measurement has low economic cost compared to physiological and genetical approaches (Costa and Nomura 2016). However, the most common responses associated to growth and morphology of larvae and metamorphs exposed to pesticide contamination are unknown, as well as the geographic distribution and publication tendencies of these studies. Thus, quantitative studies of the scientific production (i.e., scientometrical analysis) can contribute to the understanding of how pesticides affect growth and morphology of anuran larvae and metamorphs. They can also show patterns, tendencies and how the actual status of the scientific knowledge is developing (Vanti 2002). Herein, we performed a scientometrical review of evidence from experimental studies that investigated pesticide effects on growth and external morphology in anuran larvae and metamorphs. The effects on growth and morphology of tadpoles were summarized for the most common pesticides and species tested, in addition to the trends of scientific production.

2. Data Survey

Electronic databases Web of Science and Google Scholar using the key words "tadpole", "larvae" and "metamorph" applying different combinations with 8 keywords (pesticide, agrochemical, growth, length, mass, morphology, deformities, and malformation) were used for searching. Abstracts were evaluated and included those that applied an experimental approach to investigate the effects of pesticides in anurans' larvae and metamorphs between Gosner's stage 25 and 46 (i.e., corresponding to the stages of larvae and metamorphs; Gosner 1960). Effects were evaluated only at these stages because they are the exclusively aquatic stages and represent the critical moment during the metamorphosis process with hormone-regulated developmental stages (Hayes et al. 2006, Mann et al. 2009). Experimental studies carried out in laboratory, mesocosms and field, restricted to the period between January 2000 and October 2022 were considered. Studies that used embryos, hatchlings and post-metamorphs were not considered. The effects of fertilizers or other type of contaminant were excluded.

From each included study, we extracted information about (see Table 1) the (i) species name, (ii) pesticide tested—alone or mixtures—and commercial formulation, if available, (iii) pesticide class, (iv) specific effects on growth and/or morphology, (v) stage tested—larvae and/or metamorph, (vi) type of experimental approach—laboratory, outdoor mesocosm or field experiment, (vii) country where the study was conducted, (viii) year of publication and (ix) journal of publication. In studies where the authors tested the effects of different types of pesticides (alone or mixed) on different species, we reported each one as a different entry (i.e., the relationship of each species tested with each pesticide type or mixtures—the term "cases" was used throughout the text to define each observation). To understand the main effects of pesticides on growth and morphology, species-specific cases were considered in relation to 5 distinct groups: (a) morphological changes—including malformations, deformities and any type of shape changes; (b) length reduction—including body and tail; (c) length increase—including body and tail; (d) mass reduction and (e) mass increase.

3. Main Findings

A total of 153 studies were found evaluating the effects on growth and external morphology in anuran larvae and metamorphs exposed to pesticide contamination, totaling to 441 species-specific cases (Table 1). These studies evaluated 69 anuran species, in which *Lithobates pipiens* was the most tested, appearing in 69 cases (Fig. 1). A total of 65 different active ingredients were recorded, in which carbaryl, malathion and glyphosate were the most common pesticides tested, respectively (Fig. 2A). The most tested class of pesticides was insecticides (considering growth attributes), followed by herbicides and fungicides (Fig. 2B).

Effects on growth and external morphology were detected in 66.5% of the cases, while in 33.5% no effects were detected. The main effect detected was associated with length reduction (37.7%), suggesting that pesticide contamination slows the growth rate of larvae and metamorphs. The second most frequently reported effect was morphological changes, including malformations and deformities (23,4%), followed by mass reduction (23.2%) (Fig. 3). From the total, 62.5% of the cases tested the larval stage, 8.8% the metamorph stage, and 29.9% both stages. Most of the cases were evaluated under laboratory conditions (69.1%), followed by outdoor/mesocosm (29.2%) and outdoor/large-scale experimental ponds (1.3%).

Studies were performed in 25 different countries, lead by the United States (74 studies), followed by China (27 studies) and Brazil (20 studies) (Fig. 4A). We observed peaks of publications in the years 2022 (26 studies), 2021 (19 studies) and 2019 (12 studies) (Fig. 4B). Studies were published in 53 different journals. The majority of studies were published in the journal "Environmental Toxicology and Chemistry" (31 studies), followed by "Archives of Environmental Contamination and Toxicology" (12 studies) and "Aquatic Toxicology" (10 studies) (Table 2).

4. Effects on Growth and External Morphology

Growth changes (e.g., mass and body size) and external malformations (e.g., hind-limb anomalies, axial and tail deformities), are among the most dramatic effects of pesticide exposure (Mann et al. 2009). Most experiments reported the growth reduction as the main effect of pesticides on larval development. However, larvae and metamorphs can eventually respond to pesticides with a growth increase, with a few studies reporting both effects (increase or reduction in growth) under similar experimental conditions. For example, Figueiredo and Rodrigues (2014) reported increases and

Table 1. Species-specific cases of the pesticide effects on growth and external morphology in tadpoles and metamorphs [stage 25 to 46 (Gosner 1960)]. Species names were updated according to Frost (2021). St. = stage; L = Larvae; M = Metamorph; B = Both.

Species	St.	Country	Experimental site	Active ingredient/ formulation (if there)	Class	Summary of effects on growth and/or external morphology	References
Acris crepitans	L	United States	Laboratory	Chlorpyrifos/ Dursban TC®	Termiticide	Snout-vent length reduction	Widder and Bidwell 2008
Agalychnis callidryas	L	Costa Rica	Laboratory	Chlorothalonil/ Daconil 50SC	Fungicide	Mass reduction	Alza et al. 2016
Anaxyrus americanus	L	United States	Laboratory	Atrazine	Herbicide	Malformations (wavy tail, lateral tail flexure, facial edema, axial shortening, dorsal tail flexure and blistering)	Allran and Kasarov 2001
Anaxyrus americanus	L	United States	Outdoor/ mesocosm	Carbaryl/Sevin®	Insecticide	No effects on growth or morphology	Boone 2008
Anaxyrus americanus	L	United States	Outdoor/ mesocosm	Malathion/ Malathion®	Insecticide	No effects on growth or morphology	Boone 2008
Anaxyrus americanus	L	United States	Outdoor/ mesocosm	Permethrin/ Cutter's Bug Free Back Yard®	Insecticide	No effects on growth or morphology	Boone 2008
Anaxyrus americanus	L	United States	Outdoor/ mesocosm	Carbaryl + malathion - mixture/Sevin® + Malathion®	Insecticides	Mass reduction	Boone 2008
Anaxyrus americanus	L	United States	Outdoor/ mesocosm	Carbaryl/Sevin®	Insecticide	Mass reduction; mass increase	Boone and James 2003
Anaxyrus americanus	L	United States	Outdoor/ mesocosm	Atrazine/Aatrex®	Herbicide	Mass reduction	Boone and James 2003
Anaxyrus americanus	B	United States	Outdoor/ mesocosm	Carbaryl/Sevin®	Insecticide	Mass reduction	Boone et al. 2007
Anaxyrus americanus	L	United States	Outdoor/ mesocosm	Carbaryl/Sevin®	Insecticide	No effects on growth or morphology	Bulen and Distel 2011
Anaxyrus americanus	M	United States	Outdoor/ mesocosm	Carbaryl/Sevin®	Insecticide	Mass reduction	Distel and Boone 2009
Anaxyrus americanus	M	United States	Outdoor/ mesocosm	Carbaryl/Sevin®	Insecticide	No effects on growth or morphology	Distel and Boone 2010
Anaxyrus americanus	L	Canada	Laboratory	Glyphosate/ Vision®	Herbicide	No effects on growth or morphology	Edginton et al. 2004
Anaxyrus americanus	B	Canada	Laboratory	Endosulfan/ Thiodan®50WP	Insecticide	Deformities (eye deformities, luxation of the right front limb)	Harris et al. 2000

Table 1 contd. ...

...Table 1 contd.

Species	St.	Country	Experimental site	Active ingredient/ formulation (if there)	Class	Summary of effects on growth and/or external morphology	References
Anaxyrus americanus	B	Canada	Laboratory	Mancozeb/ Dithane® DG	Fungicide	Deformities (eyes missing)	Harris et al. 2000
Anaxyrus americanus	B	Canada	Laboratory	Azinphos-methyl/ Guthion®50WP	Insecticide	No effects on growth or morphology	Harris et al. 2000
Anaxyrus americanus	B	United States	Outdoor/ mesocosm	Chlorpyrifos	Insecticide	No effects on growth or morphology	Hua and Relyea 2014
Anaxyrus americanus	B	United States	Outdoor/ mesocosm	Diazinon	Insecticide	No effects on growth or morphology	Hua and Relyea 2014
Anaxyrus americanus	B	United States	Outdoor/ mesocosm	Malathion	Insecticide	No effects on growth or morphology	Hua and Relyea 2014
Anaxyrus americanus	B	United States	Outdoor/ mesocosm	Endosulfan	Insecticide	No effects on growth or morphology	Hua and Relyea 2014
Anaxyrus americanus	B	United States	Outdoor/ mesocosm	Chlorpyrifos + diazinon + malathion + endosulfan – mixture	Insecticides	No effects on growth or morphology	Hua and Relyea 2014
Anaxyrus americanus	L	United States	Outdoor/ mesocosm	Glyphosate/ Roundup Original MAX®	Herbicide	Mass reduction	Jones et al. 2010
Anaxyrus americanus	L	United States	Laboratory	Malathion/ Matathion®	Insecticide	Deformities (diamond-shaped body and stiff-tail)	Krishnamurthy and Smith 2010
Anaxyrus americanus	L	United States	Laboratory	Carbaryl/Sevin®	Insecticide	Growth reduction	Relyea 2004a
Anaxyrus americanus	L	United States	Laboratory	Diazinon	Insecticide	Growth reduction	Relyea 2004a
Anaxyrus americanus	L	United States	Laboratory	Malathion	Insecticide	Growth reduction	Relyea 2004a
Anaxyrus americanus	L	United States	Laboratory	Glyphosate/ Roundup®	Herbicide	Growth reduction	Relyea 2004a
Anaxyrus americanus	L	United States	Laboratory	Carbaryl + diazinon - mixture/ Sevin®	Insecticides	Growth reduction	Relyea 2004a
Anaxyrus americanus	L	United States	Laboratory	Carbaryl + malathion - mixture/Sevin®	Insecticides	Growth reduction	Relyea 2004a
Anaxyrus americanus	L	United States	Laboratory	Carbaryl + glyphosate - mixture/Sevin® + Roundup®	Insecticide and herbicide	Growth reduction	Relyea 2004a

Table 1 contd. ...

...Table 1 contd.

Species	St.	Country	Experimental site	Active ingredient/ formulation (if there)	Class	Summary of effects on growth and/or external morphology	References
Anaxyrus americanus	L	United States	Laboratory	Diazinon + malathion – mixture	Insecticides	Growth reduction	Relyea 2004a
Anaxyrus americanus	L	United States	Laboratory	Diazinon + glyphosate - mixture/ Roundup®	Insecticide and herbicide	Growth reduction	Relyea 2004a
Anaxyrus americanus	L	United States	Laboratory	Malathion + glyphosate - mixture/ Roundup®	Insecticide and herbicide	Growth reduction	Relyea 2004a
Anaxyrus americanus	L	United States	Outdoor/ mesocosm	Glyphosate/ Roundup Original MAX®	Herbicide	No effects on growth or morphology	Relyea 2012
Anaxyrus americanus	B	United States	Laboratory	Atrazine/Atrazine 4l®	Herbicide	No effects on growth or morphology	Willians and Semlitsch 2010
Anaxyrus americanus	B	United States	Laboratory	S-metolachlor/ Dual II Magnum®	Herbicide	No effects on growth or morphology	Willians and Semlitsch 2010
Anaxyrus americanus	B	United States	Laboratory	Glyphosate/ Roundup Original MAX®	Herbicide	No effects on growth or morphology	Willians and Semlitsch 2010
Anaxyrus americanus	B	United States	Laboratory	Glyphosate/ Roundup WeatherMax®	Herbicide	No effects on growth or morphology	Willians and Semlitsch 2010
Anaxyrus woodhousii	B	United States	Outdoor/ mesocosm	Carbaryl/Sevin®	Insecticide	No effects on growth or morphology	Boone and Semlitsch 2001
Anaxyrus woodhousii	B	United States	Outdoor/ mesocosm	Carbaryl/Sevin®	Insecticide	Mass increase	Boone and Semlitsch 2002
Anaxyrus woodhousii	B	United States	Outdoor/ large-scale experimental ponds	Carbaryl/Sevin®	Insecticide	Mass increase	Boone et al. 2004
Boana faber	L	Brazil	Laboratory	Glyphosate/ Roundup Original DI® + 2,4–D/ NORTOX®	Herbicides	Damage in body structures (mouth and intestine)	Pavan et al. 2021
Boana pulchella	L	Argentina	Laboratory	Pirimicarb/ Aficida®	Insecticide	Length reduction; morphological abnormalities	Natale et al. 2018
Boana pulchella	L	Argentina	Laboratory	Cypermethrin	Insecticide	Body length reduction; malformations (axial, eyes, gut, head and face abnormalities)	Agostini et al. 2010

Table 1 contd. ...

...Table 1 contd.

Species	St.	Country	Experimental site	Active ingredient/ formulation (if there)	Class	Summary of effects on growth and/or external morphology	References
Boana pulchella	L	Argentina	Laboratory	Cypermethrin/ Sherpa®	Insecticide	Body length reduction; malformations (axial, eyes, gut, head and face abnormalities)	Agostini et al. 2010
Boana xerophylla	L	Colombia	Laboratory	Chlorpyrifos/ Lorsban™	Insecticide	Length reduction	Henao et al. 2022
Boana xerophylla	L	Colombia	Laboratory	Diazinon/Diazol®	Insecticide	Length reduction	Henao et al. 2021
Boana xerophylla	L	Colombia	Laboratory	Monocrotophos/ Monocrotophos 600	Insecticide	Length reduction	Henao et al. 2021
Bombina variegata	L	Germany	Laboratory	Cypermethrin	Insecticide	Physical abnormalities	Greulich and Pflugmacher 2004
Bufo bufo	B	Hungary	Laboratory	Terbuthylazine	Herbicide	Mass increase	Bokony et al. 2020
Bufo bufo	L	Austria	Laboratory	Glyphosate/ Roundup PowerFlex®	Herbicide	Length reduction; malformations	Baier et al. 2016
Bufo bufo	B	Italy	Laboratory	Endosulfan	Insecticide	Mass reduction; malformations (axis, skeletal, tail and mouth malformations, edemas and lateral kink at the base of the tale)	Brunelli et al. 2009
Bufo bufo	L	Spain	Laboratory	Cooper sulfate	Fungicide	Total length reduction	García-Munõz et al. 2010
Bufotes viridis	L	Albania	Laboratory	Cooper sulfate	Algicide, fungicide and molluscicide	Total length reduction; malformations (spinal cord deformity and edemas)	Aıko et al. 2014
Bufotes viridis	L	Turkey	Laboratory	Copper sulfate	Fungicide and algicide	Reduction in body size, body width, and tail length; anomalies (edemas)	Gürkan and Hayretdağ 2012
Discoglossus galganoi	L	Spain	Laboratory	Cooper sulfate	Fungicide	Total length reduction	García-Munõz et al. 2010
Dryophytes chrysoscelis	M	United States	Laboratory	Carbaryl/Sevin®	Insecticide	Mass increase	Gaietto et al. 2014
Dryophytes chrysoscelis	M	United States	Laboratory	Cooper sulfate	Fungicide	No effects on growth or morphology	Gaietto et al. 2014

Table 1 contd. ...

...*Table 1 contd.*

Species	St.	Country	Experimental site	Active ingredient/ formulation (if there)	Class	Summary of effects on growth and/or external morphology	References
Dryophytes chrysoscelis	M	United States	Outdoor/ mesocosm	Malathion/ Malathion®	Insecticide	Mass increase	Mackey and Boone 2009
Dryophytes chrysoscelis	L	United States	Laboratory	Chlorpyrifos/ Dursban TC®	Termiticide	Mass reduction	Widder and Bidwell 2008
Dryophytes versicolor	B	United States	Outdoor/ mesocosm	Carbaryl/Sevin®	Insecticide	Mass increase	Boone and Bridges-Britton 2006
Dryophytes versicolor	B	United States	Outdoor/ mesocosm	Atrazine/Aatrex®	Herbicide	No effects on growth or morphology	Boone and Bridges-Britton 2006
Dryophytes versicolor	B	United States	Outdoor/ mesocosm	Carbaryl + Atrazine – mixture	Insecticide and herbicide	No effects on growth or morphology	Boone and Bridges-Britton 2006
Dryophytes versicolor	B	United States	Outdoor/ mesocosm	Carbaryl/Sevin®	Insecticide	Mass reduction; mass increase	Boone and Semlitsch 2001
Dryophytes versicolor	B	United States	Outdoor/ mesocosm	Malathion	Insecticide	Mass increase	Cothran et al. 2011
Dryophytes versicolor	B	United States	Outdoor/ mesocosm	Chlorpyrifos	Insecticide	Mass increase	Hua and Relyea 2014
Dryophytes versicolor	B	United States	Outdoor/ mesocosm	Diazinon	Insecticide	No effects on growth or morphology	Hua and Relyea 2014
Dryophytes versicolor	B	United States	Outdoor/ mesocosm	Malathion	Insecticide	No effects on growth or morphology	Hua and Relyea 2014
Dryophytes versicolor	B	United States	Outdoor/ mesocosm	Endosulfan	Insecticide	No effects on growth or morphology	Hua and Relyea 2014
Dryophytes versicolor	B	United States	Outdoor/ mesocosm	Chlorpyrifos + diazinon + malathion + endosulfan – mixture	Insecticides	Mass increase	Hua and Relyea 2014
Dryophytes versicolor	L	United States	Outdoor/ mesocosm	Glyphosate/ Roundup Original MAX®	Herbicide	No effects on growth or morphology	Jones et al. 2011
Dryophytes versicolor	L	United States	Outdoor/ mesocosm	Glyphosate/ Roundup Power Max®	Herbicide	Size reduction; smaller bodies; deeper tails	Katzenberger et al. 2014
Dryophytes versicolor	B	United States	Laboratory	Atrazine	Herbicide	No effects on growth or morphology	LaFiandra et al. 2008
Dryophytes versicolor	L	United States	Laboratory	Carbaryl/Sevin®	Insecticide	Growth reduction	Relyea 2004a
Dryophytes versicolor	L	United States	Laboratory	Diazinon	Insecticide	Growth reduction	Relyea 2004a

Table 1 contd. ...

...Table 1 contd.

Species	St.	Country	Experimental site	Active ingredient/ formulation (if there)	Class	Summary of effects on growth and/or external morphology	References
Dryophytes versicolor	L	United States	Laboratory	Malathion	Insecticide	No effects on growth or morphology	Relyea 2004a
Dryophytes versicolor	L	United States	Laboratory	Glyphosate/ Roundup®	Herbicide	No effects on growth or morphology	Relyea 2004a
Dryophytes versicolor	L	United States	Laboratory	Carbaryl + diazinon - mixture/ Sevin®	Insecticides	Growth reduction	Relyea 2004a
Dryophytes versicolor	L	United States	Laboratory	Carbaryl + malathion - mixture/Sevin®	Insecticides	Growth reduction	Relyea 2004a
Dryophytes versicolor	L	United States	Laboratory	Carbaryl + glyphosate - mixture/Sevin® + Roundup®	Insecticide and herbicide	No effects on growth or morphology	Relyea 2004a
Dryophytes versicolor	L	United States	Laboratory	Diazinon + malathion – mixture	Insecticides	Growth reduction	Relyea 2004a
Dryophytes versicolor	L	United States	Laboratory	Diazinon + glyphosate - mixture/ Roundup®	Insecticide and herbicide	No effects on growth or morphology	Relyea 2004a
Dryophytes versicolor	L	United States	Laboratory	Malathion + glyphosate - mixture/ Roundup®	Insecticide and herbicide	No effects on growth or morphology	Relyea 2004a
Dryophytes versicolor	L	United States	Outdoor/ mesocosm	Carbaryl	Insecticide	No effects on growth or morphology	Relyea 2009
Dryophytes versicolor	L	United States	Outdoor/ mesocosm	Malathion	Insecticide	No effects on growth or morphology	Relyea 2009
Dryophytes versicolor	L	United States	Outdoor/ mesocosm	Diazinon	Insecticide	No effects on growth or morphology	Relyea 2009
Dryophytes versicolor	L	United States	Outdoor/ mesocosm	Chlorpyrifos	Insecticide	No effects on growth or morphology	Relyea 2009
Dryophytes versicolor	L	United States	Outdoor/ mesocosm	Endosulfan	Insecticide	No effects on growth or morphology	Relyea 2009
Dryophytes versicolor	L	United States	Outdoor/ mesocosm	Carbaryl + malathion + chlorpyrifos + diazinon + endosulfan – mixture	Insecticides	Mass increase	Relyea 2009

Table 1 contd. ...

...*Table 1 contd.*

Species	St.	Country	Experimental site	Active ingredient/ formulation (if there)	Class	Summary of effects on growth and/or external morphology	References
Dryophytes versicolor	L	United States	Outdoor/ mesocosm	Acetochlor	Herbicide	No effects on growth or morphology	Relyea 2009
Dryophytes versicolor	L	United States	Outdoor/ mesocosm	Metolachlor	Herbicide	No effects on growth or morphology	Relyea 2009
Dryophytes versicolor	L	United States	Outdoor/ mesocosm	Glyphosate	Herbicide	No effects on growth or morphology	Relyea 2009
Dryophytes versicolor	L	United States	Outdoor/ mesocosm	2,4-D	Herbicide	No effects on growth or morphology	Relyea 2009
Dryophytes versicolor	L	United States	Outdoor/ mesocosm	Atrazine	Herbicide	Mass increase	Relyea 2009
Dryophytes versicolor	L	United States	Outdoor/ mesocosm	Acetochlor + metolachlor + glyphosate + 2,4-D + atrazine – mixture	Herbicides	No effects on growth or morphology	Relyea 2009
Dryophytes versicolor	L	United States	Outdoor/ mesocosm	Carbaryl + malathion + chlorpyrifos + diazinon + endosulfan + acetochlor + metolachlor + glyphosate + 2,4-D + atrazine – mixture	Herbicides and insecticides	Mass increase	Relyea 2009
Dryophytes versicolor	L	United States	Laboratory	Carbaryl	Insecticide	Growth reduction	Relyea and Mills 2001
Dryophytes versicolor	B	United States	Laboratory	Atrazine/Atrazine 4l®	Herbicide	No effects on growth or morphology	Willians and Semlitsch 2010
Dryophytes versicolor	B	United States	Laboratory	S-metolachlor/ Dual II Magnum®	Herbicide	No effects on growth or morphology	Willians and Semlitsch 2010
Dryophytes versicolor	B	United States	Laboratory	Glyphosate/ Roundup Original MAX®	Herbicide	No effects on growth or morphology	Willians and Semlitsch 2010
Dryophytes versicolor	B	United States	Laboratory	Glyphosate/ Roundup WeatherMax®	Herbicide	No effects on growth or morphology	Willians and Semlitsch 2010
Duttaphrynus melanostictus	L	India	Laboratory	Malathion	Insecticide	Malformations (scoliosis, lordosis, kyphosis, fin blistering)	David and Kartheek 2015

Table 1 contd. ...

...Table 1 contd.

Species	St.	Country	Experimental site	Active ingredient/ formulation (if there)	Class	Summary of effects on growth and/or external morphology	References
Duttaphrynus melanostictus	L	India	Laboratory	Cypermethrin	Insecticide	Malformations (deformities in coiled intestine, twisting of tail, changes in axial region, loss of conveyance of tail fin and deformities in the head)	David et al. 2012
Duttaphrynus melanostictus	B	India	Laboratory	Acephate/Not specified	Insecticide	Growth reduction; malformations (crooked tails, drooped trunk, edema and split at the tail terminal)	Ghodageri and Pancharatna 2011
Duttaphrynus melanostictus	B	India	Laboratory	Cypermethrin/Not specified	Insecticide	Length reduction; malformations (drooped trunks, tail distortions and head deformities)	Ghodageri and Pancharatna 2011
Duttaphrynus melanostictus	B	Sri Lanka	Laboratory	Chlorpyrifos/ Lorsban EC 40® or Pattas®	Insecticide	Snout-vent length increase; mass increase; spine malformations (kyphosis, lordosis, edemas and skin ulcers)	Jayawardena et al. 2011
Duttaphrynus melanostictus	B	Sri Lanka	Laboratory	Dimethoate / Dimethoate EC®	Insecticide	Snout-vent length increase; mass increase; spine malformations (kyphosis, lordosis, edemas and skin ulcers)	Jayawardena et al. 2011
Duttaphrynus melanostictus	B	Sri Lanka	Laboratory	Glyphosate/ Roundup® or Glyphosate®	Herbicide	Snout-vent length increase; mass increase; spine malformations (kyphosis, lordosis, edemas and skin ulcers)	Jayawardena et al. 2011
Duttaphrynus melanostictus	B	Sri Lanka	Laboratory	Propanil/3,4 DPA®	Herbicide	Snout-vent length increase; mass increase	Jayawardena et al. 2011
Duttaphrynus melanostictus	L	India	Laboratory	Edifenphos/ Hinosan EC®	Fungicide	Mass and length reduction	Mathew and Andrews 2003
Duttaphrynus melanostictus	L	India	Laboratory	Endosulfan/ Endosulfan 3EC®	Insecticide	Mass and length reduction	Mathew and Andrews 2003
Duttaphrynus melanostictus	L	Sri Lanka	Outdoor	Diazinon	Insecticide	Body length reduction	Sumanadasa et al. 2008a

Table 1 contd. ...

...Table 1 contd.

Species	St.	Country	Experimental site	Active ingredient/ formulation (if there)	Class	Summary of effects on growth and/or external morphology	References
Duttaphrynus melanostictus	L	Sri Lanka	Laboratory	Diazinon	Insecticide	Size reduction; abnormalities (bent tails, curved tails and slanted bodies)	Sumanadasa et al. 2008b
Duttaphrynus melanostictus	L	Sri Lanka	Laboratory	Chlorpyrifos/ Lorsban 40 EC®or Pattas®	Insecticide	Malformations - Scoliosis (vertebral column curvature, laterally deviated spine), kyphosis (hunched back, abnormal convexed spine), and edema	Jayawardena et al. 2017
Duttaphrynus melanostictus	L	Sri Lanka	Laboratory	Dimethoate/ Dimethoate 40EC®	Insecticide	Malformations - Scoliosis (vertebral column curvature, laterally deviated spine), kyphosis (hunched back, abnormal convexed spine), and edema	Jayawardena et al. 2017
Duttaphrynus melanostictus	L	Sri Lanka	Laboratory	Glyphosate/ Round Up® or Glyphosate®	Herbicide	Malformations - Scoliosis (vertebral column curvature, laterally deviated spine), kyphosis (hunched back, abnormal convexed spine), and edema	Jayawardena et al. 2017
Duttaphrynus melanostictus	L	Sri Lanka	Laboratory	Propanil/3, 4-DPA®	Herbicide	Malformations - Scoliosis (vertebral column curvature, laterally deviated spine), kyphosis (hunched back, abnormal convexed spine), and edema	Jayawardena et al. 2017
Engystomops pustulosus	L	Colombia	Laboratory	Chlorpyrifos/ Lorsban™	Insecticide	Length reduction; malformations	Henao et al. 2021
Engystomops pustulosus	L	Colombia	Laboratory	Diazinon/Diazol®	Insecticide	Length reduction	Henao et al. 2021
Engystomops pustulosus	L	Colombia	Laboratory	Monocrotophos/ Monocrotophos 600	Insecticide	Length reduction	Henao et al. 2021
Epidalea calamita	L	Spain	Laboratory	Cooper sulfate	Fungicide	Total length reduction	García-Munõz et al. 2009
Epidalea calamita	L	Spain	Laboratory	Cooper sulfate	Fungicide	Total length reduction	García-Munõz et al. 2010
Fejervarya limnocharis	L	China	Laboratory	Chlorantraniliprole	Insecticide	Mass reduction; length reduction	Wei et al. 2022

Table 1 contd. ...

...Table 1 contd.

Species	St.	Country	Experimental site	Active ingredient/ formulation (if there)	Class	Summary of effects on growth and/or external morphology	References
Fejervarya limnocharis	L	China	Laboratory	Penoxsulam	Herbicide	Mass reduction; length reduction	Wei et al. 2022
Fejervarya limnocharis	L	China	Laboratory	Pymetrozine	Insecticide	Mass reduction; length reduction	Wei et al. 2022
Fejervarya limnocharis	L	China	Laboratory	Haloxyfop-P-methyl	Herbicide	Mass reduction; length reduction	Wei et al. 2022
Fejervarya limnocharis	L	China	Laboratory	Chlorantraniliprole + Penoxsulam	Insecticide + herbicide	Mass reduction; length reduction	Wei et al. 2022
Fejervarya limnocharis	L	China	Laboratory	Chlorantraniliprole + Pymetrozine	Insecticide	Mass reduction; length reduction	Wei et al. 2022
Fejervarya limnocharis	L	China	Laboratory	Chlorantraniliprole + Haloxyfop-P-methyl	Insecticide + herbicide	Mass reduction; length reduction	Wei et al. 2022
Fejervarya limnocharis	L	China	Laboratory	Penoxsulam + Pymetrozine	Herbicide + insecticide	Mass reduction; length reduction	Wei et al. 2022
Fejervarya limnocharis	L	China	Laboratory	Penoxsulam + Haloxyfop-P-methyl	Herbicide	Mass reduction; length reduction	Wei et al. 2022
Fejervarya limnocharis	L	China	Laboratory	Pymetrozine + Haloxyfop-P-methyl	Insecticide + herbicide	Mass reduction; length reduction	Wei et al. 2022
Fejervarya limnocharis	L	India	Laboratory	Malathion	Insecticide	Mass reduction; reduction in total lenght, body lenght and tail length	Gurushankara et al. 2007
Fejervarya sp.1	B	India	Laboratory	Endosulfan/Hildan 35 EC®	Insecticide and acaricide	Morphological deformities (fore-limb deformities, axial malformation and hind-limbs deformities)	Devi and Gupta 2013
Fejervarya sp.2	B	India	Laboratory	Endosulfan/Hildan 35 EC®	Insecticide and acaricide	Morphological deformities (axial malformation)	Devi and Gupta 2013
Gastrophryne olivacea	L	United States	Laboratory	Chlorpyrifos/ Dursban TC®	Termiticide	Mass reduction; snout-vent length reduction	Widder and Bidwell 2008
Hoplobatrachus rugulosus	L	Thailand	Laboratory	Atrazine/Not specified	Herbicide	Body length reduction; asymmetrical limbs	Trachantong et al. 2013
Hyla arborea	L	Turkey	Laboratory	Dimethoate	Insecticide	Total length reduction; tail deformities (bent tail)	Sayim and Kaya 2006

Table 1 contd. ...

...*Table 1 contd.*

Species	St.	Country	Experimental site	Active ingredient/ formulation (if there)	Class	Summary of effects on growth and/or external morphology	References
Hyla intermedia	B	Italy	Laboratory	Pyrimethanil	Fungicide	Irregular profile of the tail fin, axial and tail malformations (lateral/dorsal flexure, and wavy tail), abnormal mouth, limb deformities, skeletal malformations; mass increase	Bernabò et al. 2016
Hyla intermedia	B	Italy	Laboratory	Tebuconazole	Fungicide	Irregular profile of the tail fin, axial and tail malformations (lateral/dorsal flexure, and wavy tail), abnormal mouth, limb deformities, skeletal malformations; mass increase	Bernabò et al. 2016
Leptodactylus fuscus	L	Brazil	Laboratory	2,4-D/DMA® 806	Herbicide	Length reduction	Freitas et al. 2019
Leptodactylus latrans	L	Brazil	Laboratory	Glyphosate/ Roundup Original DI® + 2,4–D/ NORTOX®	Herbicides	Mass reduction; growth reduction; damage in body structures (mouth and intestine)	Pavan et al. 2021
Leptodactylus luctator	L	Brazil	Laboratory	Imidacloprid/ Nortox	Insecticide	Lenght reduction; mass reduction; malformations (mouth and intestine)	Samojeden et al. 2022
Limnodynastes peronii	L	Australia	Laboratory	Endodulfan/ Thiodan®	Insecticide and acaricide	Total length reduction	Broomhall 2004
Lithobates blairi	B	United States	Outdoor/ mesocosm	Carbaryl/Sevin®	Insecticide	No effects on growth or morphology	Boone and Semlitsch 2002
Lithobates blairi	L	United States	Laboratory	Glyphosate/ Kleeraway®	Herbicide	No effects on growth or morphology	Smith 2001
Lithobates catesbeianus	L	Brazil	Laboratory	Abamectin/Kraft® 36EC	Vermicide	Malformations (lowest scores for mandibular pigmentation and structures)	Amaral et al. 2018
Lithobates catesbeianus	L	Brazil	Laboratory	2,4-D/DMA® 806	Herbicide	Length reduction	Freitas et al. 2019

Table 1 contd. ...

...Table 1 contd.

Species	St.	Country	Experimental site	Active ingredient/ formulation (if there)	Class	Summary of effects on growth and/or external morphology	References
Lithobates catesbeianus	L	United States	Outdoor/ mesocosm	Carbaryl/Sevin®	Insecticide	Mass increase	Boone and Semlitsch 2003
Lithobates catesbeianus	B	United States	Outdoor/ mesocosm	Carbaryl/Sevin®	Insecticide	Mass increase	Boone et al. 2007
Lithobates catesbeianus	B	United States	Outdoor/ mesocosm	Malathion	Insecticide	Growth increase	Cothran et al. 2011
Lithobates catesbeianus	L	United States	Laboratory	Malathion	Insecticide	Body length increase	Fordham et al. 2001
Lithobates catesbeianus	L	United States	Outdoor/ mesocosm	Glyphosate/ Roundup Original MAX®	Herbicide	No effects on growth or morphology	Jones et al. 2011
Lithobates catesbeianus	L	United States	Laboratory	Carbaryl/Sevin®	Insecticide	Growth reduction	Relyea 2004a
Lithobates catesbeianus	L	United States	Laboratory	Diazinon	Insecticide	Growth reduction	Relyea 2004a
Lithobates catesbeianus	L	United States	Laboratory	Malathion	Insecticide	Growth reduction	Relyea 2004a
Lithobates catesbeianus	L	United States	Laboratory	Glyphosate/ Roundup®	Herbicide	Growth reduction	Relyea 2004a
Lithobates catesbeianus	L	United States	Laboratory	Carbaryl + diazinon - mixture/ Sevin®	Insecticides	Growth reduction	Relyea 2004a
Lithobates catesbeianus	L	United States	Laboratory	Carbaryl + malathion - mixture/Sevin®	Insecticides	Growth reduction	Relyea 2004a
Lithobates catesbeianus	L	United States	Laboratory	Carbaryl + glyphosate - mixture/Sevin® + Roundup®	Insecticide and herbicide	Growth reduction	Relyea 2004a
Lithobates catesbeianus	L	United States	Laboratory	Diazinon + malathion – mixture	Insecticides	Growth reduction	Relyea 2004a
Lithobates catesbeianus	L	United States	Laboratory	Diazinon + glyphosate - mixture/ Roundup®	Insecticide and herbicide	Growth reduction	Relyea 2004a
Lithobates catesbeianus	L	United States	Laboratory	Malathion + glyphosate - mixture/ Roundup®	Insecticide and herbicide	Growth reduction	Relyea 2004a
Lithobates catesbeianus	L	United States	Outdoor/ mesocosm	Carbaryl/Sevin®	Insecticide	Growth increase	Relyea 2006
Lithobates clamitans	L	United States	Outdoor/ mesocosm	Carbaryl/Sevin®	Insecticide	Mass increase	Boone 2008
Lithobates clamitans	L	United States	Outdoor/ mesocosm	Malathion/ Malathion®	Insecticide	No effects on growth or morphology	Boone 2008

Table 1 contd. ...

...Table 1 contd.

Species	St.	Country	Experimental site	Active ingredient/ formulation (if there)	Class	Summary of effects on growth and/or external morphology	References
Lithobates clamitans	L	United States	Outdoor/ mesocosm	Permethrin/ Cutter's Bug Free Back Yard®	Insecticide	Mass increase	Boone 2008
Lithobates clamitans	L	United States	Outdoor/ mesocosm	Permethrin + malathion - mixture/Cutter's Bug Free Back Yard® + Malathion®	Insecticides	No effects on growth or morphology	Boone 2008
Lithobates clamitans	L	United States	Outdoor/ mesocosm	Permethrin + carbaryl - mixture/ Cutter's Bug Free Back Yard® + Sevin®	Insecticides	No effects on growth or morphology	Boone 2008
Lithobates clamitans	L	United States	Outdoor/ mesocosm	Malathion + carbaryl - mixture/ Malathion® + Sevin®	Insecticides	No effects on growth or morphology	Boone 2008
Lithobates clamitans	L	United States	Outdoor/ mesocosm	Permethrin + malathion + carbaryl - mixture/ Cutter's Bug Free Back Yard® + Malathion® + Sevin®	Insecticides	No effects on growth or morphology	Boone 2008
Lithobates clamitans	L	United States	Outdoor/ mesocosm	Carbaryl/Sevin®	Insecticide	Mass reduction	Boone and Bridges 2003
Lithobates clamitans	B	United States	Outdoor/ mesocosm	Carbaryl/Sevin®	Insecticide	No effects on growth or morphology	Boone and Semlitsch 2001
Lithobates clamitans	B	United States	Outdoor/ mesocosm	Carbaryl/Sevin®	Insecticide	Mass reduction	Boone and Semlitsch 2002
Lithobates clamitans	B	United States	Outdoor/ mesocosm	Carbaryl/Sevin®	Insecticide	Mass reduction	Boone et al. 2001
Lithobates clamitans	L	United States	Outdoor/ mesocosm	Carbaryl/Sevin®	Insecticide	No effects on growth or morphology	Boone et al. 2005
Lithobates clamitans	L	United States	Laboratory	Carbaryl/Sevin®	Insecticide	No effects on growth or morphology	Boone et al. 2013
Lithobates clamitans	B	United States	Outdoor/ mesocosm	Malathion	Insecticide	Growth increase	Cothran et al. 2011
Lithobates clamitans	L	Canada	Outdoor/ large-scale experimental ponds	Glyphosate/ VisionMAX®	Herbicide	No effects on growth or morphology	Edge et al. 2012
Lithobates clamitans	L	Canada	Laboratory	Glyphosate/ Vision®	Herbicide	No effects on growth or morphology	Edginton et al. 2004

Table 1 contd. ...

...*Table 1 contd.*

Species	St.	Country	Experimental site	Active ingredient/ formulation (if there)	Class	Summary of effects on growth and/or external morphology	References
Lithobates clamitans	B	United States	Outdoor/ mesocosm	Chlorpyrifos	Insecticide	No effects on growth or morphology	Hua and Relyea 2014
Lithobates clamitans	B	United States	Outdoor/ mesocosm	Diazinon	Insecticide	No effects on growth or morphology	Hua and Relyea 2014
Lithobates clamitans	B	United States	Outdoor/ mesocosm	Malathion	Insecticide	No effects on growth or morphology	Hua and Relyea 2014
Lithobates clamitans	B	United States	Outdoor/ mesocosm	Endosulfan	Insecticide	No effects on growth or morphology	Hua and Relyea 2014
Lithobates clamitans	B	United States	Outdoor/ mesocosm	Chlorpyrifos + diazinon + malathion + endosulfan – mixture	Insecticides	No effects on growth or morphology	Hua and Relyea 2014
Lithobates clamitans	L	United States	Outdoor/ mesocosm	Glyphosate/ Roundup Original MAX®	Herbicide	Mass increase	Jones et al. 2011
Lithobates clamitans	M	United States	Outdoor/ mesocosm	Malathion/ Malathion®	Insecticide	Mass increase	Mackey and Boone 2009
Lithobates clamitans	L	United States	Laboratory	Carbaryl/Sevin®	Insecticide	No effects on growth or morphology	Relyea 2004a
Lithobates clamitans	L	United States	Laboratory	Diazinon	Insecticide	Growth reduction	Relyea 2004a
Lithobates clamitans	L	United States	Laboratory	Malathion	Insecticide	Growth reduction	Relyea 2004a
Lithobates clamitans	L	United States	Laboratory	Glyphosate/ Roundup®	Herbicide	Growth reduction	Relyea 2004a
Lithobates clamitans	L	United States	Laboratory	Carbaryl + diazinon - mixture/ Sevin®	Insecticides	Growth reduction	Relyea 2004a
Lithobates clamitans	L	United States	Laboratory	Carbaryl + malathion - mixture/Sevin®	Insecticides	Growth reduction	Relyea 2004a
Lithobates clamitans	L	United States	Laboratory	Carbaryl + glyphosate - mixture/Sevin® + Roundup®	Insecticide and herbicide	Growth reduction	Relyea 2004a
Lithobates clamitans	L	United States	Laboratory	Diazinon + malathion – mixture	Insecticides	Growth reduction	Relyea 2004a
Lithobates clamitans	L	United States	Laboratory	Diazinon + glyphosate - mixture/ Roundup®	Insecticide and herbicide	Growth reduction	Relyea 2004a

Table 1 contd. ...

...Table 1 contd.

Species	St.	Country	Experimental site	Active ingredient/ formulation (if there)	Class	Summary of effects on growth and/or external morphology	References
Lithobates clamitans	L	United States	Laboratory	Malathion + glyphosate - mixture/ Roundup®	Insecticide and herbicide	Growth reduction	Relyea 2004a
Lithobates clamitans	L	United States	Outdoor/ mesocosm	Carbaryl/Sevin®	Insecticide	No effects on growth or morphology	Relyea 2006
Lithobates pipiens	M	United States	Outdoor/ mesocosm	Chlorpyrifos	Insecticide	No effects on growth or morphology	McClelland et al. 2018
Lithobates pipiens	L	United States	Laboratory	Triclopyr/ Renovate®	Herbicide	Mass increase; growth increase; morphological changes	Curtis and Bidart 2021
Lithobates pipiens	B	United States	Outdoor/ mesocosm	Atrazine/Atrazine 4L	Herbicide	Deformities	Strasburg and Boone 2021
Lithobates pipiens	B	United States	Outdoor/ mesocosm	Bacillus thuringiensis israelensis (Bti)/ Mosquito Dunk®	Insecticide	No effects on growth or morphology	Strasburg and Boone 2021
Lithobates pipiens	L	Canada	Laboratory	Clothianidin	Insecticide	No effects on growth or morphology	Gavel et al. 2021
Lithobates pipiens	L	Canada	Laboratory	Thiamethoxam	Insecticide	Standard body width reduction	Gavel et al. 2021
Lithobates pipiens	L	Canada	Laboratory	Clothianidin/ Reward®	Insecticide	No effects on growth or morphology	Robinson et al. 2021
Lithobates pipiens	B	United States	Laboratory	Atrazine	Herbicide	No effects on growth or morphology	Allran and Kasarov 2000
Lithobates pipiens	L	United States	Laboratory	Atrazine	Herbicide	Malformations (wavy tail, lateral tail flexure, facial edema, axial shortening, dorsal tail flexure and blistering)	Allran and Kasarov 2001
Lithobates pipiens	M	United States	Laboratory	Atrazine + Carbaryl – mixture	Herbicide and insecticide	Mass reduction; increase the number of bony triangles; increase the incidence of skin webbings	Bridges et al. 2004
Lithobates pipiens	L	United States	Outdoor/ mesocosm	Carbaryl/Sevin®	Insecticide	No effects on growth or morphology	Bulen and Distel 2011
Lithobates pipiens	M	United States	Outdoor/ mesocosm	Carbaryl/Sevin®	Insecticide	Mass increase	Distel and Boone 2010

Table 1 contd. ...

...Table 1 contd.

Species	St.	Country	Experimental site	Active ingredient/ formulation (if there)	Class	Summary of effects on growth and/or external morphology	References
Lithobates pipiens	L	Canada	Laboratory	Glyphosate/ Vision®	Herbicide	No effects on growth or morphology	Edginton et al. 2004
Lithobates pipiens	M	United States	Outdoor/ mesocosm	Malathion/ Malathion Plus®	Insecticide	Mass reduction; mass increase	Groner and Relyea 2011
Lithobates pipiens	M	United States	Outdoor/ mesocosm	Carbaryl/Sevin®	Insecticide	Mass reduction	Groner and Relyea 2011
Lithobates pipiens	B	Canada	Laboratory	Endosulfan/ Thiodan®50WP	Insecticide	No effects on growth or morphology	Harris et al. 2000
Lithobates pipiens	B	Canada	Laboratory	Mancozeb/ Dithane®DG	Fungicide	No effects on growth or morphology	Harris et al. 2000
Lithobates pipiens	B	Canada	Laboratory	Azinphos-methyl/ Guthion®50WP	Insecticide	Deformities (eyes missing)	Harris et al. 2000
Lithobates pipiens	M	United States	Laboratory	Atrazine	Herbicide	Snout-vent length reduction; mass reduction	Hayes et al. 2006
Lithobates pipiens	M	United States	Laboratory	Alachlor	Herbicide	No effects on growth or morphology	Hayes et al. 2006
Lithobates pipiens	M	United States	Laboratory	Nicosulfuron	Herbicide	No effects on growth or morphology	Hayes et al. 2006
Lithobates pipiens	M	United States	Laboratory	Cyfluthrin	Insecticide	Snout-vent length reduction	Hayes et al. 2006
Lithobates pipiens	M	United States	Laboratory	λ- cyhalothrin	Insecticide	No effects on growth or morphology	Hayes et al. 2006
Lithobates pipiens	M	United States	Laboratory	Tebupirimphos	Insecticide	Snout-vent length reduction; mass reduction	Hayes et al. 2006
Lithobates pipiens	M	United States	Laboratory	Metalaxyl	Fungicide	No effects on growth or morphology	Hayes et al. 2006
Lithobates pipiens	M	United States	Laboratory	Propiconizole	Fungicide	No effects on growth or morphology	Hayes et al. 2006
Lithobates pipiens	M	United States	Laboratory	S-metolachlor	Herbicide	No effects on growth or morphology	Hayes et al. 2006
Lithobates pipiens	M	United States	Laboratory	Atrazine-metolachlor/Bicep II Magnum®	Herbicide	Snout-vent length reduction; mass reduction	Hayes et al. 2006
Lithobates pipiens	M	United States	Laboratory	Atrazine + s-metolachlor – mixture	Herbicides	Snout-vent length reduction; mass reduction	Hayes et al. 2006

Table 1 contd. ...

...Table 1 contd.

Species	St.	Country	Experimental site	Active ingredient/ formulation (if there)	Class	Summary of effects on growth and/or external morphology	References
Lithobates pipiens	M	United States	Laboratory	Atrazine + alachlor + nicosulfuron + cyfluthrin + λ- cyhalothrin + tebupirimphos + metalaxyl + propiconizole + s-metolachlor – mixture	Insecticides, herbicides and fungicides	Snout-vent length reduction; mass reduction	Hayes et al. 2006
Lithobates pipiens	L	Canada	Laboratory	Glyphosate	Herbicide	Snout–vent length reduction	Howe et al. 2004
Lithobates pipiens	L	Canada	Laboratory	Polyethoxylated tallowamine surfactant (POEA)/ Surfactante substance	Herbicide	Snout–vent length reduction; tail lenght reduction; tail damage (necrosis of the tail tip, flexure of the tail tip, fin damage, abnormal growths on the tail tip and blistering on the tail fin)	Howe et al. 2004
Lithobates pipiens	L	Canada	Laboratory	Glyphosate/ Roundup Original®	Herbicide	Snout–vent length reduction; tail lenght reduction; tail damage (necrosis of the tail tip, flexure of the tail tip, fin damage, abnormal growths on the tail tip and blistering on the tail fin)	Howe et al. 2004
Lithobates pipiens	L	Canada	Laboratory	Glyphosate/ Roundup Transorb®	Herbicide	Snout–vent length reduction; tail lenght reduction; tail damage (necrosis of the tail tip, flexure of the tail tip, fin damage, abnormal growths on the tail tip and blistering on the tail fin)	Howe et al. 2004
Lithobates pipiens	B	United States	Outdoor/ mesocosm	Chlorpyrifos	Insecticide	No effects on growth or morphology	Hua and Relyea 2014
Lithobates pipiens	B	United States	Outdoor/ mesocosm	Diazinon	Insecticide	No effects on growth or morphology	Hua and Relyea 2014

Table 1 contd. ...

...*Table 1 contd.*

Species	St.	Country	Experimental site	Active ingredient/ formulation (if there)	Class	Summary of effects on growth and/or external morphology	References
Lithobates pipiens	B	United States	Outdoor/ mesocosm	Malathion	Insecticide	No effects on growth or morphology	Hua and Relyea 2014
Lithobates pipiens	B	United States	Outdoor/ mesocosm	Endosulfan	Insecticide	No effects on growth or morphology	Hua and Relyea 2014
Lithobates pipiens	B	United States	Outdoor/ mesocosm	Chlorpyrifos + diazinon + malathion + endosulfan – mixture	Insecticides	Mass increase	Hua and Relyea 2014
Lithobates pipiens	L	United States	Laboratory	Atrazine	Herbicide	No effects on growth or morphology	Orton et al. 2006
Lithobates pipiens	L	United States	Laboratory	Carbaryl/Sevin®	Insecticide	Growth reduction	Relyea 2004a
Lithobates pipiens	L	United States	Laboratory	Diazinon	Insecticide	No effects on growth or morphology	Relyea 2004a
Lithobates pipiens	L	United States	Laboratory	Malathion	Insecticide	Growth reduction	Relyea 2004a
Lithobates pipiens	L	United States	Laboratory	Glyphosate/ Roundup®	Herbicide	No effects on growth or morphology	Relyea 2004a
Lithobates pipiens	L	United States	Laboratory	Carbaryl + diazinon - mixture/ Sevin®	Insecticides	Growth reduction	Relyea 2004a
Lithobates pipiens	L	United States	Laboratory	Carbaryl + malathion - mixture/Sevin®	Insecticides	Growth reduction	Relyea 2004a
Lithobates pipiens	L	United States	Laboratory	Carbaryl + glyphosate - mixture/Sevin® + Roundup®	Insecticide and herbicide	Growth reduction	Relyea 2004a
Lithobates pipiens	L	United States	Laboratory	Diazinon + malathion – mixture	Insecticides	Growth reduction	Relyea 2004a
Lithobates pipiens	L	United States	Laboratory	Diazinon + glyphosate - mixture/ Roundup®	Insecticide and herbicide	No effects on growth or morphology	Relyea 2004a
Lithobates pipiens	L	United States	Laboratory	Malathion + glyphosate - mixture/ Roundup®	Insecticide and herbicide	Growth reduction	Relyea 2004a
Lithobates pipiens	L	United States	Outdoor/ mesocosm	Carbaryl	Insecticide	No effects on growth or morphology	Relyea 2009

Table 1 contd. ...

...Table 1 contd.

Species	St.	Country	Experimental site	Active ingredient/ formulation (if there)	Class	Summary of effects on growth and/or external morphology	References
Lithobates pipiens	L	United States	Outdoor/ mesocosm	Malathion	Insecticide	No effects on growth or morphology	Relyea 2009
Lithobates pipiens	L	United States	Outdoor/ mesocosm	Diazinon	Insecticide	Mass reduction	Relyea 2009
Lithobates pipiens	L	United States	Outdoor/ mesocosm	Chlorpyrifos	Insecticide	No effects on growth or morphology	Relyea 2009
Lithobates pipiens	L	United States	Outdoor/ mesocosm	Endosulfan	Insecticide	Mass increase	Relyea 2009
Lithobates pipiens	L	United States	Outdoor/ mesocosm	Carbaryl + malathion + chlorpyrifos + diazinon + endosulfan – mixture	Insecticides	No effects on growth or morphology	Relyea 2009
Lithobates pipiens	L	United States	Outdoor/ mesocosm	Acetochlor	Herbicide	No effects on growth or morphology	Relyea 2009
Lithobates pipiens	L	United States	Outdoor/ mesocosm	Metolachlor	Herbicide	No effects on growth or morphology	Relyea 2009
Lithobates pipiens	L	United States	Outdoor/ mesocosm	Glyphosate	Herbicide	No effects on growth or morphology	Relyea 2009
Lithobates pipiens	L	United States	Outdoor/ mesocosm	2,4-D	Herbicide	No effects on growth or morphology	Relyea 2009
Lithobates pipiens	L	United States	Outdoor/ mesocosm	Atrazine	Herbicide	No effects on growth or morphology	Relyea 2009
Lithobates pipiens	L	United States	Outdoor/ mesocosm	Acetochlor + metolachlor + glyphosate + 2,4-D + atrazine – mixture	Herbicides	No effects on growth or morphology	Relyea 2009
Lithobates pipiens	L	United States	Outdoor/ mesocosm	Carbaryl + malathion + chlorpyrifos + diazinon + endosulfan + acetochlor + metolachlor + glyphosate + 2,4-D + atrazine – mixture	Herbicides and insecticides	No effects on growth or morphology	Relyea 2009

Table 1 contd. ...

...Table 1 contd.

Species	St.	Country	Experimental site	Active ingredient/ formulation (if there)	Class	Summary of effects on growth and/or external morphology	References
Lithobates pipiens	L	United States	Outdoor/ mesocosm	Glyphosate/ Roundup Original MAX®	Herbicide	Mass reduction; morphological changes (deeper tails, deeper bodies)	Relyea 2012
Lithobates pipiens	M	United States	Outdoor/ mesocosm	Malathion/ Malathion Plus®	Insecticide	Mass reduction	Relyea and Diecks 2008
Lithobates pipiens	L	United States	Outdoor/ mesocosm	Malathion/ Malathion Plus®	Insecticide	Mass reduction	Relyea and Hoverman 2008
Lithobates pipiens	L	United States	Laboratory	Endosulfan/ Endosulfan 3EC®	Insecticide	No effects on growth or morphology	Shenoy et al. 2009
Lithobates pipiens	L	United States	Laboratory	Mancozeb/ Manzate 75DF®	Fungicide	Total length reduction	Shenoy et al. 2009
Lithobates pipiens	B	United States	Outdoor/ mesocosm	Chlorpyrifos	Insecticide	Mass reduction; body shape alterations (wider body, deeper body, deeper tail, longer body, longer forelimbs, longer thighs, longer legs and longer feet)	Woodley et al. 2015
Lithobates sphenocephalus	L	United States	Laboratory	Atrazine	Herbicide	Growth reduction (smallest surface area measurements)	Adelizzi et al. 2019
Lithobates sphenocephalus	L	United States	Outdoor/ mesocosm	Carbaryl/Sevin®	Insecticide	Mass increase	Boone and James 2003
Lithobates sphenocephalus	L	United States	Outdoor/ mesocosm	Atrazine/Aatrex®	Herbicide	Mass reduction	Boone and James 2003
Lithobates sphenocephalus	B	United States	Outdoor/ mesocosm	Carbaryl/Sevin®	Insecticide	No effects on growth or morphology	Boone and Semlitsch 2002
Lithobates sphenocephalus	B	United States	Outdoor/ large-scale experimental ponds	Carbaryl/Sevin®	Insecticide	No effects on growth or morphology	Boone et al. 2004
Lithobates sphenocephalus	B	United States	Outdoor/ mesocosm	Carbaryl/Sevin®	Insecticide	No effects on growth or morphology	Boone et al. 2007

Table 1 contd. ...

...Table 1 contd.

Species	St.	Country	Experimental site	Active ingredient/ formulation (if there)	Class	Summary of effects on growth and/or external morphology	References
Lithobates sphenocephalus	B	United States	Laboratory	Carbaryl	Insecticide	Mass reduction; visceral and limb malformations (failure of the ventral surface of the integument, only one hind limb, bends in tail near the trunk and formed of three front limbs)	Bridges 2000
Lithobates sphenocephalus	M	United States	Outdoor/ mesocosm	Carbaryl/Sevin®	Insecticide	Mass increase	Bridges and Boone 2003
Lithobates sphenocephalus	B	United States	Laboratory	Thiophanate-methyl	Fungicide	Mass increase, snout-vent length increase	Hanlon et al. 2012
Lithobates sphenocephalus	L	United States	Laboratory	Glyphosate/ Roundup Pro Concentrate®	Herbicide	No effects on growth or morphology	Hanlon et al. 2013
Lithobates sphenocephalus	M	United States	Outdoor/ mesocosm	Carbaryl/Sevin®	Insecticide	Mass increase; mass reduction	Mills and Semlistch 2004
Lithobates sphenocephalus	L	United States	Laboratory	Chlorpyrifos/ Dursban TC®	Termiticide	Mass reduction	Widder and Bidwell 2006
Lithobates sphenocephalus	L	United States	Outdoor/ mesocosm	Chlorpyrifos/ Dursban TC®	Termiticide	Mass reduction	Widder and Bidwell 2006
Lithobates sphenocephalus	L	United States	Laboratory	Chlorpyrifos/ Dursban TC®	Termiticide	Mass reduction	Widder and Bidwell 2008
Lithobates sylvaticus	L	Canada	Outdoor/ mesocosm	Imidacloprid/ Admire	Insecticide	No effects on growth or morphology	Robinson et al. 2017
Lithobates sylvaticus	L	Canada	Outdoor/ mesocosm	Thiamethoxam/ Actara	Insecticide	No effects on growth or morphology	Robinson et al. 2017
Lithobates sylvaticus	L	United States	Laboratory	Atrazine	Herbicide	Malformations (wavy tail, lateral tail flexure, facial edema, axial shortening, dorsal tail flexure and blistering)	Allran and Kasarov 2001
Lithobates sylvaticus	L	United States	Outdoor/ mesocosm	Cooper Ethalonamine Complex/Cultrine Plus®	Algicide and herbicide	Mass reduction	Cothran et al. 2011
Lithobates sylvaticus	M	Canada	Laboratory	Glyhosate/ Roundup WeatherMAX®	Herbicide	Mass increase	Gahl et al. 2011

Table 1 contd. ...

...Table 1 contd.

Species	St.	Country	Experimental site	Active ingredient/ formulation (if there)	Class	Summary of effects on growth and/or external morphology	References
Lithobates sylvaticus	B	United States	Outdoor/ mesocosm	Chlorpyrifos	Insecticide	No effects on growth or morphology	Hua and Relyea 2014
Lithobates sylvaticus	B	United States	Outdoor/ mesocosm	Diazinon	Insecticide	No effects on growth or morphology	Hua and Relyea 2014
Lithobates sylvaticus	B	United States	Outdoor/ mesocosm	Malathion	Insecticide	No effects on growth or morphology	Hua and Relyea 2014
Lithobates sylvaticus	B	United States	Outdoor/ mesocosm	Endosulfan	Insecticide	No effects on growth or morphology	Hua and Relyea 2014
Lithobates sylvaticus	B	United States	Outdoor/ mesocosm	Chlorpyrifos + diazinon + malathion + endosulfan – mixture	Insecticides	No effects on growth or morphology	Hua and Relyea 2014
Lithobates sylvaticus	L	United States	Outdoor/ mesocosm	Glyphosate/ Roundup Original MAX®	Herbicide	Mass reduction	Jones et al. 2010
Lithobates sylvaticus	B	United States	Outdoor/ large-scale experimental ponds and laboratory	Atrazine	Herbicide	Mass reduction; limb abnormalities	Kiesecker 2002
Lithobates sylvaticus	B	United States	Outdoor/ large-scale experimental ponds and laboratory	Malathion	Insecticide	Mass reduction; limb abnormalities	Kiesecker 2002
Lithobates sylvaticus	L	United States	Laboratory	Malathion	Insecticide	Mass reduction; total length alterations (larger and smaller tadpoles); abnormalities (increase of swollen bodies, increase of diamond shape bodies and increase of stiff tails)	Krishnamurthy and Smith 2011
Lithobates sylvaticus	L	United States	Laboratory	Malathion	Insecticide	Mass reduction	Krishnamurthy and Smith 2011
Lithobates sylvaticus	L	Canada	Outdoor/ large-scale experimental ponds	Glyphosate/ Roundup WeatherMax®	Herbicide	Body length increase	Lanctôt et al. 2013
Lithobates sylvaticus	L	Canada	Laboratory	Glyphosate/ Roundup WeatherMax®	Herbicide	Mass increase; snout-vent-length reduction	Lanctôt et al. 2014

Table 1 contd. ...

...Table 1 contd.

Species	St.	Country	Experimental site	Active ingredient/ formulation (if there)	Class	Summary of effects on growth and/or external morphology	References
Lithobates sylvaticus	L	Canada	Laboratory	Glyphosate/ Vision®	Herbicide	Mass increase; snout-vent-length reduction	Lanctôt et al. 2014
Lithobates sylvaticus	B	Canada	Laboratory	Glyphosate/ VisionMax®	Herbicide	Mass, tail and snout-vent length increase	Navarro-Martín et al. 2014
Lithobates sylvaticus	L	United States	Outdoor/ mesocosm	Glyphosate/ Roundup Original MAX®	Herbicide	Mass reduction; morphological changes (deeper tail muscles, deeper tails and deeper bodies)	Relyea 2012
Lithobates sylvaticus	M	United States	Outdoor/ mesocosm	Malathion/ Malathion Plus®	Insecticide	No effects on growth or morphology	Relyea and Diecks 2008
Lithobates sylvaticus	L	United States	Outdoor/ mesocosm	Malathion/ Malathion Plus®	Insecticide	No effects on growth or morphology	Relyea and Hoverman 2008
Lithobates sylvaticus	L	United States	Outdoor/ mesocosm	Atrazine	Herbicide	No effects on growth or morphology	Rohr and Crumrine 2005
Lithobates sylvaticus	L	United States	Outdoor/ mesocosm	Endosulfan	Insecticide	Mass increase; mass reduction	Rohr and Crumrine 2005
Litoria freycineti	L	Australia	Laboratory	Endosulfan	Insecticide	Total length reduction	Broomhall and Shine 2003
Litoria peronii	L	Australia	Laboratory	Endosulfan	Insecticide	Total length reduction	Broomhall and Shine 2003
Microhyla fissipes	L	China	Laboratory	Chlorantraniliprole	Insecticide	Mass reduction; length reduction	Wei et al. 2022
Microhyla fissipes	L	China	Laboratory	Penoxsulam	Herbicide	Mass reduction; length reduction	Wei et al. 2022
Microhyla fissipes	L	China	Laboratory	Pymetrozine	Insecticide	Mass reduction; length reduction	Wei et al. 2022
Microhyla fissipes	L	China	Laboratory	Haloxyfop-P-methyl	Herbicide	Mass reduction; length reduction	Wei et al. 2022
Microhyla fissipes	L	China	Laboratory	Chlorantraniliprole + Penoxsulam	Insecticide + herbicide	Mass reduction; length reduction	Wei et al. 2022
Microhyla fissipes	L	China	Laboratory	Chlorantraniliprole + Pymetrozine	Insecticide	Mass reduction; length reduction	Wei et al. 2022
Microhyla fissipes	L	China	Laboratory	Chlorantraniliprole + Haloxyfop-P-methyl	Insecticide + herbicide	Mass reduction; length reduction	Wei et al. 2022
Microhyla fissipes	L	China	Laboratory	Penoxsulam + Pymetrozine	Herbicide + insecticide	Mass reduction; length reduction	Wei et al. 2022
Microhyla fissipes	L	China	Laboratory	Penoxsulam + Haloxyfop-P-methyl	Herbicide	Mass reduction; length reduction	Wei et al. 2022

Table 1 contd. ...

...*Table 1 contd.*

Species	St.	Country	Experimental site	Active ingredient/ formulation (if there)	Class	Summary of effects on growth and/or external morphology	References
Microhyla fissipes	L	China	Laboratory	Pymetrozine + Haloxyfop-P-methyl	Insecticide + herbicide	Mass reduction; length reduction	Wei et al. 2022
Microhyla fissipes	L	China	Laboratory	Glyphosate/ Kissun®	Herbicide	Length reduction	Wang et al. 2019
Minervarya teraiensis	B	India	Laboratory	Endosulfan/Hildan 35 EC®	Insecticide and acaricide	Morphological deformities (fore-limb deformity)	Devi and Gupta 2013
Odontophrynus americanus	L	Argentina	Laboratory	Pyriproxyfen/ Dragon®	Insecticide	No effects on growth or morphology	Lajmanovich et al. 2019
Osteopilus septentrionalis	B	United States	Outdoor/ mesocosm	Atrazine	Herbicide	Snout-vent length reduction	Rohr et al. 2013
Pelobates cultripes	L	Spain	Laboratory	Cooper sulfate	Fungicide	Total length reduction	García-Munõz et al. 2010
Pelophylax nigromaculata	L	China	Laboratory	Triadimefon	Fungicide	Mass reduction; length reduction	Zhang et al. 2018
Pelophylax nigromaculata	L	China	Laboratory	Triadimenol	Fungicide	Mass reduction; length reduction	Zhang et al. 2018
Pelophylax nigromaculata	L	China	Laboratory	α-cypermethrin	Insecticide	Mass reduction; length reduction	Xu and Huang 2017
Pelophylax nigromaculata	L	China	Laboratory	Cyproconazole	Fungicide	Mass reduction; length reduction	Zhang et al. 2019
Pelophylax perezi	L	Spain	Laboratory	Cooper sulfate	Fungicide	Total length reduction	García-Munõz et al. 2010
Pelophylax ridibundus	L	Turkey	Laboratory	Malathion	Insecticide	Deformities in head and trunk (tail deformations, abnormal gut coiling and generalized edema)	Sayim 2008
Physalaemus centralis	B	Brazil	Laboratory	Glyphosate/ Agripec®	Herbicide	Size increase	Figueiredo and Rodrigues 2014
Physalaemus centralis	B	Brazil	Laboratory	U46 D-FLUID 2,4-D/Nufarm®	Herbicide	Size increase	Figueiredo and Rodrigues 2014
Physalaemus centralis	B	Brazil	Laboratory	Picloram/Padron®	Herbicide	Size reduction	Figueiredo and Rodrigues 2014
Physalaemus centralis	B	Brazil	Laboratory	Picloram + 2,4-D - mixture/Tordon®	Herbicide	Size increase	Figueiredo and Rodrigues 2014
Physalaemus cuvieri	L	Brazil	Laboratory	Glyphosate/ Roundup Original®	Herbicide	Morphological asymmetries (asymmetry in nostril-snout distance and eyes width)	Costa and Nomura 2016

Table 1 contd. ...

...Table 1 contd.

Species	St.	Country	Experimental site	Active ingredient/ formulation (if there)	Class	Summary of effects on growth and/or external morphology	References
Physalaemus cuvieri	L	Brazil	Laboratory	Imidacloprid/ Nortox	Insecticide	Lenght reduction; mass reduction; malformations (mouth and intestine)	Samojeden et al. 2022
Physalaemus cuvieri	L	Brazil	Laboratory	Glyphosate/Not informed	Herbicide	Length reduction; mass reduction; malformations (mouth)	Almeida et al. 2019
Physalaemus cuvieri	L	Brazil	Laboratory	Glyphosate/ Roundup Original®	Herbicide	Length reduction; mass reduction; malformations (mouth, epithelium colour and bowel oedema)	Herek et al. 2020
Physalaemus gracilis	L	Brazil	Laboratory	Glyphosate/ Roundup Original®	Herbicide	Length reduction; mass reduction; malformations (mouth, body morphology and epithelium colour)	Herek et al. 2020
Physalaemus gracilis	L	Brazil	Laboratory	Cypermethrin/ Cipertrin EC	Insecticide	Malformations in oral morphology	Vanzetto et al. 2019
Physalaemus gracilis	L	Brazil	Laboratory	Deltamethrin/ Decis 25 EC	Insecticide	No effects on growth or morphology	Vanzetto et al. 2019
Physalaemus gracilis	L	Brazil	Laboratory	Atrazine/Siptran ®SC	Insecticide	Malformations (mouth, intestine shape, and edemas)	Rutkoski et al. 2018
Physalaemus nattereri	L	Brazil	Laboratory	2,4-D/DMA® 806	Herbicide	Length reduction	Freitas et al. 2019
Polypedates cruciger	B	Sri Lanka	Laboratory	Chlorpyrifos/ Lorsban 40 EC®or Pattas®	Insecticide	Mass reduction; growth reduction; malformations - scoliosis (tail curvature), kyphosis (hunched back), edema, and skin ulcers	Jayawardena et al. 2016
Polypedates cruciger	B	Sri Lanka	Laboratory	Dimethoate/ Dimethoate 40EC®	Insecticide	Mass reduction; growth reduction; malformations - scoliosis (tail curvature), kyphosis (hunched back), edema, and skin ulcers	Jayawardena et al. 2016

Table 1 contd. ...

...Table 1 contd.

Species	St.	Country	Experimental site	Active ingredient/ formulation (if there)	Class	Summary of effects on growth and/or external morphology	References
Polypedates cruciger	B	Sri Lanka	Laboratory	Glyphosate/ Round Up® or Glyphosate®	Herbicide	Mass reduction; growth reduction; malformations - scoliosis (tail curvature), kyphosis (hunched back), edema, and skin ulcers	Jayawardena et al. 2016
Polypedates cruciger	B	Sri Lanka	Laboratory	Propanil/3, 4-DPA®	Herbicide	Mass reduction; growth reduction; malformations - scoliosis (tail curvature), kyphosis (hunched back), edema, and skin ulcers	Jayawardena et al. 2016
Polypedates cruciger	B	Sri Lanka	Laboratory	Chlorpyrifos/ Lorsban EC 40® or Pattas®	Insecticide	Snout-vent length reduction; mass reduction; spine malformations (kyphosis, scoliosis, edema and skin ulcers)	Jayawardena et al. 2010
Polypedates cruciger	B	Sri Lanka	Laboratory	Dimethoate / Dimethoate EC®	Insecticide	Snout-vent length reduction; mass reduction; spine malformations (kyphosis, scoliosis, edema and skin ulcers)	Jayawardena et al. 2010
Polypedates cruciger	B	Sri Lanka	Laboratory	Glyphosate/ Roundup® or Glyphosate®	Herbicide	Snout-vent length reduction; mass reduction; spine malformations (kyphosis, scoliosis, edema and skin ulcers)	Jayawardena et al. 2010
Polypedates cruciger	B	Sri Lanka	Laboratory	Propanil/3,4 DPA®	Herbicide	Snout-vent length reduction; mass reduction; spine malformations (kyphosis, scoliosis, edema and skin ulcers)	Jayawardena et al. 2010
Polypedates cruciger	L	Sri Lanka	Outdoor	Diazinon	Insecticide	Body length reduction	Sumanadasa et al. 2008a
Pseudacris crucifer	B	United States	Outdoor/ mesocosm	Chlorpyrifos	Insecticide	Mass increase	Hua and Relyea 2014

Table 1 contd. ...

...Table 1 contd.

Species	St.	Country	Experimental site	Active ingredient/ formulation (if there)	Class	Summary of effects on growth and/or external morphology	References
Pseudacris crucifer	B	United States	Outdoor/ mesocosm	Diazinon	Insecticide	No effects on growth or morphology	Hua and Relyea 2014
Pseudacris crucifer	B	United States	Outdoor/ mesocosm	Malathion	Insecticide	No effects on growth or morphology	Hua and Relyea 2014
Pseudacris crucifer	B	United States	Outdoor/ mesocosm	Endosulfan	Insecticide	No effects on growth or morphology	Hua and Relyea 2014
Pseudacris crucifer	B	United States	Outdoor/ mesocosm	Chlorpyrifos + diazinon + malathion + endosulfan – mixture	Insecticides	Mass increase	Hua and Relyea 2014
Pseudacris regilla	L	United States	Laboratory	Cypermethrin	Insecticide	No effects on growth or morphology	Biga and Blaustein 2013
Pseudacris regilla	B	United States	Outdoor/ mesocosm	Carbaryl	Insecticide	Growth increase	Buck et al. 2012
Pseudacris regilla	B	United States	Laboratory	Diazinon	Insecticide	No effects on growth or morphology	Kleinhenz et al. 2012
Pseudacris regilla	B	United States	Laboratory	Malathion	Insecticide	No effects on growth or morphology	Kleinhenz et al. 2012
Pseudacris regilla	B	United States	Laboratory	Chlorpyrifos	Insecticide	No effects on growth or morphology	Kleinhenz et al. 2012
Pseudacris regilla	B	United States	Laboratory	Endosulfan	Insecticide and acaricide	No effects on growth or morphology	Kleinhenz et al. 2012
Pseudacris regilla	B	United States	Laboratory	Diazinon + malathion + chlorpyrifos + endosulfan – mixture	Insecticide	No effects on growth or morphology	Kleinhenz et al. 2012
Pseudacris regilla	B	United States	Laboratory	Chlorpyrifos	Insecticide	No effects on growth or morphology	Sparling and Fellers 2009
Pseudacris regilla	B	United States	Laboratory	Endosulfan	Insecticide	Snout-vent length reduction; mass reduction; body abnormalities	Sparling and Fellers 2009
Pseudacris regilla	L	Canada	Laboratory	Endosulfan/ Thiodan®	Insecticide	Abnormalities (kink tail); lost pigmentation	Westman et al. 2010

Table 1 contd. ...

...Table 1 contd.

Species	St.	Country	Experimental site	Active ingredient/ formulation (if there)	Class	Summary of effects on growth and/or external morphology	References
Pseudacris regilla	L	Canada	Laboratory	Diazinon/ Diazinon®	Insecticide	No effects on growth or morphology	Westman et al. 2010
Pseudacris regilla	L	Canada	Laboratory	Azinphosmethyl/ Guthion®	Insecticide	No effects on growth or morphology	Westman et al. 2010
Pseudacris regilla	L	Canada	Laboratory	Endosulfan + diazinon + azinphosmethyl - mixture/ Thiodan®+ Diazinon® + Guthion®	Insecticides	No effects on growth or morphology	Westman et al. 2010
Pseudacris triseriata	L	United States	Laboratory	Glyphosate/ Kleeraway®	Herbicide	No effects on growth or morphology	Smith 2001
Pseudacris triseriata	B	United States	Laboratory	Atrazine/Atrazine 4l®	Herbicide	No effects on growth or morphology	Willians and Semlitsch 2010
Pseudacris triseriata	B	United States	Laboratory	S-metolachlor/ Dual II Magnum®	Herbicide	No effects on growth or morphology	Willians and Semlitsch 2010
Pseudacris triseriata	B	United States	Laboratory	Glyphosate/ Roundup Original MAX®	Herbicide	No effects on growth or morphology	Willians and Semlitsch 2010
Pseudacris triseriata	B	United States	Laboratory	Glyphosate/ Roundup WeatherMax®	Herbicide	No effects on growth or morphology	Willians and Semlitsch 2010
Rana arvalis	B	Germany	Laboratory	α-cypermethrin	Insecticide	Deformities (tail kinking); Length reduction; mass increase	Greulich and Pflugmacher 2003
Rana arvalis	L	Germany	Laboratory	Cypermethrin	Insecticide	Physical abnormalities	Greulich and Pflugmacher 2004
Rana aurora	L	United States	Laboratory	Cypermethrin	Insecticide	No effects on growth or morphology	Biga and Blaustein 2013
Rana boylii	B	United States	Laboratory	Chlorpyrifos	Insecticide	Snout-vent length and mass reduction	Sparling and Fellers 2009
Rana boylii	B	United States	Laboratory	Endosulfan	Insecticide	Mass reduction; body abnormalities	Sparling and Fellers 2009
Rana cascadae	L	United States	Laboratory	Cypermethrin	Insecticide	No effects on growth or morphology	Biga and Blaustein 2013
Rana cascadae	B	United States	Laboratory	Glyphosate/ Roundup®	Herbicide	Mass reduction; abnormalities (bent tails)	Cauble and Wagner 2005

Table 1 contd. ...

...Table 1 contd.

Species	St.	Country	Experimental site	Active ingredient/ formulation (if there)	Class	Summary of effects on growth and/or external morphology	References
Rana chensinensis	L	China	Laboratory	Trichlorfon	Insecticide	Length reduction; mass reduction; malformations (axial flexures, skeletal malformations and lateral kinks)	Ma et al. 2018
Rana dalmatina	B	Hungary	Laboratory	Terbuthylazine	Herbicide	Mass reduction	Bokony et al. 2020
Rana dalmatina	B	Italy	Laboratory	Chlorpyrifos	Insecticide	Mass reduction; abnormalities (skeletal defect, abnormal tail lateral flexure, bloated heads and edema)	Bernabò et al. 2011
Rana dalmatina	L	Italy	Laboratory	Endosulfan	Insecticide	Mass reduction; snou-vent length reduction; malformations (bloated heads and skeletal malformations)	Lavorato et al. 2013
Rana temporaria	M	Denmark	Laboratory	Prochloraz	Fungicide	Mass increase	Brande-Lavridsen et al. 2010
Rana temporaria	L	Belgica	Laboratory	Endosulfan	Insecticide	Body mass reduction	Denoël et al. 2012
Rana temporaria	B	Sweden	Laboratory	Azoxystrobin	Fungicide	Body length reduction	Johansson et al. 2006
Rana temporaria	B	Sweden	Laboratory	Cyanazine	Herbicide	Reduction in body length, tail length and dry weight	Johansson et al. 2006
Rana temporaria	B	Sweden	Laboratory	Esfenvalerate	Insecticide	No effects on growth or morphology	Johansson et al. 2006
Rana temporaria	B	Sweden	Laboratory	MCPA	Herbicide	No effects on growth or morphology	Johansson et al. 2006
Rana temporaria	B	Sweden	Laboratory	Permethrin	Insecticide	Increase in body length, tail length and wet weight	Johansson et al. 2006
Rana temporaria	B	Sweden	Laboratory	Pirimicarb	Insecticide	Reduction in tail length and dry weight	Johansson et al. 2006
Rana temporaria	L	France	Laboratory	Amitrole	Herbicide	Body and tail length increase; mass increase; shallower bodies	Mandrillon and Saglio 2009

Table 1 contd. ...

...Table 1 contd.

Species	St.	Country	Experimental site	Active ingredient/ formulation (if there)	Class	Summary of effects on growth and/or external morphology	References
Rana temporaria	B	Sweden	Laboratory	Fenpropimorph	Fungicide	Reduction in mass, tail length, tail fin depth and body length	Teplitsky et al. 2005
Rhinella arenarum	L	Argentina	Laboratory	Glyphosate/ Roundup Ultra-Max®	Herbicide	Abnormalities (general body alterations, bilateral asymmetry, swollen body or edemas, eye shape, mouth and gill chamber structure, visceral organization and tail orientation and shape)	Boccioni et al. 2021
Rhinella arenarum	L	Argentina	Laboratory	Cypermethrin/ Glextrin 25®	Insecticide	No effects on growth or morphology	Svartz et al 2016
Rhinella arenarum	L	Argentina	Laboratory	Mancozeb/ Candil®	Fungicide	Deformities (edema)	Asparch et al. 2019
Rhinella fernandezae	L	Argentina	Laboratory	Chlorpyrifos/ Lorsban® 48E	Insecticide	Growth reduction; deformities (slight and severe lateral flexure of the tail from its normal position)	Arcaute et al. 2012
Rhinella horribilis	L	Colombia	Laboratory	Chlorpyrifos/ Lorsban™	Insecticide	Length reduction	Henao et al. 2021
Rhinella horribilis	L	Colombia	Laboratory	Diazinon/Diazol®	Insecticide	Length reduction	Henao et al. 2021
Rhinella horribilis	L	Colombia	Laboratory	Monocrotophos/ Monocrotophos 600	Insecticide	Length reduction	Henao et al. 2021
Rhinella icterica	L	Brazil	Laboratory	Glyohosate/Not informed	Herbicide	Length reduction; mass reduction; malformations (body, mouth)	Almeida et al. 2020
Rhinella icterica	L	Brazil	Laboratory	Atrazine/ Primoleo®	Herbicide	Mass increase	Reichert et al. 2022
Rhinella icterica	L	Brazil	Laboratory	Glyphosate/ Roundup®	Herbicide	Condition factor K increase	Reichert et al. 2022
Rhinella icterica	L	Brazil	Laboratory	Quinclorac/Facet®	Herbicide	No effects on growth or morphology	Reichert et al. 2022
Rhinella marina	B	Brazil	Laboratory	Glyphosate/ Agripec®	Herbicide	No effects on growth or morphology	Figueiredo and Rodrigues 2014

Table 1 contd. ...

...Table 1 contd.

Species	St.	Country	Experimental site	Active ingredient/ formulation (if there)	Class	Summary of effects on growth and/or external morphology	References
Rhinella marina	B	Brazil	Laboratory	U46 D-FLUID 2,4-D/Nufarm®	Herbicide	No effects on growth or morphology	Figueiredo and Rodrigues 2014
Rhinella marina	B	Brazil	Laboratory	Picloram/Padron®	Herbicide	Size reduction	Figueiredo and Rodrigues 2014
Rhinella marina	B	Brazil	Laboratory	Picloram + 2,4-D - mixture/Tordon®	Herbicide	Size increase	Figueiredo and Rodrigues 2014
Scinax nasicus	L	Argentina	Laboratory	Glyphosate/ Glyphos®	Herbicide	Cranialfacial and mouth deformities; eye abnormalities; bent curved tails	Lajmanovich et al. 2003
Sclerophrys arabica	B	Saudi Arabia	Laboratory	Methomyl/Lannate 90% SP	Insecticide	Tail deformations, abnormal gut coiling, kinking tail, asymmetrical tail, curvature of the spinal cord, scoliosis, tail torsion, incomplete ossification of the vertebral column, retardation of tail regression, incomplete and reduction ossification of the phalanges of digits for forelimbs	Seleem 2019
Spea intermontana	L	Canada	Laboratory	Endosulfan/ Thiodan®	Insecticide	Abnormalities (kink tail)	Westman et al. 2010
Spea intermontana	L	Canada	Laboratory	Diazinon/ Diazinon®	Insecticide	No effects on growth or morphology	Westman et al. 2010
Spea intermontana	L	Canada	Laboratory	Azinphosmethyl/ Guthion®	Insecticide	No effects on growth or morphology	Westman et al. 2010
Spea intermontana	L	Canada	Laboratory	Endosulfan + diazinon + azinphosmethyl - mixture/Thiodan® + Diazinon® + Guthion®	Insecticides	No effects on growth or morphology	Westman et al. 2010
Spea multiplicata	L	United States	Laboratory	Chlorothalonil	Fungicide	Snout-vent-length reduction; tail growth inhibition	Yu et al. 2013b
Sclerophrys regularis	B	Egypt	Laboratory	Atrazine	Herbicide	Deformities - missing and misshapen limbs, tail deformities	Said et al. 2022
Xenopus laevis	L	China	Laboratory	Metamifop	Herbicide	Mass reduction	Liu et al. 2021

Table 1 contd. ...

...Table 1 contd.

Species	St.	Country	Experimental site	Active ingredient/ formulation (if there)	Class	Summary of effects on growth and/or external morphology	References
Xenopus laevis	L	Italy	Laboratory	Carbaryl	Insecticide	Malformations (microphthalmia, cardiac or abdominal edema, irregular shape of the intestinal loops, abnomial tail flexure and dorsal flexure)	Bachetta et al. 2008
Xenopus laevis	L	Italy	Laboratory	Chlorpyrifos	Insecticide	Malformation (ventral and lateral tail flexure coupled with abnormal gut coiling)	Bonfanti et al. 2004
Xenopus laevis	L	Italy	Laboratory	Malathion	Insecticide	Malformation (abnormal tail flexure)	Bonfanti et al. 2004
Xenopus laevis	M	United States	Laboratory	Atrazine	Herbicide	No effects on growth or morphology	Carr et al. 2003
Xenopus laevis	L	United States	Laboratory	Malathion	Insecticide	Length reduction; axis deformities	Chemotti et al. 2006
Xenopus laevis	B	United States	Laboratory	Atrazine	Herbicide	No effects on growth or morphology	Coady et al. 2005
Xenopus laevis	L	Italy	Laboratory	Chlorpyrifos	Insecticide	Abnormalities (abnormal tail flexure)	Colombo et al. 2005
Xenopus laevis	L	Canada	Laboratory	Acetochlor	Herbicide	Body area reduction (sculpting of the head and reduced tail fin quality)	Crump et al. 2002
Xenopus laevis	L	Canada	Laboratory	Glyphosate/ Vision®	Herbicide	No effects on growth or morphology	Edginton et al. 2004
Xenopus laevis	L	United States	Laboratory	Methoxychlor	Insecticide	Malformations (visceral edema, notochord lesions and shortening of the femur)	Fort et al. 2004
Xenopus laevis	B	United States	Laboratory	Atrazine	Herbicide	Mass increase; mass decrease	Freeman and Rayburn 2005
Xenopus laevis	M	United States	Laboratory	Atrazine	Herbicide	No effects on growth or morphology	Hayes et al. 2002
Xenopus laevis	L	United States	Laboratory	Chlorpyrifos	Insecticide	Body length reduction; spinal malformations	Richards and Kendall 2003

Table 1 contd. ...

...Table 1 contd.

Species	St.	Country	Experimental site	Active ingredient/ formulation (if there)	Class	Summary of effects on growth and/or external morphology	References
Xenopus laevis	L	United States	Laboratory	Atrazine	Herbicide	Mass decrease; reduction and increase in snout-vent-length	Sullivan and Spence 2003
Xenopus laevis	L	United States	Laboratory	Malathion/ Malathion Plus®	Insecticide	Abnormalities (bent tails at their bases)	Webb and Crain 2006
Xenopus laevis	L	Korea	Laboratory	Carbendazim/ Benomyl®	Fungicide	Growth inhibition; malformations (optic hernia and dysplasia, narrow head, fin, notochord and tail abnormalities, cephalis edema, optic edema, abdominal edema, gut dysplasia and atrophy)	Yoon et al. 2008
Xenopus laevis	L	Korea	Laboratory	N-butyl isocyanate/ Benomyl®	Fungicide	Growth inhibition; malformations (blisters, optical hernia and dysplasia, narrow head, fin, notochord and tail abnormalities, optic edema, abdominal edema, gut dysplasia and atrophy)	Yoon et al. 2008
Xenopus laevis	L	United States	Laboratory	Malathion	Insecticide	Total length reduction; malformations (edemas, axil and tail deformities)	Yu et al. 2013a
Xenopus laevis	L	United States	Laboratory	Endosulfan	Insecticide	Total length reduction; malformations (edemas, axil and tail deformities)	Yu et al. 2013a
Xenopus laevis	L	United States	Laboratory	α-cypermethrin	Insecticide	Total length reduction; malformations (edemas, axil and tail deformities)	Yu et al. 2013a
Xenopus laevis	L	United States	Laboratory	Chlorothalonil	Fungicide	Snout-vent-length reduction; tail growth inhibition	Yu et al. 2013b

reductions in the body size of *Rhinella marina* and *Physalaemus centralis* exposed to 5 types of herbicides. Groner and Relyea (2011) observed that *Lithobates pipiens* metamorphs exposed to different concentrations of malathion (Malathion Plus®) had either an increase or a reduction in

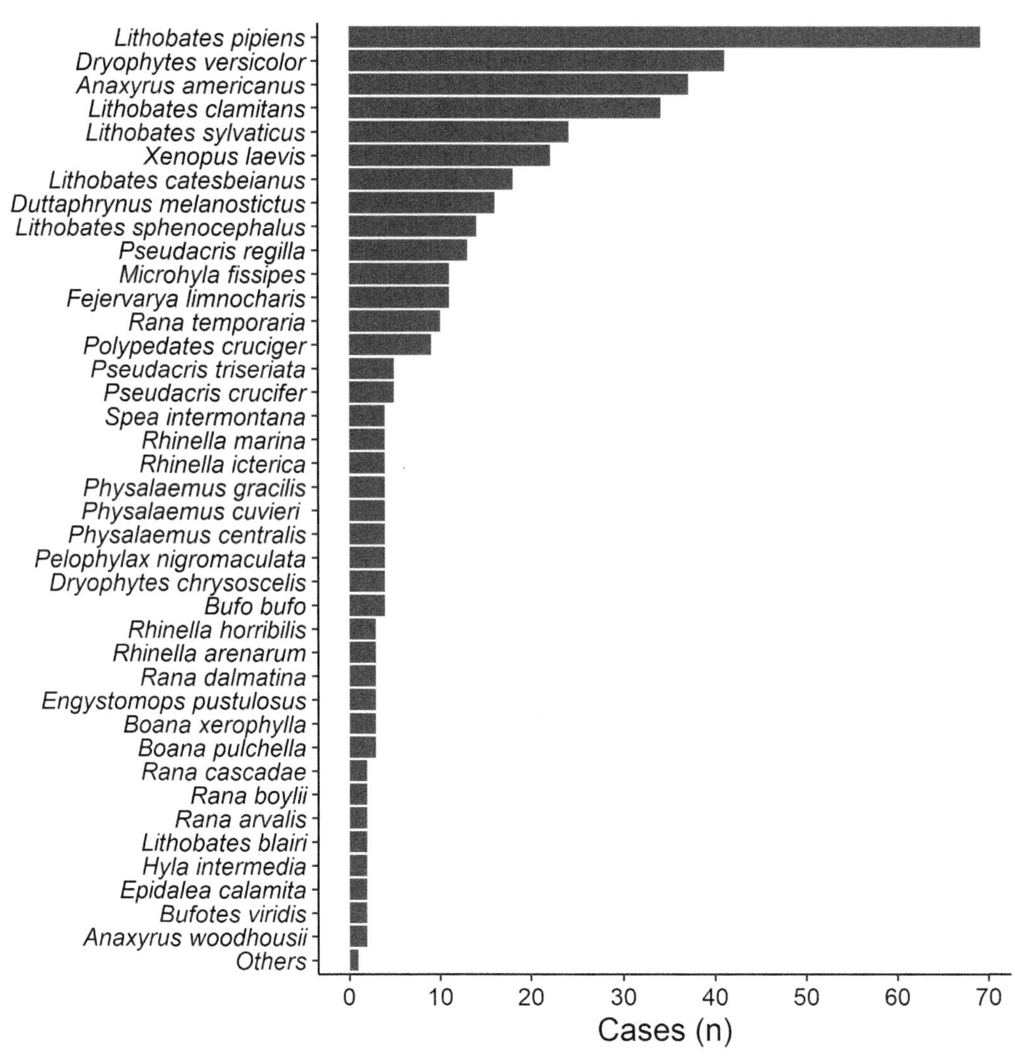

Figure 1. Species considered in ecotoxicological studies with pesticides and the number of cases found for each species. Others: species with one case—*Acris crepitans, Agalychnis callidryas, Boana faber, Bombina variegata, Discoglossus galganoi, Fejervarya* sp.1, *Fejervarya* sp.2, *Gastrophryne olivácea, Hoplobatrachus chinensis, Hyla arborea, Leptodactylus fuscus, Leptodactylus latrans, Leptodactylus luctator, Limnodynastes peronii, Litoria freycineti, Litoria peronii, Minervarya teraiensis, Odontophrynus americanos, Osteopilus septentrionalis, Pelobates cultripes, Pelophylax perezi, Pelophylax ridibundus, Physalaemus nattereri, Rana aurora, Rana chensinensis, Rhinella dorbignyi, Scinax nasicus, Sclerophrys arábica, Sclerophrys regularis, Spea multiplicata.*

their body mass. Also, these contradictory responses on growth and mass can be induced by the same pesticide type [e.g., carbaryl can be associated to mass reduction (e.g., Boone and Bridge 2003, Boone et al. 2007, Groner and Relyea 2011) or mass increase (Boone and Semlitsch 2002, Boone et al. 2004, Boone 2008)]. There are different explanations for these contradictory responses. Pesticides can act as endocrine disruptors, especially on thyroid gland affecting thyroid hormones (TH), which are primarily responsible for regulating metamorphosis in amphibians (see Chapter 5) (Brown and Cai 2007). This same pathway on metamorphosis can be affected by other stressors, such as pond drying, diseases, or other contaminants (e.g., Denver 1997, Buck et al. 2012, Figueiredo and Rodrigues 2014). Thus, depending on the developmental stage (see Chapter 2), pesticides can (i) disrupt the early metamorphosis, resulting in smaller, poorer competitors and/or individuals more susceptible to predation; or (ii) disrupt later metamorphosis, resulting in delayed metamorphosis,

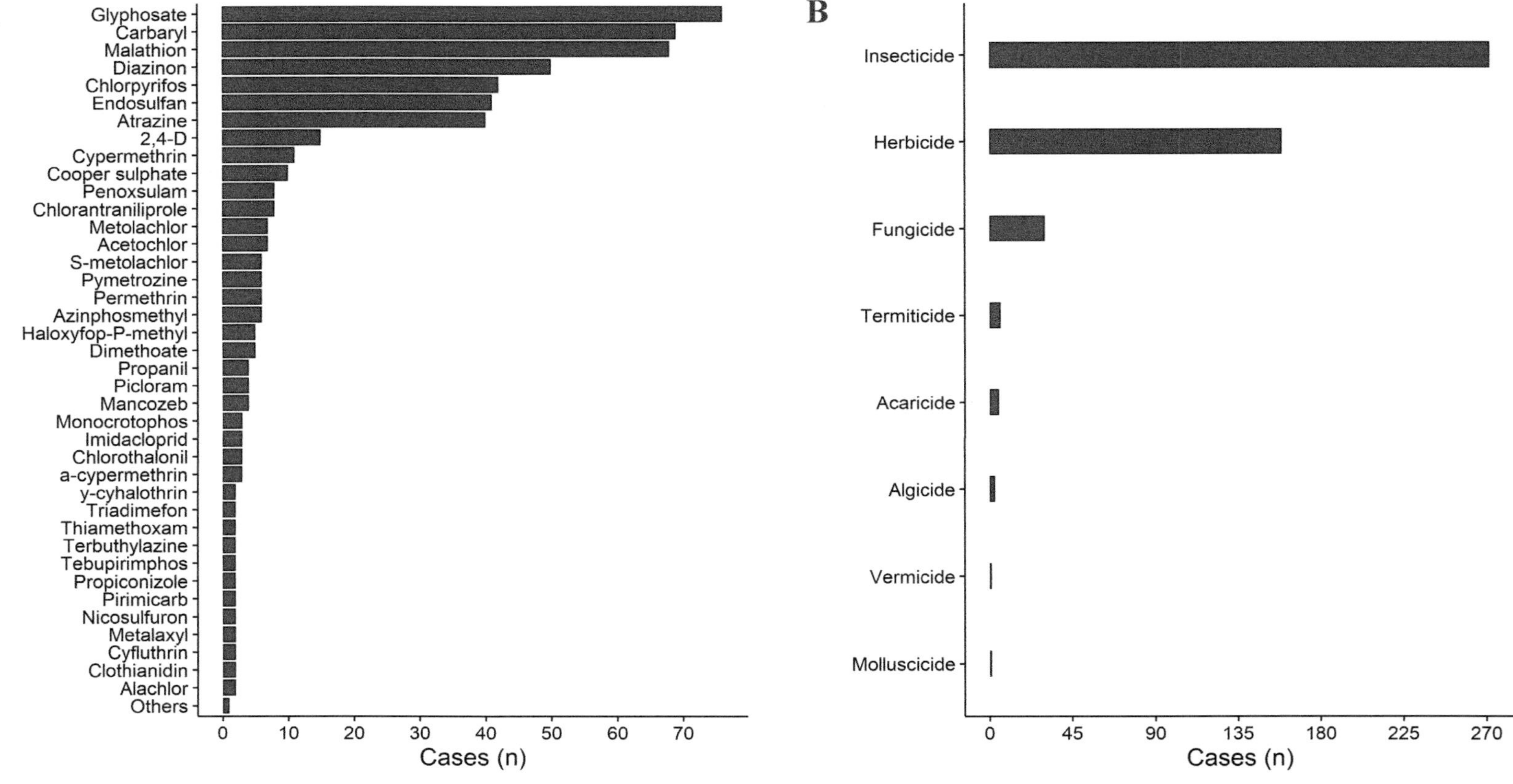

Figure 2. (A) Active ingredients considered in the experimental studies and the number of cases with each active ingredient (including pesticide mixtures). Others: pesticides with one case—*Abamectin, Acephate, Amitrole, Azoxystrobin, Bacillus thuringiensis israelenses, Carbendazim, Cyanazine, Cyproconazole, Deltamethrin, Edifenphos, Esfenvalerate, Fenpropimorph*, MCPA, Metamifop, Methomyl, Methoxychlor, N-butyl isocyanate, POEA, Prochloraz, Pyrimethanil, Pyriproxyfen, Quinclorac, Tebuconazole, Thiophanate-methyl, Trichlorfon, Triclopyr. (B) Number of cases that evaluated each pesticide class (including pesticide mixtures).

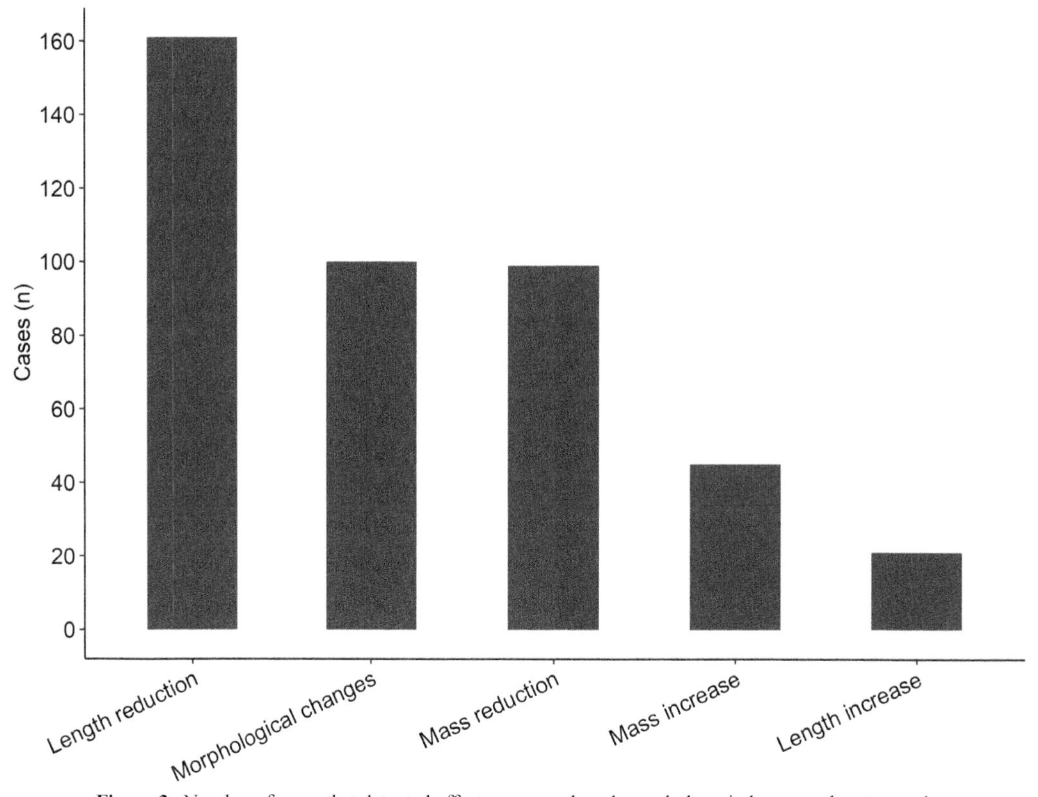

Figure 3. Number of cases that detected effects on growth and morphology in larvae and metamorphs.

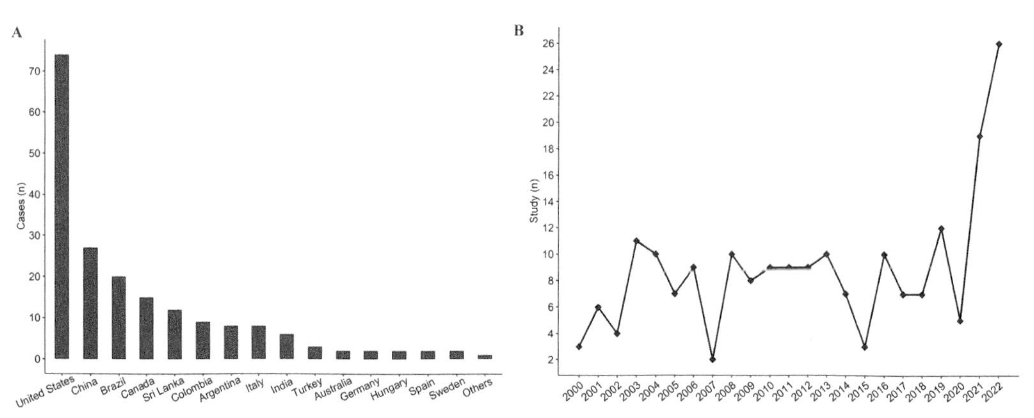

Figure 4. (A) Number of studies conducted in each country. Others: countries with one study—Albania, Denmark, Belgium, France, Thailand, Korea, Saudi Arabia, Costa Rica, Egypt, Austria. (B) Number of studies published from January 2000 to October 2022.

which increases time of exposure to contamination and, consequently, the susceptibility to adverse environmental effects, like pond drought, diseases and/or other sporadic or stochastic events (e.g., Berrill et al. 1993, Kiesecker 2002, Cauble and Wagner 2005, Hayes et al. 2006, Sayim and Kaya 2006, Bulen and Distel 2011, Buck et al. 2012).

The growth rates are not only related to metamorphosis time. Contamination by pesticides can alter the nutrient dynamic and the primary production of ponds, which, in turn, affects the foraging rates, mediates food web disruption and/or trophic cascade effects, and increases competition (Boone and Semlitsch 2001, Boone and Bridges 2003, Relyea 2006, Whiles et al. 2006,

Table 2. Number of studies published in each journal.

Journal of publication	Number of studies
Environmental Toxicology and Chemistry	32
Archives of Environmental Contamination and Toxicology	12
Aquatic Toxicology	10
Ecological Applications	9
Ecotoxicology	7
Ecotoxicology and Environmental Safety	7
Environmental Science and Pollution Research	6
Bulletin of Environmental Contamination and Toxicology	5
Oecologia	4
Environmental Pollution	4
Proceedings of the National Academy of Science USA	3
Turkish Jounal of Zoology	3
Chemosphere	3
Environmental Health Perspectives	2
Freshwater Biology	2
Functional Ecology	2
Journal of the National Science Foundation of Sri Lanka	2
Plos ONE	2
Environmental Toxicology	2
Hydrobiologia	2
Journal of Herpetology	2
Environmental Toxicology And Pharmacology	2
Journal of Toxicology, Environment and Health	2
4th ICE Conference	1
Science of the Total Environment	1
Conservation Biology	1
Biological Conservation	1
BIOS	1
Bioscene	1
Ecological Economics	1
Environmental Science & Technology	1
Herpetological Journal	1
International Journal of Agriculture and Biology	1
Journal of Applied Ecology	1
Journal of the North American Benthological Society	1
Maejo International Journal of Science and Technology	1
Oikos	1
Pesticide Biochemistry and Physiology	1
Proceedings of the Royal Society B	1
The Journal of Basic and Applied Zoology	1
The Open Zoology Journal	1
Toxicological & Environmental Chemistry	1

Table 2 contd. ...

...Table 2 contd.

Toxicological Sciences	1
Zoo's Print Journal	1
BMC Zoology	1
Ecohealth	1
Engenharia Sanitaria e Ambiental	1
Neurotoxicology and Teratology	1
Pakistan Journal of Zoology	1
Parasitology Research	1
Peerj	1
Pollution	1
Scientific Reports	1

Relyea and Diecks 2008, Hua and Relyea 2014). Thus, growth variations can be associated to different behavioral responses related to food acquisition (e.g., increasing or inhibiting the foraging rate), as well as to different levels of species susceptibility to contamination and species-specific capacity to consume food and convert it to growth (e.g., Bridges 1999, Boone and Semlitsch 2001, Relyea 2004a, Whiles et al. 2006, Denoël et al. 2012). Furthermore, growth changes may influence species-specific detoxification mechanisms, which require energy allocation mediated by a trade-off budget (Greulich and Pflugmacher 2004, Venturino and D'Angelo 2005, Durant et al. 2007). Regardless of the source, these responses on growth of larvae and metamorphs can negatively affect biological processes observed in adults, such as reduction of overwinter survival and reproductive potential (Smith 1987, Semlitsch et al. 1988, Berven 1990).

Malformations and deformities (grouped as morphological changes) were the second most common responses observed. These morphological changes triggered by chemicals negatively affect the individual performance, but can also affect different body parts, as the tail [deeper tails (Relyea 2012, Katzenberger et al. 2014), lateral flexure of the tail from its normal position (Bonfanti et al. 2004, Bernabò et al. 2011, Arcaute et al. 2012, Boccioni et al. 2021), twisting of tail (Greulich and Pflugmacher 2003, David et al. 2012), increase of stiff tails (Krishnamurthy and Smith 2011), curved tails (Lajmanovich et al. 2003, Seleem 2019), necrosis of the tail tip, flexure of the tail tip, fin damage, abnormal growth and blistering on the tail fin (Howe et al. 2004, David and Kartheek 2015)] and fore and hind-limbs [formation of 1 hind-limb only, formation of 3 front-limbs (Bridges 2000, Said et al. 2022), limb deformities (Devi and Gupta 2013) and femoral shortening (Fort et al. 2004)]. Furthermore, several types of skeletal defects, variations in intestinal shape, mouth and body deformities, and edemas have been reported (e.g., Bridges 2000, Bridges et al. 2004, Bachetta et al. 2008, Mandrillon and Saglio 2009, Bernabò et al. 2011, Krishnamurthy and Smith 2011, David et al. 2012, Relyea 2012, Aiko et al. 2014, Jayawardena et al. 2017, Herek et al. 2020, Samojeden et al. 2022). As subtler effects, pesticides affect the symmetry of morphological traits associated to sensory capabilities (e.g., eyes and nares) (Costa and Nomura 2016), causing deformities in the mouthparts, eyes (Lajmanovich et al. 2003, Herek et al. 2020), and spine malformations (e.g., scoliosis, lordosis, and kyphosis) (Jayawardena et al. 2010, 2011, Devi and Gupta 2013, Aiko et al. 2014, David and Kartheek 2015, Jayawardena et al. 2017). These morphological traits are related to swimming activity, food acquisition, and/or predator and food detection and deformities or developmental deviations (see Chapter 14 for developmental alterations induced by emerging pollutants), which can result in increased risk of predation and reduced competitive potential (e.g., Van Buskirk and Relyea 1998, Van Buskirk and McCollum 2000, Relyea 2004b, Van Buskirk 2009, Arendt 2009). Clearly, larvae and metamorphs respond to pesticide stress through variations in their body traits, as well as physical abnormalities. These responses can be better understood by evaluating a fitness-phenotype relationship, especially due to the range of effects observed on behavior and morphological traits (Van Buskirk and McCollum 2000, Arendt 2009, Allentoft and O'Brien 2010, Woodley et al. 2015).

Most of the studies revised in this chapter had a good accuracy regarding the effects of pesticides (alone or combined with another stressor), since they were generally carried out under controlled laboratory conditions. However, it also brings a low degree of realism when compared to the experiments carried out in the field, where abiotic variables are taken into account (Boone and James 2005). On the other hand, precisely by considering environmental factors, field experiments that are conducted in a large-scale environment have a cost of accuracy due to the large number of uncontrolled variables (Boone and James 2005, Mann et al. 2009). Thus, the use of mesocosms can be a very interesting alternative, as it allows better control of some variables within an experimental system that allows greater realism (see review in Boone and James 2005). Therefore, mesocosm studies should favor the evaluation of growth and morphological responses in a context closer to natural conditions. Thus, pesticide effects can be strongly mediated by species-specific susceptibility and by the different ecological contexts found in a contamination scenario, trying to simulate the conditions of freshwater ecosystems.

5. Distribution and Publication Trends

Geographical gaps are common in ecological studies (Martin et al. 2012, Trimble and Van-Arde 2012), limiting the knowledge about negative impacts on species and ecossystems in the world. We observed that most studies that experimentally evaluated the effects of pesticides on growth and morphology of larvae and metamorphs were carried out in the United States of America and, consequently, the most tested species are from North American (*Lithobates pipiens*, *Dryophytes versicolor* and *Anaxyrus americanus*). The high number of publications from the United States can reflect the high investment in science and researcher formations, besides the financial support obtained from public and private institutions (Mugnaini et al. 2004). However, we emphasize that while countries like Brazil have shown a good scientific production within the scope of this review, scientific investments in the country are still low. This may be due to the interest in conserving the large number of neotropical amphibian species.

Worldwide, there are 7516 anuran species (Frost 2021). Considering that only 69 species were identified in this review (0.91% of species), we can conclude that there is a great knowledge gap about the negative effects of pesticides on the growth and external morphology of larvae and metamorphs. Schiesari et al. (2007) show that regions with the highest numbers of amphibian species (i.e., Neotropical region) have the lowest numbers of species tested regarding the effects of pesticides. On the one hand, this makes sense considering the high number of species in these areas and the difficulty of sampling a large portion of them. On the other hand, it is clear that despite advances in research in neotropical countries, this pattern can still be observed, mainly if we consider that most of the species tested on ecotoxicological studies occur in temperate regions.

An increasing trend in scientific production within the scope of this review in the recent years was observed, mainly in 2021 and 2022. However, a larger temporal sample is needed to assess whether it is a trend or whether they are random observations. Once we only focus on studies that measure changes in growth and morphology, it is possible that studies that measure lethality, behavioral, physiological and/or genetical changes, as well as ecotoxicological tests with different contaminants (e.g., heavy metals, fertilizers), could explain the variation in the number of studies observed over time.

The publications were directed to ecotoxicological journals (e.g., Environmental Toxicology and Chemistry, Archives of Environmental Contamination and Toxicology, Aquatic Toxicology), with less publications in ecology, herpetology and/or conservation journals (Table 2). Manuscript publication in specialized ecotoxicological journals can favor the communication among researchers and the search of comparative studies. In addition, sharing scientific data in specialized journals increases the outreach of new discoveries, potentially increasing the number of citations.

6. Concluding Remarks

Growth and morphological responses in tadpoles are easily identified and can directly affect fitness and survival of species, increasing their susceptibility to stochastic events, the risk of predation and decreasing competitive ability (Brodie Jr. et al. 1991, Steiner and Van Buskirk 2008, Van Buskirk 2009, Relyea 2012). In addition to expanding information about isolated effects of pesticides on growth and morphology, it is important that future studies seek to evaluate how pesticides interact with different stressors, especially those resulting from human actions (i.e., global warming, pollution, habitat changes, invasive species).

It is also necessary to evaluate how the different species traits, especially those related to individual fitness, respond to stressor interactions and how chronic responses contribute to amphibian species loss. Studies that examined the interaction of pesticides with different stressors, such as diseases, predators, climate changes and habitat alterations, can contribute to mitigate ecological impacts and optimize the efforts on species conservation (Boone et al. 2007, Baker et al. 2013). Finally, increasing the number of ecotoxicological studies from threatened areas and those with greater biodiversity, as well as studies with rare, endemic and/or threatened species, would be important to discriminate between general and specific responses of amphibians to pesticide contamination.

Acknowledgments

We are grateful to A. Morais, R. de Mello, L. Signorelli, N. Marques, F. Juncá, P. Eterovick and D. Rossa-Feres for their review and suggestions on the manuscript. RNC thanks to Pesquisador Produtividade da UEMG – PQ/UEMG - edital 10/2022 e edital 08/2021. FN (# 301232/2018-0) and MS (# 309365/2019-8) are CNPq fellows.

References Cited

Adelizzi, R., J. Portmann and R. Van Meter. 2019. Effect of individual and combined treatments of pesticide, fertilizer, and salt on growth and Corticosterone Levels of Larval Southern Leopard Frogs (*Lithobates sphenocephala*). Archives of Environmental Contamination and Toxicology 77(1): 29–39.

Agostini, M.G., G.S. Natale and A.E. Ronco. 2010. Lethal and sublethal effects of cypermethrin to *Hypsiboas pulchellus* tadpoles. Ecotoxicology 19: 1545–1550.

Aiko, V., M. Qirjo and E. Nuna. 2014. Hematological and morphological effects of cooper sulfate on the larval development of Green Toad, *Bufo viridis*. Conference paper, 4th ICE Conference: 736–743.

Allentoft, M.E. and J. O'Brien. 2010. Global amphibian declines, loss of genetic diversity and fitness: A review. Diversity 2. 47–71.

Allran, J.W. and W.H. Karasov. 2000. Effects of atrazine and nitrate on northern leopard frog (*Rana pipiens*) larvae exposed in the laboratory from posthatch through metamorphosis. Environmental Toxicology and Chemistry 19: 2850–2855.

Allran, J.W. and W.H. Karasov. 2001. Effects of Atrazine on embryos, larvae, and adults of anuran amphibians. Environmental Toxicology and Chemistry 20(4): 769–775.

Almeida, P.R., M.V. Rodrigues and A.M. Imperador. 2019. Acute Toxicity (CL50) and behavioral and morphological effects of a commercial formulation with glyphosate active ingredient in tadpoles of *Physalaemus cuvieri* (Anura, Leptodactylidae) and *Rhinella icterica* (Anura, Bufonidae). Engenharia Sanitária e Ambiental 24(6): 1115–1125.

Alroy, J. 2015. Current extinction rates of reptiles and amphibians. Proc. Natl. Acad. Sci. USA (PNAS) 112(42): 13003–13008.

Alza, C.M., M.A. Donnelly and S.M. Whitfield. 2016. Additive Effects of mean temperature, temperature variability, and chlorothalonil to red-eyed treefrog (*Agalychnis callidryas*) Larvae. Environmental Toxicology and Chemistry 35(12): 2998–3004.

Amaral, D.F., M.F. Montalvão, B.D. Mendes, A.L.D. Castro and G. Malafaia. 2018. Behavioral and mutagenic biomarkers in tadpoles exposed to different abamectin concentrations. Environmental Science and Pollution Research 25: 12932–12946.

Arcaute, C.R., C.S. Costa, P.M. Demetrio, G.S. Natale and A.E. Ronco. 2012. Influence of existing site contamination on sensitivity of *Rhinella fernandezae* (Anura, Bufonidae) larvaes to Lorsban® 48E formulation of chlorpyrifos. Ecotoxicology 21: 2338–2348.
Arendt, J.D. 2009. Influence of sprint speed and body size on predator avoidance in New Mexican spadefoot toads (*Spea multiplicata*). Oecologia 159: 455–461.
Asparch, Y., G. Svartz and C.P. Coll. 2019. Toxicity characterization and environmental risk assessment of Mancozeb on the South American common toad *Rhinella arenarum*. Environmental Science and Pollution Research 27: 3034–3042.
Bacchetta, R., P. Mantecca, M. Andrioletti, C. Vismara and G. Vailati. 2008. Axialskeletal defects caused by carbaryl in *Xenopus laevis* embryos. Sci. Total Environment 392: 110–118.
Baier, F., E. Gruber, T. Hein, E. Bondar-Kunze, M. Ivankovic, A. Mentler, C.A. Bruhl, B. Spangl and J.G. Zaller. 2016. Non-target effects of a glyphosate-based herbicide on Common toad larvae (*Bufo bufo*, Amphibia) and associated algae are altered by temperature. PeerJ. Doi: 10.7717/peerj.2641.
Baker, N.J., B.A. Bancroft and T.S. Garcia. 2013. A meta-analysis of the effects of pesticides and fertilizers on survival and growth of amphibians. Sci. Total Environ. 449: 150–156.
Bancroft, B.A., N.J. Baker and A.R. Blaustein. 2008. A meta-analysis of the effects of ultraviolet B radiation and its synergistic interactions with pH, contaminants, and disease on amphibian survival. Conserv. Biol. 22: 987–96.
Bernabò, I., E. Sperone, S. Tripedi and E. Brunelli. 2011. Toxicity of chlorpyrifos to larval *Rana dalmatina*: Acute and chronic effects on survival, development, growth and gill apparatus. Arch. Environ. Contam. Toxicol. 61: 704–718.
Bernabò, I., A. Guardia, R. Macirella, S. Sesti, A. Crescente and E. Brunelli. 2016. Effects of long-term exposure to two fungicides, pyrimethanil and tebuconazole, on survival and life history traits of Italian tree frog (*Hyla intermedia*). Aquatic Toxicology 172: 56–66.
Berrill, M., S. Bertram, A. Wilson and S. Louis. 1993. Lethal and sublethal impacts of pyrethroid insecticides on amphibian embryos and larvaes. Environ. Toxicol. Chem. 12: 525–39.
Berven, K.A. 1990. Factors affecting population fluctuations in larval and adult stages of the wood frog (*Rana sylvatica*). Ecology 71: 1599–1608.
Biga, L.M. and A.R. Blaustein. 2013. Variations in lethal and sublethal effects of Cypermethrin among aquatic stages and species of anuran amphibians. Environmental Toxicology and Chemistry 32(12): 2855–2860.
Bishop, C.A., N.A. Mahony, J. Struger, P. Ng and K.E. Pettit. 1999. Anuran development, density and diversity in relation to agricultural activity in the Holland River Watershed, Ontario, Canada (1990–1992). Environ. Monit. Assess. 57: 21–43.
Blaustein, A.R. and J.M. Kiesecker. 2002. Complexity in conservation: Lessons from the global decline of amphibian populations. Ecol. Lett. 5(4): 597–608.
Boccioni, A.P.C., R.C. Lajmanovich, P.M. Peltzer, A.M. Attademo and C.S. Martinuzzi. 2021. Toxicity assessment at different experimental scenarios with glyphosate, chlorpyrifos and antibiotics in *Rhinella arenarum* (Anura: Bufonidae) tadpoles. Chemosphere 273: 128475.
Bokony, V., V. Verebelyi, N. Ujhegyi, Z. Miko, E. Nemeshazi, M. Szederkenyi, S. Orf, E. Vitanyi and A.M. Moricz. 2020. Effects of two little-studied environmental pollutants on early development in anurans. Environmental Pollution 260: 114078.
Bombardi, L.M. 2017. Geografia do Uso de Agrotóxicos no Brasil e Conexões com a União Europeia. São Paulo: FFLCH – USP. 296p.
Bonfanti, P., A. Colombo, F. Orsi, I. Nizzetto, M. Andrioletti, R. Bacchetta, P. Mantecca, U. Fascio, G. Vailati and C. Vismara. 2004. Comparative teratogenicity of chlorpyrifos and malathion on *Xenopus laevis* development. Aquat. Toxicol. 70: 189–200.
Boone, M., C.M. Bridges and B.B. Rothermel. 2001. Growth and development of larval green frogs (*Rana clamitans*) exposed to multiple doses of an insecticide. Oecologia 129: 518–524.
Boone, M.D. and R.D. Semlitsch. 2001. Interactions of an insecticide with larval density and predation in experimental amphibian communities. Conserv. Biol. 15: 228–238.
Boone, M.D. and R.D. Semlitsch. 2002. Interactions of an insecticide with competition and pond drying in amphibian communities. Ecological Applications 12: 307–316.
Boone, M.D. and R.D. Semlitsch. 2002. Interactions of an insecticide with competition and pond drying in amphibian communities. Ecol. Appl. 12: 307–316.
Boone, M.D. and S.M. James. 2003. Interactions of an insecticide, herbicide, and natural stressors in amphibian community mesocosms. Ecological Applications 13: 829–841.
Boone, M.D. and R.D. Semlitsch. 2003. Interactions of bullfrog tadpole predators and an insecticide: Predation release and facilitation. Oecologia 442: 610–616.

Boone, M.D. and C.M. Bridges. 2003. Effects of carbaryl on green frog (*Rana clamitans*) larvaes: Timing of exposure versus multiple exposures. Environ. Toxicol. Chem 22: 2695–2702.

Boone, M.D., R.D. Semlitsch, J.F. Fairchild and B.B. Rothermel. 2004. Effects of an insecticide on amphibians in large-scale experimental ponds. Ecol. Appl. 14: 685–691.

Boone, M.D. and S.M. James. 2005. Use of aquatic and terrestrial mesocosms in ecotoxicology. Appl. Herpetol. 2: 231–257.

Boone, M.D., C.M. Bridges, J.F. Fairchild and E.E. Little. 2005. Multiple sublethal chemicals negatively affect tadpoles of the green frog *Rana clamitans*. Environmental Toxicology and Chemistry 24: 1267–1272.

Boone, M.D. and C.M. Bridges-Britton. 2006. Examining multiple sublethal contaminants on the gray treefrog (*Hyla versicolor*): Effects of an Insecticide, Herbicide, and Fertilizer. Environmental Toxicology and Chemistry 25(12): 3261–3265.

Boone, M.D., R.D. Semlitsch, E.E. Little and M.C. Doyle. 2007. Multiple stressors in amphibian communities: Effects of chemical contamination, bullfrogs, and fish. Ecol. Appl. 17: 291–301.

Boone, M.D. 2008. Examining the single and interactive effects of three insecticides on amphibian metamorphosis. Environ. Toxicol. Chem. 27: 1561–1568.

Boone, M.D., S.A. Hammond, N. Veldhoen, M. Youngquist and C.C. Helbing. 2013. Specific time of exposure during tadpole development influences biological effects of the insecticide carbaryl in green frogs (*Lithobates clamitans*). Aquatic Toxicology 130-131: 139–148.

Brande-Lavridsen, N., J. Christensen-Dalsgaard and B. Korsgard. 2010. Effects of ethinylestradiol and the fungicide prochloraz on metamorphosis and thyroid gland morphology in *Rana temporaria*. The Open Zoology Journal 3: 7–16.

Bridges, C., E. Little, D. Gardiner, J. Petty and J. Huckins. 2004. Assessing the toxicity and teratogenicity of pond water in North-Central minnesota to amphibians. Environ. Sci. Pollut Res. 11(4): 233–239.

Bridges, C.M. 1999. Predator-prey interactions between two amphibian species: Effects of insecticide exposure. Aquat. Ecol. 33: 205–211.

Bridges, C.M. 2000. Long-term effects of pesticide exposure at various life stages of the Southern Leopard Frog (*Rana sphenocephala*). Arch. Environ. Contam. Toxicol. 39: 91–96.

Bridges, C.M. and M.D. Boone. 2003. The interactive effects of UV-B and insecticide exposure on tadpole survival, growth and development. Biological Conservation 113: 49–54.

Brodie, Jr., E.D., D.R. Formanowicz Jr. and E.D. Brodie III. 1991. Predator avoidance and antipredator mechanisms: Distinct pathways to survival. Ethol. Ecol. Evol. 3: 73–77.

Broomhall, S. and R. Shine. 2003. Effects of the insecticide endosulfan and presence of congeneric tadpoles on Australian Treefrog (*Litoria freycineti*) Tadpoles. Archives of Environmental Contamination and Toxicology 45: 221–226.

Broomhall, S.D. 2004. Egg temperature modifies predator avoidance and the effects of the insecticide endosulfan on tadpoles of an Australian frog. Journal of Applied Ecology 41: 105–113.

Brown, D.D. and L.Q. Cai. 2007. Amphibian metamorphosis. Develop. Biol. 306: 20–33.

Brunelli, E., I. Bernabò, C. Berg, K. Lundstedt-Enkel, A. Bonacci and S. Tripepi. 2009. Environmentally relevant concentrations of endosulfan impair development, metamorphosis and behaviour in *Bufo bufo* tadpoles. Aquatic Toxicology 91: 135–142.

Buck, J.C., E.A. Scheessele, R.A. Relyea and A.R. Blaustein. 2012. The effects of multiple stressors on wetland communities: Pesticides, pathogens and competing amphibians. Freshw. Biol. 57: 61–73.

Bulen, B.J. and C.A. Distel. 2011. Carbaryl concentration gradients in realistic environments and their influence on our understanding of the larvae food web. Arch. Environ. Contam. Toxicol. 60: 343–350.

Carneiro, F.F., L.G.S. Augusto, R.M. Rigotto, K. Friedrich and A.C. Búrigo. 2015. Dossiê ABRASCO: Um alerta sobre os impactos dos agrotóxicos. Escola Politécnica de Saúde Joaquim Venâncio: Rio de Janeiro. Expressão Popular: São Paulo. 624 p.

Carr, J.A., A. Gentles, E.E. Smith, W.L. Goleman, L.J. Urquidi, K. Thuett, R.J. Kendall, J.P. Giesy, T.S. Gross, K.R. Solomon and G.V.D. Kraak. 2003. Response of Larval *Xenopus laevis* to Atrazine: Assessment of Growth, Metamorphosis, and Gonodal and Laryngeal Morphology. Environmental Toxicology and Chemistry 22(2): 396–405.

Carson, R.L. 1962. Silent Spring. Boston: Houghton Mifflin. 328 p.

Cauble, K. and R.S. Wagner. 2005. Sublethal effects of the herbicide glyphosate on amphibian metamorphosis and development. Bull. Environ. Contam. Toxicol. 75: 429–435.

Chemotti, D.C., S.N. Davis, LW. Cook, I.R. Willoughby, C.J. Paradise and B. Lom. 2006. The pesticide malathion disrupts *Xenopus* and Zebrafish Embryogenesis: An investigative laboratory exercise in developmental toxicology. Bioscene 32(3): 4–18.

Coady, K.K., M.B. Murphy, D.L. Villeneuve, M. Hecker, P.D. Jones, J.A. Carr, J.A., K.R. Solomon, E.E. Smith, G. Van der Kraak, R.J. Kendall and J.P. Giesy. 2005. Effects of atrazine on metamorphosis, growth, laryngeal and gonadal development, aromatase activity, and sex steroid concentrations in *Xenopus laevis*. Ecotoxicology and Environmental Safety 62: 160–173.

Colombo, A., F. Orsi and P. Bonfanti. 2005. Exposure to the organophosphorus pesticide chlorpyrifos inhibits acetylcholinesterase activity and affects muscular integrity in *Xenopus laevis* larvae. Chemosphere 61: 1665–1671.

Costa, R.N. and F. Nomura. 2016. Measuring the impacts of Roundup Original® on fluctuating asymmetry and mortality in a Neotropical tadpole. Hydrobiologia 765: 85–96.

Cothran, R.D., F. Radarian and R.A. Relyea. 2011. Altering aquatic food webs with a global insecticide: Arthropod–amphibian links in mesocosms that simulate pond communities. J. N. Am. Benthol. Soc. 30(4): 893–912.

Crump, D., K. Werry, N. Veldhoen, G.V. Aggelen and C.C. Helbing. 2002. Exposure to the herbicide acetochlor alters thyroid hormone-dependent gene expression and metamorphosis in *Xenopus Laevis*. Environmental Health Perspectives 110(12): 1199–1205.

Curtis, A.N. and M.G. Bidart. 2021. Increased temperature influenced growth and development of *Lithobates pipiens* tadpoles exposed to leachates of the invasive plant European Buckthorn (*Rhamnus cathartica*) and a Triclopyr Herbicide. Environmental Toxicology and Chemistry 40(9): 2547–2558.

David, M., S.R. Marigoudar, V.K. Patil and R. Halappa. 2012. Behavioral, morphological deformities and biomarkers of oxidative damage as indicators of sublethal cypermethrin intoxication on the larvaes of *D. melanostictus* (Schneider, 1799). Pestic. Biochem. Physiol. 103: 127–134.

David, M. and R.M. Kartheek. 2015. Malathion acute toxicity in larvaes of *Duttaphrynus melanostictus*, morphological and behavioural study. J. Basic Appl. Zool. 72: 1–7.

Denoël, M., B. D'Hooghe, G.F. Ficetola, C. Brasseur, E. De Pauw, J.P. Thomé and P. Kestemont. 2012. Using sets of behavioral biomarkers to assess short-term effects of pesticide: A study case with endosulfan on frog larvaes. Ecotoxicology 21: 1240–1250.

Denoël, M., S. Libon, P. Kestemont, C. Brasseur, J.F. Focant and E.D. Pauw. 2013. Effects of a sublethal pesticide exposure on locomotor behavior: A video-tracking analysis in larval amphibians. Chemosphere 90: 945–951.

Denver, R.J. 1997. Proximate mechanisms of phenotypic plasticity in amphibian metamorphosis. Am. Zool. 37: 172–184.

Devi, N.N. and A. Gupta. 2013. Toxicity of endosulfan to larvaes of *Fejervarya* spp. (Anura: Dicroglossidae): mortality and morphological deformities. Ecotoxicology 22: 1395–1402.

Devine, G.J. and M.J. Furlong. 2007. Insecticide use: contexts and ecological consequences. Agric. Human Values 24: 281–306.

Distel, C.A. and M.D. Boone. 2009. Effects of aquatic exposure to the insecticide carbaryl and density on aquatic and terrestrial growth and survival in American toads. Environ. Toxicol. Chem. 28(9): 1963–1969.

Distel, C.A. and M.D. Boone. 2010. Effects of aquatic exposure to the insecticide carbaryl are species-specific across life stages and mediated by heterospecific competitors in anurans. Functional Ecology 24: 1342–1352.

Durant, S.E., W.A. Hopkins and L.G. Talent. 2007. Energy acquisition and allocation in an ectothermic predator exposed to a common environmental stressor. Comp. Bioche. Physiol. C Toxicol. Pharmacol. 145: 442–448.

Edge, C.B., D.G. Thompson, C. Hao and J.E. Houlahan. 2012. A silviculture application of the glyphosate-based herbicide Visonmax to wetlands has limited directed effects on amphibian larvae. Environmental Toxicology and Chemistry 31: 1–9.

Edginton, A.N., P.M. Sheridan, G.R. Stenphenson, D.G. Thompson and H.J. Boermans. 2004. Comparative effects of pH and Vision® herbicide on two life stages of four anuran amphibian species. Environmental Toxicology and Chemistry 23(4): 815–822.

Figueiredo, J. and D.J. Rodrigues. 2014. Effects of four types of pesticides on survival, time and size to metamorphosis of two species of larvaes (*Rhinella marina* and *Physalaemus centralis*) from the southern Amazon, Brazil. Herpetol. J. 24: 1–9.

Foley, J.A., R. Defries, G.P. Asner, C. Barford, G. Bonan, S.R. Carpenter, F.S. Chapin, M.T. Coe, G.C. Daily, H.K. Gibbs, J.H. Helkowski, T. Holloway, E.A. Howard, C.J. Kucharik, C. Monfreda, J.A. Patz, C. Prentice, N. Ramankutty and P.K. Snyder. 2005. Global Consequences of Land Use. Science 309: 570–574.

Fordham, C.L., J.D. Tessari, H.S. Ramsdell and T.J. Keefe. 2001. Effects of malathion on survival, growth, development, and equilibrium posture of bullfrog tadpoles (*Rana catesbeiana*). Environmental Toxicology and Chemistry 20: 179–184.

Fort, D.J., P.D. Guiney, J.A. Weeks, J.H. Thomas, R.L. Rogers, A.M. Noll and C.D. Spaulding. 2004. Effect of methoxychlor on various life stages of *Xenopus laevis*. Toxicol. Sci. 81: 454–466.

Freeman, J.L. and A.L. Rayburn. 2005. Developmental impact of atrazine on metamorphing *Xenopus laevis* as revealed by nuclear analysis and morphology. Environmental Toxicology and Chemistry 24(7): 1648–1653.

Freitas, J.S., L. Girotto, B.V. Goulart, L.D.G. Alho, R.C. Gebara, C.C. Montagner, L. Schiesari and E.L.G. Espindola. 2019. Effects of 2,4-D-based herbicide (DMA® 806) on sensitivity, respiration rates, energy reserves and behavior of tadpoles. Ecotoxicology and Environmental Safety 182: 109446.

Frost, D.R. 2021. Amphibian Species of the World: An Online Reference. Version 6.1 (Acessed on 19/10/2022). Electronic Database accessible at https://amphibiansoftheworld.amnh.org/index.php. American Museum of Natural History, New York, USA. doi.org/10.5531/db.vz.0001.

Fusco, G. and A. Minelli. 2010. Phenotypic plasticity in development and evolution: Facts and concepts. Philos. Trans. R. Soc. B 365: 547–556.

Gahl, M.K., B.D. Pauli and J.E. Houlahan. 2011. Effects of chytrid fungus and a glyphosate-based herbicide on survival and growth of wood frogs (*Lithobates sylvaticus*). Ecological Applications 21(7): 2521–2529.

Gaietto, K.M., S.L. Rumschlag and M.D. Boone. 2014. Effects of pesticide exposure and the amphibian chytrid fungus on gray treefrog (*Hyla chrysoscelis*) metamorphosis. Environmental Toxicology and Chemistry 33(10): 2358–2362.

Gallant, A.L., R.W. Klaver, G.S. Casper and M.J. Lannoo. 2007. Global rates of habitat loss and implications for amphibian conservation. Copeia 4: 967–979.

García-Muñoz, E., F. Guerreo and G. Parra. 2009. Effects of copper sulfate on growth, development, and escape behavior in *Epidalea calamita* embryos and larvae. Archives of Environment Contamination and Toxicology 56: 557–565.

García-Muñoz, E., F. Guerreo and G. Parra. 2010. Intraspecific and interspecific tolerance to copper sulphate in five iberian amphibian species at two developmental stages. Archives of Environment Contamination and Toxicology 59: 312–321.

Gavel, M.J., S.D. Young, N. Blais, M.R. Forbes and S.A. Robinson. 2021. Trematodes coupled with neonicotinoids: Effects on blood cell profiles of a model amphibian. Parasitology Research 120: 2135–2148.

Ghodageri, M.G. and K. Pancharatna. 2011. Morphological and behavioral alterations induced by endocrine disrupters in amphibian tadpoles. Toxicological & Environmental Chemistry 93(10): 2012–2021.

Giesy, J.P., S. Dobson and K.R. Solomon. 2000. Ecotoxicological risk assessment for Roundup® herbicide. Rev. Environ. Contam. Toxicol. 167: 35–120.

Gosner, K.L. 1960. A simplified table for staging anuran ambryos and larvae with notes on identification. Herpetologica 16: 183–190.

Greulich, K. and S. Pflugmacher. 2003. Differences in susceptibility of various life stages of amphibians to pesticide exposure. Aquat. Toxicol. 65: 329–336.

Greulich, K. and S. Pflugmacher. 2004. Uptake and effects on detoxication enzymes of cypermethrin in embryos and tadpoles of amphibians. Archives of Environmental Contamination and Toxicology 47: 489–495.

Groner, M.L. and R.A. Relyea. 2011. A tale of two pesticides: How common insecticides affect aquatic communities. Freshw. Biol. 56(11): 2391–2404.

Gürkan, M. and S. Hayretdağ. 2012. Morphological and histological effects of copper sulfate on the larval development of green toad, *Bufo viridis*. Turkish Jounal of Zoology 36(2): 231–240.

Gurushankara, H.P., S.V. Krishnamurthy and V. Vasudev. 2007. Effect of malathion on survival, growth, and food consumption of indian cricket frog (*Limnonectus limnocharis*) Tadpoles. Archives of Environment Contamination and Toxicology 52: 251–256.

Hanlon, S.M., J.L. Kerby and M.J. Parris. 2012. Unlikely remedy: Fungicide clears infection from pathogenic fungus in larval southern leopard frogs (*Lithobates sphenocephalus*). PLoS ONE 7(8): 1–8.

Hanlon, S.M., K.J. Lynch and M.T. Parris. 2013. Mouthparts of southern leopard frog, *Lithobates sphenocephalus*, tadpoles not affected by exposure to a formulation of glyphosate. Bulletin of Environmental Contamination and Toxicology 91: 611–615.

Harris, M.L., L. Chora, C.A. Bishop and J.P. Bogart. 2000. Species- and age-related differences in susceptibility to pesticide exposure for two amphibians, *Rana pipiens*, and *Bufo americanus*. Bulletin of Environment Contamination and Toxicology 64: 263–270.

Hayes, T.B., A. Collins, M. Lee, M. Mendoza, N. Noriega, S.A. Stuart and A. Vonk. 2002. Hermaphroditic, demasculinized frogs after exposure to the herbicide atrazine at low ecologically relevant doses. Proceedings of the National Academy of Sciences of the United States of America 99: 5476–5480.

Hayes, T.B., P. Case, S. Chui, D. Chung, C. Haeffele, K. Haston, M. Lee, V.P. Mai, Y. Marjuoa, J. Parker and M. Tsui. 2006. Pesticide mixtures, endocrine disruption, and amphibian declines: Are we underestimating the impact? Environ. Health Perspect. 114: 40–50.

Hayes, T.B., P. Falso, S. Gallipeau and M. Stice. 2010. The cause of global amphibian declines: A developmental endocrinologist's perspective. J. Exp. Biol. 213: 921–933.

Henao, L.M., J.J. Mendez and M.H. Bernal. 2022. UVB radiation enhances the toxic effects of three organophosphorus insecticides on tadpoles from tropical anurans. Hydrobiologia 849(1): 141–153.

Herek, J.S., L. Vargas, S.A.R. Trindade, C.F. Rutkoski, N. Macagnan, P.A. Hartmann and M.T. Hartmann. 2020. Can environmental concentrations of glyphosate affect survival and cause malformation in amphibians? Effects from a glyphosate-based herbicide on *Physalaemus cuvieri* and P. *gracilis* (Anura: Leptodactylidae). Environmental Science and Pollution Research 27: 22619–22630.

Howe, C.M., M. Berrill, B.D. Pauli, C.C. Helbing, K. Werry and N. Veldhoen. 2004. Toxicity of glyphosate-based pesticides to four North American frog species. Environ. Toxicol. Chem. 23: 1928–1938.

Hua, J. and R.A. Relyea. 2014. Chemical cocktails in aquatic systems: Pesticide effects on the response and recovery of > 20 animal taxa. Environ. Pollut. 189: 18–26.

Hua, J., D.K. Jones, B.M. Mattes, R.D. Cothran, R.A. Relyea and T.J. Hoverman. 2015. The contribution of phenotypic plasticity to the evolution of insecticide tolerance in amphibian populations. Evol. Applic. 8(6): 586–596.

Jayawardena, U.A., P.S. Rajakaruna, A.N. Navaratne and PH. Amerasinghe. 2010. Toxicity of agrochemicals to common hourglass tree frog (*Polypedates cruciger*) in acute and chronic exposure. Int. J. Agric. Biol. 12: 641–648.

Jayawardena, U.A., A.N. Navaratne, P.H. Amerasinghe and R.S. Rajakaruna. 2011. Acute and chronic toxicity of four commonly used agricultural pesticides on the Asian common toad, *Bufo melanostictus* Schneider. J. Natl. Sci. Found. 39(3): 267–276.

Jayawardena, U.A., J.R. Rohr, A.N. Navaratne, P.H. Amerasinghe and R.S. Rajakaruna. 2016. Combined effects of pesticides and trematode infections on hourglass tree frog *Polypedates cruciger*. Ecohealth 13: 111–122.

Jayawardena, U.A., J.R. Rohr, P.H. Amerasinghe, A.N. Navaratne and R.S. Rajakaruna. 2017. Effects of agrochemicals on disease severity of *Acanthostomum burminis* infections (Digenea: Trematoda) in the Asian common toad, *Duttaphrynus melanostictus* BMC Zoology 2(13). https://doi.org/10.1186/s40850-017-0022-1.

Johansson, M., H. Piha, H. Kylin and J. Merilä. 2006. Toxicity of six pesticides to common frog (*Rana temporaria*) tadpoles. Environmental Toxicology and Chemistry 25(12): 3164–3170.

Jones, D.K., J.I. Hammond and R.A. Relyea. 2010. Roundup® and Amphibians: The Importance of concentration, application time, and stratification. Environmental Toxicology and Chemistry 29(9): 2016–2025.

Jones, D.K., J.I. Hammond and R.A. Relyea. 2011. Competitive stress can make the herbicide Roundup® more deadly to larval amphibians. Environ. Toxicol. Chem. 30(2): 446–454.

Katzenberger, M., J. Hammond, H. Duarte, M. Tejedo, C. Calabuig and R.A. Relyea. 2014. Swimming with predators and pesticides: How environmental stressors affect the thermal physiology of tadpoles. PloS one 9(5): 1–11.

Kiesecker, J. 2002. Synergism between trematode infection and pesticide exposure: A link to amphibian limb deformities in nature? Proc. Natl. Acad. Sci. USA 99: 9900–9904.

Kleinhenz, P., M.D. Boone and G. Fellers. 2012. Effects of the amphibian chytrid fungus and four insecticides on pacific treefrogs (*Pseudacris regilla*). Journal of Herpetology 46(4): 625–631.

Krishnamurthy, S.V. and G.R. Smith. 2010. Growth, abnormalities, and mortality of tadpoles of american toad exposed to combinations of malathion and nitrate. Environmental Toxicology and Chemistry 29(12): 2777–2782.

Krishnamurthy, S.V. and G.R. Smith. 2011. Combined effects of malathion and nitrate on early growth, abnormalities, and mortality of wood frog (*Rana sylvatica*) larvaes. Ecotoxicology 20: 1361–1367.

LaFiandra, E.M., K.J. Babbitt and S.A. Sower. 2008. Effects of atrazine on anuran development are altered by the presence of a nonlethal predator. Journal of Toxicology and Environmental Health-Part A 71: 505–511.

Lajmanovich, R.C., M.T. Sandoval and P.M. Peltzer. 2003. Induction of mortality and malformation in *Scinax nasicus* larvaes exposed to glyphosate formulations. Bull. Environ. Contam. Toxicol. 70: 612– 618.

Lajmanovich, R.C., P.M. Peltzer, C.S. Martinuzzi, A.M. Attademo, A. Basso and C.L. Colussi. 2019. Insecticide pyriproxyfen (Dragon®) damage biotransformation, thyroid hormones, heart rate, and swimming performance of *Odontophrynus americanus* tadpoles. Chemosphere 220: 714–722.

Lanctôt, C., C. Robertson, L. Navarro-Martín, C. Edge, S.D. Melvin, J. Houlahan and V.L. Trudeau. 2013. Effects of the glyphosate-based herbicide Roundup WeatherMax® on metamorphosis of wood frogs (*Lithobates sylvaticus*) in natural wetlands. Aquatic Toxicology 140-141: 48–57.

Lanctôt, C., L. Navarro-Martín, C. Robertson, B. Park, P. Jackman, B.D. Pauli and V.L. Trudeau. 2014. Effects of glyphosate-based herbicides on survival, development, growth and sex ratios of wood frog (*Lithobates sylvaticus*) tadpoles. II: Agriculturally relevant exposures to Roundup WeatherMax® and Vision® under laboratory conditions. Aquatic Toxicology 154: 291–303.

Lavorato, M., I. Bernabò, A. Crescente, M. Denoël, S. Tripedi and E. Brunelli. 2013. Endosulfan effects on *Rana dalmatina* tadpoles: Quantitative developmental and behavioural analysis. Archives of Environmental Contamination and Toxicology 64: 253–262.

Liu, R., Y.A. Qin, J.L. Diao and H.J. Zhang. 2021. *Xenopus laevis* tadpoles exposed to metamifop: Changes in growth, behavioral endpoints, neurotransmitters, antioxidant system and thyroid development. Ecotoxicology and Environmental Safety 220: 112417.

Ma, Y., B. Li, Y. Ke and Y.H. Zhang. 2018. Effects of low doses Trichlorfon exposure on *Rana chensinensis* tadpoles. Environmental Toxicology 34(1): 30–36.

Mackey, M.J. and M. Boone. 2009. Single and interactive effects of malathion, overwintered green frog tadpoles, and cyanobacteria on gray treefrog tadpoles. Environmental Toxicology and Chemistry 28: 637–643.

Mandrillon, A. and P. Saglio. 2009. Effects of single and combined embryonic exposures to herbicide and conspecific chemical alarm cues on hatching and larval traits in the common frog (*Rana temporaria*). Arch. Environ. Contam. Toxicol. 56: 566–576.

Mann, M.R., R.V. Hyne, C.B. Choung and S.P. Wilson. 2009. Amphibians and agricultural chemicals: Review of the risks in a complex environment. Environ. Pollut. 157: 2903–2927.

MAPA. 2022. Ministério da agricultura, pecuária e abastecimento. Registros de agrotóxicos concedidos entre 2005 a 2021. Available at: https://www.gov.br/agricultura/pt-br/assuntos/insumos-agropecuarios/insumos-agricolas/agrotoxicos/informacoes-tecnicas. Acsess date: 28/01/2022.

Martin, L.J., B. Blossey and E. Ellis. 2012. Mapping where ecologists work: Biases in the global distribution of terrestrial ecological observations. Front. Ecol. Environ. 10(4): 195–201.

Mathew, M. and M.I. Andrews. 2003. Impact of some pesticides on the growth of tadpoles of commom Indian toad *Bufo melanosticus* Schneider. Zoo's Print Journal 18 (2): 1007–1010.

McClelland, S.J., R.J. Bendis, R.A. Relyea and S.K. Woodley. 2018. Insecticide-induced changes in amphibian brains: How sublethal concentrations of chlorpyrifos directly affect neurodevelopment. Environmental Toxicology and Chemistry 37(10): 2692–2698.

Mills, N.E. and R.D. Semlitsch. 2004. Competition and predation mediate the indirect effects of an insecticide on southern leopard frogs. Ecol. Appl. 14: 1041–1054.

Mugnaini, R., P.M. Jannuzzi and L. Quoniam. 2004. Indicadores bibliométricos da produção científica brasileira: uma análise a partir da base Pascal. Ci. Inf. 33(2): 123–131.

Natale, G.S., J. Vera-Candioti, C.R. de Arcaute, S. Soloneski, M.L. Larramendy and A.E. Ronco. 2018. Lethal and sublethal effects of the pirimicarb-based formulation Aficida® on *Boana pulchella* (Dumeril and Bibron, 1841) tadpoles (Anura, Hylidae). Ecotoxicology and Environmental Safety 147: 471–479.

Navarro-Martín, L., C. Lanctôt, P. Jackman, B.J. Park, K. Doe, B.D. Pauli and V.L. Trudeau. 2014. Effects of glyphosate-based herbicides on survival, development, growth and sex ratios of wood frogs (*Lithobates sylvaticus*) tadpoles. I: Chronic laboratory exposures to VisionMax®. Aquatic Toxicology 154: 278–290.

Orton, F., J.A. Carr and R.D. Handy. 2006. Effects of nitrate and atrazine on larval development and sexual differentiation in the northern leopard frog *Rana pipiens*. Environmental Toxicology and Chemistry 25: 65–71.

Pavan, F.A., C.G. Samojeden, C.F. Rutkoski, A. Folador, S.P. da Fre, C. Muller, P.A. Hartmann and M.T. Hartmann. 2021. Morphological, behavioral and genotoxic effects of glyphosate and 2,4-D mixture in tadpoles of two native species of South American amphibians. Environmental Toxicology and Pharmacology 85: 103637.

Pedlowski, M.A., M.C. Canela, M.A.C. Terra and R.M.R. de Faria. 2012. Modes of pesticides utilization by Brazilian smallholders and their implications for human health and the environment. Crop Prot. 31: 113–118.

Pietrzak, D., J. Kania, G. Malina, E. Kmiecik and K. Wątor. 2019. Pesticides from the EU First and Second watch lists in the water environment. Clean-Soil Air Water 47(7): 1–13.

Pimentel, D. 2009. Environmental and economic costs of the application of pesticides primarily in the United States. pp. 89–111. *In*: Peshin, R. and A.K. Dhawan [eds.]. Integrated Pest Management: Innovation-Development Process. Springer Science Business Media B.V.

Reichert, L.M.M., D.R. de Oliveira, J.L. Papaleo, A.A.N. Valgas and G.T. Oliveira. 2022. Biochemical and body condition markers in *Rhinella icterica* tadpoles exposed to atrazine, glyphosate, and quinclorac based herbicides in ecologically relevant concentrations. Environmental Toxicology and Pharmacology 93: 103884.

Relyea, R.A. and N.E. Mills. 2001. Predator-induced stress makes the pesticide carbaryl more deadly to gray treefrog tadpoles (*Hyla versicolor*). Proceedings of the National Academy of Science USA 98: 2491–2496.

Relyea, R.A. 2004a. Growth and survival of five amphibian species exposed to combinations of pesticides. Environ. Toxicol. Chem. 23: 1737–1742.

Relyea, R.A. 2004b. Fine-tuned phenotypes: Larvae plasticity under 16 combinations of predators and competitors. Ecology 85(1): 172–179.

Relyea, R.A. 2005. The impact of insecticides and herbicides on the biodiversity and productivity of aquatic communities. Ecol. Appl. 15: 618–627.

Relyea, R.A. 2006. The effects of pesticides, pH, and predatory stress on amphibians under mesocosm conditions. Ecotoxicology 15: 503–511.

Relyea, R.A. and J. Hoverman. 2006. Assessing the ecology in ecotoxicology: A review and synthesis in freshwater systems. Ecol. Lett. 9: 1157–1171.

Relyea, R.A. and J.T. Hoverman. 2008. Interactive effects of predators and a pesticide on aquatic communities. Oikos 117: 1647–1658.

Relyea, R.A. and N. Diecks. 2008. An unforeseen chain of events: Lethal effects of pesticides at sublethal concentrations. Ecol. Appl. 18: 1728–1742.

Relyea, R.A. 2009. A cocktail of contaminants: How mixtures of pesticides at low concentrations affect aquatic communities. Oecologia 159: 363–376.

Relyea, R.A. and D.K. Jones. 2009. The toxicity of Roundup® Max to 13 species of larval amphibians. Environ. Toxicol. Chem. 28(9): 2004–2008.

Relyea, R.A. 2010. Multiple stressors and indirect food web effects of contaminants on herptofauna. pp. 475–485. *In*: Sparling, D.W., G. Linder, C.A. Bishop and S.K. Krest [eds.]. Ecotoxicology of amphibians and reptiles, 2nd edn. CRC, Boca Raton.

Relyea, R.A. 2012. New effects of Roundup® on amphibians: Predators reduce herbicide mortality; herbicides induce antipredators morphology. Ecol. Appl. 22(2): 634–647.

Richards, S.M. and R.J. Kendall. 2003. Physical effects of chlorpyrifos on two stages of *Xenopus laevis*. Journal of Toxicology, Environment and Health 66: 75–91.

Robinson, S.A., R.J. Chlebak, S.D. Young, R.L. Dalton, M.J. Gavel, R.S. Prosser, A.J. Bartlett and S.R. de Solla. 2021. Clothianidin alters leukocyte profiles and elevates measures of oxidative stress in tadpoles of the amphibian, *Rana pipiens*. Environmental Pollution 284: 117149.

Robinson, S.A., S.D. Richardson, R.L. Dalton, F. Maisonneuve, V.L. Trudeau, B.D. Pauli and S.S.Y. Lee-Jenkins. 2017. Sublethal effects on wood frogs chronically exposed to environmentally relevant concentrations of two neonicotinoid insecticides. Environmental Toxicology and Chemistry 36(4): 1101–1109.

Rohr, J.R. and P.W. Crumrine. 2005. Effects of an herbicide and an insecticide on pond community structure and processes. Ecological Applications 15(4): 1135–1147.

Rohr, J.R., T.R. Raffel, N.T. Halstead, T.A. McMahon, S.A. Johnson, R.K. Boughton and L.B. Martin. 2013. Early-life exposure to a herbicide has enduring effects on pathogen-induced mortality. Proc. R. Soc. B 280: 1–7.

Rutkoski, C.F., N. Macagnan, C. Kolcenti, G.V. Vanzetto, P.F.P.A. Hartmann and M.T. Hartmann. 2018. Lethal and sublethal effects of the herbicide atrazine in the early stages of development of *Physalaemus gracilis* (Anura: Leptodactylidae). Archives Of Environmental Contamination and Toxicology 74: 587–593.

Said, R.E.M., A.S. Said, S.A.L. Saber and B.A. ElSalkh. 2022. Biomarker Rresponses in *Sclerophrys regularis* (Anura: Bufonidae) exposed to atrazine and nitrate. Pollution 8(4): 1387–1397.

Samojeden, C.G., F.A. Pavan, C.F. Rutkoski, A. Folador, S.P. da Fre, C. Muller, P.A. Hartmann and M.T. Hartmann. 2022. Toxicity and genotoxicity of imidacloprid in the tadpoles of *Leptodactylus luctator* and *Physalaemus cuvieri* (Anura: Leptodactylidae). Scientific Reports 12: 11926.

Sayim, F. 2008. Acute toxic effects of malathion on the 21st stage larvae of the marsh frog. Turkish Journal of Zoology 32: 99–106.

Sayim, F. and U. Kaya. 2006. Effects of dimethoate on tree frog (Hyla arborea) larvae. Turk. J. Zool. 30: 261–266.

Schiesari, L., B. Grillitsch and H. Grillitsch. 2007. Biogeographic biases in research and their consequences for linking amphibian declines to pollution. Conserv. Biol. 21(2): 465–471.

Schiesari, L. and B. Grillitsch. 2011. Pesticides meet megadiversity in the expansion of biofuel crops. Front. Ecol. Environ. 9: 215–221.

Schiesari, L., A. Waichman, T. Brock, C. Adams and B. Grillitsch. 2013. Pesticide use and biodiversity conservation in the Amazonian agriculture frontier. Philos. Trans. R. Soc. Lond. B 368: 1–9.

Seleem, A.A. 2019. Teratogenicity and neurotoxicity effects induced by methomyl insecticide on the developmental stages of *Bufo arabicus*. Neurotoxicology and Teratology 72: 1–9.

Semlitsch, R.D., D.E. Scott and J.H.K. Pechmann. 1988. Time and size at metamorphosis related to adult fitness in *Ambystoma talpoideum*. Ecology 69: 184–192.

Shenoy, K., B.T. Cunningham, J.W. Renfroe and P.H. Crowley. 2009. Growth and survival of northern leopard frog (*Rana pipiens*) tadpoles exposed to two Commom Pesticides. Environmental Toxicology and Chemistry 28(7): 1469–1474.

Smith, D.C. 1987. Adult recruitment in chorus frogs: Effects of size and date at metamorphosis. Ecology 68: 344–350.

Smith, G.R. 2001. Effects of acute exposure to a commercial formulation of glyphosate on the tadpoles of two species of anurans. Bulletin of Environmental Contamination and Toxicology 67: 483–488.

Sparling, D.W., G.M. Fellers and L.L. McConnell. 2001. Pesticides and Amphibian Population Declines in California, USA. Environ. Toxicol. Chem. 20(7): 1591–1595.

Sparling, D.W. and G.M. Fellers. 2009. Toxicity of two insecticides to California, USA, anurans and its relevance to declining amphibian populations. Environmental Toxicology and Chemistry 28(8): 1696–1703,

Steiner, U.K. and J. Van Buskirk. 2008. Environmental stress and the costs of whole-organism phenotypic plasticity in larvaes. J. Evol. Biol. 21: 97–103.

Strasburg, M. and M.D. Boone. 2021. Effects of trematode parasites on Snails and Northern Leopard Frogs (*Lithobates pipiens*) in pesticide-exposed mesocosm communities. Journal of Herpetology 55(3): 229–236.

Stuart, S.N., J.S. Chanson, N.A. Cox, B.E. Young, A.S.L. Rodrigues, D.L. Fischman and R.W. Waller. 2004. Status and trends of amphibian declines and extinctions worldwide. Science 306: 1783–1786.

Sullivan, K.B. and K.M. Spence. 2003. Effects of sublethal concentrations of atrazine and nitrate on metamorphosis of the African clawed frog. Environmental Toxicology and Chemistry 22(3): 627–635.

Sumanadasa, D.M., M.R. Wijesinghe and W.D. Ratnasooriya. 2008a. Effects of diazinon on survival and growth of two amphibian larvae. Journal of the National Science Foundation of Sri Lanka 36(2): 165–169.

Sumanadasa, D.M., M.R. Wijesinghe and W.D. Ratnasooriya. 2008b. Effects of diazinon on larvae of the asian commom toad (*Bufo melanostictus*, Schneider 1799). Environmental Toxicology and Chemistry 27(11): 2320–2325.

Svartz, G., C. Aronzon and C.P. Coll. 2016. Comparative sensitivity among early life stages of the South American toad to cypermethrin-based pesticide. Environmental Science and Pollution Research 23: 2906–2913.

Teplitsky, C., H. Piha, A. Laurila and J. Merila. 2005. Common pesticide increases costs of antipredator defenses in *Rana temporaria* tadpoles. Environmental Science & Technology 39: 6079–6085.

Trachantong, W., J. Promya, S. Saenphet and K. Saenphet. 2013. Effects of atrazine herbicide on metamorphosis and gonadal development of *Hoplobatrachus rugulosus*. Maejo International Journal of Science and Technology 7: 60–71.

Trimble, M.J. and R.J. Van-Aarde. 2012. Geographical and taxonomic biases in research on biodiversity in human-modified landscapes. Ecosphere. 3(12): 1–16.

Van Buskirk, J. and R.A. Relyea. 1998. Selection for phenotypic plasticity in *Rana sylvatica* larvaes. Biol. J. Linn. Soc. 65: 301–328.

Van Buskirk, J. and S.A. McCollum. 2000. Influence of tail shape on larvae swimming performance. J. Exp. Biol. 203: 2149–2158.

Van Buskirk, J. 2009. Natural variation in morphology of larval amphibians: Phenotypic plasticity in nature? Ecol. Monogr. 79(4): 681–705.

Vanti, N.A.P. 2002. Da bibliometria à webometria: uma exploração conceitual dos mecanismos utilizados para medir o registro da informação e a difusão do conhecimento. Ci. Inf. 31(2): 152–162.

Vanzetto, G.V., J.G. Slaviero, P.F. Sturza, C.F. Rutkoski, N. Macagnan, C. Kolcenti, P.A. Hartmann, C.M. Ferreira and M.T. Hartmann. 2019. Toxic effects of pyrethroids in tadpoles of *Physalaemus gracilis* (Anura: Leptodactylidae). Ecotoxicology 28: 1105–1114.

Venturino, A. and A.M.P. D'Angelo. 2005. Biochemical targets of xenobiotics: Biomarkers in amphibian ecotoxicology. App. Herpetol. 2: 335–353.

Wang, X.G., L.M. Chang, T. Zhao, L.S. Liu, M.J. Zhang, C. Li, F. Xie, J.P. Jiang and W. Zhu. 2019. Metabolic switch in energy metabolism mediates the sublethal effects induced by glyphosate-based herbicide on tadpoles of a farmland frog *Microhyla fissipes*. Ecotoxicology and Environmental Safety 186: 109794.

Webb, C.M. and A. Crain. 2006. Effects of ecologically relevant doses of malathion on developing *Xenopus laevis* tadpoles. BIOS 77(1): 1–6.

Wei, L., W.W. Shao and Z.H. Lin. 2022. Effects of four individual pesticides and their pairwise combinations on the survival and growth of the tadpoles of two Anuran species. Pakistan Journal of Zoology 54(2): 791–800.

Westman, A.D.J., J. Elliot, K. Cheng, G. Van Aggelen and C.A. Bishop. 2010. Effects of environmentally relevant concentrations of Endosulfan, Azinphosmethyl, and Diazinon on great basin spadefoot (*Spea intermontana*) and Pacific Treefrog (*Pseudacris regilla*). Environmental Toxicology and Chemistry 29(7): 1604–1612.

Whiles, M.R., K.R. Lips, C.M. Pringle, S.S. Kilham, R.J. Bixby, R. Brenes, S. Connelly, J.C. Colon-Gaud, M. Hunte-Brown, A.D. Huryn, C. Montgomery and S. Peterson. 2006. The effects of amphibian population declines on the structure and function of neotropical stream ecosystems. Front. Ecol. Environ. 4: 27–34.

Widder, P.D. and J.R. Bidwell. 2006. Cholinesterase activity and behavior in chloppyrifos-exposed *Rana sphenocephala* Tadpoles. Environmental Toxicology and Chemistry 25(9): 2446–2454.

Widder, P.D. and J.R. Bidwell. 2008. Tadpole size, cholinesterase activity, and swim speed in four frog species after exposure to sub-lethal concentrations of chlorpyrifos. Aquatic Toxicology 88: 9–18.

Williams, B.K. and R.D. Semlitsch. 2010. Larval responses of three midwestern Anurans to chronic, low-dose exposures of four herbicides. Archives of Environment Contamination and Toxicology 58: 819–827.

Wilson, C. and C. Tisdell. 2001. Why farmers continue to use pesticides despite environmental, health and sustainability costs. Ecol. Econ. 39: 449–462.

Woodley, S.K., B.M. Mattes, E.K. Yates and R.A. Relyea. 2015. Exposure to sublethal concentrations of a pesticide or predator cues induces changes in brain architecture in larval amphibians. Oecologia 179(3): 655–665.

Xu, P. and L.D. Huang. 2017. Effects of alpha-cypermethrin enantiomers on the growth, biochemical parameters and bioaccumulation in *Rana nigromaculata* tadpoles of the anuran amphibians. Ecotoxicology and Environmental Safety 139: 431–438.

Yoon, C.S., J.H. Jin, J.H. Park, C.Y. Yeo, S.J. Kim, Y.G. Hwang, S.J. Hong and S.W. Cheong. 2008. Toxic effects of carbendazim and n-butyl isocyanate, metabolites of the fungicide benomyl, on early development in the African clawed frog, *Xenopus laevis*. Environmental Toxicology 23: 131–144.

Yu, S., M.R. Wages, Q. Cai, J.D. Maul and G.P. Cobb. 2013a. Lethal and sublethal effects of three insecticides on two developmental stages of *Xenopus laevis* and comparison with other amphibians. Environmental Toxicology and Chemistry 32(9): 2056–2064.

Yu, S., M.R. Wages, G.P. Cobb and J.D. Maul. 2013b. Effects of chlorothalonil on development and growth of amphibian embryos and larvae. Environmental Pollution 181: 329–334.

Zhang, W.J., Y.L. Lu, L.D. Huang, C. Cheng, S.S. Di, L. Chen, Z.Q. Zhou and J.L. Diao. 2018. Comparison of triadimefon and its metabolite on acute toxicity and chronic effects during the early development of *Rana nigromaculata* tadpoles. Ecotoxicology and Environmental Safety 156: 247–254.

Zhang, W.J., L. Chen, Y.Y. Xu, Y. Deng, L.Y. Zhang, Y.N. Qin, Z.K. Wang, R. ZK, Z.Q. Zhou and L.J. Diao. 2019. Amphibian (*Rana nigromaculata*) exposed to cyproconazole: Changes in growth index, behavioral endpoints, antioxidant biomarkers, thyroid and gonad development. Aquatic Toxicology 208: 62–70.

13

Ecotoxicological Impacts of Metals on Amphibian Tadpoles

Cleoni dos Santos Carvalho[1,*] and
Felipe Augusto Pinto-Vidal[2,*]

1. Introduction

Metals are natural elements that share fundamental properties such as high density, malleability, ductility, electrical and thermal conduction (Wittmann 1981). As part of nature, these elements occur naturally in aquatic environments as they are released from natural sources, for example, by weathering of geological matrices. Once in water, the levels of these metals are under physicochemical equilibrium; however, properties such as pH, hardness, redox-potential, and dissolved oxygen play an important role in their bioavailability to aquatic organisms. At limited concentrations in water, some metal species such as Fe, Cu, Se and Mn are benefic to most organisms, since they are essential for numerous physiological and biochemical processes. However, if they exceed natural background, surpassing the physiological needs, they can be toxic to the exposed organism (Aigberua et al. 2018). Usually, the natural levels of metals in aquatic environments are surpassed due to anthropogenic influence, mostly as a result of agricultural and industrial activities, in addition to domestic sewage and air emissions (Velusamy et al. 2014, Veronez et al. 2016).

Excessive concentrations of metals in water may disrupt homeostasis in resident organisms, including those under development, such as tadpoles. These metallic contaminants can be taken up by tadpoles through skin (Loumbourdis et al. 1999), gills (Alvarado and Moody 1970, Baldwin and Bentley 1980) and intestines (Lefcort et al. 1998, Dovick et al. 2020, Girotto et al. 2020), becoming potentially bioavailable to exert either physiological functions or toxicity. However, most aquatic animals, including tadpoles, are equipped with physiological and biochemical responses to deal

[1] Laboratory of Biochemistry and Microbiology (LaBioM), Program of Biotechnology and Environmental Monitoring (PPGBMA) - Federal University of São Carlos (UFSCar), Sorocaba, São Paulo, Brazil.
[2] RECETOX, Faculty of Science, Masaryk University, Kotlarska 2, 611 37 Brno, Czech Republic.
* Corresponding authors: san-cleo@ufscar.br; felipe.vidal@recetox.muni.cz

with variations above the natural background levels, for example by the expression of metal-binding proteins, such as metallothioneins (Coyle et al. 2002). Negative consequences of metal intoxication on development and metamorphosis, as well as on biochemical and physiological processes, have been shown in amphibian tadpoles (Kelepertzis et al. 2012, Fernando et al. 2016, Veronez et al. 2016, Carlsson and Tydén 2018, Boiarski et al. 2020, Zhang et al. 2020, Pinto-Vidal et al. 2021a, Pinto-Vidal et al. 2021b). These negative outcomes have been associated with the declines of amphibians' population observed over the recent decades. The rising interest on the toxicity of metals in water bodies, along with its consequences to the general health and development of amphibians, has brought important information to elucidate the potential consequences that metal pollution can have on communities and populations. Some important studies have been assembled here to provide a comprehensive framework on the knowledge of the toxic effects of diverse metallic species in water-dwelling tadpoles.

1.1 Influence of Physicochemical Properties of Water on Metal Toxicity

The risk assessment of aquatic environments should consider the evaluation of the bioavailability of metals in water and sediments, in addition to the peculiarities of each location, considering their natural backgrounds levels (Väänänen et al. 2018). The water matrix offers a set of favorable conditions for chemical reactions, but it is under a complex chemical and physicochemical equilibrium. Properties such as pH, dissolved oxygen (DO), temperature, redox potential, hardness, and the presence of organic and inorganic matters might have an impact on speciation (oxidation state, ionic form, stoichiometry, complexation, etc.) of metals, rising the complexity for understanding how they can interact with the biological phase. In turn, the toxicity of these elements can also get substantially impacted by all these properties, influencing their bioavailability in the water columns (Fig. 1) and their interaction with biological structures (Prosi 1981, Davies et al. 1993, Mastrángelo et al. 2011, de Paiva Magalhães et al. 2015, Adams et al. 2019). Efforts have been made over the last 90 years to understand how metal toxicity can vary according to different characteristics of the aquatic environment (Adams et al. 2019). Since then, the influence of physicochemical properties of water and metal toxicity have been studied in correlation (Adams et al. 2019).

Sediment is also an important matrix for the risk assessment of metals on tadpoles. From the sediment, metals can return to water due to changes in the physicochemical conditions (Tayab 1991). Another aspect of metals' mobility in water is the fact that these compounds can be absorbed not only by filtering organisms (Rainbow 2002), but also by primary producers, which are a food source for a variety of herbivores. In addition, when tadpoles die, they can be consumed by detritivore organisms, resulting in the incorporation of metals into the food chain (Mishra et al. 2008, Rowe et al. 2011). A study by Dovick et al. (2020), for example, recorded high concentrations of As and Sb in larvae and tadpoles of *Anaxyrus boreas* living in contaminated areas with metalloids. The authors reported the presence of As and Sb in the intestine and, due to redistribution resulting from metamorphosis, in other tissues as well. Furthermore, the metal retained by the soil can be present in different chemical forms: soluble, exchangeable, linked to organic matter, linked to Fe/Al, manganese oxides (MnO)/hydroxides (OH), carbonates (CO_3^{2}), phosphates (PO_4^{3-}), sulfates (SO_4^{2-}), or bound to silicates (Devesa-Rey et al. 2010). Therefore, the determination of metals in the soil and/or sediment is fundamental for understanding their bioavailability and release in the water (Karbassi et al. 2010, Peltzer et al. 2013).

As stated before, the interaction between metals, organisms, and the aquatic environment are regulated by physicochemical properties which are intrinsically related to each other. Some of these physicochemical properties will be discussed below separately, but it is important to bear in mind that all the conditions will exert influence on the metal toxicity together. This gives us an idea of the complex context behind the metal's toxicity in aquatic environments.

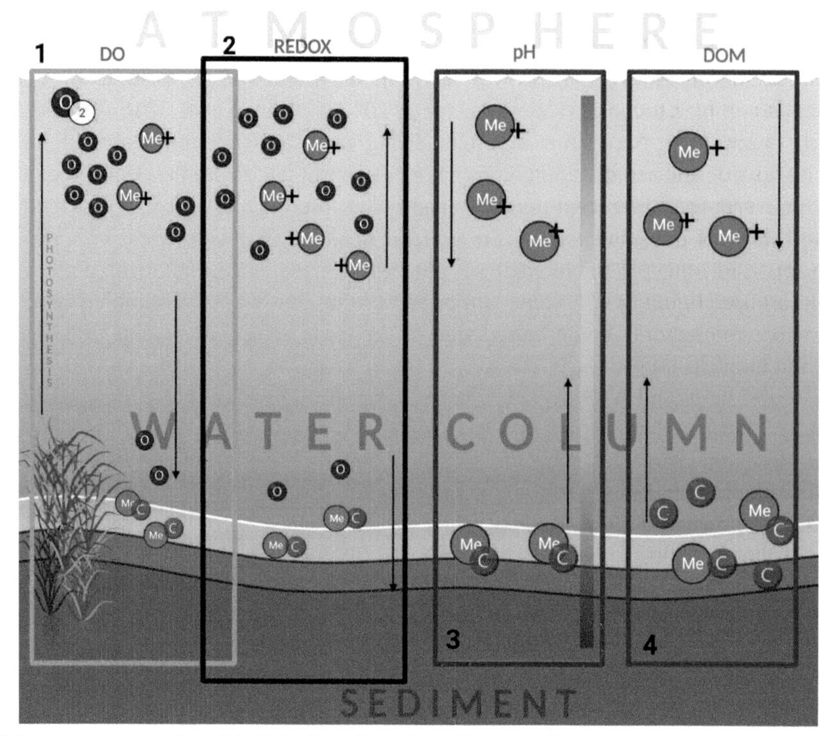

Figure 1. Schematic representation of the behaviour of metal species under different physicochemical scenarios. Dissolved Oxygen (DO); Oxygen in solution (O); Potential Redox (REDOX), pH, and Dissolved Organic Matter (DOM) Me+ = metal species in solution (increased bioavailability). MeC = metal complexed (decreased bioavailability). (↑) indicates that the increase in a given parameter will result in increase in the bioavailability of a given metal species. (↓) indicates that the decrease in the respective parameter will lead to a decrease in the metal species bioavailability. The overlapping in the DO and REDOX is due to the intrinsic relationship between these two physicochemical parameters. It is noteworthy to notice that this scheme represents the parameters considered isolated, and the interaction between them may lead to different levels of metals release. Created with BioRender.com.

1.1.1 Water pH

The pH of water indicates the balance between H^+ and OH^- at a given temperature, and its value is represented by a scale ranging from 1 (acidic) to 14 (basic). The balance between H^+ and OH^- affects metals' toxicity, since the solubility of these compounds is influenced by the pH of the medium. In general, the more soluble a metal species , higher its toxicity as it can be more bioavailable (de Paiva Magalhães et al. 2015, Väänänen et al. 2018). This explains why, in general, acid environments are related with greater toxicity of metallic contaminants: under low pH ranges, the dissociation of metals (cations) is favored in the water column, increasing its solubility and potential toxicity.

Although metal's bioavailability can be influenced by pH, and, in most cases, toxicity decreases as the pH increases, biological responses must also be considered (Wang et al. 2016a). Low pH ranges will also induce faster release of metals from the sediment to the water column (Playle 1998, Li et al. 2013), which could increase the natural levels in a given water environment. When these bioavailable metals interact with aquatic fauna, another variant needs to be considered: the excess of H^+, at low pH, competes with cationic metals at uptake surfaces, which may protect the animal from absorbing metallic species (de Paiva Magalhães et al. 2015). In the opposite way, the competition between H^+ and the cationic metals may be less significant at high pH, which implies that even less bioavailable metals can interact with the absorption sites. Alkaline or basic pH is also related to the presence of bicarbonate - HCO_3^- and carbonate - CO_3^{-2} - in water (Davies et al. 1993). This condition will lead to precipitation of metallic species as oxides or hydroxides, which makes them less bioavailable in the water column (de Paiva Magalhães et al. 2015). However, precipitates accumulate in the sediment, making metals available to benthonic tadpoles through ingestion (Lefcort et al. 1998).

The pH itself can induce changes in physiological, developmental and genotoxic profiles of tadpoles, as shown by Lu et al. (2021), which exposed tadpoles of *Rana zhenhaiensis* to Cd under different pH ranges (low: 5, neutral: 7.23, and high: 9). In the study, the hepatic content of Cd, Zn, Cu, and Fe was not altered when tadpoles were exposed at low (5.0), neutral (7.2), or high (9,0) pH. Mg levels seemed to be marginally decreased in low and high pH groups compared to the animals from neutral pH group. The survival and metamorphic rates were decreased for animals from both low and high pH treatments and a higher incidence of abnormalities was reported in the low pH group. Higher body mass and snout-vent length in the low and high pH groups were observed, when compared to the neutral groups. Moreover, swimming speed and jumping distance were also impaired in both treatments, in addition to a higher percentage of abnormalities in the peripheral blood cells and a general decrease in the antioxidant defense system. Another study performed by Marques et al. (2013) evaluated the effects of *in situ* exposure of *Pelophylax perezi* tadpoles in 3 effluent ponds from a deactivated uranium mine. These 3 lagoons presented increasing levels of metal concentrations, and the death rate of the tadpoles was higher in the effluent with a higher concentration of metals, despite metal not accumulating in the animals. The high mortality was attributed to the extremely low water pH ($\approx$ 3.77), which reduced the accumulation of metal probably due to the damage to the integument, affecting the ions uptake.

1.1.2 Water Hardness

Water hardness refers to the amount of dissolved Ca^{2+} and Mg^{2+} in the system (Davies et al. 1993, USGS 2020b). The levels of hardness are usually reported as mg/L of equivalent calcium carbonate ($CaCO_3$) (Pascoe et al. 1986). The variations in hardness generally do not affect the bioavailability of metals in water, but the increase in water hardness is often related to a decrease in metal toxicity (Ebrahimpour et al. 2010, Pascoe et al. 1986) (Table 1). This happens because Ca^{2+} and Mg^{2+} can compete for the divalent binding sites of the target organism, which may block the uptake of the metals through the cell's transporters (de Paiva Magalhães et al. 2015, Ebrahimpour et al. 2010). The protective effects of high Ca^{2+} levels have been associated with the decreased Cd toxicity in *Rhinella arenarum* after 24 and 96 h of exposure (Mastrángelo et al. 2011).

A study published by Home and Dunson (1995) verified the interactions and the influence of pH (4.5 and 5.5) and metals (Al, Cu, Fe, Pb, and Zn) along with 2 different water hardness (low hardness: 1700 and 470 mg/L of Ca^{2+} and Mg^{2+}, respectively; high hardness: 3300 and 1000 mg/L of Ca^{2+} and Mg^{2+}, respectively) on *Rana sylvatica* tadpoles. The concentrations of metals used were environmentally relevant and caused varied effects on survival after acute (7 days) and chronic (28 days) exposure. For example, acute exposure to Al and Cu reduced survival, but it was later increased with increasing water hardness. Chronic exposure to Al and Cu at pH 5.5 and lower water hardness also reduced the animal survival. In fact, pH and hardness are important determinants of metal toxicity in amphibians; their sensitivity also depends on their stage of development, as shown in Table 1. Even though aquatic animals can tolerate different levels of hardness, mortality and morbidity can be observed under sudden changes in these parameters (Diggs and Parker 2009, USGS 2020).

1.1.3 Dissolved Oxygen (DO)

Tadpoles present an interesting plasticity regarding their respiratory behavior; they can uptake oxygen dissolved in water (dissolved oxygen, DO) by their surface structures, such as skin, gills, oral surfaces (buccal and pharyngeal, including gill filters), but can also capture atmospheric air when the lungs are developing (Wassersug and Seibert 1975). Streams have higher levels of DO compared to stagnant waters, in which most tadpoles develop. The incidence of sunlight (which increases the photosynthesis of aquatic plants) will also increase DO levels, whereas the increase in the temperature leads to DO decreases (USGS 2020b). DO in a range of 7–9 mg/L may increase

Table 1. Mean Lethal Concentrations (LC50) for amphibians at the larval stage.

	Mean values for LC 50 mg/L									
Species		12 hr	24 hr	48 hr	72 hr	96 hr	pH range	Hardness (mg/L)	DO (mg/L)	
D. melanostictus	Aluminium (Al)		7.7	3.4	2.2	1.9	6.5 ± 1	18.6 ± 1.8	6.3 ± 0.1	Shuhaimi-Othman et al. (2012)
R. hexadactyla	Arsenic (As)		0 4	0.3	0.3	0.25	6.2 – 6.7	13 – 80	6.2 – 7.0	Khangarot et al. (1985)
B. metanosticus	Cadmium (Cd)	22	20	12		8.2	7.1 – 7.6	165 – 215	5.8 – 7.8	Khangarot and Ray (1987)
D. melanostictus			1	1	0.5	0.3	6.5 ± 1	18.6 ± 1.8	6.3 ± 0.1	Shuhaimi-Othman et al. (2012)
*M. ornata**			3	2.5	2	1.6	6.86 – 6.94	142 – 145	8.2 – 8.4	Jayaprakash and Madhyastha (1987)
*M. ornata***			3	2.7	2.2	1.8	6.86 – 6.94	142 – 145	8.2 – 8.4	Jayaprakash and Madhyastha (1987)
B. metanosticus	Chromium (Cr)	74	58	53.4		49.3	7.1 – 7.6	165 – 215	5.8 – 7.8	Khangarot and Ray (1987)
R. hexadactyla			75	51	46.8	42.6	6.2 – 6.7	13 – 80	6.2 – 7.0	Khangarot et al. (1985)
R. hexadactyla	Cobalt (Co)		30	26	26	17.6	6.2 – 6.7	13 – 80	6.2 – 7.0	Khangarot et al. (1985)
B. metanosticus	Copper (Cu)	2	0.8	0.45		0.32	7.1 – 7.6	165 – 215	5.8 – 7.8	Khangarot and Ray (1987)
D. melanostictus			0.1	0.05	0.04	0.03	6.5 ± 1	18.6 ± 1.8	6.3 ± 0.1	Shuhaimi-Othman et al. (2012)
*M. ornata***			5.6	5.3	5.1	5.04	6.86 – 6.94	142 – 145	8.2 – 8.4	Jayaprakash and Madhyastha (1987)
*M. ornata***			6.0	5.7	5.5	5.4	6.86 – 6.94	142 – 145	8.2 – 8.4	Jayaprakash and Madhyastha (1987)
R. hexadactyla			0.4	0.04	0.04	0.04	6.2 – 6.7	13 – 80	6.2 – 7.0	Khangarot et al. (1985)
D. melanostictus	Iron (Fe)		1.1	0.6	0.5	0.4	6.5 ± 1	18.6 ± 1.8	6.3 ± 0.1	Shuhaimi-Othman et al. (2012)
R. hexadactyla			25	22.4	19.7	17.6	6.2 – 6.7	13 – 80	6.2 – 7.0	Khangarot et al. (1985)
D. melanostictus	Lead (Pb)		8.2	3.5	2.2	1.5	6.5 ± 1	18.6 ± 1.8	6.3 ± 0.1	Shuhaimi-Othman et al. (2012)
R. hexadactyla			100	66.6	41.3	33.3	6.2 – 6.7	13 – 80	6.2 – 7.0	Khangarot et al. (1985)

Table 1 contd. ...

...Table 1 contd.

	Mean values for LC 50 mg/L									
Species		12 hr	24 hr	48 hr	72 hr	96 hr	pH range	Hardness (mg/L)	DO (mg/L)	
D. melanostictus	Manganese (Mn)		138	92	53	39	6.5 ± 1	18.6 ± 1.8	6.3 ± 0.1	Shuhaimi-Othman et al. (2012)
M. ornata			17	16	15.6	15	6.86 – 6.94	142 – 145	8.2 – 8.4	Jayaprakash and Madhyastha (1987)
M. ornata			18	16.5	15.7	14.3	6.86 – 6.94	142 – 145	8.2 – 8.4	Jayaprakash and Madhyastha (1987)
B. metanosticus	Mercury (Hg)	0.07	0.1	0.05		0.05	7.1 – 7.6	165 – 215	5.8 – 7.8	Khangarot and Ray (1987)
*M. ornata**			2.0	1.7	1.2	1.12	6.86 – 6.94	142 – 145	8.2 – 8.4	Jayaprakash and Madhyastha (1987)
*M. ornata***			2.4	2.1	1.7	1.43	6.86 – 6.94	142 – 145	8.2 – 8.4	Jayaprakash and Madhyastha (1987)
R. hexadactyla			0.8	0.12	0.07	0.05	6.2 – 6.7	13 – 80	6.2 – 7.0	Khangarot et al. (1985)
B. metanosticus	Nickel (Ni)	61.4	53	34.3		25.3	7.1 – 7.6	165 – 215	5.8 – 7.8	Khangarot and Ray (1987)
D. melanostictus			30	26	14	8.8	6.5 ± 1	18.6 ± 1.8	6.3 ± 0.1	Shuhaimi-Othman et al. (2012)
B. metanosticus	Silver (Ag)	0.01	0.003	0.006		0.004	7.1 – 7.6	165 – 215	5.8 – 7.8	Khangarot and Ray (1987)
R. hexadactyla			0.03	0.024	0.03	0.026	6.2 – 6.7	13 – 80	6.2 – 7.0	Khangarot et al. (1985)
B. metanosticus	Zinc (Zn)	50	48	25.65		19.9	7.1 – 7.6	165 — 215	5.8 – 7.8	Khangarot and Ray (1987)
D. melanostictus			18	14	9.2	4.2	6.5 ± 1	18.6 ± 1.8	6.3 ± 0.1	Shuhaimi-Othman et al. (2012)
*M. ornata**			24	23.4	23.0	22.4	6.8 — 6.9	142 – 145	8.2 – 8.4	Jayaprakash and Madhyastha (1987)
*M. ornata***			25	24.4	23.5	23.1	6.8 – 6.9	142 – 145	8.2 – 8.4	Jayaprakash and Madhyastha (1987)
R. hexadactyla			8	3.7	3.2	2.1	6.2 – 6.7	13 – 80	6.2 – 7.0	Khangarot et al. (1985)

* (1week old) ** (4 weeks old)

the release of metals from the sediment, as it can influence the oxidation rate of metals that are bound to organic matter. The same trend has been observed regarding the mobility of metals from the sediment to the water column. In contrast, low oxygen concentration, i.e., between < 1 and 3 mg/L, might increase the adsorption of metals to the sediment (Kang et al. 2019, Li et al. 2013). According, the toxicity of zinc sulphate to fish (*Salmo gairdnerii*) is increased in environments with higher DO (Lloyd 1960).

To the best of our knowledge, there are no studies on toxic effects of metals associated with different levels of DO in tadpoles. However, considering that tadpoles hold a branchial performance comparable to teleost fish (Taylor et al. 2010), similar effects of metal toxicity combined with DO levels would be expected for tadpoles. This is especially relevant when comparing species tolerant or not to hypoxia. Hypoxia-tolerant tadpoles can spend more time foraging in sediment at the bottom of the aquatic environment, where DO concentrations are usually lower. This behavior is also advantageous to protect tadpoles against hypoxia-intolerant predators, as shown by McIntyre and McCollum (2000) in bullfrog tadpoles. However, these authors also showed that the presence of predators increased the swimming activity of tadpoles, which may favor the uptake of metals due to the increased ventilation as a consequence of higher energetic demands.

1.1.4 Redox Potential

Reduction-oxidation reactions involves the flow of electrons from reducing (reductant) to an oxidizing (oxidant) agent (DeLaune and Reddy 2005, Søndergaard 2009). These two forms represent the redox couple (Søndergaard 2009). The propension of redox reaction to water is influenced by the pH and oxygen levels in the solution (Søndergaard 2009). Once inside the body, metals can interact with molecules and participate in oxidation-reduction reactions (Ossana et al. 2013, Veronez et al. 2016, Pérez-Iglesias et al. 2019). The redox potential has been shown to influence the bioavailability of metal species in aquatic environments, such as Cr, Cu, Hg, Mn, and Fe, that under oxidizing conditions and pH > 5.5 produce low soluble oxides and hydroxides. When they are under reducing circumstances, e.g., under low concentrations of DO and pH < 7, the solubility increases, especially due to the formation of soluble cations (de Paiva Magalhães et al. 2015). Fe^{3+} is also reduced to Fe^{2+}, increasing Fe^{2+} levels in water and favoring the release of metals bound to the Fe-Mn oxyhydroxides in sediments (Kang et al. 2019).

1.1.5 Organic and Inorganic Ligands Matters

Different organic and inorganic ligand compounds present in water can bind to metallic ions forming soluble or insoluble complexes (Davies et al. 1993). Inorganic compounds, such as oxygen, nitrate, nitrite, sulfate, arsenates, phosphates, etc., as well as organic compounds, such as ethylenediaminetetraacetic acid (EDTA), hydroxyl, carboxylate, cyano, amino groups, humic substances and dissolved organic carbon (DOC), may offer binding sites to metals, resulting in changes in their availabilities and toxicities (Benedetti et al. 1996, Tao et al. 2001, DeLaune and Reddy 2005, Baken et al. 2011, de Paiva Magalhães et al. 2015). When DOC is present in the medium, more Cu is needed to exert toxicity in the exposed animals, which means that Cu complexation with DOC makes it non-bioavailable (Di-Toro et al. 2001). Organic matter is well known to be a ligand to metals in water, decreasing their bioavailability (Di-Toro et al. 2001) and toxicity (Freda et al. 1990, Mastrángelo et al. 2011). The bioavailability of trace metals is strongly influenced by the presence of dissolved organic matter (DOM). Once DOM presents affinity for binding to trace metals, it will decrease their mobility in aquatic medium, possibly decreasing the bioavailability of these metallic elements in the water column (Baken et al. 2011). However, once the metallic species are bound to organic matter, they may present increased lipophilicity, which may favor a passive diffusion across biological membranes, potentially increasing chemical bioavailability (de Paiva Magalhães et al. 2015).

Other classes of organic matter, namely humic and fulvic substances, are rendered from decomposition of organic matter and can impact the bioavailability of metals. The interaction between metal and humic substances is influenced by the load and chemical heterogeneity of the humic material. It can occur through adsorption, cation exchange reactions, and complexation (Bezerra et al. 2009). Complexation reactions are considered more important as they affect the geochemistry of metal ions, changing their solubility, charge, and redox potential; it can potentially influence the bioavailability, transport, and migration, with consequent effects on metal toxicity. For example, Cd toxicity was decreased in pre-metamorphic tadpoles of *R. arenarum* when humic substances were present in water with variable hardness (54–407 mg as $CaCO_3$) and Ca excess (3.5 mM as $CaSO_4$), after 96 h (Mastrángelo et al. 2011).

2. Mechanisms of Metal Absorption by Tadpoles

Despite many studies on metals and their effects, information on the absorption mechanisms in tadpoles is still scarce when compared to other taxa. Previous studies suggested that metals and metal-containing species enter cells by ionic and/or molecular mimicry using transporters of essential ions and/or organic molecules (Ballatori 2002, Bridges and Zalups 2005). In ionic mimicry, the cationic metal mimics an essential element or a cationic species of an element at the carrier site, whereas in molecular mimicry, the metallic ion bind to an organic molecule to form a type of organometallic compound that would act as a functional or structural mimic of essential molecules at the transport sites. In the review by Bridges and Zalups (2005), they classified molecular and ionic mimics as structural (similar in size and shape to another element or molecule) and/or functional (elicit the same effect as the native element or molecule).

Several carrier proteins, such as amino acid and organic anion transporters, have been implicated in the transport of certain toxic metals (Ballatori 2002, Bridges and Zalups 2005), while cationic forms of some metals can mimic anionic complexes (i.e., oxyanions) as a form of molecular mimicry. In ionic mimicry, cationic metal species can serve as a structural and/or functional homolog or mimic of another element (usually essential) at the site of a carrier protein, ion channel, enzyme, structural protein, transcription factor and/or protein metal binding sites. For example, Cd^{2+} and Hg^{2+} can use ion channels (such as Ca^{2+}, Fe^{2+} and Zn^{2+} channels) and some membrane transporters to access mammalian target cells, competing with ligands for cell-binding sites (Tao et al. 2001, Zalups and Ahmad 2003, Farina et al. 2011). In addition, experiments have shown that Cd^{2+} and Hg^{2+} may form covalent complexes with biomolecules containing sulfhydryl groups, such as glutathione (GSH), cysteine (Cys), and homocysteine (Hcy) (Rabenstein et al. 1983, Zalups and Ahmad 2004, Bridges and Zalups 2005, Farina et al. 2011), which means that they can potentially interact with these proteins and their related roles in biochemical process, such as in antioxidant defense responses.

2.1 Oral Apparatus, Gill, and Skin as uptake Sites in Tadpoles

Chemical uptake refers to the migration of compounds across biological membranes from the environment to the internal system of the exposed organism. The amount of chemicals available for uptake or absorption is roughly defined as bioavailable fraction. Thus, once a certain amount of metal is bioavailable in the water, it can interact with the organism and be absorbed into the body, where it can exert either physiological or toxic role. Regarding aquatic animals, the gills are accepted as one of the main sites for the uptake of bioavailable metals. This respiratory organ represents an important target for the interaction between the animal and the metallic species in its surroundings.

Gill membranes are negatively charged due to the presence of phospholipids that act as a Lewis base (electron pair-donor), while the metals in solutions (cations) behave as Lewis acid (electron pair-acceptor) (Wittmann 1981, Pagenkopf 1983). Due to these inherently shared properties, the bioavailable metals will be inherently attracted by the organisms' membranes, leading to a chemical interaction between these two phases. In general, excessive levels of metals in the medium can interact with the gill's membrane causing mechanical damage and mucus accumulation in the branchial arches. Low concentrations, in turn, may impair ionoregulatory processes due to competition between the metals in the solution and the other elements needed for keeping the homeostasis (Lloyd 1960, Playle 1998) (Fig. 2). It is worth mentioning that, as in fish, tadpoles' gills are the most important site for the ionic regulation (Alvarado and Moody 1970, Mcdonald et al. 1984). The branchial system is accountable for the active transport of Na^+ and Cl^- (Alvarado and Moody 1970), nitrogenous waste excretion (exchanges Na^+ for H^+/NH_4^+ and Cl^- for OH^- or HCO_3^-), acid-base regulation (Dietz and Alvarado 1974, Mcdonald et al. 1984), and homeostasis of Ca^{2+}, which is responsible for 70% of the total calcium influx in tadpoles (Baldwin and Bentley 1980). This ionic regulation exerted by gills might alter the pH of the branchial surface by the movement of NH_3, NH_4^+, CO_2 and HCO_3^-. Once the pH in the gill is different from the surrounding water, the speciation of the metal interaction with this organ will change, potentially increasing its bioavailability/toxicity, as well as the speciation of the metal within this microenvironment. In addition, mucus produced in response to the presence of these contaminants comprise several types of glycoproteins acting as an ion exchange system, which might affect kinetic parameters such as accumulation and/or elimination of toxic metals (Tao et al. 2001, Guardiola et al. 2015, Dubaissi et al. 2018, Reverter et al. 2018).

Dubaissi et al. (2018) described the tissue architecture of the epidermis of *Xenopus tropicalis* tadpoles as similar to the respiratory mucosal epithelium of mammals, presenting as the main structural barrier component a mucin glycoprotein with the sequence, domain organization, and structural properties similar to the gel forming mucins of human. Thus, the interaction of supraphysiologic levels of metals with the biological phase may impair both ionoregulatory and barrier functions (McDonald and Wood 1993).

Skin is also pointed out as an important route for metal uptake in tadpoles (Wei et al. 2015, Dubaissi et al. 2018). Walentek et al. (2014) showed that the embryonic skin of *Xenopus* tadpoles constitutes of mucus-secreting cells, multiciliary cells (MCCs) and a third cell type, the ion-secreting cells (ISCs). Thus, it can be assumed that metals present in the aquatic environment cross the tadpole skin and gill barriers through contact with mucus proteins and/or through ion exchange channels.

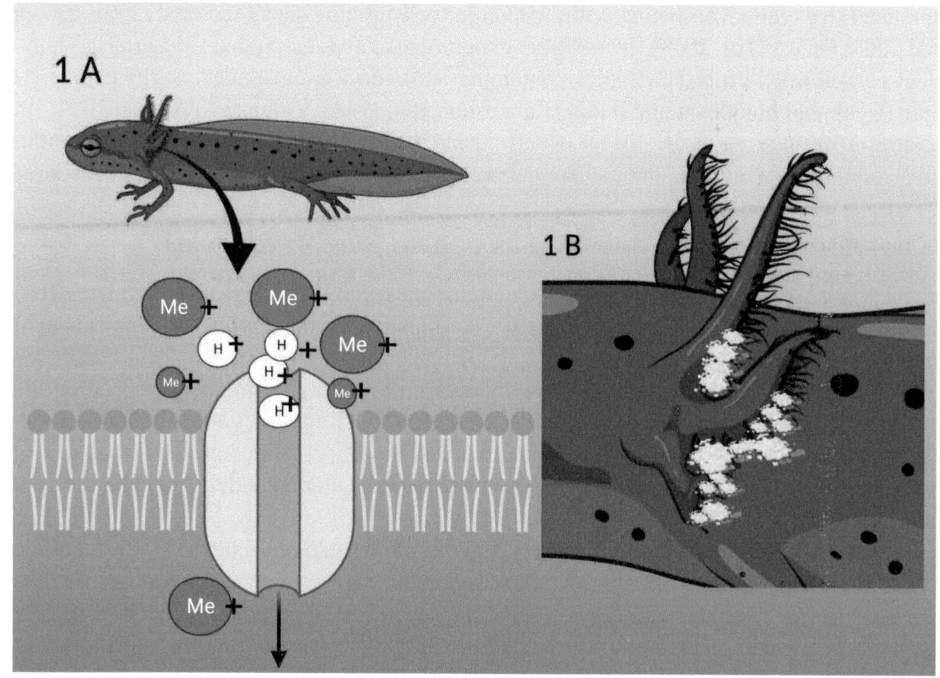

Figure 2. Schematic representation of the mode of action of metals on tadpoles. 1A. Competition of bioavailable metals with H+ in uptake sites. 1B. Accumulation of mucus on the gills triggered by high levels of metals. Created with BioRender.com.

The oral route for exposure to metals is also worth considering. Girotto et al. (2020) exposed tadpoles of *L. catesbeianus* to different concentrations of mining waste (10–100% dilutions) from the Fundão dam (Mariana - MG, Brazil), containing Zn, Fe, Mn, Pb, Cd and Al. The authors reported the presence of tailings in the mouth and intestine of the tadpoles, indicating the ingestion of metals by the organisms. Depending on the form of exposure, metals such as Cd^{2+}, Hg^{2+} and Pb^{2+}, can be absorbed by the intestine and reach the liver, where they are absorbed from sinusoidal blood by the hepatocytes. The uptake of these metals by erythrocytes can be mediated by an anion exchanger; however, the mechanisms of how this happens still needs to be clarified. As previously described, these metals can enter through mimicry and/or can bind to proteins that are absorbed in the hepatocytes, kidney, and intestine by receptor-mediated endocytosis (Bridges and Zalups 2005). Metals (such as Cd, Hg and Pb) can be taken up by different transporters, including the divalent metal transporter 1 (DMT1) (Trinder et al. 2000, Ohta and Ohba 2020). The transport of these metals seems to involve an ionic mimicry mechanism (both passive and active mechanisms), whereby Cd^{2+}, Hg^{2+} and Pb^{2+} mimic the ferrous form of Fe (Fe^{2+}) and zinc (Zn^{2+}) to access the cytosolic compartment of the hepatocytes (Bridges and Zalups 2005). These mentioned studies are in mammals, but Okubo et al. (2003) demonstrated a Cd^{2+} transporter by DMT1 in *X. laevis* oocytes. The oocytes were microinjected with DMR1 encoding mRNA and, subsequently, exposed to $CdCl_2$. The result was that the extracellular medium promoted an internal Cd^{2+} flow.

Metallothioneins (MTs) are low-weight, thiol-rich proteins, known to be specific markers for metal exposure (Loumbourdis et al. 2007, Monserrat et al. 2007). For example, MT concentration increased in embryos of *X. laevis* exposed to Cu, Cd and Pb (Yologlu and Ozmen 2015) and in tadpoles of *L. catesbeianus* (mainly in liver and muscle) exposed to Zn, Cu and Cd at concentrations considered safe by Brazilian legislation (BRAZIL 2005) (Carvalho et al. 2017). These studies showed a positive correlation between MT and metal-contaminated environments, which corroborate with the uptake of metals by the organism. In addition, complexes formed by metal-MT can be transported to cells via receptor-mediated endocytosis (Bridges and Zalups 2005).

The CdMT complex (or other metal-metallothionein - MT complex) can be released from the liver to the kidneys by hepatocellular necrosis and/or apoptosis. Afterwards, the complex will be filtered by the glomeruli in the kidneys and absorbed by the epithelial cells of the proximal tubule (Dudley et al. 1985, Webb 1986, Bridges and Zalups 2005); it can also be transported from hepatocytes back to sinusoidal blood (Bridges and Zalups 2005). As previously described, several transport proteins in the sinusoidal plasma membrane may be involved, such as the organic anion transport polypeptides, organic cation transporters, metal transport proteins and amino acid or peptide transporters. In addition, a fraction of Cd^{2+} conjugated to reduced glutathione (GSH) (such as GS-Cd-SG) in hepatocytes can be secreted into the bile and later delivered to the duodenum by a membrane transporter, so it is excreted together with the feces (Cherian and Vostal 1977, Leslie et al. 2001). Thus, the GS-Cd-SG complex, due to its structural similarity, can act as a GSSG mimic. According to the studies by Heijn et al. (1997), Keppler et al. (1998), and Leier et al. (1996), which were conducted with mammals, this transporter is located in the canalicular membrane and performs the transport of GSH, glutathione disulfide (GSSG) and GSH S-conjugates out of hepatocytes. The tripeptide glutathione (g-L-glutamyl-L-cysteinyl-glycine) exists in the body as reduced (GSH) and oxidized disulfide (GSSG) forms. Both are important in regulating the redox state of cells and play a role in many biological processes (Anderson 1998).

Tadpoles' kidneys are also one of the organs that can be affected following exposure to metals, especially Cr (Monteiro et al. 2018) and Cd (Carvalho et al. 2020), which may cause injury and renal dysfunction. Cd cannot undergo biodegradation or biotransformation as it has a prolonged biological half-life (15–30 years) and low excretion rate, which is why it accumulates mainly in the liver and kidneys of the animals (Medina et al. 2012). Cd^{2+} was reported in the epithelial cells lining the proximal tubule of the rabbit kidney (Robinson et al. 1993) and in cultured proximal tubular cells (Endo 2002). The mentioned studies showed that Cd^{2+} is taken up in a cationic form by these cells in complexed form (GS-Cd-SG or Cys-S-Cd-SCys), as previously mentioned.

These complexes are formed due to the high affinity of Cd^{2+} for biomolecules containing thiol groups, and are mediated by the action of γ-glutamyltransferase (GGT) (that cleaves the g-glutamylcysteine bond in GSH molecules) and cysteinylglycinase (which degrades the cysteinylglycine derivative formed) on the luminal plasma membrane of proximal tubular cells.

A study carried out in rats by Zalups (2000) showed that GSH or Cys injected (intravenous) simultaneously with Cd^{2+} increased renal Cd^{2+} uptake, demonstrating that there was a luminal and basolateral mechanism involved in the renal uptake of Cd^{2+}. Monteiro et al. (2018) also observed more intense effects of Cr on tadpole's kidneys. These authors found that potassium dichromate has more affinity for renal epithelial cells than for liver cells. The study showed that the glomerular filtration of this metal followed by its resorption by the proximal tubules leads to its accumulation in the renal parenchyma. The presence of Cr allows it to interact with kidney cells contributing to changes that can be irreversible. To the best of our knowledge, to date, no study has identified a specific mechanism for tadpoles, however, we can infer that the mechanisms described for mammals, such as Cys-S-Cd-S-Cys and Cys-S-Hg-S-Cys can also occur in tadpoles exposed to other metals. As reviewed by Bridges and Zalups (2005), the carriers used by metals are structurally similar and, therefore, can also mediate the absorption of other metals by similar conjugates. Thus, in the same way as for Cd, Hg, Cu, and Zn (or other metals), they may be bound to MT and/or other metal-binding proteins, resulting in detoxification, tolerance, and homeostasis.

3. Metal Effects and Biomarkers

3.1 Bioaccumulation

Ionic homeostasis is maintained under very controlled conditions by an efficient transport system, which keeps the intracellular concentrations of many metals lower than the extracellular concentration. However, when the net balance between the metal influx and efflux results in increased uptake and decreased excretion, the metals will accumulate in the exposed animal. Bioaccumulation is an important parameter because this is one of the required scenarios for the first signs of toxicity (Wang and Tan 2019). Tadpoles accumulate metals in different organs, progressively from their diet and by absorbing them from water and sediments. Studies have shown that sediments and water contaminated with metals exert negative effects on amphibians at several developmental stages of life. Berzins and Bundy (2002) indicated progressive lead (Pb) absorption in tadpoles (*Xenopus laevis*) as exposure period and concentration increased. Similar results were observed in *Rana sphenocephala* tadpoles exposed to several concentrations of Pb in sediment and water (Sparling et al. 2006). Boczulak et al. (2017) found that more developed and massive tadpoles (*Rana sylvatica*) in the North American Prairie Pothole region had higher concentrations of Hg than the less developed ones (*Pseudacris maculata*). The study demonstrated that Hg biomagnification was influenced by the life history, stage of development, and dietary changes as the body approaches metamorphosis. Another study (Kelepertzis et al. 2012) was carried out to correlate the presence of Pb, Zn, Cu and Cd in water and sediments with their levels detected in tadpole tissues. In fact, the accumulation of metals such as Cd and Pb (Perez-Coll and Herkovits 1990, 1996, Herkovits and Perez-Coll 1993, Berzins and Bundy 2002, Dobrovoljc et al. 2003), Cu (Chen et al. 2007), Ni (Klemish et al. 2018), Cu, Pb, Zn, Mn, Cd, Ni and Cr (Kelepertzis et al. 2012, Girotto et al. 2020), iron ore, Fe and Mn (Veronez et al. 2016), Hg (Edwards et al. 2013, Boczulak et al. 2017), Zn, Cu and Cd (Carvalho et al. 2017) and V, Mo and Cd (Krohn et al. 2020) have been documented in tissues and organs of embryos and tadpoles.

The study of Lanctôt et al. (2016) with tadpoles of the striped marsh frog (*Limnodynastes peronii*) exposed (four weeks) to different concentrations of wastewater from coal mines in central Queensland, Australia, reported high levels of Se, Co, Mn, and As in the tail and liver. This study showed that hepatic tissue accumulated 8–9 times higher concentrations of Co, Mn and Se than tail tissue. This approach is essential to investigate possible relationship between metal bioaccumulation and morphophysiological effects during development, as well as the effects related to individual and population performance. In another example, tadpoles of *L. peronii* were exposed to Cd (0.5 μg/L), Se (0.7 μg/L) and Zn (25 μg/L) alone and in mixture, followed by depuration and feeding in aquaria with clean water. Tadpoles showed a higher bioconcentration factor and retention for Se, and there was greater Cd accumulation when the metals were combined, suggesting antagonistic effects of Se and/or Zn against Cd uptake. The results also showed that tadpoles fed in aquaria containing metal solution accumulated more Cd and Zn compared to tadpoles fed in clean water, indicating that the presence of food particles was an important factor influencing the absorption of metals (Lanctôt et al. 2017). Rowe et al. (2011) verified accumulation and depuration in *Hyla versicolor* larvae and tadpoles exposed to Se (1.0, 7.5 and 32.7 μg/g dw) and V (3.0, 132.1 and 485.7 μ/g dw) through diet. Accumulation of Se was higher in larvae, followed by premetamorphs and metamorphs, whereas V accumulation was lower in larvae and metamorphs compared to premetamorphs. No effects were observed in development, growth, and metabolism, suggesting that there was depuration in the presence of V in the animals during metamorphosis.

The element As is highly toxic according to the Agency for Toxic Substances and Disease Registry (ATSDR 2003). When combined with anionic species of oxygen, chlorine, or sulfur, it is referred to as inorganic As. Organic As, the least toxic form, is formed by combining As ions with carbon and hydrogen (Barra et al. 2000). The mechanisms of As absorption and elimination in cells can occur as described for other metals, i.e., absorption through co-transport of a GSH S-conjugate and the elimination of this conjugate by multidrug resistance proteins (MRPs) to outside the cells. MRPs act in detoxification and transport of a vast number of substrates. However, the molecular mimicry in the As transport mediated by MRP is not yet clear (Bridges and Zalups 2005).

Bioaccumulation of As has also been reported in tadpoles. The content of As in the tissues of frog embryos (*Silurana tropicalis*) increased when its concentration (0.5–1.0 mg/L) augmented in water and the prevalent metal species observed was the organic trimethylarsine oxide (TMAO) (Koch et al. 2015). The study also verified the expression of the DNA methyltransferases AS3MT and DNMT1. AS3MT, which is an enzyme that acts on As metabolism by facilitating the addition of a methyl group to the trivalent metabolites of As, was decreased in animals exposed to the lowest As concentration. This enzyme maintains the epigenomic integrity and, therefore, the methylation patterns of genomic DNA during DNA replication. An important study on As speciation was carried out by Dovick et al. (2020) in larvae and tadpoles of *Anaxyrus boreas* in an environment contaminated by high concentrations of As and Sb. This study suggested that these animals have the ability to biomethylate inorganic As (and perhaps Sb) due to the presence of organic forms such as arsenic TMAO, monomethylsonic acid (MMA), and dimethylsinic acid (DMA) in tadpoles' tissues. In mammals, inorganic As methylation (detoxification) occurs in the kidneys, reducing the compound's affinity for the tissue (Barra et al. 2000). These results demonstrate the biological and molecular mechanisms involved in the speciation of As in tadpoles and the ability they have to tolerate metals that are detrimental to their development.

3.2 Nanoecotoxicology

Research in the nanoecotoxicology field has gained ground in the face of the potential threat that nanoparticles pose to the environment. However, few studies have reported the effects of metal nanoparticles on tadpoles. Birhanli et al. (2014) exposed *X. laevis* embryos and tadpoles to nano-TiO_2 (titanium dioxide) (5 – 320 mg/L, 96 h) and showed that embryos were more susceptible to the nanoparticles than tadpoles. Another study by Al Mahrouqi et al. (2018) in *Sclerophrys arabica* exposed tadpoles to nano-TiO_2 and zinc oxide (Zn-NPs) (1, 10 and 100 mg/L) in 3 different conditions: directly to water, in food or in a combined way (NPs and food). The results showed that tadpoles with lower body mass were those exposed only to NPs in water, while those exposed to Ti-NPs+food had higher body mass. The study suggests that the exposure route affected tadpoles' growth, with the possibility of synergistic interaction at different trophic levels. Nations et al. (2015) found alterations in growth and development in tadpoles of *X. laevis* following a subchronic exposure

(0.156 – 2.5 mg/L, 14 days) to CuO (copper oxide nanoparticle) at the highest concentration tested. Moreover, according to the authors, when tadpoles were exposed to the lowest CuO concentrations, metamorphosis and growth were stimulated, suggesting that in lower doses, CuO-NPs may present hormesis effects. Gold nanoparticles (AuNPs) (0.05, 0.5 and 5 pM in particles, 55 days) reduced time to metamorphosis by up to 3 days in *Lithobates sylvaticus* tadpoles (Fong et al. 2016). Tadpole exposure to NPs also revealed oxidative and metabolic stress (Salvaterra et al. 2013, Pattanayak et al. 2018, Murthy et al. 2022), in addition to bioaccumulation of Ag and Zn in the blood, liver, kidney and bones (Murthy et al. 2022). Svartz et al. (2020) reported oxidative stress and genotoxicity of alumina (Al) nanoparticles (5 and 25 mg/L, 96 h) in *R. arenarum* tadpoles.

3.3 Oxidative Stress

Reactive oxygen species (ROS) are radical or molecular species produced mainly in mitochondria from molecular oxygen. About 85% of O_2 is completely reduced to water in the mitochondrial respiratory chain and partially reduced O_2 intermediates are also produced in small amounts (Collin 2019). ROS include superoxide anions ($O_2^{\bullet-}$), hydrogen peroxyl (H_2O_2), hydroxyl ($^{\bullet}OH$), alkoxyl ($RO^{\bullet}$) and peroxyl ($RO_2^{\bullet}$) radicals (see more details in Chapter 6). Reactive nitrogen species (RNS) can also be produced due to oxygen metabolism in cells, including nitric oxide ($NO^{\bullet}$), peroxynitrite ($ONOO^-$) and nitrogen dioxide ($NO_2^{\bullet}$) (Van der Oost et al. 2003, Collin 2019). Increases on ROS and RNS are also often a result of toxicity caused by metals in aquatic organisms (Dusse et al. 2003, Van der Oost et al. 2003, Vasconcelos et al. 2007, Viarengo et al. 2007, Atli and Canli 2010, Barhoumi et al. 2012), and this status can lead to oxidative stress in the cell, when its generation overcomes the antioxidant capacity. Transition metals such as Fe and Cu, in the presence of $O_2^{-\bullet}$ and H_2O_2, can catalyze the generation $^{\bullet}OH$ through the Fenton and Haber-Weiss reactions (Storey 1996). However, other metal ions also generate ROS (Stohs et al. 2000), such as Cr and V, which undergo redox cycling, while Cd, Hg, and Ni, as well as Pb, deplete GSH and protein-bound SH-groups, reducing the cellular protection against ROS (Stohs and Bagchi 1995, Stohs et al. 2000, Wang et al. 2019).

The antioxidant defense system is responsible for intercepting or avoiding excessive ROS generation (Halliwell and Gutteridge 2007). They include enzymes and non-enzymatic compounds, such as GSH, carotenoids and tocopherols (Fig. 3). Superoxide dismutase (SOD, Cu/ZnSOD - in the cytosol and MnSOD - in the mitochondria) is one of the first-line cellular antioxidant enzyme that reduces $O_2^{\bullet-}$ into H_2O_2, which in turn can then be decomposed into O_2 and H_2O by catalase (CAT) (Van Der Oost et al. 2003). Hydrogen peroxide and other organic peroxides can also be decomposed by cytosolic and mitochondrial glutathione peroxidases (GPx), using electrons from GSH, and then converting it into GSSG (Gaté et al. 1999). Glutathione reductase (GR) restores GSH levels from GGSG in the presence of nicotinamide adenine dinucleotide phosphate (NADPH). GSH is also important as a cofactor for the xenobiotic biotransformation enzyme glutathione S-transferase (GST) (Halliwell and Gutteridge 2007). GST catalyzes the conjugation of GSH with various substances in phase II detoxification and plays a role in preventing oxidative damage by conjugating the decomposition products of lipid peroxides, such as 4-hydroxynonenal, to GSH (Van der Oost et al. 2003, Carvalho et al. 2015, Veronez et al. 2016).

CAT is proved to participate in controlling oxidative stress and in several physiological and metabolic reactions (Van der Oost et al. 2003, Hermes-Lima 2005), which makes it a sensitive biomarker of metal exposure in tadpoles (Hermes-Lima 2005, Boiarski et al. 2020, Carvalho et al. 2020, Fernandes et al. 2021). Numerous studies with metals and oxidative stress have been performed in tadpoles (Ossana et al. 2010, 2013, Mardirosian et al. 2015, Veronez et al. 2016, Boiarski et al. 2020, Carvalho et al. 2020, Peluso et al. 2020, Zhang et al. 2020, Fernandes et al. 2021) (Table 2). For example, CAT and GST increased in tadpoles of *Trachycephalus typhonius* exposed to a sediment containing Fe, Mn, Zn, Cu, Cd, Cr, Ni, Pb, As and Hg (Peltzer et al. 2013), and in *L. catesbeianus* exposed to iron ore, Fe and Mn (Veronez et al. 2016), as well as to Zn, Cu

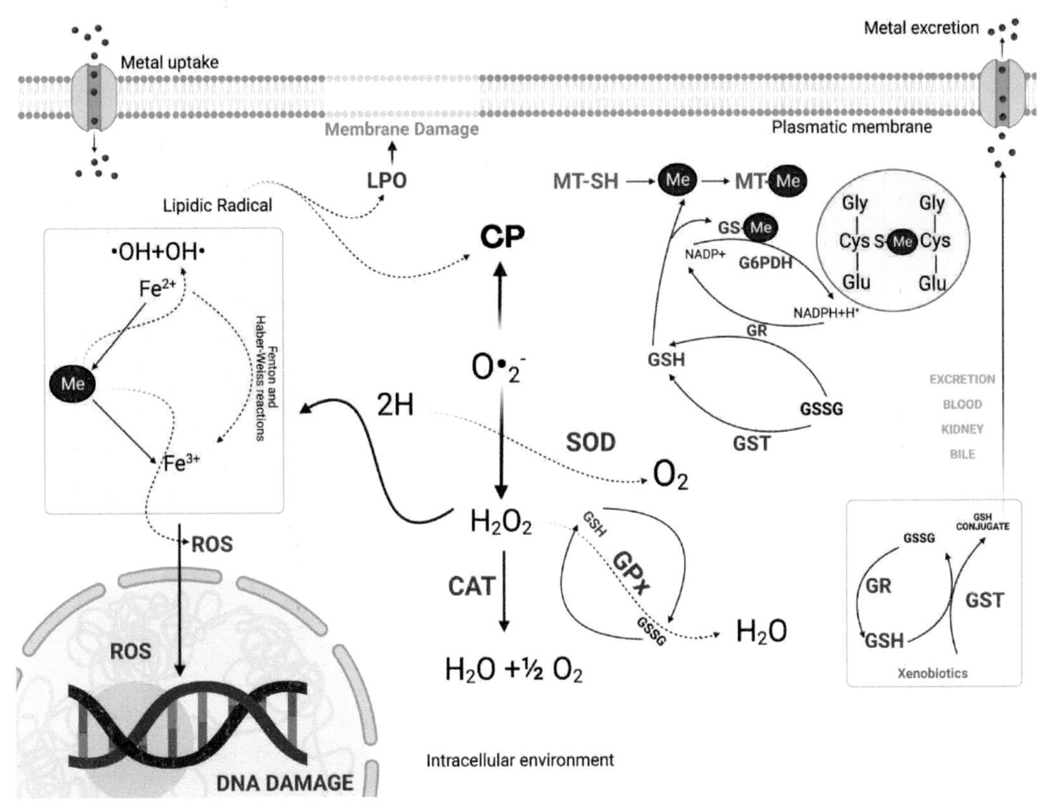

Figure 3. Schematic representation of the major effects of metals in cells and the action of enzymatic and non-enzymatic antioxidants: CAT = catalase; LPO = lipoperoxidation; GSH = reduced glutathione; GSSG = oxidized glutathione; GR = glutathione reductase; GPx = glutathione peroxidase; GST = glutathione S-transferase; G6PDH = glucose-6-phosphate dehydrogenase; Me = metals; MT = metallothionein; NADPH + H^+ = nicotinamide adenine dinucleotide reduced phosphate; $NADP^+$ = nicotinamide adenine dinucleotide oxidized phosphate; CP = protein carbonylated; SOD = superoxide desmutase; OH = hydroxyl radical; OH^- = hydroxyl anion; $O_2^{\cdot-}$ = anion superoxide; H_2O_2 = hydrogen peroxide; ROS = reactive oxygen species; Cys = cysteine; Glu = glutamate; Gly = glycine. The Fenton reaction is a catalytic process (by Fe (3+/2+) that converts organic substrates (RH), H_2O_2, a product of mitochondrial oxidative respiration, into a highly toxic ·OH free radical. The Haber–Weiss reaction is generation of ·OH and OH^- ions from the reaction of H_2O_2 and $O_2^{\cdot-}$ catalyzed by Fe. Created with BioRender.com.

and Cd (Carvalho et al. 2020). Peluso et al. (2020) reported alterations in CAT and GST activities, in addition to GSH and thiobarbituric acid (TBARs) levels, in *R. arenarum* embryos and tadpoles exposed (96 h) to water samples from two water bodies containing metals (As, Cr, Cu, Pb, Ni and Zn) from Paraná River (Argentina).

Increase in GST activity and GSH levels after As exposure (10–25 mg/L, 7 days) was observed by Mardirosian et al. (2015) in *R. arenarum* tadpoles, suggesting a protective response of embryos at the end of their embryonic development by preventing oxidative stress-induced damage. However, decreased GST activity observed by Samanta et al. (2020) in *Euphlyctis cyanophlyctis* tadpoles exposed to different As concentrations (10–100 mg/L, 96 h) were associated to oxidative stress, cytotoxicity, and higher mortality of the animals. In contrast, Güngördü et al. (2010) did not observe any changes in GST or GR in *X. laevis* embryos after acute exposure to Cd, Pb and Cu.

Changes in CAT and SOD activities have been associated with the presence of metals in tadpoles of *L. catesbeianus* exposed to river water (Boiarski et al. 2020, Fernandes et al. 2021) and under laboratory conditions, as *Bufo raddei* exposed to Cd and Pb (Zhang et al. 2007), *X. laevis* exposed to Cu, Cd and Pb (Yologlu and Ozmen 2015) and *Bufo gargarizans* exposed to Cd, Pb and Cd+Pb (Zheng et al. 2021). Metals may result in decreased antioxidant enzyme activity due to their binding with amino acids, causing changes in the enzymatic structure and therefore altering its activity

Table 2. Studies of metals and oxidative stress in tadpoles.

Species	Metals	Exposure	Concentration	Biomarkers	References
B. raddei	Cd, Pb	45, 60, 75d	Cd: 0.0015, 0.03, and 0.15 mg/L	+ SOD	Zhang et al. (2007)
X. laevis	Cu, Cd, Pb	96 hr	Cu: 0.01, 0.085, 0.425, 0.85 mg/L Cd: 0.005, 0.52, 2.59, 5.18 mg/L Pb: 0.01, 12.3, 61.53, 123.05 mg/L	+ CAT (Cu, Cd); – CAT (CuCd, PbCd, + PbCu, PbCuCd) + GR (Cd, Cu), – GR (Pb); – GR (CuCd, + PbCuCd) + GPx (Cd), – GPx (Pb, Cu); + GPx (PbCu) + GST (Cd, Pb), +/– GST (Cu); + GST (PbCu, PbCd, PbCuCd) + MT (Cu, Cd, Pb)	Yologlu and Ozmen (2015)
L. catesbeianus	As, Cu, Cr, Cd, Pb, Zn	96 hr		– CAT (gills), –/+ GST, + GSH (liver), + SOD	Ossana et al. (2013)
L. catesbeianus	Iron ore, Fe, Mn	30 d	Iron ore: 3.79 mg/L Fe: 0.5063 mg/L Mn: 5.23 mg/L	+ CAT, + GST	Veronez et al. (2016)
L. catesbeianus	Cu, Fe, Mn, Pb	7 d	ND	+ CAT, – LPO	Boiarski et al. (2020)
L. catesbeianus	Zn, Cu, Cd	96 hr	1 μg/L	+ CAT, + GST + LPO, + GSH	Carvalho et al. (2020)
L. catesbeianus	Al, Cu, Mn, Zn	96 hr		– CAT, – SOD + LPO, + GSH, + GST, – GPx, + CP	Fernandes et al. (2021)
B. raddei	Cd, Pb	*in situ*		+ CAT, + SOD, + MDA	Zhang et al. (2020)
T. typhonius	Fe, Mn, Zn, Cu, Cd, Cr, Ni, Pb, As, Hg	*Gs 2,3–38*		+ CAT, + GST	Peltzer et al. (2013)
E. cyanophlyctis	As	96 hr	10-100 mg/L	– GST	Samanta et al. (2020)
B. gargarizans	Cd	7 d	5, 50, 100, 200, 500 g/L	+ SOD, + GPx (mRNA)	Ya et al. (2017)
B. gargarizans	Cd, Pb, Cd/Pb	*Gs 2,3–38*	1 μM Cd, Pb	– SOD, – GPx (mRNA)	Zheng et al. (2021)
R. arenarum	As	7 d	10, 20, 25 mg/L	+ GST, + GSH	Mardirosian et al. (2015)
R. arenarum *	As, Cr, Cu, Pb, Ni, Zn	96 hr		+ CAT, + GST + LPO (by TBARs)	Peluso et al. (2020)
R. arenarum	Al-NPs	96 hr	5, 25 mg/L	+ LPO	Svartz et al. (2020)
P. perezi	TiSiO4-NPs	96 hr	8.2, 10.2, 12.8, 16, 20 mg/L	+ CAT	Salvaterra et al. (2013)
P. perezi	Be, Al, Ca, Mn, Fe, Co, U	48, 96 hr		+ GPx, + LPO (by TBARs)	Marques et al. (2013)
P. maculatus	Ag-NPs	60 d	1, 5 mg/L	+ SOD	Pattanayak et al. (2018)

Table 2 contd. ...

...*Table 2 contd.*

Species	Metals	Exposure	Concentration	Biomarkers	References
P. maculatus	Ag-NP, ZnO-NP	60 d	Ag-NPs: 1, 5, 10 mg/L	– CAT, – SOD	Murthy et al. (2022)
			ZnO-NPs: 1, 10, 50 mg/L	+ LPO	
R. temporaria	Cr, Cu, Zn	*in situ*		+ GSH, + GST, + MT	Johansen (2013)

CAT = catalase; GSH = reduced glutathione; GSSG = oxidized glutathione; GR = glutathione reductase; GPx = glutathione peroxidase; GST = glutathione S-transferase; LPO = lipoperoxidation; TBARs = thiobarbituric acid; MDA = malondialdehyde, MT = metallothionein; CP = protein carbonylated; SOD = superoxide dismutase; NPs = nanoparticules; Al-NPs = alumina nanoparticules; TiSiO4-NP = titanium silicate nanoparticles; Ag-NP = silver nanoparticles; ZnO-NP = zinc oxide nanoparticles; ND = not determined; + indicates an increase; – indicates a decrease. * (embryos and larvae). Gs = Gosner stage.

(Rowe et al. 2011, Carvalho et al. 2020, Fernandes et al. 2021). In addition, *B. gargarizans* embryos exposed to different concentrations of Cd (5, 50, 100, 200 and 500 g/L for 7 days) had upregulated expression of SOD and GPx mRNA (Wu et al. 2017). In contrast, CAT and SOD activities decreased in the serum of *Polypedates maculatus* tadpoles exposed to Ag-NPs (1–10 mg/L) and ZnO-NPs (1–50 mg/L) after 60 days, leading to enhanced malondialdehyde (MDA) levels (Murthy et al. 2022). GPx, CAT and SOD are metalloproteins and CAT contains heme protein as a prosthetic group and Fe linked to its active center (Putnam et al. 2000), whose biosynthesis can be inhibited by metals. Metals can sequester and compete with Fe for transporters and, in addition, can bind to proteins, impairing enzymatic activity and integration of Fe into protoporphyrin (Schauder et al. 2010). Furthermore, it should be considered that enzyme activities may vary in each organ due to their specificity.

A study by Chen et al. (2018) demonstrated an *in vitro* interaction of trivalent chromium (Cr^{3+}) and hexavalent chromium (Cr^{6+}) with CAT by spectroscopy and computer simulations. The fluorescence analysis showed that both metal forms extinguished the fluorescence of CAT by a dynamic and static extinction mechanism, as well as inducing conformational changes in CAT; the degree of influence between the two forms was different. The SOD-CAT system represents the first line of defense in the clearance of free radicals (Yildirim et al. 2011, Gusso-Choueri et al. 2015) and the maintenance of high constitutive levels of these antioxidant enzymes is essential to prevent the harmful action of radicals.

ROS and RNS are capable of altering structures and functions of biomolecules. In lipids, they can cause a chain reaction that results in peroxidation or lipoperoxidation (LPO) (Van der Oost et al. 2003, Vasconcelos et al. 2007, Marques et al. 2013). Several studies report elevations in LPO in amphibian tadpoles after exposure to metals (Marques et al. 2013, Carvalho et al. 2020, Peluso et al. 2020, Fernandes et al. 2021) and nanomaterials (NMs), such as alumina (Al) nanoparticles in animals treated with 25 mg/L at 96 h (Svartz et al. 2020). These effects are usually associated with metal accumulation, high ROS production and oxidative damage (Marques et al. 2013, Peluso et al. 2020). Formation of carbonylated proteins (CP) is another effect of metals that must be considered in studies with tadpoles. CP occur through direct oxidation of amino acid side chains by metals or ROS, but also by the action of LPO by-products (Stadtman and Oliver 1991, Vasconcelos et al. 2007, Farina et al. 2011), which is an irreversible process. Some studies show and relate the presence of contaminants to increased protein degradation (Grune et al. 2003, Almroth et al. 2005), however, studies regarding metals and the effects of CP in tadpoles are still lacking. We highlight the study by Fernandes et al. (2021) which verified increased CP associated with the presence of metals in *L. catesbeianus* tadpoles after exposure to water from the Sorocaba River in Brazil. Higher CP levels are usually a sign of oxidative stress (Dalle-Donne et al. 2003, Monteiro et al. 2010) and have an integrated result with oxidative damage (Stadman and Levine 2000). Therefore, it can be used as a reference for oxidative stress index in tadpoles exposed to metals.

As previously mentioned, in addition to enzymatic antioxidants, tadpoles also have non-enzymatic antioxidant mechanisms, such as carotenoids (vitamin A), ascorbic acid (vitamin C), tocopherol (lipid-soluble vitamin E), ubiquinol and carotene, that act to combat ROS and prevent, or reduce, the deleterious effects of oxidative stress. In this group, we highlight the tripeptide glutamyl-cysteine-glycine and GSH, with the latter representing the main antioxidant and redox regulator in cells to combat oxidation of cellular constituents.

GSH plays a key role in detoxifying peroxides, RNS, and xenobiotic compounds (such as electrophilic reactive molecules) in cells (Han et al. 2006). This tripeptide serves as a substrate for GPx and GST. The GSH is oxidized generating GSSG and can be recycled back in a NADPH-dependent reaction catalyzed by GR or by thioredoxin reductase systems. In fact, there are several reports in relevant literature on the effect of metals on GSH levels in tadpole tissues. Johansen (2013) found increased GSH levels in tadpoles (*Rana temporaria*) from two sedimentation ponds with high concentrations of Cr, Cu and Zn. Carvalho et al. (2020) and Fernandes et al. (2021) found an increase in GSH in tissues of *L. catesbeianus* tadpoles after exposure to Al, Zn, Cu, Cd and Mn. Although high concentrations of GSH may indicate exposure to metals and detoxification by GSH, its decrease can be a result of the direct interaction between metal and GSH to prevent the contaminant from interacting with cellular targets (Cuypers et al. 2010, Peluso et al. 2020). However, depletion of intracellular GSH may also affect the detoxification processes with worsening oxidative stress and cellular changes, interfering with several biological processes (Papadimitriou and Loumbourdis 2002, De Boeck et al. 2003, Authman et al. 2012).

3.4 Metallothionein (MT)

Metals are neither metabolizable nor biodegradable. For this reason, the organisms can not break them down into less toxic substances. Therefore, mechanisms of protection against excessive levels of metals consist in the synthesis of proteins, such as MT and ferritin. These proteins are specialized in binding to the metal species decreasing their bioavailability (Coyle et al. 2002, Sayre et al. 2005, Loumbourdis et al. 2007, Monserrat et al. 2007, Yologlu and Ozmen 2015, Veronez et al. 2016, Carvalho et al. 2017). Metals can also be stored in intracellular granules (Fairbrother et al. 2007) as mechanism to keep the homeosthasis. These mechanisms imply in increases of proteins that bind to metals and store them, such as MT and ferritin (Coyle et al. 2002, Sayre et al. 2005, Loumbourdis et al. 2007, Monserrat et al. 2007, Yologlu and Ozmen 2015, Veronez et al. 2016, Carvalho et al. 2017). Metals can also be stored in intracellular granules (Fairbrother et al. 2007).

MTs are low molecular weight cytosolic proteins containing thiol groups (–SH) that form metal-thiolate complexes involved in metal homeostasis, such as Zn and Cu. They also act in the detoxification of toxic metals, such as Cd and Hg, and participate in ROS elimination (Monteiro et al. 2010, Carvalho et al. 2017, Šulinskienė et al. 2019). This biological characteristic makes MT a useful biomarker of metal exposure in many aquatic organisms, including tadpoles (Johansen 2013, Yologlu and Ozmen 2015, Carvalho et al. 2017, Carlsson and Tydén 2018). Studies with tadpoles have associated increases in MT to the presence of metals in the environment (Johansen 2013, Yologlu and Ozmen 2015, Carvalho et al. 2017, Carlsson and Tydén 2018). Higher MT contents may prevent metals from interacting and reacting with cellular molecules, which can be important for conferring tadpole tolerance to metals.

3.5 Neurotoxicity

Acetylcholinesterase (AChE) interrupts the transmission of nerve impulses at cholinergic synapses by the hydrolysis of the neurotransmitter acetylcholine (ACh). Thus, both an increase and a decrease in its activity can impair the functioning of the body, leading to behavioral changes, or negative effects on development, nutrition and reproduction (Nunes 2011, Ossana et al. 2013, Boiarski et al. 2020). However, little is known about the action and effects of metals and their mixtures on amphibians' AChE. As previously mentioned, metals can have a direct action on the enzyme by binding to its free SH groups, causing reversible inhibition, enzymatic denaturation, and protein aggregates. Some studies have reported disrupting effects of metals on AChE, thus considering this enzyme as an important indicator to attest neurotoxicity of these compounds in tadpoles (Güngördü et al. 2010, Birhanli et al. 2014, Yologlu and Ozmen 2015, Carvalho et al. 2020).

Mercury is a well-known neurotoxic agent; when Hg^{2+} enters the aquatic ecosystems, it can be readily transformed into the neurotoxic organic methylmercury, which can be taken up by organisms and biomagnified as the trophic level increases (Farina et al. 2011, Boczulak et al. 2017). All forms of Hg (metallic-Hg^0, inorganic - mercury - Hg^{1+} or mercury - Hg^{2+} ions and/or organic forms - methylmercury - CH_3Hg^+) are toxic to almost all organisms (Farina et al. 2011).

3.6 Genotoxicity

Erythrocytes in amphibians are nucleated and undergo cell division in the circulation, especially during the larval stages (Patar et al. 2016). Nucleated erythrocytes allow genotoxic studies with environmental pollutants through the evaluation of morphological alterations of the nuclei (Zhang et al. 2007, Ossana et al. 2013, Veronez et al. 2016, Hu et al. 2019, Benvindo-Souza et al. 2020, Peluso et al. 2020, Svartz et al. 2020). A reliable, relatively simple and sensitive test used to assess nuclei damage is the micronuclei (MN) assay (Benvindo-Souza et al. 2020). The MN corresponds to chromosomal fragments as a result of chromosomal breaks (clastogenic action) or abnormal chromosome segregation (aneugenic effects) (Al-Sabti and Metcalfe 1995, Cavas et al. 2005). Analyzes of erythrocytes for diameter, refringence and color, and the assessment of DNA damage, such as MN (biomarker of chromosome breakage and/or loss), nuclear buds (gene amplification and/or elimination of DNA repair complexes), and nuclear abnormalities (such as lobe and notch) (Alimba et al. 2018) allow the evaluation of cytotoxicity of environmental contaminants.

DNA is the key-target for genotoxic stress and the main types of damage commonly observed are base damage, single-strand breaks or double-strand breaks (Gastaldo et al. 2008). Some studies have proposed the occurrence of MN and DNA breaks (by using the comet assay) as biomarkers for DNA damage in tadpoles exposed to metals in the environment (Ferreira et al. 2003, Mouchet et al. 2006, 2007, Zhang et al. 2007, Peltzer et al. 2013, Patar et al. 2016, Veronez et al. 2016, Monteiro et al. 2018, Hu et al. 2019, Benvindo-Souza et al. 2020, Peluso et al. 2020, Svartz et al. 2020, Patar et al. 2021).

Peltzer et al. (2013) identified nuclear abnormalities such as MN and increased levels of immature and anucleated erythrocytes in *Trachycephalus typhonius* tadpoles exposed to metals in the sediment. Different studies correlate the exposure of bullfrog tadpoles (*L. catesbeianus*) with genotoxic alterations in response to the presence of metals in the medium. For example, higher frequency of MN was found in the erythrocytes after exposure to river water samples containing Cu, Cd, Ni, Pb at lower levels, and Mn, Zn and Cr at levels above those accepted by local legislation for aquatic life protection (Ossana et al. 2013). DNA damage and MN were verified after exposure to iron ore, Fe, Mn, (Veronez et al. 2016) and Cr^{6+} (4, 12 and 36 mg/L) in different stages of larval development (Monteiro et al. 2018). Mutagenic effects and inhibition of the cell cycle, such as cell retention in the Sub-G1 phase and decreased cells in the S and phases G2/M, were also observed effects (Monteiro et al. 2018). Cd, for example, can have direct effects on DNA by binding to it and/or by inhibiting enzymes that act in DNA repair, or indeed by induction of oxidative stress (Lutzen et al. 2004, Patar et al. 2016, Hu et al. 2019). The study by Peluso et al. (2020) in embryos and larvae of *R. arenarum* exposed (96h) to water samples from the Paraná River (Argentina) containing As, Cr, Cu, Pb, Ni and Zn also showed a higher frequency of MN. *Fejervarya limnocharis* tadpoles showed increases in MN frequency with the raising on $ZnCl_2$ concentration (1.5 to 2.0 mg/L, 24–96 h), in addition to DNA strand breaks at lower concentrations (Patar et al. 2021). These studies highlight the reliability of MN assays used as a proxy to assess genotoxicity in amphibians exposed to metals.

3.7 Effects on Metamorphosis

Metamorphosis is a process of in-depth organ remodeling of amphibians that undergo the transition from the aquatic larval stage toward the adult terrestrial/semiterrestrial life. During this process, many adjustments of physiological and biochemical functions occur. Effects of metals on the time of metamorphosis seem to present a hormesis effect (U-shape dose-response), i.e., an acceleration of the metamorphosis at low concentrations and a delay when high levels of toxic metals are present in the medium. Thus, the influence of metals during the metamorphic period have been studied to elucidate the potential effects of these contaminants in an animal that is undergoing remarkable readaptations.

The effects of metals on metamorphic processes of tadpoles have been assessed by external characteristics, such as the emergence of hind-limb, front-limb, tails resorption, snout-vent length in function of time, among others. Moreover, histopathological evaluation of the thyroid gland has also been performed. The responses following the exposure to soil burdened with several metals were linked to delayed metamorphosis in tadpoles (Lefcort et al. 1998). Tadpoles exposed to Fe and Mn also required a longer time to reach the metamorphic climax (Veronez et al. 2016). Other metals such as Hg (24 and 30 µg/L) (Shi et al. 2018) and Cd (100 500 µg/L) (Sun et al. 2017, Wu et al. 2017, Ya et al. 2021) were shown to decelerate the metamorphic process in *B. gargarizans*. The presence of Al (2000 µg/L) and Cu (50, 100 and 200 µg/L) also delayed the time required for metamorphic development in *R. sylvatica* (Peles 2013). This effect on development was discussed as a result of thyroid architecture damage induced by exposure to high levels of metals in tadpoles (Huang et al. 2014, Wu et al. 2017, Shi et al. 2018).

Nations et al. (2011) showed that *X. laevis* exposed to high concentrations of ZnO nanoparticles (0.513 and 0.799 mg/L) prolonged their larval life, whereas low levels of the same compound (0.067 mg/L) accelerated the metamorphosis. Studies using low and high levels of CuO nanoparticles resulted in the same response, regarding the time required to reach metamorphic climax (Nations et al. 2015). Tadpoles of *Rana luteiventris* experienced decreased weight and delays in metamorphosis when exposed to Pb and Zn (Lefcort et al. 1998). Exposure to Pb (40–1280 mg/L) caused several effects in *Pelophylax* (formerly *Rana*) *nigromaculata* tadpoles (Gosner stage 19–46) (Huang et al. 2014). Increases in Pb concentration affected metamorphosis time and survival rate, caused spinal malformations (scoliosis, lordosis and kyphosis), arched femurs and shortened the forelimbs (causing animals to lean to one side and making it difficult to move), and also affected the jumping and swimming speed of tadpoles.

Gross et al. (2007) observed that *Rana pipens* exposed to 0.25 and 5 µg/L of Cd had their metamorphic spam shortened, considering their trend in reaching the emergence of the forelimb significantly earlier than the control group. Similar effects were seen in *Bufo americanus*, where tadpoles exposed to Cd at concentrations of 5 µg and 54 µg/L also reached the metamorphic climax sooner. Exposure to Li (2.5 mg/L) and Se (10 µg/L), both isolated and in mixture, induced decreases in the number and size of the thyroid follicles in *L. catesbeianus* after 21 days (Pinto-Vidal et al. 2021a), which is indicative of accelerated development, even in the absence of significant manifestations concerning the external development. Low levels of Cd (5 µg/L) were also shown to shorten the metamorphic period in *B. gargarizans* (Sun et al. 2017). This moderate stress, elicited by low levels of metals, seems to trigger escaping strategies in the exposing animal, which is probably a strategy of pushing them away from contaminated water towards a potentially safer environment (James and Little 2003, Gross et al. 2007, Sun et al. 2017, Pinto-Vidal et al. 2021a).

3.8 Effects of Metals on Energetic Metabolism

Metabolism of tadpoles can be impaired by metal exposure. Increases in ATPase activity in the intestine caused by Cd (0.0015–0.15 mg/L), and in liver and skin caused by Pb (0.7–70 mg/L) were observed in *B. raddei* tadpoles after 60 and 75 days, indicating a higher energy demand required for detoxification processes following metal exposure (Zhang et al. 2007).

In studies with *B. gargarizans* at larval stage, animals expressed disturbance in their hepatic lipidic metabolism after chronically exposed to high levels of Cd (Wu et al. 2017, Ju et al. 2020, Ya et al. 2021), Cr^{6+} (Yang et al. 2020) and Hg (Shi et al. 2018). In these experiments, authors observed a downregulation of β-oxidation related genes (probably due to hepatotoxicity followed by degeneration of the mitochondria), whereas genes related to the synthesis and elongation of fatty acid to long chain fatty acid were upregulated, which could improve fatty acid accumulation in liver.

The effects of Li (2.5 mg/L) and Se (10 μg/L) at concentrations allowed by Brazilian legislation (BRASIL 2005) were tested on premetamorphic *L. catesbeianus* (American bullfrog) tadpoles, over acute (7 days) and chronic (21 days) sets, to assess the effects on the mobilization of hepatic triglycerides (Pinto-Vidal et al. 2021b). The study showed raises in lipid mobilization, which could be indicative of increased energy demand to deal with water contamination. A latter study also observed that the presence of Li, in addition to its mixture with Se, was linked to the increased mobilization of glucose after 7 days of exposure, whereas the presence of Se decreased the mobilization of this carbohydrate after 21 days (Pinto-Vidal et al. 2021b). In the aforementioned study, authors reported no difference in the total protein content in premetamorphic *L. catesbeianus* exposed to Li, Se, or their mixture, neither after 7 nor 21 days.

Exposure of *L. catesbeianus* tadpoles to Zn, Cu and Cd activated the anaerobiosis (by lactate dehydrogenase—LDH) and aerobiosis (by malate dehydrogenase—MDH) in liver and kidney after short-term exposure (2 days) (Chagas et al. 2020). In the kidney, the aerobic metabolism (MDH) was also activated after chronic exposure (16 days). The same study reported that these metals altered the glucose, protein, and triglycerides levels in liver, kidney, and muscle of tadpoles. Li (2.5 mg/L) and Se (10 μg/L) in mixture decreased lactate dehydrogenase (LDH) activity from caudal muscle of premetamorphic *L. catesbeianus* (Pinto-Vidal et al. 2022), whereas the isolated elements caused no significant changes, indicating a synergic toxicity induced by the combined compounds.

3.9 Effects of Metals on Morphological Parameters

The influence of metals in tadpoles have been assessed by morphological changes in target organs, especially in liver (De Oliveira et al. 2017, Ju et al. 2020, Yang et al. 2020, Krohn et al. 2020, Fernandes et al. 2021). Tadpoles of *B. gargarizans* exposed to different concentrations of Cd (5 to 200 μg/L) at Gosner stages 26–42 showed histological alterations such as hepatocyte deformation, nuclear pyknosis, increases in melanomacrophage centers (MMCs), and lipid aggregates (Ju et al. 2020). Similar results were observed in this same species exposed to different concentrations (13–416 mg/L) of Cr^{6+} at larval stages 2–42 of Gosner (Yang et al. 2020). Cr^{6+} (416 mg/L, 70 d) also altered intestinal tissue structure and reduced the total body length, wet body weight, intestinal length, and tadpole wet weight (Yao et al. 2019). A recent study reported growth retardation, intestinal histological injury, and decreased activity of digestive enzymes (lipase, amylase, trypsin and pepsin) in *B. gargarizans* tadpoles co-exposed to Cd and Pb (1 μM) (Zheng et al. 2021). The study also suggested that combined metals may have more severe effects than their isolated exposure.

MMCs increase was also noticed by Pinto-Vidal et al. (2021b) when tadpoles were exposed to Li (2.5 mg/L) and Se (10 μg/L). This same exposure elicited signs of hepatotoxicity, namely microvesicular steatosis and cytoplasmic vacuolation, however, without nuclear damage or signs of cell death. Barni et al. (2002) and Fenoglio et al. (2005) suggest that histological changes, mainly MMCs, can be used as biomarkers of water contaminants, being related to stress and chronic inflammation. MMCs are melanin-pigmented phagocytes with catabolic functions (De Oliveira et al. 2017) and non-enzymatic antioxidant action (Fenoglio et al. 2005). Exposure to Ag/ZnO-NPs of different concentrations increased serum alanine transaminase, aspartate transaminase, and alkaline phosphatase activities in *Polypedates maculatus* tadpoles, which may be indicative of liver damage caused by NPs with the release of enzymes into the blood (Murthy et al. 2022).

Monteiro et al. (2018) assessed the responses of *L. catesbeianus* tadpoles (Gosner stages 25–31) exposed to 4, 12 and 36 mg/L of $K_2Cr_2O_7$. Authors found severe histopathological changes, both in liver and kidney, which were dependent on Cr concentration. Congestion and inflammation of sinusoid areas, hemorrhages in liver parenchyma, hepatocytes with pyknotic nuclei, loss of cell integrity, and necrosis were the main hepatic effects found by the study. In the kidney, the effects were as intense as those in the liver, presenting areas with inflammatory infiltrate, granulomas, tubular and glomerular hypertrophy, and tubular necrosis. The tropism of metals to liver and kidney can represent a potential impairment of metabolic and excretion functions.

Peltzer et al. (2013) assessed effects on hatching (stage Gosner 23) of *Trachycephalus thyphonius* exposed to sediment containing metals such as Fe, Mn, Zn, Cu, Cd, Cr, Ni, Pb, As and Hg until they reached the metamorphic climax (stage Gosner 46). The presence of Al, Cu, Fe, Pb and Zn induced sublethal effects added to developmental and morphological abnormalities, such as swollen bodies and diamond shape, bowel unfolding, deviated intestine, rigid tails, polydactyly, and visceral and posterior hemorrhage in *T. typhonius* tadpoles. This study demonstrated that metals were teratogenic and suggested that the abnormalities were related to the high concentrations of Al, Cu, Fe, Pb and Zn in the sediment.

Exposure to Hg induced hepatotoxicity in *B. gargarizans* (Shi et al. 2018) expressed as a reduction in the number of mitochondria, mitochondrial vacuolation, degeneration, disintegration of hepatocytes, and endoplasmic reticulum breakdown. High levels of Cd also induced histopathological responses in the thyroid gland of *B. gargarizans* tadpoles (100 and 500 μg/L), such as cell hyperplasia and deformation (Sun et al. 2017). In *Pseudepidalea variabilis* tadpoles, the exposure to Cd at 5, 10, and 25 μg/L elicited gill lamellar fusions, deformations of pronephric tubules in the kidneys, or hepatic hemorrhage and deformation, which were associated with visceral edema (Gürkan et al. 2014). Deformation in gills and mouth, the presence of edema, and underdevelopment (Chen et al. 2007), degeneration of hepatocytes, reduced size and diameter of thyroid gland, and increased number of follicle (Chai et al. 2014) were described in tadpoles exposed to Cu. Malformations as scoliosis, microphthalmia, and micromelia were linked to the exposure to Hg (Unrine et al. 2004). Thyroid gland of *B. gargarizans* presented several histopathological responses in presence of Hg, as enlarged thyroid follicular space, severe deformation in follicular cell (12 μg/L of Hg), follicular cells hyperplasia, and general deformations (18, 24, and 30 μg/L) (Shi et al. 2018). Morphological changes were also reported in *L. catesbeianus* exposed to Cd at 1 μg/L (Abdalla et al. 2013). In this study, authors described changes in somatic cells and gonadal architecture, gonadal tissue degeneration, and cell death followed by absorption after 96 h of exposure. Developmental malformations, such as polygonadism, sex reversal or incomplete metamorphism, in addition to gonad reabsorption, were also detected following 16-days exposure.

3.10 Effects Growth and Development

Metals can cause physiological dysfunctions that affect growth and development of tadpoles. Impairment in development, such as prolonged larval life or reduced body size in the metamorphic period are factors with a significant negative potential to impact amphibian ecology (Berven 1990, Rumrill et al. 2016, Meindl et al. 2020). Disturbances in development will also affect size and age at first reproduction (Semlitsch et al. 1998). Larvae that spend longer time before reaching metamorphic climax are associated with larger adults. Expanding the larval life in water seems to be rather beneficial for *Ambystoma talpoideum* (Semlitsch et al. 1998). To summarize, the larval life strongly affects the fitness of the adult individual (Berven 1990, Semlitsch et al. 1998).

Rana sphenocephala tadpoles exposed to Pb in sediment and water from the initial free-swimming stage to metamorphosis had decreased development and growth and, mainly, skeletal development, with severe skeletal malformations at high concentrations of Pb. The surviving animals exhibited spinal alterations, reduced femur and humerus lengths, deformed fingers, among other bone malformations (Sparling et al. 2006). Reduced survival to the free-swimming stage was observed in embryos and larvae of the southern frog *Anaxyrus* (*Bufo*) *terrestris* after exposure to 10 mg/L of Cu, while at concentrations above 15 mg/L no larvae reached the metamorphic climax (Lance et al. 2013). Another study developed by Rumrill et al. (2016), who exposed *A. terrestris* larvae to Cu (30 μg/L) from parents belonging to metal contaminated sites, observed a reduction in metamorphosis and a delayed larval development when compared to tadpoles from parents collected at non-contaminated sites.

The presence of Cu also had a negative impact in *B. gargarizans* tadpole growth, expressed by length and weight measurements (Chai et al. 2014, Wang et al. 2016b). Moreover, *Duttaphrynus melanostictus* tadpoles in the Gosner stages 24–26 exposed to Cr^{6+} (0.002–2.0 mg/L) for 4 and 21 days presented mortality, growth retardation, developmental delays and structural aberrations, depending on the concentration and duration of the exposure (Fernando et al. 2016). In addition, none of the tadpoles exposed to concentrations of 1 and 2 mg/L of Cr^{6+} completed metamorphosis (Gosner 46) after 21 days. Furthermore, around 25% of tadpoles exhibited abnormal curvatures in their tails and lateral spine deviation.

The exposure of *Bufo gargarizans* tadpoles to Cd at concentrations of 50, 100, and 500 μg/L decreased skeleton ossification, but 10 μg/L of Cd conversely increased the process (Sun et al. 2017). Tadpoles exposed to Cd at 100 and 500 μg/L also presented a reduction in biomarkers related to development (e.g., total length, hind-limb length and tail length) and body mass. Similar results were shown by Ya et al. (2021), i.e., a growth reduction expressed by decreases in the total length, tail length, hindlimb length, and body weight of *B. gargarizans* exposed to 100 and 200 μg/L of Cd. The presence of Cd at 0.20 mg/L decreased *D. melanostictus* size (Ranatunge et al. 2012) and impacted its development, with over 80% of the tadpoles not completing metamorphosis. In contrary, exposure to Cd at 0.20 mg/L for 14 days increased the body size of *Microhyla fissipes* tadpoles, which was attributed to delayed metamorphosis and slower locomotion (Hu et al. 2019). Hg in several concentrations also affected the total length, average tail length, and body mass of *B. gargarizans* tadpoles (Shi et al. 2018).

Chen et al. (2007) showed deformities in tadpoles of *Rana pipiens* (Northern leopard frogs) exposed to 100 μg/L of $CuSO_4$ on the tenth day in an experiment of 154 days. In this same experiment, concentrations of 25 and 100 μg/L decreased growth rate of the animals, while survival and swimming performance were negatively affected at 100 μg/L. Tadpoles of *X. laevis* treated with different concentrations of Pb (3.7–18.7 mg/L) for 3–6 weeks had increased body weight at higher concentrations, but with additional metal bioaccumulation (Berzins and Bundy 2002).

3.11 Metal Effects on Other Physiological Parameters

Metals can cause ionic and respiratory disorders in fish, for example by inhibiting branchial Na^+-K^+-ATPase (Martinez et al. 2004, Simonato et al. 2013). Thus, similar chang es are also expected to occur in tadpoles exposed to metals, with possible impacts on blood composition. For example, tadpoles of *L. catesbeianus* exposed to different concentrations of Cu (0.2, 1.2 and 2.4 mg/L) presented immature erythrocytes and alterations in neutrophils and lymphocytes number at the highest concentrations (Ferreira et al. 2003). Another study with *L. catesbeianus* tadpoles exposed to isolated and combined Zn, Cu and Cd showed an increase in hemoglobin (Hb), red blood cell count (RBCs) and mean corpuscular hemoglobin (MCH) in most treated groups after 2 days; after 16 days, RBC increased, and MHC decreased in some groups exposed to the combined metals (Carvalho et al. 2017). In addition, the animals also had their lymphocytes increased and the neutrophils, eosinophils, basophils, and monocytes decreased. The results suggested that the raises in Hb, RBCs and MCH in tadpoles was associated to a maintenance of oxygen capacity to meet the metabolic demand caused by metal stress.

Another study worth mentioning was carried out by Dal-Medico et al. (2014), which demonstrated a marked bradycardic response correlated with incomplete cardiac relaxation after exposure to 1 μg/L of Cd in tadpoles of *L. catesbeianus*. Moreover, *Polypedates maculatus* tadpoles exposed to Ag/ZnO NPs (1–50 mg/L, 60 days) had lower blood parameters (RBCs, hematocrit, white blood cells, monocytes, lymphocytes, and neutrophils) and different serum immune characteristics (alternative complement activity, lysozyme activity, total immunoglobulin content, serum total protein, serum albumin and globulin levels) (Murthy et al. 2022).

3.12 Behavioral Biomarkers

Behavioral changes can indicate direct or indirect impacts of pollutants on aquatic organisms, some of which can be ecologically significant. For example, Lefcort et al. (1998) verified changes in the interactions between *Rana luteiventris* tadpoles and their predators (rainbow trout). The animals were exposed to Zn, Cd and Pb and challenged to respond to predatory stimuli, resulting in a reduced fright response. Araújo et al. (2014), evaluated the responses of 3 tadpole's species (*Leptodactylus latrans, L. catesbeianus,* and *P. perezi*) to Cu. They used a non-forced exposure system with a Cu gradient (110–650 µg/L) in which animals could move around and escape from the environment with the metal. At the lowest concentration, all species avoided Cu, while at the highest concentrations (> 450 µg/L), higher mortality was observed. The study pointed out that metals can disrupt the capacity of tadpoles to sense the presence of the metal in the environment and avoid it. A study by Zhang et al. (2020) found altered swimming performance in *B. raddei* tadpoles collected from 2 sites with different concentrations of Cu, Zn, Cd and Pb, with animals decreasing their swimming speed or resisting to swim. These responses can have ecological consequences, as slower tadpoles are disadvantaged in obtaining food and evading predators.

The element Li has been linked with significant decreases in the activity of premetamorphics *L. catesbeianus* (Pinto-Vidal et al. 2021a) and Cd was associated with a reduced swimming behavior (Chen et al. 2007, Ranatunge et al. 2012, Hu et al. 2019). Moreover, *L. catesbeianus* tadpoles exposed (acutely and chronically) to Zn, Fe, Mn, Pb, Cd and Al from mining waste from the Fundão dam (Mariana, Brazil) presented lower swimming activity and oxygen consumption than animals from the reference (non-contaminated) area (Girotto et al. 2020). The study also showed that the animals avoided areas with the higher concentrations of tailings.

3.13 Mortality

Exposures to Cd (20 µg/L) and Zn (50 µg/L) were described as highly toxic to *Rana luteiventris* tadpoles, leading animals to death after a few weeks at low concentrations. Cd was also associated to mortality in *Pelophylax ridibundus* (syn. for *Rana ridibunda*) tadpoles after exposure to 12.5–100 mg/L over 15 and 30 days (LC_{50} of 71.8 mg/L) (Loumbourdis et al. 1999) and in *Duttaphrynus melanostictus* tadpoles exposed to 1 mg/L for 10 days (Ranatunge et al. 2012). Cu has been described as the cause of mortality in *Rana pipens* (25 and 100 µg/L at 154 days of exposure) (Chen et al. 2007) and in *X. laevis* in CuO nanoparticles exposed animals (0.156 – 2.5 mg/L at 14 days) (Nations et al. 2015).

4. Conclusions and Perspectives

Toxicity of metals in tadpoles is especially relevant due to the broad presence of these contaminants in aquatic environments; however, there is still a lack of information on their molecular, biochemical, and cellular effects on tadpoles. As described in this chapter, the variation in the physicochemical parameters of water can affect the bioavailability and, consequently, the toxicity of metals. Accumulation of metals in tadpoles is heterogeneous, depending on the route of entry, as well as the pattern of bioaccumulation of the element, which can be characterized by a marked initial increase followed by a subsequent drop in concentration. However, several studies have also shown the persistence of metals in tadpole tissues, which also depends on the life stage and concentration to which the animals were exposed to water and/or sediment. Differences in the dynamics of absorption and excretion of metals in the organs of amphibians were noted. These differences can be attributed to a direct contact of the gills and skin with water, as well as the intestinal mucosa with contaminated ingested material. The presence of metals in the body of tadpoles trigger biological changes that can compromise the functionality of the organism and, therefore, its development. Thus, the presence of metals in certain ranges in the environment and their consequent toxicity to tadpoles are of ecological relevance.

Acknowledgments

Dra Claudia Bueno dos Reis Martinez from Animal Ecophysiology Laboratory, Londrina State University - UEL, the Department of Physiological Sciences, Londrina, PR, Brazil, revised the original version of the manuscript. CSC thanks the Program of Biotechnology and Environmental Monitoring (PPGBMA) - Federal University of São Carlos (UFSCar) (Electronic Information System - SEI - No. 23112.013674/2022-26) and the Brazilian research funding institution FAPESP (Proc. 2017/23781-9) for providing the financial support. FAP-V work was supported by the Operational Programme Research, Development and Education - project "Internal Grant Agency of Masaryk University (No.CZ.02.2.69/0.0/0.0/19_073/0016943). FAP-V also thank the Research Infrastructure RECETOX RI (No LM2018121) financed by the Czech Ministry of Education, Youth and Sports, and the Operational Programme Research, Development, and Innovation-project CETOCOEN EXCELLENCE (No CZ.02.1.01/0.0/0.0/26617_043/0009632) for the supportive background. This work was supported by the European Union's Horizon 2020 research and innovation program under grant agreement No 857560. This publication reflects only the author's view, and the European Commission is not responsible for any use that may be made of the information it contains.

References Cited

Abdalla, F.C., L.P.A. Martins, E.C.M.S. Zacarin, M.J. Costa, A.L. Kalinin and D.A. Monteiro. 2013. The Impact of cadmium chloride on the gonadal morphology of the North American bullfrog tadpoles, *Lithobates catesbeianus* (Shaw, 1802). Fresenius Environ. Bull. 22: 1962–1966.

Adams, W., R. Blust, R. Dwyer, D. Mount, E. Nordheim, P.H. Rodriguez et al. 2019. Bioavailability assessment of metals in freshwater environments: A historical review. Environ. Toxicol. Chem. 39: 48–59.

Agency for Toxic Substance and Disease Registry - ATSDR. Toxicological profile for arsenic. U.S. Department of Health and Humans Services, Public Health Service, Centers for Disease Control; Atlanta, GA, 2003.

Aigberua, A., J. Tarawou and C. Abasi. 2018. Effect of oxidation-reduction fluctuations on metal mobility of speciated metals and arsenic in bottom sediments of Middleton River, Bayelsa State, Nigeria. J. Appl. Sci. Environ. Manag. 22: 1511.

Alimba, C.G., A.M. Aladeyelu, I.A. Nwabisi and A.A. Bakare. 2018. Micronucleus cytome assay in the differential assessment of cytotoxicity and genotoxicity of cadmium and lead in *Amietophrynus regularis*. EXCLI J. 17: 89–101.

Almroth, B.C., J. Sturve, A. Berglund and L. Forlin. 2005. Oxidative damage in eelpout (*Zoarces viviparus*), measured as protein carbonyls and TBARS, as biomarkers. Aquat. Toxicol. 73: 171–180.

Al Mahrouqi, D., S. Al Riyami and M.J. Barry. 2018. Effects of Zn and Ti nanoparticles on the survival and growth of *Sclerophrys arabica* tadpoles in a two level trophic system. Bull. Environ. Contam. Toxicol.

Al-Sabti, K. and C.D. Metcalfe. 1995. Fish micronuclei for assessing genotoxicity in water. Mutat Res. 343(2-3): 121–135.

Alvarado, R.H. and A. Moody. 1970. Sodium and chloride transport in tadpoles of the bullfrog *Rana catesbeiana*. Am. J. Physiol. 218: 1510–1516.

Anderson, M.E. 1998. Glutathione: An overview of biosynthesis and modulation. Chem. Biol. Interact. 111-112: 1–14.

Araújo, C.V.M., C. Shinn, M. Moreira-Santos, I. Lopes, E.L.G. Espíndola and R. Ribeiro. 2014. Copper-driven avoidance and mortality in temperate and tropical tadpoles. Aquat. Toxicol. 146: 70–75.

Atli, G. and M. Canli. 2010. Response of antioxidant system of freshwater fish *Oreochromis niloticus* to acute and chronic metal (Cd, Cu, Cr, Zn, Fe) exposures. Ecotoxicol. Environ. Saf. 73: 1884–1889.

Authman, M.M.N., W.T. Abbas and A.Y. Gaafar. 2012. Metals concentrations in Nile tilapia *Oreochromis niloticus* (Linnaeus, 1758) from illegal fish farm in Al-Minufiya Province, Egypt, and their effects on some tissues structures. Ecotoxicol. Environ. Saf. 84: 163–172.

Baken, S., F. Degryse, L. Verheyen, R. Merckx and E. Smolders. 2011. Metal complexation properties of freshwater dissolved organic matter are explained by its aromaticity and by anthropogenic ligands. Environ. Sci. Technol. 45: 2584–2590.

Baldwin, G.F. and P.J. Bentley. 1980. Calcium metabolism in bullfrog tadpoles (*Rana catesbeiana*). J. Exp. Biol. 88: 357–365.

Ballatori, N. 2002. Transport of toxic metals by molecular mimicry. Environ. Health Perspect. 110 (Suppl 5): 689–694. [PubMed: 12426113].

Barhoumi, S., I. Messaoudi, F. Gagné and A. Kerkeni. 2012. Spatial and seasonal variability of some biomarkers in *Salaria basilisca* (Pisces: Blennidae): Implication for biomonitoring in Tunisian coasts. Ecol. Indic. 14(1): 222–228.

Barni, S., R. Vaccarone, V. Bertone, A. Fraschini, F. Bernini and C. Fenoglio. 2002. Mechanisms of changes to the liver pigmentary component during the annual cycle (activity and hibernation) of *Rana esculenta* L. J. Anat. 200: 185–194.

Barra, C.M., R.E. Santelli, J.J. Abrão and M. de la Guardia. 2000. Especiação de arsênio - uma revisão. Quím. Nova, 23 (1): 58–70.

Benedetti, M.F., W.H. Van Riemsdijk, L.K. Koopal, D.G. Kinniburgh, D.C. Gooddy and Milne. 1996. Metal ion binding by natural organic matter: From the model to the field. Geochim. Cosmochim. Acta 60(14): 2503–2513.

Benvindo-Souza, M., E.A.S. Oliveira, R.A. Assis, C.G.A. Santos, R.E. Borges, D.L. Melo e Silva and L.R.S. Santos. 2020. Micronucleus test in tadpole erythrocytes: Trends in studies and new paths. Chemosphere 240: 124910.

Berven, K.A. 1990. Factors affecting population fluctuations in larval and adult stages of the wood frog (*Rana Sylvatica*). Ecology 71(4): 1599–1608.

Berzins, D. and K.J. Bundy. 2002. Bioaccumulation of lead in *Xenopus laevis* tadpoles from water and sediment. Environ. Int. 28: 69–77.

Bezerra, P.S.S., L.R. Takiyama and C.W.B. Bezerra. 2009. Complexação de íons de metais por matéria orgânica dissolvida: modelagem e aplicação em sistemas reais. Acta Amaz. 39(3): 639–648.

Birhanli, A., F.B. Emre, F. Sayilkan and A. Gungordu. 2014. Effect of nanosized TiO_2 particles on the development of *Xenopus laevis* embryos. Turk. J. Biol. 38: 283–288.

Boczulak, S.A., M.C. Vanderwel and B.D. Hall. 2017. Survey of mercury in boreal chorus frog (*Pseudacris maculata*) and wood frog (*Rana sylvatica*) tadpoles from wetland ponds in the Prairie Pothole Region of Canada. FACETS 2: 315–329.

Boiarski, D.R., C.M. Toigo, T.M. Sobjak, A.F.P. Santos, S. Romão and A.T.B. Guimarães. 2020. Assessment of antioxidant system, cholinesterase activity and histopathology in *Lithobates catesbeianus* tadpoles exposed to water from an urban stream. Ecotoxicology 29(3): 314–326.

Brasil. 2005. Resolução CONAMA nº 357, de 17 de março de 2005. Conselho Nacional de Meio Ambiente. Available in: http://pnqa.ana.gov.br/Publicacao/RESOLUCAO_CONAMA_n_357.pdf.

Bridges, C.M., F.J. Dwyer, D.K. Hardesty and D.W. Whites. 2002. Comparative contaminant toxicity: Are amphibian larvae more sensitive than fish? Bull. Environ. Contam. Toxicol. 69(4): 562–569.

Bridges, C.C. and R.K. Zalups. 2005. Molecular and ionic mimicry and the transport of toxic metals. Toxicol. Appl. Pharm. 204(3): 274–308.

Carlsson, G. and E. Tydén. 2018. Development and evaluation of gene expression biomarkers for chemical pollution in common frog (*Rana temporaria*) tadpoles. Environ. Sci. Poll. Res. 25(1): 33131–33139.

Carvalho, C.S., V.A., Bernusso and M.N. Fernandes. 2015. Copper levels and changes in pH induce oxidative stress in the tissue of curimbata (*Prochilodus lineatus*). Aquat Toxicol. 167: 220–227.

Carvalho, C.S., H.S.M. Utsunomiya, T. Pasquoto, R. Lima, M.J. Costa and M.N.F. Fernandes. 2017. Blood cell responses and metallothionein in the liver, kidney and muscles of bullfrog tadpoles, *Lithobates catesbeianus*, following exposure to different metals. Environ. Poll. 221: 445–452.

Carvalho, C.S., H.S.M. Utsunomiya, T. Pasquoto, M.J. Costa and M.N. Fernandes. 2020. Biomarkers of the oxidative stress and neurotoxicity in tissues of the bullfrog, *Lithobates catesbeianus* to assess exposure to metals. Ecotoxicol. Environ. Saf. 196: 110560.

Cavas, T., N.N. Garanko and V.V. Arkhipchuk. 2005. Induction of micronuclei and binuclei in blood, gill and liver cells of fishes subchronically exposed to cadmium chloride and copper sulphate. Food Chem. Toxicol. 43(4): 569–574.

Chagas, B.R.C., H.S.M. Utsunomiya, M.N. Fernandes and C.S. Carvalho. 2020. Metabolic responses in bullfrog, *Lithobates catesbeianus* after exposure to zinc, copper and cadmium. Comp. Biochem. Physiol. C. 233: 108768.

Chai, L., H. Wang, H. Deng, H. Zhao and W. Wang. 2014. Chronic exposure effects of copper on growth, metamorphosis and thyroid gland, liver health in Chinese toad, *Bufo gargarizans* tadpoles. Chemistry and Ecology 30(7): 589–601.

Chen, L., J. Zhang, Y. Zhu and Y. Zhang. 2018. Interaction of chromium (III) or chromium (VI) with catalase and its effect on the structure and function of catalase: An *in vitro* study. Food Chem. 244: 378–385.

Chen, T.-H., J.A. Gross and W.H. Karasov. 2007. Adverse effects of chronic copper exposure in larval northern leopard frogs (*Rana pipiens*). Environ. Toxicol. Chem. 26(7): 1470.

Cherian, M.G. and J.J. Vostal. 1977. Biliary excretion of cadmium in rat. I. Dose-dependent biliary excretion and the form of cadmium in the bile. J. Toxicol. Environ. Health 2: 945–954.

Collin, F. 2019. Chemical basis of reactive oxygen species reactivity and involvement in neurodegenerative diseases. Int. J. Mol. Sci. 20(10): 2407.

Colombo, A., P. Bonfanti, F. Orsi and M. Camatini. 2003. Differential modulation of cytochrome P-450 1A and P-glycoprotein expression by aryl hydrocarbon receptor agonists and thyroid hormone in *Xenopus laevis* liver and intestine. Aquat. Toxicol. 63: 173–186.

Coyle, P., J.C. Philcox, L.C. Carey and A.M. Rofe. 2002. Metallothionein: The multipurpose protein. Cell. Mol. Life Sci. 59(4): 627–647.

Cuypers, A., M. Plusquin, T. Remans, M. Jozefczak, E. Keunen, H. Gielen et al. 2010. Cadmium stress: An oxidative challenge. Biometals. 23(5): 927–940.

Dalle-Donne, I., D. Giustarini, R. Colombo, R. Rossi and A. Milzani. 2003. Protein carbonylation in human diseases. Trends Mol. Med. 9(4): 169–176.

Dal-Medico, S.E., R.Z. Rissoli, F.U. Gamero, J.A. Victório, R.F. Salla, F.C. Abdalla et al. 2014. Negative impact of a cadmium concentration considered environmentally safe in Brazil on the cardiac performance of bullfrog tadpoles. Ecotoxicol. Environ. Saf. 104: 168–174.

Davies, P.H., W.C. Gorman, S.F. Brinkman and C.A. Carlson. 1993. Effect of hardness on bioavailability and toxicity of cadmium to rainbow trout. Chem. Speciat. Bioavailab. 5: 67–77.

De Boeck, G., T.T. Nago, K. Van Campenhout and R. Blust. 2003. Differential metallothionein induction pattern in three freshwater fish during sublethal copper exposure. Aquat. Toxicol. 65(4): 413–424.

De Oliveira, C., L. Franco-Belussi, L.Z. Fanali and L.R. Santos. 2017. Use of melanin-pigmented cells as a new tool to evaluate effects of agrochemicals and other emerging contaminants in Brazilian anurans. Ecotoxicology Genotoxicology: Non-traditional Terrestrial Models 32: 125.

de Paiva Magalhães, D., M.R. da Costa Marques, D.F. Baptista and D.F. Buss. 2015. Metal bioavailability and toxicity in freshwaters. Environ. Chem. Lett. 13: 69–87.

DeLaune, R.D. and K.R. Reddy. 2005. Redox Potential. pp. 366–371. *In:* Hillel, D. [ed.]. Encyclopedia of Soils in the Environment. Academic Press, New York, USA.

Devesa-Rey, R., F. Díaz-Fierros and M.T. Barral. 2010. Trace metals in river bed sediments: an assessment of their partitioning and bioavailability by using multivariate exploratory analysis. J. Environ. Manage. 9: 2471–2477.

Dietz, T.H. and R.H. Alvarado. 1974. Na and Cl transport across gill chamber epithelium of *Rana catesbeiana* tadpoles. Am. J. Physiol. 226: 764–770.

Diggs, H.E. and J.M. Parker. 2009. Aquatic facilities. Plan. Des. Res. Anim. Facil. 323–331.

Di-Toro, D.M., H.E. Allen, H.L. Bergman, J.S. Meyer, P.R. Paquin and R.C. Santore. 2001. Hazard/risk assessment: Biotic ligand model of the acute toxicity of metals. Environ. Toxicol. Chem. 20: 2383–2396.

Dobrovoljc, K., Z. Jeran and B. Bulog. 2003. Uptake and elimination of cadmium in *Rana dalmatina* (Anura, Amphibia) tadpoles. Bull. Environ. Contam. Toxicol. 70: 78–84.

Dovick, M.A., R.S. Arkle, T.R. Kulp and D.S. Pilliod. 2020. Extreme arsenic and antimony uptake and tolerance in toad tadpoles during development in highly contaminated Wetlands. Environ. Sci. Technol. 54: 7983–7991.

Dubaissi, E., K. Rousseau, G.W. Hughes, C. Ridley, R.K. Grencis, I.S. Roberts et al. 2018. Functional characterization of the mucus barrier on the *Xenopus tropicalis* skin surface Proc. Natl. Acad. Sci. Unit. States Am. 115(4): 726–731.

Dudley, R.E., L.M. Gammal and C.D. Klaassen. 1985. Cadmium-induced hepatic and renal injury in chronically exposed rats: likely role of hepatic cadmium-metallothionein in nephrotoxicity. Toxicol. Appl. Pharmacol. 77: 414–426.

Dusse, L.M.S., L.M. Vieira and M.G. Carvalho. 2003. Revisão sobre óxido nítrico. J. Bras. Patol. Med. Lab. 39(4): 343–350.

Ebrahimpour, M., H. Alipour and S. Rakhshah. 2010. Influence of water hardness on acute toxicity of copper and zinc on fish. Toxicol. Ind. Health 26: 361–365.

Edwards, P.G., K.F. Gaines, A.L. Bryan, J.M. Novak and S.A. Blas. 2013. Trophic dynamics of U, Ni, Hg and other contaminants of potential concern on the Department of Energy's Savannah River Site. Environ. Monit. Assess. 186(1): 481–500.

Endo, T. 2002. Transport of cadmium across the apical membrane of epithelial cell lines. Comp. Biochem. Physiol., C Toxicol. Pharmacol. 131: 223–239.

Fairbrother, A., R. Wenstel, K. Sappington and W. Wood. 2007. Framework for metals risk assessment. Ecotoxicol. Environ. Saf. 68(2): 145–227.

Farina, M., M. Aschner and J.B.T. Rocha. 2011. Special issue: Environmental chemicals and neurotoxicity oxidative stress in MeHg-induced neurotoxicity. Toxicol. Appl. Pharmacol. 256(3): 405–417.

Fenoglio, C., E. Boncompagni, M. Fasola, C. Gandini, S. Comizzoli, G. Milanesi et al. 2005. Effects of environmental pollution on the liver parenchymal cells and Kupffer-melanomacrophagic cells of the frog *Rana esculenta*. Ecotoxicol. Environ. Saf. 60(3): 259–268.

Fernandes, I.F., H.S.M. Utsunomiya, B.S.L. Valverde, J.V.C. Ferraz, G.H. Fujiwara, D.M. Gutierres et al. 2021. Ecotoxicological evaluation of water from the Sorocaba River using an integrated analysis of biochemical and morphological biomarkers in bullfrog tadpoles, *Lithobates catesbeianus* (Shaw, 1802). Chemosphere 275: 130000.

Fernando, V.A., J. Weerasena, G.P. Lakraj, I.C. Perera, C.D. Dangalle, S. Handunnetti et al. 2016. Lethal and sub-lethal effects on the Asian common toad *Duttaphrynus melanostictus* from exposure to hexavalent chromium. Aquat. Toxicol. 177: 98–105.

Ferreira, C.M., H.M. Bueno-Guimarães, M.J.T. Ranrani-Paiva, S.R.C. Soares, D.H.R.F. Rivero and P.H.N. Saldiva. 2003. Hematological markers of copper toxicity in *Rana catesbeiana* tadpoles (Bullfrog). Rev. Bras. Toxicol. 16(2): 83–88.

Freda, J., V. Cavdek and D.G. McDonald. 1990. Role of organic complexation in the toxicity of aluminum to *Rana pipiens* embryos and *Bufo americanus* tadpoles. Can. J. Fish. Aquat. Sci. 47(1): 217–224.

Fong, P.P., L.B. Thompson, G.L.F. Carfagno and A.J. Sitton. 2016. Long-term exposure to gold nanoparticles accelerates larval metamorphosis without affecting mass in wood frogs (*Lithobates sylvaticus*) at environmentally relevant concentrations. Environ. Toxicol. Chem. 35(9): 2304–2310.

Gastaldo, J., M. Viau, M. Bouchot, A. Joubert, A.-M. Charvet and N. Foray. 2008. Induction and repair rate of DNA damage: A unified model for describing effects of external and internal irradiation and contamination with heavy metals. J. Theor. Biol. 251: 68–81.

Gaté, L., J. Paul, G.N. Ba, K.D. Tew and H. Tapiero. 1999. Oxidative stress induced in pathologies: The role of antioxidants. Biomed. Pharmacother. 53(4): 169–180.

Girotto, L., E.L.G. Espíndola, R.C. Gebara and J.S. Freitas. 2020. Acute and chronic effects on tadpoles (*Lithobates catesbeianus*) exposed to mining tailings from the Dam rupture in Mariana, MG (Brazil). Water Air Soil Pollut. 231(7).

Gross, J.A., T.-H. Chen and W.H. Karasov. 2007. Lethal and sublethal effects of chronic cadmium exposure on northern leopard frog (*Rana pipiens*) tadpoles. Environ. Toxicol. Chem. 26(6): 1192.

Grune, T., K. Merker, G. Sandig and K.J.A. Davies. 2003. Selective degradation of oxidatively modified protein substrates by the proteasome. Biochem. Biophys. Res. Commun. 305(3): 709–718.

Guardiola, F.A., M. Dioguardi, M.G. Parisi, M.R. Trapani, J. Meseguer, A. Cuesta et al. 2015. Evaluation of waterborne exposure to heavy metals in innate immune defences present on skin mucus of gilthead seabream (*Sparus aurata*). Fish Shellfish Immunol. 45(1): 112–123.

Güngördü, A., A. Birhanlı and M. Ozmen. 2010. Assessment of embryotoxic effects of cadmium, lead and copper on *Xenopus laevis*. Fresenius Environ. Bull. 19: 2528–2535.

Gürkan, M., A. Çetin and S. Hayretdağ. 2014. Acute toxic effects of cadmium in larvae of the green toad, *Pseudepidalea variabilis* (Pallas, 1769) (Amphibia: Anura). Arch. Ind. Hyg. Toxicol. 65(3): 301–309.

Gusso-Choueri, P.K., R.B. Choueri, G.S. de Araújo, A.C.F. Cruz, T. Stremel, S. Campos et al. 2015. Assessing pollution in marine protected areas: The role of a multi-biomarker and multi-organ approach. Environ. Sci. Pollut. Res. 22: 18047–18065.

Halliwell, B. and J.M.C. Gutteridge. 2007. Free Radicals in Biology and Medicine. Oxford University Press, Oxford.

Han, D., N. Hanawa, B. Saberi and N. Kaplowitz. 2006. Mechanisms of Liver Injury. III. Role of glutathione redox status in liver injury. Am. J. Physiol. Gastrointest. Liver Physiol. 291(1): G1–G7.

Heijn, M., J.H. Hooijberg, G.L. Scheffer, G. Szabo, H.V. Westerhoff and J. Lankelma. 1997. Anthracyclines modulate multidrug resistance protein (MRP) mediated organic anion transport. Biochim. Biophys. Acta 1326: 12–22.

Herkovits, J. and C.S. Perez-Coll. 1993. Stage dependent susceptibility of *Bufo arenarum* embryos to cadmium. Bull. Environ. Contam. Toxicol. 50: 608–611.

Hermes Lima, M. 2005. Oxygen in biology and biochemistry. Role of Free Radicals. pp. 319–368. *In*: Storey, K.B. [ed.]. Functional Metabolism: Regulation and Adaptation. John Wiley & Sons, Inc., Hoboken, USA.

Hinojosa-Garro, D., J.R.-V. Ostenb and R. Dzul-Caamal. 2020. Banded tetra (*Astyanax aeneus*) as bioindicator of trace metals in aquatic ecosystems of the Yucatan Peninsula, Mexico: Experimental biomarkers validation and wild populations biomonitoring. Ecotoxicol. Environ. Saf. 195: 110477.

Hofer, R., R. Lackner and G. Lorbeer. 2005. Accumulation of toxicants in tadpoles of the common frog (*Rana temporaria*) in high mountains. Arch. Enviro. Contam. Toxicol. 49(2): 192–199.

Home, M.T. and W.A. Dunson. 1995. Effects of low pH, metals, and water hardness on larval Amphibians. Arch. Environ. Contam. Toxicol. 29: 500–505.

Huang, M.-Y., R.-Y. Duan and X. Ji. 2014. Chronic effects of environmentally-relevant concentrations of lead in *Pelophylax nigromaculata* tadpoles: Threshold dose and adverse effects. Ecotoxicol. Environ. Saf. 104: 310–316.

Hu, Y.-C., Y. Tang, Z.-Q. Chen, J.-Y. Chen and G.-H. Ding. 2019. Evaluation of the sensitivity of *Microhyla fissipes* tadpoles to aqueous cadmium. Ecotoxicology.

James, S.M. and E.E. Little. 2003. The effects of chronic cadmium exposure on American toad (*Bufo americanus*) tadpoles. Environ. Toxicol. Chem. 22(2): 377–380.

Javed, M., N. Usmani, I. Ahmad and M. Ahmad. 2015. Studies on the oxidative stress and gill histopathology in *Channa punctatus* of the canal receiving heavy metal loaded effluent of Kasimpur Thermal Power Plant. Environ. Monit. Assess. 187: 1–11.

Javed, M., I. Ahmad, N. Usmani and M. Ahmad. 2016. Studies on biomarkers of oxidative stress and associated genotoxicity and histopathology in *Channa punctatus* from heavy metal polluted canal. Chemosphere 151: 210–219.

Jayaprakash Rao, I. and M.N. Madhyastha. 1987. Toxicities of some heavy metals to the tadpoles of frog, *Microhyla ornata* (dumeril & bibron). Toxicol. Lett. 36(2): 205–208.

Johansen, S. 2013. Element accumulation and levels of four biomarkers in common frog (*Rana temporaria*) tadpoles in two sedimentation ponds and a naturally occurring pond. Master Theses.

Ju, Z., J. Ya, X. Li, H. Wang and H. Zhao. 2020. The effects of chronic cadmium exposure on *Bufo gargarizans* larvae: Histopathological impairment, gene expression alteration and fatty acid metabolism disorder in the liver. Aquat. Toxicol. 222: 105470.

Kang, M., Y. Tian, S. Peng and M. Wang. 2019. Effect of dissolved oxygen and nutrient levels on heavy metal contents and fractions in river surface sediments. Sci. Total Environ. 648: 861–870.

Karbassi, A., G. Bidhendi, A. Pejman and M.E. Bidhendi. 2010. Environmental impacts of desalination on the ecology of Lake Urmia. Journal of Great Lakes Research 36: 419–424.

Kelepertzis, E., A. Argyraki, E. Valakos and E. Daftsis. 2012. Distribution and accumulation of metals in tadpoles inhabiting the metalliferous streams of eastern Chalkidiki, northeast Greece. Arch. Environ. Contam. Toxicol. 63: 409–420.

Keppler, D., I. Leier, G. Jedlitschky and J. Konig. 1998. ATP-dependent transport of glutathione S-conjugates by the multidrug resistance protein MRP. Chem. Biol. Interact. 24: 111-112, 153–161.

Khangarot, B.S., A. Sehgal and M.K. Bhasin. 1985. Man and Biosphere - Studies on the sikkim himalayas. Part 6: Toxicity of selected pesticides to frog tadpole *Rana hexadactyla* (Lesson). Acta Hydrochim. Hydrobiol. 13(3): 391–394.

Khangarot, B.S. and P.K. Ray. 1987. Sensitivity of toad tadpoles, *Bufo melanostictus* (Schneider), to heavy metals. Bull. Environ. Contam. Toxicol. 38(3): 523–527.

Klemish, J.L., S.J. Bogart, A. Luek, M.J. Lannoo and G.G. Pyle. 2018. Nickel toxicity in wood frog tadpoles: Bioaccumulation and sublethal effects on body condition, food consumption, activity, and chemosensory function. Environ. Toxicol. Chem. 37(9): 2458–2466.

Koch, I., J. Zhang, M. Button, L.A. Gibson, G. Caumette, V.S. Langlois et al. 2015. Arsenic (+3) and DNA methyltransferases, and arsenic speciation in tadpole and frog life stages of western clawed frogs (*Silurana tropicalis*) exposed to arsenate. Metallomics. (8): 1274–1284.

Krohn, R.M., V. Palace and J.E.G. Smits. 2020. Metal changes in pre- and post-metamorphic wood frog (*Lithobates sylvaticus*) tadpoles: Implications for ecotoxicological studies. Arch. Environ. Contam. Toxicol. 80: 760–768.

Lance, S.L., R.W. Flynn, M.R. Erickson and D.E. Scott. 2013. Within- and among-population level differences in response to chronic copper exposure in southern toads, *Anaxyrus terrestris*. Environ. Poll. 177: 135–142.

Lanctôt, C., W. Bennett, S. Wilson, L. Fabbro, F.D.L. Leusch and S.D. Melvin. 2016. Behaviour, development and metal accumulation in striped marsh frog tadpoles (*Limnodynastes peronii*) exposed to coal mine wastewater. Aquat. Toxicol. 173: 218–227.

Lanctôt, C.M., T. Cresswell and S.D. Melvin. 2017. Uptake and tissue distributions of cadmium, selenium and zinc in striped marsh frog tadpoles exposed during early post-embryonic development. Ecotoxicol. Environ. Saf. 144: 291–299.

Lefcort, H., R.A. Meguire, L.H. Wilson and W.F. Ettinger. 1998. Heavy metals alter the survival, growth, metamorphosis, and antipredatory behavior of Columbia spotted frog (*Rana luteiventris*) tadpoles. Arch. Environ. Contam. Toxicol. 35(3): 447–456.

Leier, I., G. Jedlitschky, U. Buchholz, M. Center, S.P.C. Cole, R.G. Deeley et al. 1996. ATP-dependent glutathione disulfide transport mediated by the MRP gene-encoded conjugate export pump. Biochem. J. 314: 433–437.

Leslie, E.M., R.G. Deeley and S.P. Cole. 2001. Toxicological relevance of the multidrug resistance protein 1, MRP1 (ABCC1) and related transporters. Toxicology 167: 3–23.

Li, H., A. Shi, M. Li and X. Zhang. 2013. Effect of pH, temperature, dissolved oxygen, and flow rate of overlying water on heavy metals release from storm sewer sediments. J. Chem. 2013: 1–11.

Lloyd, R. 1960. The toxicity of zinc sulphate to rainbow trout. Ann. App. Biol. 48(1): 84–94.

Lloyd, R. 1961. Effect of dissolved oxygen concentrations on the toxicity of several poisons to rainbow trout (*Salmo gairdnerii* Richardson). Exp. Biol. 38(2): 447–455.

Lopez, E.L., J. EliasSedeño-Díaz, C. Soto and L. Favari. 2011. Responses of antioxidant enzymes, lipid peroxidation, and Na+/K+- ATPase in liver of the fish *Goodea atripinnis* exposed to Lake Yuriria water. Fish Physiol. Biochem. 37: 511–522.

Loumbourdis, N.S., P. Kyriakopoulou-Sklavounou and G. Zachariadis. 1999. Effects of cadmium exposure on bioaccumulation and larval growth in the frog *Rana ridibunda*. Environ. Poll. 104(3): 429–433.

Loumbourdis, N.S., I. Kostaropoulos, B. Theodoropoulou and D. Kalmanti. 2007. Heavy metal accumulation and metallothionein concentration in the frog *Rana ridibunda* after exposure to chromium or a mixture of chromium and cadmium. Environ. Poll. 145: 787–792.

Lu, H., Y. Hu, C. Kang, Q. Meng and Z. Lin. 2021. Cadmium-induced toxicity to amphibian tadpoles might be exacerbated by alkaline not acidic pH level. Ecotoxicol. Environ. Saf. 218: 112288.

Lutzen, A., S.E. Liberti and L.J. Rasmussen. 2004. Cadmium inhibits human DNA mismatch repair *in vivo*. Biophys. Res. Commun. 321: 21–25.

Mardirosian, M.N., C.I. Lascano, A. Ferrari, G.A. Bongiovanni and A. Venturino. 2015. Acute toxicity of arsenic and oxidative stress responses in the embryonic development of the common South American toad *Rhinella arenarum*. Environ. Toxicol. Chem. 34(5): 1009–1014.

Marques, S.M., S. Chaves, F. Gonçalves and R. Pereira. 2013. Evaluation of growth, biochemical and bioaccumulation parameters in *Pelophylax perezi* tadpoles, following an in-situ acute exposure to three different effluent ponds from a uranium mine. Sci. Total. Environ. 445-446: 321–328.

Martinez, C.B.R., M.Y. Nagae, C.T.B.V. Zaia and D.A.A. Zaia. 2004. Acute morphological and physiological effects of lead in the neotropical fish *Prochilodus lineatus*. Braz. J. Biol. 64: 797–807.

Mastrángelo, M., M. Dos Santos Afonso and L. Ferrari. 2011. Cadmium toxicity in tadpoles of *Rhinella arenarum* in relation to calcium and humic acids. Ecotoxicology 20: 1225–1232.

McDonald, D.G., J.L. Ozog and B.P. Simons. 1984. The influence of low pH environments on ion regulation in the larval stages of the anuran amphibian, *Rana clamitans*. Can. J. Zool. 62(11): 2171–2177.

McDonald D.G. and C.M. Wood. 1993. Branchial mechanisms of acclimation to metals in freshwater fish. *In:* Rankin, J.C. and F.B. Jensen [eds.]. Fish Ecophysiology. Chapman & Hall Fish and Fisheries Series, vol 9. Springer, Dordrecht.

McIntyre P.B. and S.A. McCollum. 2000. Responses of bullfrog tadpoles to hypoxia and predators. Oecologia 125(2): 301–308.

Medina, M.F., A. Cosci, S. Cisint, C.A. Crespo, I. Ramos, A.L.I. Villagra et al. 2012. Histopathological and biological studies of the effect of cadmium on *Rhinella arenarum* gonads. Tissue Cell 44: 418–426.

Meindl, G.A., N. Schleissmann, B. Sander, M. Lam, W. Parker, G. Fitzgerald et al. 2020. Exposure to metals (Ca, K, Mn) and road salt (NaCl) differentially affect development and survival in two model amphibians. Chem. Ecol. 1–11.

Mills, N.E. and M.C. Barnhart. 1999. Effects of hypoxia on embryonic development in two Ambystoma and two Rana species. Physiol. Biochem. Zool: Ecol. Evol. Approaches 72: 179–788.

Mishra, V.K., A.R. Upadhyay, S.K. Pandey and B.D. Tripathi. 2008. Concentrations of heavy metals and aquatic macrophytes of Govind Ballabh Pant Sagar an anthropogenic lake affected by coal mining effluent. Environ. Monit. Assess. 141: 49–58.

Monserrat, J.M., P.E. Martínez, L.A. Geracitano, L.L. Amado, C.M.G. Martins, G.L.L. Pinho et al. 2007. Pollution biomarkers in estuarine animals: Critical review and new perspectives. Comp. Biochem. Physiol. C 146: 221–234.

Monteiro, D.A., F.T. Rantin and A.L. Kalinin. 2010. Inorganic mercury exposure: Toxicological effects, oxidative stress biomarkers and bioaccumulation in the tropical freshwater fish matrinxã, *Brycon amazonicus* (Spix and Agassiz, 1829). Ecotoxicology 19: 105–123.

Monteiro, J.A.D., L.A. da Cunha, M.H.P. da Costa, H.S. dos Reis, A.C.D. Aguiar, V.R.L. de Oliveira-Bahia et al. 2018. Mutagenic and histopathological effects of hexavalent chromium in tadpoles of *Lithobates catesbeianus* (Shaw, 1802) (Anura, Ranidae). Ecotoxicol. Environ. Saf. 163: 400–407.

Mouchet, F., M. Baudrimont, P. Gonzalez, Y. Cuenot, J.P. Bourdineaud, A. Boudou et al. 2006. Genotoxic and stress inductive potential of cadmium in *Xenopus laevis* larvae. Aquat. Toxicol. 78(2): 157–166.

Mouchet, F., L. Gauthier, M. Baudrimont, P. Gonzalez, C. Mailhes, V. Ferrier, et al. 2007. Comparative evaluation of the toxicity and genotoxicity of cadmium in amphibian larvae (*Xenopus laevis* and *Pleurodeles waltl*) using the comet assay and the micronucleus test. Environ. Toxicol. 22: 422–435.

Murthy, M.K., C.S. Mohanty, P. Swain and R. Pattanayak. 2022. Assessment of toxicity in the freshwater tadpole *Polypedates maculatus* exposed to silver and zinc oxide nanoparticles: A multi-biomarker approach. Chemosphere. 293: 133511.

Nations, S., M. Long, M. Wages, J. Canas, J.D. Maul, C. Theodorakis et al. 2011. Effects of ZnO nanomaterials on *Xenopus laevis* growth and development. Ecotoxicol. Environ. Saf. 74(2): 203–210.

Nations, S., M. Long, M. Wages, J.D. Maul, C.W. Theodorakis and G.P. Cobb. 2015. Subchronic and chronic developmental effects of copper oxide (CuO) nanoparticles on *Xenopus laevis*. Chemosphere 135: 166–174.

Nunes, B. 2011. The use of cholinesterases in ecotoxicology. Rev. Environ. Contam. Toxicol. 212: 29–59.

Okubo, M., K. Yamada, M. Hosoyamada, T. Shibasaki and H. Endou. 2003. Cadmium transport by human Nramp 2 expressed in *Xenopus laevis* oocytes. Toxicol. Appl. Pharmacol. 187: 162–167.

Ossana, N.A., P.M. Castañé, P.L. Sarmiento and A. Salibián. 2010. Do *Lithobates catesbeianus* tadpoles acclimatise to sub-lethal copper? Int. J. Environ. Health 4(4): 342–354.

Ossana, N.A., P.M. Castané and A. Salibán. 2013. Use of *Lithobates catesbeianus* tadpoles in a multiple biomarker approach for the assessment of water quality of the Reconquista River (Argentina). Arch. Environ. Contam. Toxicol. 65(3): 486–497.

Ohta, H. and K. Ohba. 2020. Involvement of metal transporters in the intestinal uptake of cadmium. J. Toxicol. Sci. 45(9): 539–548.

Pagenkopf, G.K. 1983. Gill surface interaction model for trace metal toxicity of fish. Role of complexation, pH, water hardness. Environ. Sci. Technol. 17: 342–347.

Papadimitriou, E. and N.S. Loumbourdis. 2002. Exposure of the frog *Rana ridibunda* to copper: Impact on two biomarkers, lipid peroxidation, and glutathione. Bull. Environ. Contam. Toxicol. 69(6): 885–891.

Pascoe, D., S.A. Evans and J. Woodworth. 1986. Heavy metal toxicity to fish and the influence of water hardness. Arch. Environ. Contam. Toxicol. 15: 481–487.

Patar, A., A. Giri, F. Boro, K. Bhuyan, U. Singha and S. Giri. 2016. Cadmium pollution and amphibians—Studies in tadpoles of *Rana limnocharis*. Chemosphere 144: 1043–1049.

Patar, A., I. Das, S. Giri and A. Giri. 2021. Zinc contamination is an underestimated risk to amphibians: Toxicity evaluation in tadpoles of *Fejervarya limnocharis*. J. Environ. Eng. Landsc. Manag. 29(4): 489–498.

Pattanayak, R., R. Das, A. Das, S.K. Padhi, S.S. Sahu, S. Pattnaik et al. 2018. Toxicological effects of silver nanoparticles (Ag-NPs) on different physiological parameters of tadpoles, *Polypedates maculatus*. Intl. J. Bio-res. Stress Manag. 9(5): 647–654.

Peles, J.D. 2013. Effects of chronic aluminum and copper exposure on growth and development of wood frog (*Rana sylvatica*) larvae. Aquat. Toxicol. 140-141: 242–248.

Peltzer, P.M., R.C. Lajmanovich, A.M. Attademo, C.M. Junges, M.C. Cabagna-Zenklusen, M.R. Repetti et al. 2013. Effect of exposure to contaminated pond sediments on survival, development, and enzyme and blood biomarkers in veined treefrog (*Trachycephalus typhonius*) tadpoles. Ecotoxicol. Environ. Saf. 98: 142–151.

Peluso, J., C.M. Aronzon, M. Ríos de Molina, M. del C., D.E. Rojas, D. Cristos and C.S. Pérez Coll. 2020. Integrated analysis of the quality of water bodies from the lower Paraná River basin with different productive uses by physicochemical and biological indicators. Environ. Poll. 263: 114434.

Perez-Coll, C.S. and J. Herkovits. 1990. Stage dependent susceptibility to lead in *Bufo arenarum* embryos. Environ. Poll. 63: 239–245.

Perez-Coll, C.S. and J. Herkovits. 1996. Stage-dependent uptake of cadmium by *Bufo arenarum* embryos. Bull. Environ. Contam. Toxicol. 56: 663–669.

Pérez-Iglesias, J.M., L. Franco-Belussi, G.S. Natale and C. de Oliveira. 2019. Biomarkers at different levels of organisation after atrazine formulation (SIPTRAN 500SC®) exposure in *Rhinellas chineideri* (Anura: Bufonidae) Neotropical tadpoles. Environ. Poll. 244: 733–746.

Pinto Vidal, F.A., F.C. Abdalla, C. dos S. Carvalho, H.S. Moraes Utsunomiya, L.A. Teixeira Oliveira, R.F. Salla et al. 2021a. Metamorphic acceleration following the exposure to lithium and selenium on American bullfrog tadpoles (*Lithobates catesbeianus*). Ecotoxicol. Environ. Saf. 207: 111101.

Pinto-Vidal, F.A., C.S. Carvalho, F.C. Abdalla, L. Ceschi-Bertoli, H.S. Moraes Utsunomiya, R.H. Silva et al. 2021b. Metabolic, immunologic, and histopathologic responses on premetamorphic American bullfrog (*Lithobates catesbeianus*) following exposure to lithium and selenium. Environ. Pollut. 270: 116086.

Pinto-Vidal, F.A., C. dos, S. Carvalho, F.C. Abdalla, H.S.M. Utsunomiya, R.F. Salla and M. Jones-Costa. 2022. Effects of lithium and selenium in the tail muscle of American bullfrog tadpoles (*Lithobates catesbeianus*) during premetamorphosis. Environ. Sci. Pollut. Res. 29: 1975–1984.

Playle, R.C. 1998. Waterborne metals can bind to gills of freshwater fish and disrupt the ionoregulatory and respiratory functions of the gills. Sci. Total Environ. 219: 147–163.

Prosi, F. 1981. Heavy metals in aquatic organisms. pp. 271–323. *In:* Metal Pollution in the Aquatic Environment. Springer Study Edition. Springer, Berlin, Heidelberg.

Putnam, C.D., A.S. Arvai, Y. Bourne and J.A. Tainer. 2000. Active and inhibited human catalase structures: Ligand and NADPH binding and catalytic mechanism. J. Mol. Biol. 296: 295–309.

Rabenstein, D.L., A.A. Isab, W. Kadima and P. Mohanakrishnan. 1983. A proton nuclear magnetic resonance study of the interaction of cadmium with human erythrocytes. Biochim. Biophys. Acta 762: 531–541.

Rainbow, P.S. 2002. Trace metal concentrations in aquatic invertebrates: why and so what? Environ. Poll. 120(3): 497–507.

Ranatunge, R.A.A.R., M.R. Wijesinghe, W.D. Ratnasooriya, H.A.S.G. Dharmarathne and R.D. Wijesekera. 2012. Cadmium-induced toxicity on larvae of the common Asian toad *Duttaphrynus Melanostictus* (Schneider 1799): Evidence from empirical trials. Bull. Environ. Contam. Toxicol. 89(1): 143–146.

Reverter, M., N. Tapissier-Bontemps, D. Lecchini, B. Banaigs and P. Sasal. 2018. Biological and ecological roles of external fish mucus: A review. Fishes 3: 41.

Ritz, G.F. and J.A. Collins. 2008, pH 6.4 (ver. 2.0, October 2008): U.S. Geological Survey Techniques of Water-Resources Investigations, book 9, chap. A6.4,

Robinson, M.K., D.W. Barfuss and R.K. Zalups. 1993. Cadmium transport and toxicity in isolated perfused segments of the renal proximal tubule. Toxicol. Appl. Pharmacol. 121(1): 103–111.

Rowe, C.L., A. Heyes and J. Hilton. 2011. Differential patterns of accumulation and depuration of dietary selenium and vanadium during metamorphosis in the gray treefrog (*Hyla versicolor*). Arch. Environ. Contam. Toxicol. 60: 336–342.

Rumrill, C.T., D.E. Scott and S.L. Lance. 2016. Effects of metal and predator stressors in larval southern toads (*Anaxyrus terrestris*). Ecotoxicology 25(6): 1278–1286.

Salvaterra, T., M.G. Alves, I. Domingues, R. Pereira, M.G. Rasteiro, R.A. Carvalho et al. 2013. Biochemical and metabolic effects of a short-term exposure to nanoparticles of titanium silicate in tadpoles of *Pelophylax perezi* (Seoane). Aquat. Toxicol. 128-129: 190–192.

Samanta, P., S. Pal, A.K. Mukherjee and A.R. Ghosh. 2020. Acute toxicity assessment of arsenic, chromium and almix 20WP in *Euphlyctis cyanophlyctis* tadpoles. Ecotoxicol. Environ. Saf. 191: 110209.

Sayre, L.M., P.I. Moreira, M.A. Smith and G. Perry. 2005. Metal ions and oxidative protein modification in neurological disease. Ann. Ist. Super Sanita. 41(2): 143–164.

Semlitsch, R.D., D.E. Scott and J.H.K. Pechmann. 1988. Time and size at metamorphosis related to adult fitness in *Ambystoma talpoideum*. Ecology 69(1): 184–192.

Seymour, R.S., J.D. Roberts, N.J. Mitchell and A.J. Blaylock. 2000. Influence of environmental oxygen on development and hatching of aquatic eggs of the Australian Frog, *Crinia georgiana*. Physiol. Biochem. Zool: Ecol. Evol. Approaches 73: 501–507.

Schauder, A., A. Avital and Z. Malik. 2010. Regulation and gene expression of heme synthesis under heavy metal exposure—Review. J. Environ. Pathol. Toxicol. Oncol. 29(2): 137–158.

Shi, Q., N. Sun, H. Kou, H. Wang and H. Zhao. 2018. Chronic effects of mercury on *Bufo gargarizans* larvae: Thyroid disruption, liver damage, oxidative stress and lipid metabolism disorder. Ecotoxicol. Environ. Saf. 164: 500–509.

Shuhaimi-Othman, M., Y. Nadzifah, N.S. Umirah and A.K. Ahmad. 2012. Toxicity of metals to tadpoles of the common Sunda toad, *Duttaphrynus melanostictus*. Toxicol. Environ. Chem. 94(2): 364–376.

Simonato, J.D., M.N. Fernandes and C.B.R. Martinez. 2013. Physiological effects of gasoline on the freshwater fish *Prochilodus lineatus* (Characiformes: Prochilodontidae). Neotrop. Ichthyol. 11: 683–691.

Solé, M., M. Antó, M. Bae, M. Carrasson, J.E. Cartes and F. Maynou. 2010. Hepatic biomarkers of xenobiotic metabolism in eighteen marine fish from NW Mediterranean shelf and slope waters in relation to some of their biological and ecological variables. Mar. Environ. Res. 70: 181–188.

Søndergaard, M. 2009. Redox potential. pp. 852–859. *In:* Likens, G. [ed.]. Encyclopedia of Inland Waters. Pergamon Press, Oxford.

Sparling, D.W., S. Krest and M. Ortiz-Santaliestra. 2006. Effects of lead-contaminated sediment on *Rana sphenocephala* tadpoles. Arch. Environ. Contam. Toxicol. 51: 458–466.

Stadtman, E. and C. Oliver. 1991. Metal-catalyzed oxidation of proteins. Physiological consequences. J. Biol. Chem. 266: 2005–2008.

Stadtman, E.R. and R.L. Levine. 2000. Protein oxidation. Annals of the New York Academy of Sciences 899: 191–208.
Stohs, S. and D. Bagchi. 1995. Oxidative mechanisms in the toxicity of metal ions. Free Radical Biology and Medicine 18(2): 321–336.
Stohs, S.J., D. Bagchi, E. Hassoun and M. Bagchi. 2000. Oxidative mechanisms in the toxicity of chromium and cadmium ions. J. Environ. Pathol. Toxicol. Oncol. 19: 201–213.
Storey, K.B. 1996. Oxidative stress: animal adaptations in nature. Braz. J. Med. Biol. Res. 29: 1715–1733.
Šulinskienė, J., R. Bernotienė, D. Baranauskienė, R. Naginienė, I. Stanevičienė, A. Kašauskas et al. 2019. Effect of zinc on the oxidative stress biomarkers in the brain of nickel-treated mice. Oxid. Med. Cell. Longev. 1–9.
Sun, N., H. Wang, Z. Ju, and H. Zhao. 2017. Effects of chronic cadmium exposure on metamorphosis, skeletal development, and thyroid endocrine disruption in Chinese toad *Bufo gargarizans* tadpoles. Environ. Toxicol. Chem. 37(1): 213–223.
Svartz, G., C. Aronzon, S. Pérez Catán, S. Soloneski and C. Pérez Coll. 2020. Oxidative stress and genotoxicity in *Rhinella arenarum* (Anura: Bufonidae) tadpoles after acute exposure to Ni-Al nanoceramics. Environ. Toxicol. Pharmacol. 80: 103508.
Tao, S., Y. Wen, A. Long, R. Dawson, J. Cao and F. Xu. 2001. Simulation of acid-base condition and copper speciation in the fish gill microenvironment. Comput. Chem. 25: 215–222.
Tayab, M.R. 1991. Environmental impact of heavy metal pollution in natural aquatic systems. Environmental pollution science, PhD Thesis. pp. 171. The University of west London.
Taylor, J.R., E.M. Mager and M. Grosell. 2010. Basolateral NBCe1 plays a rate-limiting role in transepithelial intestinal HCO_3^- secretion, contributing to marine fish osmoregulation. J. Exp. Biol. 213: 459–468.
Trinder, D., P.S. Oates, C. Thomas, J. Sadleir and E.H. Morgan. 2000. Localization of divalent metal transporter 1 (DMT1) to the microvillus membrane of rat duodenal enterocytes in iron deficiency, but to hepatocytes in iron overload. Gut 46: 270–276.
U.S. Geological Survey. 2020b. Dissolved oxygen: U.S. Geological Survey Techniques and Methods, book 9, chap. A6.2, 33 p., [Supersedes USGS Techniques of Water-Resources Investigations, book 9, chap. A6.2, version 3.0.].
Unrine, J.M., C.H. Jagoe, W.A. Hopkins and H.A. Brant. 2004. Adverse effects of ecologically relevant dietary mercury exposure in southern leopard frog (*Rana sphenocephala*) larvae. Environ. Toxicol. Chem. 23(12): 2964.
Väänänen, K., M.T. Leppänen, X.P. Chen and J. Akkanen. 2018. Metal bioavailability in ecological risk assessment of freshwater ecosystems: From science to environmental management. Ecotoxicol. Environ. Saf. 147 : 430–446.
Van der Oost, R., J. Beyer and N.P. Vermeulen. 2003. Fish bioaccumulation and biomarkers in environmental risk assessment: a review. Environ. Toxicol. Pharmacol. 13(2): 57–149.
Vasconcelos, S.M.L., M.O.F. Goulart, J.B.F. Moura, V. Manfredini, M.S. Benfato and L.T. Kubota. 2007. Espécies reativas de oxigênio e de nitrogênio, antioxidantes e marcadores de dano oxidativo em sangue humano: principais métodos analíticos para sua determinação. Quim. Nova 30(5): 1323–1338.
Velusamy, A., P. Satheesh Kumar, A. Ram and S. Chinnadurai. 2014. Bioaccumulation of heavy metals in commercially important marine fishes from Mumbai Harbor, India. Mar. Poll. Bull. 81(1): 218–224.
Veronez, A.C.S., R.V. Salla, V.D. Baroni, I.F. Barcarolli, A. Bianchini, C.B.R. Martinez et al. 2016. Genetic and biochemical effects induced by iron ore, Fe and Mn exposure in tadpoles of the bullfrog *Lithobates catesbeianus*. Aquat. Toxicol. 174: 101–108.
Viarengo, A., D. Lowe, C. Bolognesi, E. Fabbri and A. Koehler. 2007. The use of biomarkers in biomonitoring: A 2-tier approach assessing the level of pollutant-induced stress syndrome in sentinel organisms. Comp. Biochem. Physiol. C 146: 281–300.
Walentek, P., S. Bogusch, T. Thumberger, P. Vick, E. Dubaissi, T. Beyer et al. 2014. A novel serotonin-secreting cell type regulates ciliary motility in the mucociliary epidermis of *Xenopus* tadpoles. Development 141(7): 1526–1533.
Wang, C., G. Liang, L. Chai and H. Wang. 2016b. Effects of copper on growth, metamorphosis and endocrine disruption of *Bufo gargarizans* larvae. Aquat. Toxicol. 170: 24-30.
Wang, R., J. Lou, J. Fang, J. Cai, Z. Hu and P. Sun. 2019. Effects of heavy metals and metal (oxide) nanoparticles on enhanced biological phosphorus removal. Rev. Chem. Eng. 36(8): 947–970.
Wang, W.-X. and Q.-G. Tan. 2019. Applications of dynamic models in predicting the bioaccumulation, transport and toxicity of trace metals in aquatic organisms. Environ. Poll. 252(B): 1561–1573.
Wang, Z., J.P. Meador and K.M.Y. Leung. 2016a. Metal toxicity to freshwater organisms as a function of pH: A meta-analysis. Chemosphere 144: 1544–1552.
Wassersug, R.J. and E.A. Seibert. 1975. Behavioral responses of amphibian larvae to variation in dissolved oxygen. Copeia 1975(1): 86–103.

Webb, M. 1986. Role of metallothionein in cadmium metabolism. pp. 281–337. *In:* Foulkes, E.C. [ed.]. Cadmium. Springer Verlag, Berlin.

Wei, L., G. Ding, S. Guo, M. Tong, W. Chen, J. Flanders et al. 2015. Toxic effects of three heavy metallic ions on *Rana zhenhaiensis* tadpoles. Asian Herpetol. Res. 6(2): 132–142.

Wittmann, G. 1981. Toxic Metals. pp. 3–70. *In:* Metal pollution in the aquatic environment. Springer Study Edition. Springer, Berlin, Heidelberg.

Wu, C., Y. Zhang, L. Chai and H. Wang. 2017. Histological changes, lipid metabolism and oxidative stress in the liver of *Bufo gargarizans* exposed to cadmium concentrations. Chemosphere 179: 337–346.

Ya, J., Y. Xu, G. Wang and H. Zhao. 2021. Cadmium induced skeletal underdevelopment, liver cell apoptosis and hepatic energy metabolism disorder in *Bufo gargarizans* larvae by disrupting thyroid hormone signaling. Ecotoxicol. Environ. Saf. 211: 2021.

Yang, Y., W. Wang, X. Liu, X. Song and L. Chai. 2020. Probing the effects of hexavalent chromium exposure on histology and fatty acid metabolism in liver of *Bufo gargarizans* tadpoles. Chemosphere 243: 125437.

Yao, Q., H. Yang, X. Wang and H. Wang. 2019. Effects of hexavalent chromium on intestinal histology and microbiota in *Bufo gargarizans* tadpoles. Chemosphere 216: 313–323.

Yildirim, N.C., F. Benzer and D. Danabas. 2011. Evaluation of environmental pollution at Munzur River of Tunceli applying oxidative stress biomarkers in *Capoeta trutta* (Heckel, 1843). J. Anim. Plant Sci. 21(1): 66–71.

Yologlu, E. and M. Ozmen. 2015. Low concentrations of metal mixture exposures have adverse effects on selected biomarkers of *Xenopus laevis* tadpoles. Aquat. Toxicol. 168: 19–27.

Zalups, R.K. 2000. Evidence for basolateral uptake of cadmium in the kidneys of rats. Toxicol. Appl. Pharmacol. 164: 15–23.

Zalups, R.K. and S. Ahmad. 2003. Molecular handling of cadmium in transporting epithelia. Toxicol. Appl. Pharmacol. 186: 163–188.

Zalups, R.K. and S. Ahmad. 2004. Homocysteine and the renal epithelial transport of inorganic mercury: role of basolateral transporter OAT1. J. Am. Soc. Nephrol. 15: 2023–2031.

Zhang, Y., D. Huang, D. Zhao, J. Long, G. Song and A. Li. 2007. Long-term toxicity effects of cadmium and lead on *Bufo raddei* tadpoles. Bull. Environ. Contam. Toxicol. 79(2): 178–183.

Zhang, W., H. Zhi and H. Sun. 2020. Effects of heavy metal pollution on fitness and swimming performance of *Bufo raddei* tadpole. Bull. Environ. Contam. Toxicol. 105: 387–392.

Zheng, R., P. Wang, B. Cao, M. Wu, X. Li, H. Wang et al. 2021. Intestinal response characteristic and potential microbial dysbiosis in digestive tract of *Bufo gargarizans* after exposure to cadmium and lead, alone or combined. Chemosphere 271: 129511.

14

Developmental Abnormalities in Tadpoles as Biomarkers to Assess the Ecotoxicity of Traditional and Emerging Pollutants

Raquel Fernanda Salla,[1,*] *Felipe Augusto Pinto-Vidal,*[2,*] *Guilherme Andrade Neto Schmitz Boeing,*[3] *Michele Provase,*[3] *Elisabete Tsukada*[4] and *Thiago Lopes Rocha*[5]

1. Introduction

The ecosystem has been continuously impacted by chemicals from anthropogenic actions. For the sake of progress, a copious number of chemicals have been synthesized to support the growing number of apparent needs of human society, however, those compounds and/or their byproducts

[1] Laboratory of Environmental Biotechnology and Ecotoxicology, Institute of Tropical Pathology and Public Health, Federal University of Goiás (UFG), Goiânia, Goiás, Brazil; Federal University of São Carlos (UFSCar), Sorocaba, São Paulo, Brazil.

[2] RECETOX, Faculty of Science, Masaryk University, Kotlarska 2, 61137 Brno, Czech Republic.

[3] Postgraduate Program in Biotechnology and Environmental Monitoring, Federal University of São Carlos (UFSCar), Sorocaba, São Paulo, Brazil, Laboratory of Structural and Functional Biology, Sorocaba, São Paulo, Brazil.
Emails: provase.michele@gmail.com; guiboeing@gmail.com

[4] Federal University of São Carlos (UFSCar), Postgraduate Program in Biotechnology and Environmental Monitoring (PPGBMA), Laboratory Conservation Physiology (LaFisC), Sorocaba, São Paulo, Brazil.
Email: elisabete.tsukada@gmail.com

[5] Laboratory of Environmental Biotechnology and Ecotoxicology, Institute of Tropical Pathology and Public Health, Federal University of Goiás (UFG), Goiânia, Goiás, Brazil. Email: thiagorochabio20@ufg.br

* Corresponding authors: raquelsalla.ufscar@gmail.com; felipe.vidal@recetox.muni.cz

may be disposed into environmental matrices (Sodré et al. 2010, Escher et al. 2020, de Souza et al. 2021) including water, where their presence might negatively interfere in the homeostasis of aquatic organisms. Among the organisms that dwelled in water, amphibians have drawn the attention of the scientific community in several research fields, including (eco)toxicology of traditional and emerging pollutants (Venturino et al. 2003, Lee et al. 2005, Salla et al. 2016, Amaral et al. 2019, Sievers et al. 2019, Araújo et al. 2021, Pinto Vidal et al. 2021) due to their suitability as unconventional animal models (Fort et al. 2004, Helbing 2012). Most amphibians' embryonic phases are obligatorily aquatic, with a subsequent terrestrial/semi-terrestrial period after metamorphosis (Frieden 1961, Dodd and Dodd 1976, Brown and Cai 2007). Such circumstances might expose the embryos to environmental contaminants throughout their development. After hatching, the amphibian skin is still highly permeable, facilitating chemical absorption (Bentley and Yorio 1976, Toledo and Jared 1993). In sum, these features make them good indicators of the anthropogenic impacts in aquatic and terrestrial environments (Helbing 2012).

The pre-metamorphic phases and the developmental stages that occur in amphibian metamorphosis are critical for the future establishment and stability of their populations in the wild (Scott 1994, Beckerman et al. 2002, Scott et al. 2007). Therefore, any environmental changes or stress factors that harm the development of tadpoles or cause body malformations can also affect individuals' survival and reproductive success, thus compromising the fitness of these species (Searcy et al. 2015). For didactic purposes, we have illustrated this context in Fig. 1.

The presence of developmental abnormalities in amphibians, such as extra limbs, missing parts, or the altered morphology of body parts and internal organs has gained scientific attention for years (Van Valen 1974, Ouellet 1997). Such morphological changes in amphibians are defined as developmental errors that result in asymmetric or abnormal modifications of the body structure (Lannoo 2008, Sodré et al. 2022). Although a small number of abnormalities can result from baseline mutation levels, developmental variables, and occasional traumas, the expected proportion

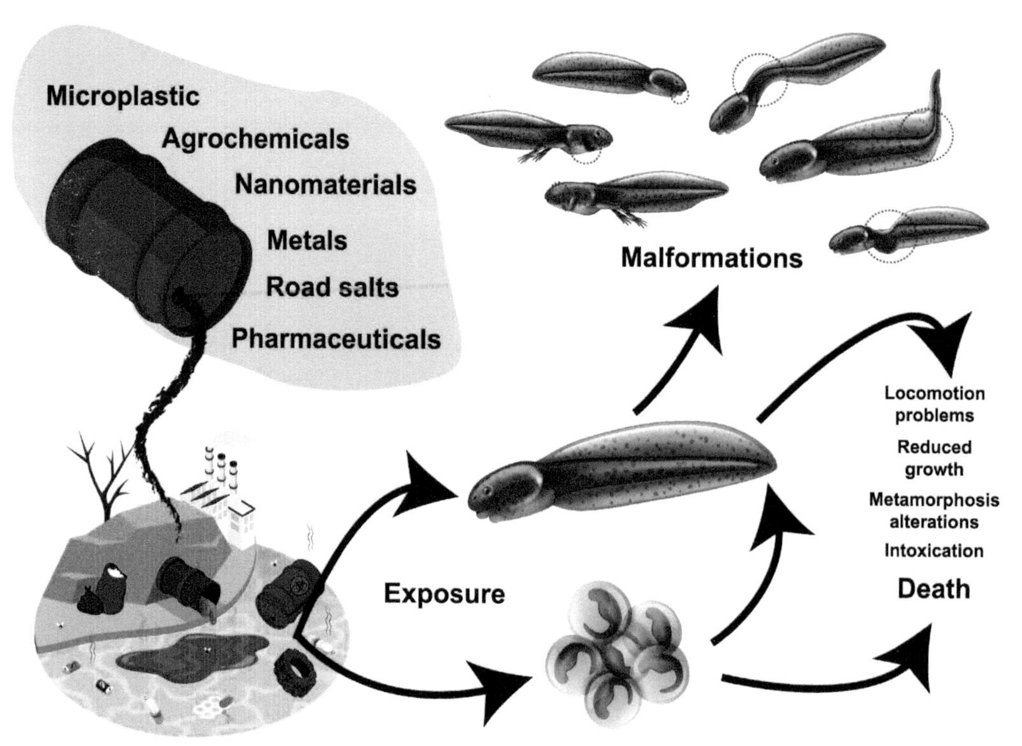

Figure 1. Illustration of the toxicological context that summarizes the interactions among anthropogenic contaminants and their effects on tadpoles. Graphics provided free of charge by Freepik Company, schematic editing and diagramming by Elisabete Tsukada.

of abnormalities is typically under 5% in a population (Berger and Johnson 2003). However, the wide increase in this phenomenon has been notable for its severity. In recent decades, many studies have drawn attention to the increased appearance of deformed amphibians in the wild (Blaustein and Johnson 2003, Hamlin and Guillette Jr. 2010). Many anthropogenic features had been associated with the increased abnormalities in tadpoles, including the UV radiation (Blaustein and Johnson 2003), road salts (Sanzo and Hecnar 2006), metals (Girotto et al. 2020, Ojha et al. 2021), coal residues (Hopkins et al. 2000), pharmaceuticals (Araújo et al. 2019, Boccioni et al. 2021), microplastics and nanomaterials (Tussellino et al. 2015, Amaral et al. 2019, Araújo et al. 2020, Venâncio et al. 2022), and especially the agrochemical products (Ouellet et al. 1997, Sodré et al. 2002, Gurushankara et al. 2007).

The identification of malformations in the early stages of an organism's life is extremely relevant for ecotoxicological studies as they represent biomarkers of toxicity at individual level. Several developmental abnormalities in tadpoles exposed to environmental chemicals have been reported (McDaniel et al. 2004, Peixoto et al. 2022), however, there is a high variability in terminologies and endpoints analyzed, which somewhat limits the comparative studies. Thus, in this chapter, we build an overview of the main reports of malformations in tadpoles exposed to traditional and emerging pollutants, such as agrochemicals, metals, microplastics and nanomaterials. Besides, we propose a new standardized system to classify the distinct types of developmental abnormalities in amphibian tadpoles, which could be used as an important approach in biomonitoring programs, and toxicological and ecotoxicological studies.

2. Developmental Abnormalities Caused by Agrochemicals

Agrochemicals are substances capable of preventing, repelling, or attenuating pests, and can be classified as insecticides, fungicides, herbicides, acaricides, molluscicides, nematicides and rodenticides (Tudi et al. 2021, USEPA 2022). We all know that pesticides play an important role in agricultural development, reducing losses of products, and improving the quality and availability of food (Aktar et al. 2008, Fenik et al. 2011, Strassemeyer et al. 2017). However, the urgency to improve food production inevitably led to the increased use of pesticides during the second World War, which increased even more after the 1940s with the development of many synthetic crop protection chemicals (Bernardes et al. 2015). Since then, the use of pesticides has reached values around three billion kilograms consumed worldwide every year (Hayes et al. 2017). The residues of these products cannot be controlled, so large amounts of remaining pesticides penetrate the soil, accumulate in non-target plants and reach aquatic environments. Consequently, the aquatic pollution by agrochemicals causes severe impacts on human and environmental health (Bernardes et al. 2015, Hayes et al. 2017).

As already expected, the agrochemical exposure has been associated with multiple impairments in amphibian homeostasis. The release of agrochemical waste in fields near breeding habitats (McDaniel et al. 2008, Rutkoski et al. 2020, Sánchez-Domene et al. 2018) can expose amphibian embryos and tadpoles to relatively uncontrolled levels of these contaminants during their most sensitive period of development. These chemicals have been proved to elicit many health impairments in amphibians at the early stages of development.

Vertebrates are non-target organisms for pesticides. These organic chemicals have shown to exert toxicity through interference in enzymatic pathways. The organophosphate chlorpyrifos, for example, is known to inhibit the enzyme acetylcholinesterase (AChE), resulting in the excess of the neurotransmitter acetylcholine in the neuromuscular system (Colombo et al. 2005), thus impacting muscular activity. On the other hand, aromatase induction has been hypothesized as the mechanism whereby atrazine causes gonadal malformations in *Xenopus laevis* (Hayes et al. 2006). Other effects elicited by pesticides include the high incidence of mortality, endocrine disruption, behavioral and development alterations, genotoxicity, among others (Bernabò et al. 2016, Clark et al. 1998,

Crump 2001, Guillette, 2006, Hu et al. 2008, Jayawardena et al. 2010, Mnif et al. 2011, Navarro-Martín et al. 2014).

There is also evidence that exposure to agrochemicals may impair development of tadpoles of several species. Cooke (1973) showed that despite the short life of the dichloro-diphenyl-trichloroethane (DDT) in water, this organochlorine insecticide caused a high prevalence of malformations in *Rana temporaria* tadpoles. Changes in hindlimbs (twisted or bent parts) were also described in *Rana clamitans, Rana pipiens, Bufo americanus*, and *Lithobates catesbeianus* exposed to pesticides' runoff in wildlife (Ouellet et al. 1997) or at laboratory conditions.

Apparently, the stage in which the tadpoles are exposed to those contaminants might play a role in the impact the chemical has. For instance, *X. laevis* embryos at neurula stage exposed to the fungicide triazole had more prominent changes in the branchial arch compared to those exposed at gastrula stage (Groppelli et al. 2005, Papis et al. 2006). Also, *Bufo arabicus* exposed in earlier stages to the carbamate insecticide methomyl presented axial malformation (i.e., changes in the body axis of the tadpole, such as twisted spine, curved trunk, etc.) more evident than animals in later stages of development (Seleem 2019). On the other hand, *X. laevis* at later stages (metamorphs stage 46 NF) were shown to be more sensitive than animals in the early stages (premetamorphs stage 14 NF) to malformations elicited by the exposure to the organophosphate pesticide chlorpyrifos (Richards and Kendall 2002). Overall, these results demonstrated that the type and frequency of abnormalities in tadpoles exposed to pesticides are dependent on the developmental stages.

Under laboratory conditions, some abnormalities following the exposure to agrochemicals proved to be reversible if the tadpoles reached more advanced stages of development. For example, axial malformations seem to be reversible prior the emergence of the forelimbs (Cooke 1981), and tail malformations may disappear after the reabsorption of this organ (Jayawardena et al. 2010). However, in nature, these conditions can limit their life, since some malformations can hamper their movement, foraging, and swimming capability (Jayawardena et al. 2010, Ghodageri and Pancharatna 2011). For example, the presence of edema or absence of limbs can impair swimming because tadpoles lose their center of gravity, leading them to twist around (Jayawardena et al. 2010, Devi and Gupta 2013). Malformations in the oral disc and abnormalities in the intestine, as observed in *Physalaemus albonotatus* exposed to the commercial herbicide 2,4-dichlorophenoxyacetic acid (2,4-D) (Curi et al. 2019), might impair the ingestion and absorption of nutrients. Even sublethal concentrations of agrochemicals may disrupt the development of the exposed animals. The fact that these chemicals are frequently released in breeding areas can lead to the exposure of amphibians at the embryonic/larval stage. Impacts on their development can result in the decline of their communities and populations, with long-term consequences to the ecosystem (Böll et al. 2013, Green et al. 2019).

Morphological changes were also reported in tadpoles after exposure to insecticides, such as organophosphates, organochlorines, carbamates, pyrazoles and pyrethroids. Recent studies highlight insecticides as one of the main chemicals inducing mortality and population declines in amphibians worldwide (Mann et al. 2009, Lopes et al. 2021). This is a worrying phenomenon, since they play an important role in ecosystems (Böll et al. 2013, Green et al. 2019). Amphibians at earlier stages are also more susceptible to the negative impacts caused by the insecticides (Table 2). Such greater sensibility can be explained by diverse biological factors: their gills and eggs can readily absorb substances from the environment (Bantle et al.1992), they still present an immature immune system (Carey and Bryant, 1995, Disner et al. 2021) and incomplete tissue and organ differentiation (Herkovits and Fernández, 1978).

According to Saka and Tada (2021), *Silurana tropicalis* was described with axial malformations and lateral and vertical curvatures in tail after exposure to the insecticide fipronil. Commonly known for its neurotoxic action on insects (Simon-Delso et al. 2015), fipronil has proved to be a teratogenic chemical. Teratogenic deformities usually end up impairing swimming movements, affecting individual survival, and consistently achieving population levels (Alvarez et al. 1995, Hopkins 2000). Similarly, *Physalaemus gracilis* tadpoles exposed to the pyrethroids cypermethrin and deltamethrin showed a partial or total absence of the mandible and denticles (Vanzetto et al. 2019). Such malformations in the oral region makes it difficult for animals to eat properly, compromising their health and integrity along the metamorphosis (Vanzetto et al. 2019).

Pseudacris regilla showed no impairment in the survival rate nor any change in body size after exposure to chlorpyrifos; however, when combined with endosulfan, the mixture of these insecticides developed axial malformations in the tadpoles (Dimitrie and Sparling 2014). Similarly, increased mortality, malformation scoliosis (curvature of the spine) and edema were induced in *Duttaphrynus melanostictus* by a combination of profenofos and abamectin (Rathanayaka and Rajakaruna 2018). Those results corroborate the hypothesis that combined formulations of agrochemicals might maximize the toxic effects in many amphibian species. In summary, all those mentioned effects can impact tadpoles' development, thus affecting mobility, fitness, foraging, as well as causing reproductive difficulties and greater vulnerability to diseases (Dimitrie and Sparling 2014, Rathanayaka and Rajakaruna 2018). A summary of these results is presented in Table 1.

3. Developmental Abnormalities Caused by Trace metals

Metals are essential for the maintenance of life (Stankovic et al. 2014, Peana et al. 2021) and consist of the most abundant group of chemical elements, accounting 25% of the earth's crust (Peana et al. 2021). Metals can be divided into essential (with biological role) and non-essential elements (without biological role) (Stankovic et al. 2014, Peana et al. 2021). Essential metals participate in cellular homeostasis, in the formation of rigid structures such as bones and teeth (by biomineralization), on structural functions of protein and nucleic acids, and can also act as enzymatic cofactors or prosthetic groups of proteins (Peana et al. 2021). Despite playing essential roles in life, concentrations of these elements, both in deficiency and in excess, can cause alterations, such as reduced biological functions, morphological abnormalities, intoxication and even death (Stankovic et al. 2014, Peana et al. 2021).

Potentially toxic metal emissions that occur in natural environments usually originate from rocks, ores, volcanic eruptions, among others (Stankovic et al. 2014). The anthropogenic pressure caused by large urban centers considerably amplifies the emission of these compounds. Mining activities have already nearly exhausted cadmium, copper, zinc, nickel, and lead mines (Stankovic et al. 2014). In addition to the ore extraction, power plant units and metallurgical activities can also contribute to the increase in such emissions. Once released into nature, metals are not easily degraded and may accumulate in soils, sediments, and in the organisms (Stankovic et al. 2014). When reaching aquatic ecosystems, such compounds also end up affecting amphibian communities and all other organisms that depend on these habitats (Stankovic et al. 2014).

Amphibian exposure to metals had already caused various sublethal effects in tadpoles, such as reduced cardiac performance (Dal-Médico et al. 2014), delayed growth and metamorphosis (Lecfort et al. 1998), alterations in the thyroid gland (Wang et al. 2021), reduced body size, hemorrhages, edema, abnormalities in various organs (Gürkan and Hayretdağ 2014), alterations on oxidative stress biomarkers, and neurotoxicity (Carvalho et al. 2020). When individuals survive to metals exposure, morphological alterations that occurred in embryonic and larval stages persist in adults (Plowman et al. 1994), affecting their adaptive and reproductive ability. Even with irreversible effects, studies on the occurrence of morphological abnormalities in tadpoles exposed to metals do not present many records (Table 3). Such fact reinforces the need for long-term studies for a more accurate assessment of the consequences of these alterations.

Abnormal spinal curvatures (scoliosis, lordosis, and kyphosis) (Sparling et al. 2006, Chen et al. 2006, Huang et al. 2014, Peixoto et al. 2022) have been observed at angles of up to 90° in *Rana pipiens* (Chen et al. 2006). In addition to the spine, abnormal curvature of the femur was also reported, affecting locomotion of *Rana sphenocephala* and *Pelophylax nigromaculata* juveniles (Sparling et al. 2006, Huang et al. 2014). Finger abnormalities in *Rana palustris* and *L. catesbeianus* (Stansley et al. 1997), abnormal digit count (polyphalangia or ectrodactyly) in

Table 1. Developmental abnormalities in embryos and tadpoles after exposure to agrochemicals.

Agrochemical	Chemical concentration	Exposure stage	Species	Biomarker expressed as malformation/ abnormality	References
2,4 Dichlorophenoxyacetic acid (2,4-D)	43.7, 87.5, 175, or 262.5 mg/L	Tadpole	*Physalaemus albonotatus*	Oral disc changes Uncoiling intestine	Curi et al. 2019
	7.29, 21.88, 36.47, 51.05 mg/L	Tadpole	*Xenopus laevis*	Intestine abnormalities Edemas Disruption of skeletal muscle	Lenkowski et al. 2010
Abamectin (Avermectins)	0.01, 0.02, 0.03, 0.04, 0.05 mg/L	Tadpoles	*Duttaphrynus melanostictus*	Edemas	Rathanayaka and Rajakaruna 2018
Acephate	0.01, 0.05, 0.1, 0.5, and 1 µg/L	Embryo	*Bufo melanostictus*	Tail distortions (laterally crooked trunk) Decreased pigmentation	Ghodageri and Pancharatna 2011
Atrazine,	0.35, 3.5, 10, 35 mg/L	Tadpole	*Xenopus laevis*	Intestine malformations Edemas Disruption of skeletal muscle	Lenkowski et al. 2010
Azinphos-methyl (Organophosphorus)	9 mg/L	Embryo	*Rhinella arenarum*	altered body size Dorsal tail flexure, Wavy tail Abdominal edema	Lascano et al. 2011
Carbaryl	0.1, 1.0 or 10 mg/L	Embryos	*Xenopus laevis*	Microcephaly Edema Occasional fusion of the optic cup	Elliott-feeley and Armstrong 1982
Carbendazin*	0–7 µM	Embryos	*Xenopus laevis*	Dysplasia of the brain, eyes, intestine, and somatic muscle Swelling of the pronephric ducts	Yoon et al. 2006
Chlorpyrifos	0, 11, 30, 90, 250 and 500 µg/L	Tadpole	*Physalaemus gracilis*	Morphological anomalies in the mouth and intestine	Rutkoski et al. 2020
	0.05, 0.10, 0.25 and 0.5 mg/L	Tadpole	*Polypedates cruciger*	Malformations in the spine (kyphosis and scoliosis) Edema and skin ulcers	Jayawardena et al. 2010
	0.025, 0.05 and 0.1 mg/L	Tadpole	*Rana dalmatina*	Tail flexure Skeletal Muscle defects	Bernabò et al. 2011
Cypermethrin	0.01, 0.05, 0.1, 0.5, and 1 µg/L	Embryo	*Bufo melanostictus*	Deformities in tail, trunk and head region	Ghodageri and Pancharatna 2011
	0.1 and 0.01 mg/L	Tadpoles	*Physalaemus gracilis*	Malformations in the mouth	Vanzetto et al. 2019

Table 1 contd. ...

...Table 1 contd.

Agrochemical	Chemical concentration	Exposure stage	Species	Biomarker expressed as malformation/ abnormality	References
Deltamethrin (Pyrethroid)	0.009 and 0.001 mg/L	Tadpoles	*Physalaemus gracilis*	M.n.d	Vanzetto et al. 2019
p,p'-Dichlorodiphenyldichloroethylene (DDE)	564, 393, 195, 13.2, 269, 199, and 214 μM	Embryos	*Xenopus laevis*	M.n.d	Saka 2004
Dichloro-diphenyl-trichloroethane (DDT)	564, 393, 195, 13.2, 269, 199, and 214 μM	Embryos	*Xenopus laevis*	Axial malformations Irregular gut coiling	Saka 2004
	0.0001 or 0.001 mg/L	Tadpole	*Rana temporaria*	Abnormal spine Straight left leg with the knee joint completely rigid abnormal position of the limbs	Cooke 1973
Dimethoate	0.25, 0.50, 0.75 and 1.00 mg/L	Tadpole	*Polypedates cruciger*	Malformations in the spine (kyphosis and scoliosis) Edema and skin ulcers	Jayawardena et al. 2010
Endosulfan	0.005, 0.01, and 0.05 mg/L	Tadpole	*Rana dalmatina*	Bloated heads Skeletal malformations	Lavorato et al. 2013
	5 and 0.5, 0.35 and 0.18, and 0.3 and 0.03 μg/L	Tadpole	*Fejervarya* spp.	Abnormal development of fore- and hind-limbs Bending and twisting of the body axis	Devi and Gupta 2013
Fenitrothion	0.1, 1.0 and 10 mg/L	Embryos	*Xenopus laevis*	Microcephaly Edema	Elliott-Feeley et al. 1982
	0.25, 0.50, 0.75 and 1.00 mg/L	Tadpole	*Polypedates cruciger*	Malformations in the spine, such as hunched back (kyphosis) and curvature (scoliosis), Edema and skin ulcers	Jayawardena et al. 2010
Fipronil (Phenylpyrazole)	0.35 – 3.5 mg/L	Tadpoles	*Silurana tropicalis*	Axial malformations Malformations in vertical- lateral curvature of the tail	Saka and Tada 2021
	0.5, 1, 2, 4, 8, 16, and 32 mg/L	Tadpole	*Bombina bombina*	Edema Wavy tail fin Improper gut coiling	Sayim 2010

Table 1 contd. ...

...Table 1 contd.

Agrochemical	Chemical concentration	Exposure stage	Species	Biomarker expressed as malformation/ abnormality	References
	15 – 600 g/L	Tadpoles	*Rana palustris*	Axial flexure Craniofacial abnormality Ventralization Tail abnormality Edema	Budischak et al. 2008
Linuron	5.0 to 25.0 mg/L	Embryos	*Pelophylax perezi*	Edemas Spinal curvature Blistering Microphtalmia	Quintaneiro et al. 2018
Malathion	1,000 mg/L	Tadpole	*Bufo americanus*	Increased frequency of diamond-shaped and stiff-tail abnormalities	Krishnamurthy and Smith 2010
	22 and 44 mg/L	Embryo	*Rhinella arenarum*	Malformation lateral of the tail Edema Axial shortening Abnormal neural tube impairment Deformed body axis	Lascano et al. 2011
Methoxychlor	1 mg/L	Embryos and tadpole	*Xenopus laevis*	Visceral edema Craniofacial dysmorphogenesis, Subtle notochord lesions	Fort et al. 2004
n-Butyl isocyanate	0–0.2 μM	Embryos	*Xenopus laevis*	Dysplasia of the brain, eyes, intestine, and somatic muscle Swelling of the pronephric ducts	Yoon et al. 2006
Profenophos (Organophosphates)	2.5, 3.0, 3.5, 4.0, 4.5 and 5.0 mg/L	Tadpoles	*Duttaphrynus melanostictus*	Malformations in the spine (scoliosis and kyphosis)	Rathanayaka and Rajakaruna 2018
Propanil	0.25, 0.50, 0.75 and 1.00 mg/L	Tadpole	*Polypedates cruciger*	Malformations in the spine (kyphosis and scoliosis) Edema and skin ulcers	Jayawardena et al. 2010
Pyrimethanil	5 and 50 μg/L	Tadpole	*Hyla intermedia*	Axial and tail malformations a specimen without the right limb (ectromelia) and with curved tail	Bernabò et al. 2016
S-metolachlor	10.0 to 50.0 mg/L	Embryos	*Pelophylax perezi*	Edemas Spinal curvature Blistering microphtalmia	Quintaneiro et al. 2018

Table 1 contd. ...

...*Table 1 contd.*

Agrochemical	Chemical concentration	Exposure stage	Species	Biomarker expressed as malformation/ abnormality	References
Tebuconazole	5 and 50 µg/L	Tadpole	*Hyla intermedia*	Skeletal malformations an individual affected by hyperextension of the hind limbs	Bernabò et al. 2016
Temephos	0.5, 1.0, and 1.5 mg/L	Tadpole	*Bufo melanostictus*	Rotation of bones Micromelia of the limb bones Hemimelia of femur Skin webbing in the hind limbs and ectrodactyly Edemas	Harischandra et al. 2012
Triadimefon	1.96, 3.91, 7.82, 15.63, 31.25, 45, 62.5, 125, 250, and 500 µM	Embryos	*Xenopus laevis*	Alteration in the gills or buccal apparatus Alteration of neural crest-derived cartilages	Papis et al. 2006
	20, 40, 60, 70 mg/L	Tadpole	*Xenopus laevis*	Intestine malformations Edemas Skeletal muscle	Lenkowski et al. 2010
	10, 30, 62.5, 125, 250 and 500 µM	Embryos	*Xenopus laevis*	craniofacial malformations	Groppelli et al. 2005
	10, 30, 62.5, 125, 250 and 500 µM	Embryos	*Xenopus laevis*	craniofacial malformations of branchial arches	Groppelli et al. 2005
Trichlorfon Organophosphates	0.01, 0.1, and 1.0 mg/L	Tadpoles	*Rana chensinensis*	Axial flexures Skeletal malformations and lateral kinks	Ma et al. 2018
Trifluralin	0.1, 0.5, 1, 3, 5, 9, and 15 mg/L	Embryos	*Bombina bombina*	Edema Axial abnormalities	Sayim 2010

*M.n.d = Malformation not detected.

R. pipiens and *Rana clamitans* (McDaniel et al. 2004), shortening and altered position of the fingers in *Rana sphenocephala* (Sparling et al. 2006) have also been observed after the exposure to metals. Most of the articles compiled in this chapter related the occurrence of malformations to lead exposure (Table 2). However, the effects were more severe in organisms exposed to low or considerably diluted concentrations (0.0002 to 0.0024 mg/L) of environmental samples (McDaniel et al. 2004, Peixoto et al. 2022). Individual eggs of *Pelophylax perezi* failed to hatch, or presented severe edema, abnormal spinal curvature, and high incidence of mortality (Peixoto et al. 2022) after the exposure to acid mine drainage. In polluted ecosystems, non-development of forelimbs and/or eyes has been reported in *R. pipiens* and *R. clamitans* (McDaniel et al. 2004) (Table 2).

These data demonstrate that metals might present distinct toxicodynamic pathways in tadpoles, which apparently vary according to the type of metallic species, its chemical nature and concentration,

Table 2. Developmental abnormalities in embryos and tadpoles after exposure to metals.

Metal	Chemical concentration	Exposure Stage	Species	Biomarker expressed as malformation/abnormality	References
Pb	25, 50, 75 and 100% Lead-Contamined Surface Water	tadpoles	*Rana palustris* and *Lithibates catesbeianus*	hind limb malformations	Stansley et al. 1997
Pb	45, 75, 180, 540, 2360 mg/Kg	tadpoles	*Rana sphenocephala*	bowing of the femurs, shortening of limb digits (brachydactyly), spinal curvature, digits twisted (clinodactyly), femurs and other long bones curvature (brachymely), absence of the development of forelimbs and other spinal abnormalities	Sparling et al. 2002
Zn Cu Pb Cd	5, 10, 15 and 20 mg/L 0.1, 0.2, 0.3, 0.4 and 0,5 mg/L 0.5, 0.6, 0.7, 0.8 and 0.9 mg/L 0.5, 0.6, 0.7, 0.8 and 0.9 mg/L	tadpoles	*Xenopus Laevis*	curling tails, bent tails, kinked tails gut and heart edema	Haywood et al. 2004
Al Cd Cr Cu Ni Pb Zn	0.009 to 1.350 mg/L 0.0001 to 0.0002 mg/L 0.0002 to 0.0024 mg/L 0.0002 to 0.0044 mg/L 0.0002 to 0.0022 mg/L 0.0002 to 0.0020 mg/L 0.0009 to 0.0164 mg/L	tadpoles	*Rana pipiens* and *Rana clamitans*	dorsal or lateral flexure of the notochord, missing eyes (anophthalmy), extra toes (polyphalangy), missing toes (ectrodactyly), lateral flexure of the spinal cord and edema	McDaniel et al. 2004
Pb	0, 3, 10 and 100 ug/L	tadpoles	*Rana pipiens*	lateral spinal curvature (scoliosis)	Chen et al. 2006
Pb	0, 40, 80, 160, 320, 640 and 1280 ug/L	tadpoles	*Pelophylax nigromaculata*	spinal malformations (scoliosis, lordosis and kyphosis)	Huang et al. 2014
Acid mine drainage (AMD)	1.39%, 1.95%, 2.73%, 3.83%, 5.36%, and 7.5%	embrio and tadpoles	*Pelophylax perezi*	body malformations (spinal curvatures), the presence of edemas (minor and severe), hatch failure	Peixoto et al. 2022

*M.n.d = Malformation not detected.

and the exposed species. In addition, the combination of compounds (a scenario which is closer to reality) seems to exert even more pronounced effects.

4. Developmental Abnormalities Caused by Microplastics and Nanomaterials

In the last two decades, the use of plastics (micro- and nanoplastics) and nanomaterials has attracted the attention of the industry and academia research fields due to their extensive application in the aeronautical, pharmaceutical, cosmetic, textile and other industries (Guarnieri et al. 2011, Guarnieri et al. 2012). As with other products generated by industries, increased use also increases disposal in the environment (Scown et al. 2010, Casado et al. 2013). This situation represents a threat because synthetic plastics and nanoparticles can potentially induce toxic effects on living organisms (Kumar et al. 2011). In fact, experimental evidence shows that nanomaterials and synthetic plastics that enter the environment are associated with the increased risk of certain diseases in living organisms as well as the development of morphological changes in early life stage organisms. Thus, studies on the exposure of amphibian embryos and tadpoles to nanomaterials (Amaral et al. 2019), microplastics and nanoplastics (Araújo et al. 2021) are essential for a better assessment of the applicability and damage caused by these materials (Krysanov et al. 2010).

The study conducted by da Costa Araujo et al. (2020) aimed to assess the toxicological potential of polyethylene microplastics (PE MPs) in *P. cuvieri* tadpoles. Tadpoles exposed to PE MPs at 60 mg/L for 7 days showed nuclear abnormalities in erythrocytes, which is indicative of mutagenicity. PE MP also induced external malformations in tadpoles, such as a reduced ratio between length and mouth-cloaca distance, differences in caudal length, ocular and mouth areas, among others. PE MPs also increased the number of melanophores and pigmentation rate in the skin in the assessed areas, which could interfere with their camouflage capacity in natural environments. Also, PE MPs were found bioaccumulated in the gills, gastrointestinal tract, liver, tail muscle, and blood of the tadpoles.

In another study, embryos, and adults of *X. leavis* were exposed to high-molecular-weight polyvinyl chloride (HMW-PVC). Six-week treatment of HMW-PVC at a concentration of 1% of body weight (offered twice each week) on parental frogs were conducted. Afterwards, *in vitro* fertilization was performed in 4 groups of adult frogs: the first group was designed as a control group, which included normal female and male frogs' embryos; the second, third, and fourth groups included normal females + assay males, assay females + normal males, and assay females + assay males' embryos, respectively. The results showed that abnormal and dead embryo ratios were significantly increased in all embryo groups whose parents were exposed to HMW-PVC, when compared to control group. Length values of the second, third, and fourth embryo groups were significantly decreased when compared to the control (Pekmezekmek et al. 2021).

The third group of embryos composed of assay females and normal males showed more decreased average body length value than the second group of embryos, which included assay males and normal females. These results showed that the embryos produced from assay females were more harmfully affected by HMW-PVC exposure than those produced from assay males. Also, it was seen that HMW-PVC exposure caused weight loss in the embryos of assay males against length decrease in the embryos of assay females. Abnormal and dead embryo ratios were significantly increased in all assay groups in comparison with controls. Malformations such as microcephaly, microftalmia, cyclopia, gut malformation, tail malformation, edema and anencephaly were found in this study. These results infer that treatment of parental frogs with HMW-PVC caused adverse effects on their gametes and subsequently on the embryos (Pekmezekmek et al. 2021).

Using the same model of anuran species, the study conducted by Venâncio et al. (2022) assessed the effects of Polymethylmethacrylate nanoplastics (PMMA-NPLs) to early life stages of *X. leavis*. Two toxicity assays were conducted by exposing embryos and tadpoles to 3 concentrations of PMMA-NPLs (1, 100 and 1000 μg/L): a 96 h embryo teratogenicity assay, where survival, total body length (BL) and malformation of embryos were evaluated, and a 48 h feeding rate assay, where survival, feeding (FR), growth rates (body weight-BW and BL) and malformations of tadpoles were assessed. PMMA-NPLs exposure had no significant effects on malformation or mortality of *X. leavis* embryos, but BL was lower at 1000 μg PMMA-NPLs/L. No effects on survival or FR were observed in tadpoles after exposure to PMMA-NPLs, but significant changes occurred in BL and BW. Moreover, anatomical changes in the abdominal region (externalization of the gut) were observed in 62.5% of the tadpoles exposed to 1000 μg PMMA-NPLs/L. Morphological abnormalities such as the curvature of the spine and the occurrence of red eyes were also found in embryos. Considering the continuous release and subsequent accumulation of PMMA, the malformations obtained in the feeding assays suggest that, in the future, these nano-polymers may constitute a risk for aquatic life stages of amphibians.

Regarding nanoparticles, the study conducted by Tussellino et al. (2015) investigated the effects of 50-nm-uncoated polystyrene nanoparticles (PSNPs) in *X. leavis* embryos to understand the suitability of the possible use of this species as model to microplastics toxicity studies. They used the standardized Frog Embryo Teratogenesis Assay-*Xenopus* test (FETAX) during the early stages of *X. laevis* and employed either contact exposure or microinjections. Results showed that the embryos' mortality rate was concentration-dependent and that embryos who survived had a high percentage of morphological changes, such as disorders in pigmentation distribution, malformations of the head, gut and tail, edema in the anterior ventral region, and a shorter body length, when compared with untreated sibling embryos. Further examination showed that the nanoparticles were localized in the cytoplasm, nucleus and periphery of the digestive gut cells. This study suggested that PSNPs are toxic and have a potential teratogenic effect on *Xenopus* larvae. The researchers hypothesize that these effects may be due either to the number of NPs that penetrate the cells and/or to the "corona" effect (concept given by the authors to represent the tendency of particles to aggregate around specific areas in the cell) caused by the interaction of PSNPs with cytoplasm components.

In contrast, the study conducted by de Felice et al. (2018) exposed *X. leavis* tadpoles to polystyrene MPs (PS MPs, size = 3 μm) under semi-static conditions at 0.125, 1.25, and 12.5 μg/mL and allowed their development from stage 36 to 46. At the end of the experiment, the digestive tract and gills from exposed and control tadpoles were microscopically examined, along with changes in the body growth and swimming activity. As expected, PS MPs were observed in tadpoles' digestive tract, but not in the gill, from each test concentration. However, neither body growth nor swimming activity was affected by PS MP exposure, unlike what was found by Tussellino et al. (2015). The results of de Felice et al. (2018) demonstrated that PS MPs can be ingested by tadpoles, but they did not alter *X. leavis* development and swimming behavior at least during early-life stages, also at high unrealistic concentrations.

The study conducted by Bacchetta et al. (2012) proposed to evaluate the lethal and teratogenic potential of carbon nanoparticles (CNPs) in their amorphous form. For this, they utilized the FETAX, a 96 h *in vitro* whole-embryo toxicity test based on the amphibian *X. laevis*. Embryos were acutely exposed to 1, 10, 100 and 500 mg/L CNP suspensions and evaluated for lethality, malformation, and growth inhibition. Also, larvae were processed for ultrastructural and histological analyses to detect the main affected organs, specific lesions at subcellular level, and to track CNPs into the tissues. The results showed that only the highest CNP suspension was lethal to *X. leavis* larvae, while malformed larvae percentages significantly increased, starting from 100 mg/L. Results from histopathological analyses indicated that malformations were observed almost entirely in the gut, but no significant malformation at developmental plan was observed. The authors suggested that the vitelline membrane works as a powerful protective barrier against NPs, and that this protection likely preserves vertebrate embryos from the potential NP-inducible teratogenic effects. As mentioned above, the stomach and gut were the preferential CNP accumulation site, on the contrary, the digestive epithelium remained intact. This study confirms the tolerance of *X. leavis* towards pure elemental carbon in its nanoparticulate amorphous form but emphasizes the capacity of CNP migration towards all body areas.

Other types of nanomaterials were also tested to evaluate their teratogenic potential. FETAX test was also used in a study conducted by Bacchetta et al. (2011) to access the effects of commercially available copper oxide (CuO), titanium dioxide (TiO_2) and zinc oxide (ZnO) nanoparticles in tadpoles. Except for CuO NPs, which was found to be weakly embryo-lethal (effects were observed only at the highest concentration), the NPs did not cause mortality at concentrations up to 500 mg/L. However, these particles induced significant malformation rates, with the gut being the main target organ. CuO NPs presented the highest significant potential, although no specific interactions among the compounds were observed. Also, ZnO NPs caused the most severe lesions to the intestinal barrier, allowing NPs to reach the underlying tissues. TiO_2 NPs presented mild embryotoxicity, suggesting that this substance could be associated with hidden biological effects. Lastly, the ions released from CuO NPs contributed greatly to the observed embryotoxic effects, but those from ZnO NPs did not, indicating that their action mechanisms may be different for amphibian larvae.

Nations et al. (2011) conducted a study to examine the effects of ZnO, TiO_2, Fe_2O_3 and CuO NPs (20–100 nm) on amphibians (*X. leavis*) utilizing the same FETAX protocol. Results showed that mortality was not increased in the static renewal exposures containing up to 1000 mg/L of TiO_2,

Fe_2O_3, CuO, and ZnO, but they induced developmental abnormalities or malformations. Spinal, gastrointestinal, and other abnormalities were observed in ZnO and CuO NP exposed tadpoles at concentrations as low as 3.16 mg/L (ZnO NPs). The minimum concentration to inhibit the growth in tadpoles exposed to CuO or ZnO NPs was 10 mg/L. ZnO and CuO NPs increased malformations. Ingestion of metal oxide nanomaterials increases contact with cells in the intestinal tract, enabling gut malformations such as edema. In this case, gut malformations were the most prominent in animals exposed to ZnO and CuO NP. However, the frequencies of each type of malformation were different between the two nanomaterials: zinc is a more effective inducer of metallothionein than Cu, while Cu would have a greater probability of being absorbed by the gastrointestinal tract and carried to other parts of a tadpole's body. Comparatively, ZnO also produced more gut malformations in tadpoles than CuO. ZnO NPs (0.1, 1.0, and 10 mg/L) also induced genotoxic (DNA damage) and mutagenic effects (nuclear abnormalities) in *L. catesbeianus* (stage 25 G) after 7 days of exposure (Motta et al. 2020). Similarly, TiO_2 NPs at the same concentrations (0.1, 1.0, and 10 mg/L) and exposure time (7 days) induced genotoxic, morphological, and behavioral changes in *Dendropsophus minutus* (Peters 1872) tadpoles (Amaral et al. 2019), confirming the risk of nanomaterials to amphibian health.

These results point out that select nanomaterial can negatively affect amphibians during their development, so the exposure to these chemicals on vertebrates should be imperative for responsible production and introduction of nanomaterials in products and industry. All these studies showed the relevance of researching microplastics and nanoparticles and their potential negative effects on environmental conservation and the safety of the amphibians. All the malformation data can be found summarized in Table 3.

Table 3. Developmental abnormalities in embryos and tadpoles after exposure to microplastics and nanomaterials.

Pollutant	Chemical concentration	Exposure stage	Species	Biomarker expressed as malformation/abnormality	References
Carbon nanoparticles	1, 10, 100 and 500 mg/L	embryo	*Xenopus laevis*	Gut malformation	Bacchetta et al. 2012
Copper oxide nanomaterial	500 mg/L	embryo	*Xenopus laevis*	Gut malformation	Bacchetta et al. 2011
	1000 mg/L	tadpole	*Xenopus laevis*	Spinal, and gastrointestinal malformations; Edema	Nations et al. 2011
Polyethylene microplastics	60 mg/L	tadpole	*Physalaemus cuvieri*	Reduced ratio between length and mouth-cloaca distance; Caudal length; Ocular area and mouth area	Araujo et al. 2019
Polymethylmethacrylate nanoplastics	1, 100 and 1000 μg/L	embryo	*Xenopus laevis*	Curvature of the spine; Occurence of red eyes	Venâncio et al. 2022
	1, 100 and 1000 μg/L	tadpole	*Xenopus laevis*	Exteriorization of the gut from the abdominal bowel cavity; Daily increase of body mass	Venâncio et al. 2022
Polystyrene nanoparticles	50-nm-uncoated	embryo	*Xenopus laevis*	Disorder in pigmentation; Head, tail, and gut malformation; Shorter body length	Tussellino et al. 2015
Polyvinyl chloride	1% of body weight	embryo	*Xenopus laevis*	Microcephaly; Microftalmy; Cyclopia; Gut malformation; Tail malformation; Odema; Anencephaly	Pekmezekmek et al. 2021
Zinc oxide nanomaterial	500 mg/L	embryo	*Xenopus laevis*	Intestinal barrier; Gut malformation	Bacchetta et al. 2012
	1000 mg/L	tadpoles	*Xenopus laevis*	Spinal, and gastrointestinal malformations; Edema	Nations et al. 2012

5. Developmental Abnormalities Caused by Other Types of Pollutants

Concerning the emerging pollutants, the effects of medicinal and pharmacological residues in amphibians are also very scarce (Peltzer et al. 2019). However, developmental abnormalities (intestine displacement to the sides of the body, and inconsistent coiling pattern of intestine cords) have already been reported in tadpoles of *L. catesbeianus* exposed to the most used anticancer drugs cyclophosphamide and 5-fluorouracil. Such anatomical anomaly seemed to result from the abnormal rotation of the gut to the abdominal cavity during the development of the animals (Araújo et al. 2019). The vastly used anti-inflammatory drug, diclofenac, also altered the rotating pattern of tadpoles' intestines, leading to asymmetrically inverted guts. Those alterations were accompanied by alterations in the body axis and chondrocranium and hyobranchial skeleton, as well as abnormalities in other organs (cardiac hypoplasia and cholecystitis) (Peltzer et al. 2019). Similar types of abnormalities (macrocephaly, microcardia, edema, gastrointestinal deviations, hypopigmentation, microcephaly, abdominal, craniofacial and heart edema, stunted body, eyes and tail, yolk tamponade and bent notochord) have also been observed in *X. laevis* at high concentrations of diclofenac (2000 μg/L) (Chae et al. 2015). Although the specific mechanisms by which these pollutants lead to the above-mentioned developmental abnormalities are still unknown, such results indicate that pharmaceuticals might exert similar toxicodynamic pathways in tadpoles.

In addition to the anthropic effects resulting from agriculture and industry, the constant modification of natural landscapes is another important consequence of human population growth (Vitousek et al. 1997, McDaniel and Borton 2002), with special attention to the vast network of roads that cover land in urbanized areas (Forman et al. 1998). Road traffic itself has already been linked to mortality of adult amphibians that collide with vehicles (Ashley and Robinson 1996, Anđelković and Bogdanović 2022). Nonetheless, the pollution of these areas (metals, hydrocarbon fragments from rubber tires, and residues of petroleum products) in addition to the frequent use of defrosting chemicals (known as road salts) generate runoff residues (Norrstrom and Jacks 1998) that can reach aquatic ecosystems where amphibians breed (Sanzo and Hecnar 2006). Indeed, studies with *R. sylvatica* reported that road de-icing salts composed of NaCl caused significantly lower survivorship, decreased time to metamorphosis, reduced weight and activity, and increased physical abnormalities (bent tails), which led them to struggle and swim in circles (Sanzo and Hecnar 2006). Dorchin and Shanas (2010) observed an exaggerated development of the eyes and mouth parts, as well as defective or broken tail and abdominal edemas in tadpoles directly exposed to environmental road runoff samples that contained $MgCl_2$. For amphibian embryonic stages, studies with NaCl and $MgCl_2$ road salts reported a high incidence of morphological abnormalities including the presence of cysts in the organs, shrunken or missing limbs and organs (gills, limb-buds, head, eyes), and spinal deformities (bent tails and bent bodies) (Hopkins et al. 2013). All these morphological alterations can directly affect the development, and consequently the establishment and survivorship of these animals in the wild. Such high prevalence of developmental abnormalities in response to road salts indicate that these biomarkers are sensitive and can be used as early diagnostic tools on similar environmental studies. A summary of the mentioned studies is available in Table 4.

Although the ecotoxicology of amphibians has gained more attention among scientific studies, knowledge about the mechanisms of action and the toxicity of the vast range of residues that are dumped into aquatic ecosystems still have large gaps to be filled. Additionally, the use of developmental biomarkers combined with a deeper comprehension of such mechanisms can provide important basis for future risk assessment approaches.

6. Toxicodynamic Mechanisms Underlying Developmental Abnormalities in Amphibians

Toxicodynamics is the area of science that studies the mechanisms of action of toxicants in living organisms, that is, their toxicity. Pollutants, once absorbed and distributed in the organism, can

Table 4. Developmental abnormalities in embryos and tadpoles after exposure to microplastics and nanomaterials

Pollutant	Chemical concentration	Exposure stage	Species	Biomarker expressed as malformation/abnormality	References
NaCL (road de-icing salt)	0.39, 77.5, and 1030 mg/L	Tadpole	*Rana Sylvatica*	Tail distortions (laterally bent tails)	Sanzo and Hecnar 2006
NaCL and $MgCl_2$ (road de-icing salts)	1.0, 1.5, and 2.0 g/L	Embryo	*Taricha granulosa*	Cysts, and shrunken or missing limbs and organs (gills, limb-buds, heads, eyes). Spinal deformities (bent tails and bent bodies)	Hopkins et al. 2013
$MgCl_2$ (road de-icing salt	0.1 mg/L	Tadpoles	*Bufo viridis*	Exaggerated development of the eyes and mouthparts, Defective or broken tail Abdominal edemas	Dorchin and Shanas 2010
Diclofenac (Pharmaceutical)	125, 250, 500, 1000 and 2000 μg/L	Tadpoles	*Trachycephalus typhonius* and *Physalaemus albonotatus*	Altered rotation of intestines Asymmetrically inverted guts altered body axis Chondrocranium and hyobranchial skeleton cardiac hypoplasia Cholecystitis	Peltzer et al. 2019
Diclofenac (Pharmaceutical)	2000 μg/L	Tadpoles	*Xenopus laevis*	Shortening of the axis, Abdominal bulging, Prominent blister formation. Altered shape of internal organs	Chae et al. 2015
Cyclophosphamide (Cyc) and 5-fluorouracil (5-FU) (Pharmaceuticals)	50 mg/L, 0.2 μg/L, 0.5 μg/L, 2.0 μg/L, 13.0 μg/L, 30.4 μg/L and 123.5 μg/L	Tadpoles	*Lithobates catesbeianus*	Intestine lateral displacement, Inconsistent coiling of the intestines	Araújo et al. 2019

interact in a specific or non-specific way with target molecules. The biological consequences of this interaction can be used as a clue to identify the mechanisms of toxicity that determine the action of a toxicant (Klaassen and Amdur 2013). In this sense, developmental biomarkers are useful tools to help us to identify the mechanisms of action of various toxicants in amphibians. Similar developmental abnormalities in amphibians in response to certain classes of pollutants may be also an indicative of similar toxicity pathways.

Considering that amphibian egg membranes are highly thin and permeable to many molecules (Hunter and de Luque 1959), many toxicants could be easily absorbed. Later, if the interaction with the target molecules causes any shrinkage of the perivitelline space of the egg (Karraker and Gibbs 2011), it can substantially decrease the area in which the embryo can develop. So, this mechanism has been suggested as one of the possible causes of curved body and tail abnormalities seen in anuran hatchlings (Padhye and Ghate 1992, Hopkins et al. 2013). Physiological stress usually increases glucocorticoids, which are also known to result in spinal deformities in fish (Eriksen et al. 2008) and could also be partly responsible for the bent bodies and tails observed in embryos developing in stressful environments (Hopkins et al. 2013). In addition, the presence and contact with pollutants during amphibian larval stages seem to induce a slower developmental rate, reduced body condition, and incomplete or altered development of some body parts (e.g., cranium abnormalities) (Peltzer et al. 2019).

Amphibian trade-offs between developmental rates, body size, and time to metamorphosis can be very complex since compensatory physiological effects might occur, and any stressful conditions could alter such biomarkers (Strong et al. 2017). Gardiner et al. (2003) also drew attention to the fact that inappropriate modulation of retinoid signaling by environmental contaminants may be the main

Table 5. Toxicity classification of distinct pollutants in relation to their mechanisms of action.

	Developmental Abnormalities															
	Cranio-facial abnormalities			**Visceral abnormalities**			**Pigmentation /Tegumentary changes**			**Members and tail abnormalities**		**Muscleskeletal disorders**				**Circulatory disorders**
Pollutant	**Oral disc changes**	**Head/Brain abnormalities**	**Orbital abnormalities**	**Uncoiling intestine**	**Intestinal abnormalities**	**Renal abnormalities**	**Edema**	**Altered pigmentation**	**Skin abnormalities**	**Tail distortions**	**Limb abnormalities bnormalities**	**Disruption of skeletal muscle**	**Altered body size**	**Skeletal abnormalities**	**Axial abnormalities**	**Cardiac abnormalities**
2,4 Dichlorophenoxyacetic acid	X			X	X		X					X				
Abamectin							X									
Acephate								X		X						
Acid mine drainage (AMD)							X							X		
Atrazine					X		X					X				
Azinphos-methyl (Organophosphorus)							X			X			X			
Carbaryl		X	X													
Carbendazin	X	X	X		X	X	X		X			X				
Carbon nanoparticles					X											
Chlorpyrifos										X		X		X		
Copper oxide nanoparticles					X		X							X		
Cyclophosphamide				X	X											
Cypermethrin	X	X								X				X		
Dichloro-diphenyl-trichloroethane (DDT)				X						X				X	X	
Diclofenac	X			X	X									X	X	X
Dimethoate							X		X							
Endosulfan		X								X				X	X	
Fenitrothion		X					X		X					X		
Fipronil		X		X			X			X					X	
Lead (Pb)										X				X		
Linuron		X					X							X		
Malathion		X					X			X					X	
Methoxychlor		X	X		X		X					X				
Sodium/Magnesium chloride (road de-icing salts)	X	X	X				X			X	X			X		
n-Butyl isocyanate						X										

Table 5 contd. ...

...Table 5 contd.

	Developmental Abnormalities															
	Cranio-facial abnormalities			**Visceral abnormalities**			**Pigmentation /Tegumentary changes**			**Members and tail abnormalities**		**Muscleskeletal disorders**				**Circulatory disorders**
Pollutant	**Oral disc changes**	**Head/Brain abnormalities**	**Orbital abnormalities**	**Uncoiling intestine**	**Intestinal abnormalities**	**Renal abnormalities**	**Edema**	**Altered pigmentation**	**Skin abnormalities**	**Tail distortions**	**Limb abnormalities bnormalities**	**Disruption of skeletal muscle**	**Altered body size**	**Skeletal abnormalities**	**Axial abnormalities**	**Cardiac abnormalities**
Polyethylene microplastics	X		X							X			X			
Polymethylmethacrylate nanoplastics			X		X								X	X		
Polystyrene nanoparticles		X			X			X		X			X			
Polyvinyl chloride		X	X		X		X			X						
Profenophos														X		
Propanil							X		X					X		
Pyrimethanil										X	X				X	
S-metolachlor			X				X							X		
Tebuconazole										X				X		
Temephos							X		X	X				X		
Triadimefon	X	X			X		X							X		
Trichlorfon														X	X	
Trifluralin							X								X	
Zinc oxide nanoporticles					X		X							X		

mechanism underlying the increased incidence of amphibian malformations. Their conclusions were built based on the evidence that retinoids can cause a wide variety of developmental abnormalities in amphibians, including craniofacial deformities, gastrointestinal and neurological defects (Gardiner et al. 2003).

With that discussion in mind, considering that different types of pollutants can generate similar effects to the target molecules/organs, it is possible to infer that the toxicodynamic mechanisms of such compounds could also be similar (Gardiner et al. 2003). Thus, we here propose a classification to the pollutants reported within the text into six types of "toxicity pathways": (1) cranio-facial abnormalities; (2) visceral abnormalities; (3) pigmentation/tegumentary changes; (4) members and tail abnormalities; (5) muscle-skeletal disorders; and (6) circulatory disorders. This toxicological classification proposal, based on tadpoles' abnormalities (Table 5), may also help future studies to identify the mechanisms of action of other pollutants in amphibians.

Although many of these mechanisms can help us to elucidate the toxicity pathways of pollutants in amphibians, it is important to consider that frog malformations in the wild can have multiple etiologies, and the complexity of environmental mixtures of contaminants still represents one of the main challenges to the development of viable risk assessment studies and conservation plans.

7. Conclusions and Perspectives

Morphological changes in amphibian embryos and tadpoles are suitable biomarkers to assess the ecotoxic effects of traditional and emerging pollutants. Revised studies confirmed that early developmental stages of amphibians are emerging model systems in toxicology and ecotoxicology. In this chapter, morphological changes in amphibian embryos and tadpoles exposed to pollutants were classified into six categories (cranio-facial abnormalities; visceral abnormalities; pigmentation/tegumentary changes; members and tail abnormalities; musculoskeletal disorder; and circulatory disorder), aiming to contribute to a better characterization of the mechanisms of action and toxicity of pollutants in amphibians.

Despite the growing use of morphological changes in amphibians as biomarkers, it is necessary to develop standard guides for identifying and describing them, as well as standardize the nomenclature of the various morphological alterations observed in embryos and tadpoles. Furthermore, several research gaps deserve further attention, such as: (i) assessment of morphological changes induced by pollutants in more environmentally relevant conditions, (ii) analysis of toxicity after multigenerational exposure, (iii) assessment of the mixture toxicity.

Acknowledgments

FAPV work was supported by the Operational Programme Research, Development and Education—project "Internal Grant Agency of Masaryk University (No.CZ.02.2.69/0.0/0.0/19_073/0016943). FAPV also thanks the Research Infrastructure RECETOX RI (No LM2018121) financed by the Czech Ministry of Education, Youth and Sports, and the Operational Programme Research, Development, and Innovation-project CETOCOEN EXCELLENCE (No CZ.02.1.01/0.0/0.0/26617_043/0009632) for supportive background. Rocha T.L. is granted with productivity scholarship from CNPq (proc. n. 306329/2020-4). This work was supported by the European Union's Horizon 2020 research and innovation programme under grant agreement No 857560. This publication reflects only the authors views, and the European Commission is not responsible for any use that may be made of the information it contains.

References Cited

Aktar, W., M. Paramasivam, D. Sengupta, S. Purkait, M. Ganguly and S. Banerjee. 2008. Impact assessment of pesticide residues in fish of *Ganga river* around Kolkata in West Bengal. Environ. Monit. Assess. 157: 97–104.

Alvarez, R., M.P. Honrubia and M.P. Herráez. 1995. Skeletal malformations induced by the insecticides ZZ-Aphox and Folidol during larval development of *Rana perezi*. Arch. Environ. Contam. Toxicol. 28: 349–356.

Amaral, D.F., V. Guerra, A.G.C. Motta, D. de Melo e Silva and T.L. Rocha. 2019. Ecotoxicity of nanomaterials in amphibians: A critical review. Sci. Tot. Environ. 686: 332–344.

Anđelković, M. and N. Bogdanović. 2022. Amphibian and reptile road mortality in special nature reserve Obedska Bara, Serbia. Animals. 12: 561.

Araújo, A.P.C., C. Mesak, M.F Montalvão, I.N. Freitas, T.Q. Chagas and G. Malafaia. 2019. Anti-cancer drugs in aquatic environments can cause cancer: Insight about mutagenicity in tadpoles. Sci. Tot. Environ. 650: 2284–2293.

Ashley, E.P. and J.T. Robinson. 1996. Road mortality of amphibians, reptiles and other wildlife on the Long Point Causeway, Lake Erie, Ontario. Canadian Field Naturalist. 110: 403–412.

Bacchetta, R., N. Santo, U. Fascio, E. Moschini, S. Freddi, G. Chirico, M. Camatini and P. Mantecca. 2011. Nano-sized CuO, TiO_2 and ZnO affect *Xenopus laevis* development. Nanotoxicol. 6: 381–398.

Bacchetta, R., P. Tremolada, C. di Benedetto, N. Santo, U. Fascio, C. Chirico et al. 2012. Does carbon nanopowder threaten amphibian development? Carbon 50: 4607–4618.

Bantle, J.A, J.N. Dumont, R.A. Finch and G. Linder. 1992. Atlas of abnormalities: A guide for the performance of FETAX. Oklahoma State University Press, Stillwater, OK, USA.

Beckerman, A., T.G. Benton, E. Ranta, V. Kaitala and P. Lundberg. 2002. Population dynamic consequences of delayed life-history effects. Trend. Ecol. Evol. 17: 263–269.

Bentley, P.J. and T. Yorio. 1976. The passive permeability of the skin of anuran amphibian: A comparison of frog (*Rana pipiens*) and toads (*Bufo marinus*). J. Physiol. 261: 603–615.

Bernabò, I., E. Brunelli, C. Berg, A. Bonacci and S. Tripepi. 2008. Endosulfan acute toxicity in *Bufo bufo* gills: Ultrastructural changes and nitric oxide synthase localization. Aquat. Toxicol. 86: 447–456.

Bernabò, I., E. Sperone, S. Tripepi and E. Brunelli. 2011. Toxicity of chlorpyrifos to larval *Rana dalmatina*: Acute and chronic effects on survival, development, growth, and gill apparatus. Arch. Environ. Contam. Toxicol. 61: 704–718.

Bernabò, I., A. Guardia, R. Macirella, S. Sesti, A. Crescente and E. Brunelli. 2016. Effects of long-term exposure to two fungicides, pyrimethanil and tebuconazole, on survival and life history traits of Italian tree frog (*Hyla intermedia*). Aquat. Toxicol. 172: 56–66.

Bernardes, M.F.F., M. Pazin, L.C. Pereira and D.J. Dorta. 2015. Impact of pesticides on environmental and human health. *In*: Andreazza C.C. and G. Scola [eds]. Toxicology studies sells, drugs and environment. IntechOpen: London, UK. pp. 198.

Blaustein, A.R. and P.T.J. Johnson. 2003. The complexity of deformed amphibians. Front. Ecol. Environ. 1: 87–94.

Boccioni, A.P.C., R.C. Lajmanovich, P.M. Peltzer, A.M. Attademo and C.S. Martinuzzi. 2021. Toxicity assessment at different experimental scenarios with glyphosate, chlorpyrifos and antibiotics in *Rhinella arenarum* (Anura: Bufonidae) tadpoles. Chemosphere 273: 128475.

Böll, S., B. Schmidt, M. Veith, N. Wagner, D. Rödder, C. Weinmann et al. 2013. Amphibians as indicators of changes in aquatic and terrestrial ecosystems following GM crop cultivation: A monitoring guideline. BioRisk 8: 39.

Brown, D.D. and L. Cai. 2007. Amphibian metamorphosis. Dev. Biol. 306: 20–33.

Budischak, S.A., L.K. Belden and W.A. Hopkins. 2008. Effects of malathion on embryonic development and latent susceptibility to trematode parasites in ranid tadpoles. Environ. Toxicol. Chem. 27: 2496–2500.

Carey, C. and C.J. Bryant. 1995. Possible interrelations among environmental toxicants, amphibian development, and decline of amphibian populations. Environ. Health Perspect. 4: 13–17.

Carvalho, C.S., H.S.M. Utsunomiya, T. Pasquoto-Stigliani, M.J. Costa and M.N. Fernandes. 2020. Biomarkers of the oxidative stress and neurotoxicity in tissues of the bullfrog, *Lithobates catesbeianus* to assess exposure to metals. Ecotoxicol. Environ. Saf. 196: 10560.

Casado, M.P., A. Macken and H.J. Byrne. 2013. Ecotoxicological assessment of silica and polystyrene nanoparticles assessed by a multitrophic test battery. Environ. Internat. 51: 97–105.

Chae, J.P., M.S. Park, Y.S. Hwang, B.H. Min, S.H. Kim and H.S. Lee. 2015. Evaluation of developmental toxicity and teratogenicity of diclofenac using *Xenopus* embryos. Chemosphere 120: 52–58.

Chen, T.H., J.A. Gross and W.H. Karasov. 2006. Sublethal effects of lead on northern leopard frog (*Rana pipiens*) tadpoles. Environ. Toxicol. Chem. 25: 1383–1389.

Clark, E.J., D.O. Norris and R.E. Jones. 1998. Interactions of gonadal steroids and pesticides (DDT, DDE) on gonaduct growth in larval tiger salamanders, *Ambystoma tigrinum*. Gen. Comp. Endocrinol. 109: 94–105.

Colombo, A., F. Orsi and P. Bonfanti. 2005. Exposure to the organophosphorus pesticide chlorpyrifos inhibits acetylcholinesterase activity and affects muscular integrity in *Xenopus laevis* larvae. Chemosphere 61: 1665–1671.

Cooke, A.S. 1981. Tadpoles as indicators of harmful levels of pollution in the field. Environ. Pollution. Ser. A Ecol. Biol. 25: 123–133.

Cooke, A.A.S. 1973. Response of *Rana temporaria* tadpoles to chronic doses of pp′ -DDT. Copeia 2: 647–652.

Crump, D. 2001. The effects of UV-B radiation and endocrine-disrupting chemicals (EDCs) on the biology of amphibians. Environ. Rev. 9: 61–80.

Curi, L.M., P.M. Peltzer, M.T. Sandoval and R.C. Lajmanovich. 2019. Acute toxicity and sublethal effects caused by a commercial herbicide formulated with 2,4-D on *Physalaemus albonotatus* tadpoles. Water Air Soil Pollut. 230.

da Costa Araújo, A.P., N.F.S. de Melo, A.G. de Oliveira Junior, F.P. Rodrigues, T. Fernandes, J.E. de Andrade Vieira et al. 2020. How much are microplastics harmful to the health of amphibians? A study with pristine polyethylene microplastics and *Physalaemus cuvieri*. J. Hazard. Mat. 382: 121066.

Dal-Medico, S.E., R.Z. Rissoli, F.U. Gamero, J.A. Victório, R.F. Salla, F.C. Abdalla, E.C.M. Silva-Zacarin, C.S. Carvalho and M.J. Costa. 2014. Negative impact of a cadmium concentration considered environmentally safe in Brazil on the cardiac performance of bullfrog tadpoles. Ecotoxicol. Environ. Saf. 104: 168–174.

de Felice, B., R. Bacchetta and N. Santo. 2018. Polystyrene microplastics did not affect body growth and swimming activity in *Xenopus laevis* tadpoles. Environ. Sci. Pollut. Res. 25: 34644–34651.

de Souza, R.C., A.A. Godoy, F. Kummrow, T.L. dos Santos, C.J. Brandão and E. Pinto. 2021. Occurrence of caffeine, fluoxetine, bezafibrate and levothyroxine in surface freshwater of São Paulo State (Brazil) and risk assessment for aquatic life protection. Environ. Sci. Pollut. Res. 28: 20751–20761.

Devi, N.N. and A. Gupta. 2013. Toxicity of endosulfan to tadpoles of *Fejervarya* spp. (Anura: Dicroglossidae): mortality and morphological deformities. Ecotoxicology 22: 1395–1402.

Dimitrie, D.A. and D.W. Sparling. 2014. Joint toxicity of chlorpyrifos and endosulfan to pacific treefrog (*Pseudacris regilla*) tadpoles. Arch. Environ. Contam Toxicol. 67: 444–452.

Disner, G.R., M. Falcão, A.I. Andrade-Barros, N.V. Leite Dos Santos, A. Soares, M. Marcolino-Souza et al. 2021. The toxic effects of glyphosate, chlorpyrifos, abamectin, and 2,4-d on animal models: A systematic review of Brazilian studies. Integr. Environ. Assess Manag. 17: 507–520.

do Amaral, D.F., V. Guerra, A.G.C. Motta, D. de Melo e Silva and T.L. Rocha. 2019. Ecotoxicity of nanomaterials in amphibians: A critical review. Sci. Tot. Environ. 686: 332–344.

Dodd, M.H.I. and J.M. Dodd. 1976. The biology of metamorphosis, physiology of the amphibia. Academic Press.

Dorchin, A. and U. Shanas. 2010. Assessment of pollution in road runoff using a *Bufo viridis* biological assay. Environ. Pollut. 158: 3626–3633.

Elliott-Feeley, E. and J.B. Armstrong. 1982. Effects of fenitrothion and carbaryl on *Xenopus laevis* development. Toxicol. 22: 319–335.

Eriksen, M.S., A.M. Espmark, R. Poppe, B.O. Braastad, R. Salte and M. Bakken. 2008. Fluctuating asymmetry in farmed Atlantic salmon (*Salmo salar*) juveniles: Also a maternal matter? Environ. Biol. of Fishes 81: 87–99.

Escher, B.I., H.M. Stapleton and E.L. Schymanski. 2020. Tracking complex mixtures in our changing environment. Science 6476: 388–392.

Fenik, J., M. Tankiewicz and M. Biziuk. 2011. Properties and determination of pesticides in fruits and vegetables. Trac. Trends Anal. Chem. 30: 814–826.

Forman, R.T.T. and L.E. Alexander. 1998. Roads and their major ecological effects. Annual Rev. of Ecol. Systematics 29: 207–231.

Fort, D.J., P.D. Guiney, J.A. Weeks, J.H. Thomas, R.L. Rogers, A.M. Noll et al. 2004. Effect of methoxychlor on various life stages of *Xenopus laevis*. Toxicol. Sci. 81: 454–466.

Frieden, E. 1961. Biochemical adaptation and anuran metamorphosis. Integr. Comp. Biol. 1: 115–149.

Gardiner, D., A. Ndayibagira, F. Grün and B. Blumberg. 2003. Deformed frogs and environmental retinoids. Pure Applied Chem. 75: 2263–2273.

Gareth, R.H., S.F. Susannah and D.E. Brodie Jr. 2013. Increased frequency and severity of developmental deformities in rough-skinned newt (*Taricha granulosa*) embryos exposed to road deicing salts (NaCl and MgCl2). Environ. Pollut. 173: 264–269.

Ghodageri, M.G. and K. Pancharatna. 2011. Morphological and behavioral alterations induced by endocrine disrupters in amphibian tadpoles. Toxicol. Environ. Chem. 93: 2012–2021.

Girotto, L.E., E.L. Gaeta, R.C. Gebara and J.S. Freitas. 2020. Acute and chronic effects on tadpoles (*Lithobates catesbeianus*) exposed to mining tailings from the dam rupture in mariana, mg (Brazil). Water, Air, and Soil Pollut. 231: 325.

Green, F.B., A.G. East and C.J. Salice. 2019. Will temperature increases associated with climate change potentiate toxicity of environmentally relevant concentrations of chloride on larval green frogs (*Lithobates clamitans*)? Sci Total Environ. 682: 282–290.

Groppelli, S., R. Pennati, F. e Bernardi, E. Menegola, E. Giavini and C. Sotgia. 2005. Teratogenic effects of two antifungal triazoles, triadimefon and triadimenol, on *Xenopus laevis* development: craniofacial defects. Aquat. Toxicol. 73: 370–381.

Guarnieri, D., A. Guaccio, S. Fusco and P.A. Netti. 2011. Effect of serum proteins on polystyrene nanoparticle uptake and intracellular trafficking in endothelial cells. J. Nanopart. Res. 13: 4295–4309.

Guarnieri, D., A. Falanga, O. Muscetti, R. Tarallo, S. Fusco, M. Galdiero et al. 2012. Shuttle-mediated nanoparticle delivery to the blood-brain barrier. Small 9: 853–862.

Guillette, L.J. 2006. Endocrine disrupting contaminants-beyond the dogma. Environ. Health Perspect. 114: 9–12.

Gürkan, M., A. Cetin and S. Hayretdağ. 2014. Acute toxic effects of cadmium in larvae of the green toad, *Pseudepidalea variabilis* (pallas, 1769) (amphibia: anura). Arh. Hig. Rada.Toksikol. 29: 301–309.

Gurushankara, H.P., S.V. Krishnamurthy and V. Vasudev. 2007. Morphological abnormalities in natural populations of common frogs inhabiting agroecosystems of the central Western Ghats. Appl. Herpetol. 4: 39–45.

Hamlin, H.J. and L.J. Guillette Jr. 2010. Birth defects in wildlife: the role of environmental contaminants as inducers of reproductive and developmental dysfunction. Systems Biol. Reproduct. Med. 56: 113–121.

Harischandra, H., S. Karunaratne and R. Rajakaruna. 2012. Effect of mosquito larvicide Abate® on the developmental stages of the Asian common toad, *Bufo melanostictus*. Ceylon J. Sci. Biological Sci. 40: 133.

Hayes, T.B., P. Case, S. Chui, D. Chung, C. Haeffele, K. Haston et al. 2006. Pesticide mixtures, endocrine disruption, and amphibian declines: Are we underestimating the impact? Environ. Health Perspect. 114: 40–50.

Hayes, T.B., M. Hansen, A.R. Kapuscinski, K.A. Locke and A. Barnosky. 2017. From silent spring to silent night: Agrochemicals and the anthropocene. Elem. Sci. Anth. 5: 1–24.

Haywood, L.K., J.A. Graham, M.J. Byrne and E. Cukrowska. 2004. *Xenopus laevis* embryos and tadpoles as models for testing for pollution by zinc, copper, lead and cadmium, African Zool. 39: 163–174.

Helbing, C.C. 2012. The metamorphosis of amphibian toxicogenomics. Front. Gen. 3: 37.

Herkovits, J. and A. Fernández. 1978. Tolerancia a noxas durante el desarrollo embrionario. Medicina Buenos Aires 39: 400–408.

Hopkins W.A., J. Congdon and J.K. Ray. 2000. Incidence and impact of axial malformations in larval bullfrogs *Rana catesbeiana* developing in sites polluted by a coal-burning power plant. Environ. Toxicol. Chem. 19: 862–868.

Hopkins, G.R., S.S. French and E.D. Brodie Jr. 2013. Increased frequency and severity of developmental deformities in rough-skinned newt (*Taricha granulosa*) embryos exposed to road deicing salts (NaCl & MgCl2). Environ. Pollut. 173: 264–269.

Hu, F., E.E. Smith and J.A. Carr. 2008. Effects of larval exposure to estradiol on spermatogenesis and *in vitro* gonadal steroid secretion in African clawed frogs, *Xenopus laevis*. Gen. Comp. Endocrinol. 155: 190–200.

Huang, M., R. Duan and J. Xiang. 2014. Chronic effects of environmentally-relevant concentrations of lead in *Pelophylax nigromaculata* tadpoles: Threshold dose and adverse effects. Ecotoxicol. Environ. Saf. 104: 310–316.

Hunter, F.R. and O. de Luque. 1959. Osmotic studies of amphibian eggs. II. Ovarian eggs. Biological Bull. 117: 468–481.

Jayawardena, U.A., R.S. Rajakaruna, A.N. Navaratne and P.H. Amerasinghe. 2010. Toxicity of agrochemicals to common hourglass tree frog (*Polypedates cruciger*) in acute and chronic exposure. Int. J. Agric. Biol. 12: 641–648.

Klaassen, C.D. and M.O. Amdur. 2013. Casarett and Doull's toxicology: the basic science of poisons. New York, McGraw-Hill. 1236: 189–190.

Karraker, N.E. and J.P. Gibbs. 2011. Road deicing salt irreversibly disrupts osmoregulation of salamander egg clutches. Environ. Pollut. 159: 833–835.

Krysanov, E.Yu., D.S. Pavlov, T.B. Demidova and Y.Y. Dgebuadze. 2010. Effect of nanoparticles on aquatic organisms. Biology Bull. 37: 406–412.

Krishnamurthy, S.V. and G.R. Smith. 2010. Growth, abnormalities, and mortality of tadpoles of American toad exposed to combinations of malathion and nitrate. Environ. Toxicol. Chem. 29: 2777–2782.

Kumar, V., A. Kumari, P. Guleria and S.K. Yadav. 2011. Evaluating the toxicity of selected types of nanochemicals. Rev. of Environ. Contam. and Toxicol. 215: 39–121.

Lannoo, M. 2008. Malformed frogs: the collapse of aquatic ecosystems. Los Angeles, University of California Press.

Lascano, C.I., A. Ferrari, L.E. Gauna, C. Cocca, A.C. Cochón, C. Verrengia et al. 2011. Organophosphorus insecticides affect normal polyamine metabolism in amphibian embryogenesis. Pest. Bioch. and Physiol. 101: 240–247.

Lavorato, M., I. Bernabò, A. Crescente, M. Denoël, S. Tripepi and E. Brunelli. 2013. Endosulfan effects on *Rana dalmatina* tadpoles: Quantitative developmental and behavioural analysis. Arch. Environ. Contam. Toxicol. 64: 253–262.

Lee, S.K., G.A. Owens and D.N.R. Veeramachaneni. 2005. Exposure to low concentrations of Di-n-butyl phthalate during embryogenesis reduces survivability and impairs development of *Xenopus laevis* frogs. J. Toxicol. Environ. Heal. - Part A 68: 763–772.

Lefcort, H., R.A. Meguire, L.H. Wilson and W.F. Ettinger. 1998. Heavy metals alter the survival, growth, metamorphosis, and antipredatory behavior of Columbia spotted frog (*Rana luteiventris*) tadpoles. Arch. Environ. Contam. Toxicol. 35: 447–456.

Lenkowski, J.R., G. Sanchez-Bravo and K.A. McLaughlin. 2010. Low concentrations of atrazine, glyphosate, 2, 4-dichlorophenoxyacetic acid, and triadimefon exposures have diverse effects on *Xenopus laevis* organ morphogenesis. J. Environ. Sci. 22: 1305–1308.

Lopes, A., M. Benvindo-Souza, W.F. Carvalho, H.F. Nunes, P.N. de Lima, M. Costa et al. 2021. Evaluation of the genotoxic, mutagenic, and histopathological hepatic effects of polyoxyethylene amine (POEA) and glyphosate on *Dendropsophus minutus* tadpoles. Environ. Pollut. 289: 117911.

Ma, Y., B. Li, Y. Ke and Y. Zhang. 2019. Effects of low doses Trichlorfon exposure on *Rana chensinensis* tadpoles. Environ. Toxicol. 34: 30–36.

Mann, R.M., R.V. Hyne, C.B. Choung and S.P. Wilson. 2009. Amphibians and agricultural chemicals: Review of the risks in a complex environment. Environ. Pollut. 157: 2903–2927.

McDaniel, C.N. and D.N. Borton. 2002. Increased human energy use causes biological diversity loss and undermines prospects for sustainability. BioScience 52: 929–936.

McDaniel, T.V., L. Megan, C.A. Harris and J.S. Bishop. 2004. Development and survivorship of northern leopard frogs (*Rana pipiens*) and green frogs (*Rana clamitans*) exposed to contaminants in the water and sediments of the St. Lawrence river near Cornwall, Ontario. Water Quality Res. J. 39: 160–174.

McDaniel, T.V., P.A. Martin, J. Struger, J. Sherry, C.H. Marvin, A. McMaster et al. 2008. Potential endocrine disruption of sexual development in free ranging male northern leopard frogs (*Rana pipiens*) and green frogs (*Rana clamitans*) from areas of intensive row crop agriculture. Aquat. Toxicol. 88: 230–242.

Mnif, W., A.I.H. Hassine, A. Bouaziz, A. Bartegi, O. Thomas and B. Roig. 2011. Effect of endocrine disruptor pesticides: A review. Int. J. Environ. Res. Public Health 8: 2265–2303.

Motta, A.G.C., D.F. do Amaral, M. Benvindo-Souza, T.L. Rocha and D.D.M. Silva. 2020. Genotoxic and mutagenic effects of zinc oxide nanoparticles and zinc chloride on tadpoles of *Lithobates catesbeianus* (Anura: Ranidae). Environ. Nanotechnol. Monitoring & Management. 14: 100356.

Muyesaier, T.H., D. Ruan, L. Wang, L. Jia, R. Sadler, D. Connell, C. Cordia and T.P. Dung. 2021. Agriculture development, pesticide application and its impact on the environment. Internat. J. Environ. Res. and Public Health 18: 1112.

Nations, S., M. Wages, J.E. Cañas, J. Maul, C. Theodorakis and G.P. Cobb. 2011. Acute effects of Fe_2O_3, TiO_2, ZnO and CuO nanomaterials on *Xenopus laevis*. Chemosphere 83: 1053–1061.

Navarro-Martín, L., C. Lanctôt, P. Jackman, B.J. Park, K. Doe, B.D. Pauli et al. 2014. Effects of glyphosate-based herbicides on survival, development, growth and sex ratios of wood frogs (*Lithobates sylvaticus*) tadpoles. I: Chronic laboratory exposures to VisionMax®. Aquat. Toxicol. 154: 278–290.

Norrström, A.C. and G. Jacks. 1998. Concentration and fractionation of heavy metals in roadside soils receiving de-icing salts. STOTEN. 218: 161–174.

Ojha, S., A. Roy and A.K. Mohapatra. 2021. Environmentally relevant concentrations of Cadmium impair morpho-physiological development and metamorphosis in *Polypedates maculatus* (Anura, Rhacophoridae) tadpoles. Environ. Chem. and Ecotoxicol. 3: 133–141.

Ouellet, M., J. Bonin, J. Rodrigue and S. Lair. 1997. Hindlimb deformities (Ectromelia, ectrodactyly) in free-living anurans from agricultural habitats. J. Wildl. Dis. 33: 95–104.

Padhye, A.D. and H.V. Ghate. 1992. Sodium chloride and potassium chloride tolerance of different stages of the frog, *Microhyla ornata*. Herpetol. J. 2: 18–23.

Papis, E., G. Bernardini, R. Gornati and M. Prati. 2006. Triadimefon causes branchial arch malformations in *Xenopus laevis* embryos. Environ. Sci. Pollut. Res. 13: 251–255.

Peana, M., A. Pelucelli, S. Medici, R. Cappai, V.M. Nurchi and M.A. Zoroddu. 2021. Metal toxicity and speciation: a review. Curr. Med. Chem. 28: 7190–7208.

Peixoto, S., B. Santos, G. Lopes, P. Dias-Pereira and I. Lopes. 2022. Differential sensitivity of aquatic life stages of *Pelophylax perezi* to an acidic metal-contaminated effluent. Environ. Sci. Pollut. Res. 4: 1402.

Pekmezekmek, A.B., M. Emre, S. Erdogan, B. Yilmaz, E. Tunc, S. Sertdemir et al. 2021. Effects of high-molecular-weight polyvinyl chloride on *Xenopus laevis* adults and embryos: The mRNA expression profiles of Myf5, Esr1, Bmp4, Pax6, and Hsp70 genes during early embryonic development. Environ. Sci. Pollut. Res. 29: 14767–14779.

Peltzer, P.M., R.C. Lajmanovich, C.A. Martinuzzi, M. Andrés, L.M. Curi and M.T. Sandoval. 2019. Biotoxicity of diclofenac on two larval amphibians: Assessment of development, growth, cardiac function and rhythm, behavior and antioxidant system. Sci. Tot. Environ. 683: 624–637.

Pinto Vidal, F.A., F.C. Abdalla, C.S. Carvalho, H.S.M. Utsunomiya, L.A.T. Oliveira, R.F. Salla et al. 2021. Metamorphic acceleration following the exposure to lithium and selenium on American bullfrog tadpoles (*Lithobates catesbeianus*). Ecotoxicol. Environ. Saf. 207: 111101.

Plowman, M.C., S. Grbac-Ivankovic, J. Martin, S.M. Hopfer and F.W. Sunderman. 1994. Malformations persist after metamorphosis of *Xenopus laevis* tadpoles exposed to Ni^{2+}, Co^{2+}, or Cd^{2+} in FETAX assays. Teratog. Carcinog. Mutagen. 14: 135–144.

Quintaneiro, C., A.M.V.M. Soares and M.S. Monteiro. 2018. Effects of the herbicides linuron and S-metolachlor on Perez's frog embryos. Chemosphere 194: 595–601.

Rathanayaka, R. and R. Rajakaruna. 2018. Cocktail effect of profenophos and abamectin on tadpoles of asian common toad (*Duttaphrynus melanostictus*). Ceyl. Jour. Scien. 47: 185–194.

Richards, S.M. and R.J. Kendall. 2002. Biochemical effects of chlorpyrifos on two developmental stages of *Xenopus laevis*. Environ. Toxicol. Chem. 21: 1826.

Rutkoski, C.F., N. Macagnan, A. Folador, V.J. Skovronski, do A.M.B. Amaral, J. Leitemperger et al. 2020. Morphological and biochemical traits and mortality in *Physalaemus gracilis* (Anura: Leptodactylidae) tadpoles exposed to the insecticide chlorpyrifos. Chemosphere 250: 126162.

Saka, M. 2004. Developmental toxicity of p,p′-dichlorodiphenyltrichloroethane, 2,4,6-trinitrotoluene, their metabolites, and benzo[a]pyrene in *Xenopus laevis* embryos. Environ. Toxicol. Chem. 23: 1065–1073.

Saka, M. and N. Tada. 2021. Acute and chronic toxicity tests of systemic insecticides, four neonicotinoids and fipronil, using the tadpoles of the western clawed frog *Silurana tropicalis*. Chemosphere 270: 12941.

Salla, R.F., F.U. Gamero, R.Z. Rissoli, S.E. Dal-Medico, L.M. Castanho, C.S. Carvalho et al. 2016. Impact of an environmental relevant concentration of 17α-ethinylestradiol on the cardiac function of bullfrog tadpoles. Chemosphere 144: 1862–1868.

Sánchez-Domene, D., A. Navarro-Lozano, R. Acayaba, K. Picheli, C. Montagner, D. de Cerqueira Rossa-Feres et al. 2018. Eye malformation baseline in *Scinax fuscovarius* larvae populations that inhabit agroecosystem ponds in southern Brazil. Amphib. Reptil. 39: 325–334.

Sanzo, D., and S.J. Hecnar. 2006. Effects of road de-icing salt (NaCl) on larval wood frogs (*Rana sylvatica*). Environ. Pollut. 140: 247–256.

Sayim, F. 2010. Toxicity of trifluralin on the embryos and larvae of the red-bellied toad, *Bombina bombina*. Turkish J. Zool. 34: 479–486.

Scott, D.E. 1994. The effect of larval density on adult demographic traits in *Ambystoma opacum*. Ecology 75: 1383–1396.

Scott, D.E., E.D. Casey, M.F. Donovan and T.K. Lynch. 2007. Amphibian lipid levels at metamorphosis correlate to post-metamorphic terrestrial survival. Oecologia 153: 521–532.

Scown, T.M., R. van Aerle and C.R. Tyler. 2010. Review: Do engineered nanoparticles pose a significant threat to the aquatic environment? Critical. Rev. Toxicol. 40: 653–670.

Searcy, C.A., H. Snaas and H.B. Shaffer. 2015. Determinants of size at metamorphosis in an endangered amphibian and their projected effects on population stability. Oikos 124: 724–731.

Seleem, A.A. 2019. Teratogenicity and neurotoxicity effects induced by methomyl insecticide on the developmental stages of *Bufo arabicus*. Neurotoxicol. Teratol. 72: 1–9.

Sievers, M., R. Hale, K.M. Parris, S.D. Melvin, C.M. Lanctôt and S.E Swearer. 2019. Contaminant-induced behavioural changes in amphibians: A meta-analysis. Sci. Total Environ. 693: 133570.

Simon-Delso, N., V. Amaral-Rogers, L.P. Belzunces, J.M. Bonmatin, M. Chagnon, C. Downs et al. 2015. Systemic insecticides (neonicotinoids and fipronil): Trends, uses, mode of action and metabolites. Environ. Sci. Pollut. Res. Int. 22: 5–34.

Sodré, D., A. D'Angiolella, C. Rocha, D. Sarmento and M. Vallinoto. 2022. A hotspot of toad malformation in the Amazon. Herpetol. Notes 15: 111–115.

Sodré, F.F., M.A.F. Locatelli and W.F. Jardim. 2010. Occurrence of emerging contaminants in Brazilian drinking waters: A sewage-to-tap issue. Water. Air. Soil Pollut. 206: 57–67.

Sparling, D.W., S. Krest and M. Ortiz-Santaliestra. 2006. Effects of lead-contaminated sediment on *Rana sphenocephala* tadpoles. Arch. Environ. Contam. Toxicol. 51: 458–466.

Stankovic, S., P. Kalaba and A.R. Stankovic. 2014. Biota as toxic metal indicators. Environ. Chem. Lett. 12: 63–84.

Stansley, W., M.A. Kosenak, J.E. Huffman and D.E. Roscoe. 1997. Effects of lead-contaminated surface water from a trap and skeet range on frog hatching and development. Environ. Pollut. 96: 69–74.

Strassemeyer, J., D. Daehmlow, A. Dominic, S. Lorenz and B. Golla. 2017. SYNOPS-WEB, an online tool for environmental risk assessment to evaluate pesticide strategies on field level. Crop. Prot. 97: 28–44.

Strong, R., F.L. Martin, K.C. Jones, R.F. Shore and C.J. Halsall. 2017. Subtle effects of environmental stress observed in the early life stages of the common frog, *Rana temporaria*. Sci. Rep. 7: 44438.

Toledo, R.C. and C. Jared. 1993. Cutaneous adaptations to waterbalance in amphibians. Comp. Biochem. Physiol. A 105: 593–608.

Toledo, R.C. and C. Jared. 1993. The calcified dermal layer in anurans. Comp. Biochem. and Physiol. Part A. 104: 443–448.

Tussellino, M., R. Ronca, F. Formiggini, N.D. Marco, S. Fusco, P.A. Netti and R. Carotenuto. 2015. Polystyrene nanoparticles affect *Xenopus laevis* development. J. Nanoparticle Res. 17: 1–17.

Tudi, M., H. Daniel Ruan, L. Wang, J. Lyu, R. Sadler, D. Connell and D.T. Phung, 2021. Agriculture development, pesticide application and its impact on the environment. Int. J. Environ. Res. Public Health 18: 1112.

USEPA United States Environmental Protection Agency. 2022. Insecticides. https://www.epa.gov/caddis-vol2/insecticides.

Van Valen, L. 1974. A natural model for the origin of some higher taxa. J. Herpetol. 109–121.

Vanzetto, G.V., J.G. Slaviero, P.F. Sturza, C.F. Rutkoski, N. Macagnan, C. Kolcenti et al. 2019. Toxic effects of pyrethroids in tadpoles of *Physalaemus gracilis* (Anura: Leptodactylidae). Ecotoxicology 28: 1105–1114.

Venâncio, C., I. Melnic, M. Tamayo-Belda, M. Oliveira, M.A. Martins and I. Lopes. 2022. Polymethylmethacrylate nanoplastics can cause developmental malformations in early life stages of *Xenopus laevis*. Sci. Tot. Environ. 806: 150491.

Venturino, A., E. Rosenbaum, A. Caballero De Castro, O.L. Anguiano, L. Gauna, T. Fonovich de Schroeder et al. 2003. Biomarkers of effect in toads and frogs. Biomarkers 8: 167–186.

Vitousek, P.M., H.A. Mooney, J. Lubchenco and J.M. Melillo. 1997. Human domination of Earth's ecosystems. Science 25: 494–499.

Wang, W., Y. Yang, A. Chen, X. Song and L. Chai. 2021. Inhibition of metamorphosis, thyroid gland, and skeletal ossification induced by hexavalent chromium in *Bufo gargarizans* larvae. Environ. Toxicol. Chem. 40: 2474–2483.

Watson, F.L., H. Schmidt, Z.K. Turman, N. Hole, H. Garcia, J. Gregg et al. 2014. Organophosphate pesticides induce morphological abnormalities and decrease locomotor activity and heart rate in *Danio rerio* and *Xenopus laevis*. Environ. Toxicol. Chem. 33: 1337–1345.

Yoon, C.S., J.H. Jin, C.Y. Yeo, S.J. Kim, Y.G. Hwang, S.J. Hong et al. 2006. First record of the ground beetle *Trechoblemus postilenatus* (Coleoptera, Carabidae) in Primorskii krai. Far East. Entomol. 165: 16.

Zagrebin, A.O., V.A. Rumyantsev and V.D. Tonkopii. 2016. Developing methods for bioidentification of xenobiotics for water quality assessment. Water Resour. 43: 141–144.

Index